Carl H. Hamann, Andrew Hamnett, and Wolf Vielstich

Electrochemistry

Carl H. Hamann, Andrew Hamnett, and Wolf Vielstich

Electrochemistry

Second, Completely Revised and Updated Edition

WILEY-VCH Verlag GmbH & Co. KGaA

The Authors

Prof. Dr. Carl H. Hamann
Institut für Angewandte Chemie
der Universität Oldenburg
Carl-von Ossietzky-Str. 9 – 11
26129 Oldenburg
Germany

Prof. Dr. Andrew Hamnett
University of Strathclyde
Principal and Vice Chancellor
16 Richmond Street McCance Building
Glasgow G1 1XQ
Great Britain

Prof. Dr. Wolf Vielstich
Instituto de Quimica, USP
P. O. Box 780
13560-970 Sao Carlos-SP
Brazil

Library of Congress Card No.:
applied for

British Library Cataloguing-in-Publication Data:
A catalogue record for this book is available from the British Library.

Bibliographic information published by the Deutsche Nationalbibliothek
The Deutsche Nationalbibliothek lists this publication in the Deutsche Nationalbibliografie; detailed bibliographic data is available in the Internet at http://dnb.d-nb.de

Composition Mitterweger & Partner Kommunikationsgesellschaft mbH, Plankstadt
Printing betz-druck GmbH, Darmstadt
Bookbinding Litges & Dopf GmbH, Heppenheim
Cover Grafik-Design Schulz, Fußgönheim
Wiley Bicentennial Logo Richard J. Pacifico

Printed in the Federal Republic of Germany.
Printed on acid-free paper

ISBN 978-3-527-31069-2

Table of Contents

Electrochemistry. Carl H. Hamann, Andrew Hamnett, Wolf Vielstich

ISBN: 978-3-527-31069-2

Preface to the Second Edition

Nearly a decade has passed since the first edition of this textbook, and developments in the discipline of electrochemistry have been rapid and on a remarkably wide front. Increasingly, electrochemistry impacts on areas of science once quite remote from the central concerns of traditional electrochemists, and in turn these other areas are having a major impact on the range and extent of knowledge with which modern electrochemists must be familiar. A generation ago, for example, aside from some optical methods, spectroscopy had little to offer electrochemistry: now it is one of the largest chapters in the current text. Fundamental theoretical studies, using ab initio methods, now allow us to probe basic molecular processes in the vicinity of the electrode surface that are finally giving us direct glimpses of behaviour that, until recently, we could only model with crude approximations. The advent of scanning surface microscopies has allowed us to study the arrangements of atoms and molecules on surfaces, and how these change with potential and electrolyte, as well as shedding new light on processes such as corrosion that depend on spatial inhomogeneities at the electrode surface. New materials, often developed for different purposes, have allowed us to develop new technologies, such as low-temperature solid-state fuel cells, and in turn new materials developed by electrochemists, such as electroactive polymers, are having far-reaching impacts on other technologies.

This ferment of ideas has somehow to be incorporated into our textbook in ways that do not cause the book to become too large to be useful to students. We have tried to bring in new material but also to remove older material where this could be safely done without imperilling the overall coverage. In the earlier chapters, the main changes have been in the introduction of new insights from ab initio theories. This is a very new area, but already exciting discoveries have been made. Chapters 4 and 5 remain the physico-chemical core of the text, and we have introduced many changes to reflect modern approaches and techniques. We have also added a section on bioelectrochemistry, emphasising the differences in behaviour associated with long-range electron transfer. Chapter 6, on mechanisms, has been completely re-written to accommodate the explosion in our understanding of such processes as electro-oxidation of methanol and CO. We have also added sections on electro-polymerisation and oscillating electrochemical reactions, reflecting enormous modern interest in these areas. Chapter 7 has been extended to include modern developments in membranes for cells, and a brief introduction to the important area of room-temperature

Electrochemistry. Carl H. Hamann, Andrew Hamnett, Wolf Vielstich

ISBN: 978-3-527-31069-2

melts. The final three chapters have all been updated as the relevant technologies have developed. We are particularly conscious of the profound impact that developments in bio-electrochemistry will have on sensors in the future, and in this, as in many other areas, we can offer signposts and the confidence that in an ever-widening group of disciplines, the skills of the electrochemist will be urgently needed.

The first edition of this textbook was based on a translation and recension by A. Hamnett on the third edition of the German textbook "Elektrochemie", written by C. H. Hamann and W. Vielstich. Both books were published in 1998. The fourth German edition was revised predominantly by C. H. Hamann and published in 2005. However, the second English edition was revised predominantly by W. Vielstich and A. Hamnett, and as a result, the two texts do differ.

It is always a pleasure to acknowledge the help of others in the preparation of a textbook. In particular, we would like to thank Prof. Teresa Iwasita and Prof. Paul Christensen, both of whom have given freely of their time and expertise to read and comment on many sections of the text. We are hugely grateful to them for all their helpful comments. We would also like to thank Prof. Velia Solis from Cordoba for her help with the bio-electrochemistry section, and Drs. Hamilton Valera, Demetrius Profeti and Roberto Batista de Lima, Bruno Carreira Batista and Eduardo Ciapina for their help as well. We have also had helpful comments from readers of the first edition, and these have been greatly appreciated.

January 2007

Carl H. Hamann
Andrew Hamnett
Wolf Vielstich

List of Symbols and Units

Symbols

a_i	activity
A	area
A_v	electrode surface per cell volume
c	concentration
C	capacitance
C_D	double layer capacitance
C_d	differential capacitance
C_s^{th}	theoretical specific capacitance
e_0	elemental charge
E	potential
E_0	open circuit potential
E^0	standard potential
E_r	rest potential
E_{pzc}	potential of zero charge
E_c	cell voltage
$E_{c,o}$	open circuit voltage, EMF
E_c^o	thermodynamic cell voltage
E_D	decomposition voltage
$\boldsymbol{E}$	electric field
F	Farad
F	Faraday constant
$\boldsymbol{F}$	force
G	molar free energy
$\Delta_f G$	molar free energy of reaction
$\Delta_f G^0$	standard molar free energy of reaction
i	current
i_0	exchange current
I	ionic strength
j	current density
j_0	exchange current density
j_{00}	standard exchange current density

Electrochemistry. Carl H. Hamann, Andrew Hamnett, Wolf Vielstich

ISBN: 978-3-527-31069-2

k_r, k_d	rate constance
k_B	Boltzmann constant
L	conductance, film thickness
m	molality
m^0	standard molality
M	molar mass
N_A	Avogadro number
Q	charge
q_i	particle charge
r	radius
$\mathbf{r}$	vector
R	Molar gas constant
R	resistance
R_i	internal resistance
R_E	electrolyte resistance
R_e	external resistance
R_{CT}	charge transfer resistance
RF	roughness factor, real surface
RHE	reversible hydrogen electrode (in same solution)
S	entropy
SCE	saturated calomel electrode
SHE	standard hydrogen electrode
t^+, t^-	transport numbers
U	energy of interaction
u	mobility
v	velocity
z	number of charges
Z	impedance
α	degree of dissociation
β	transfer coefficient (asymetry parameter)
δ_N	Nernst diffusion layer
δ_{Pr}	Prandtl layer
δ_R	reaction layer
ε_r	relative permittivity
ε_o	permittivity of free space
κ_I	ionic conductivity
κ^{-1}	radius of the ionic cloud
Λ	molar conductivity
Λ_0	limiting molar conductivity
Λ_{eq}	equivalent conductivity
λ^+, λ^-	ionic conductivity
λ_o^+, λ_o^-	limiting ionic conductivity
μ_i	chemical potential
$\tilde{\mu}_i$	electrochemical potential

$\mu_i^{0\dagger}$	chemical potential of a hypothetical solution of non-interacting ions of molality m^0
ν	kinematic viscosity
ν_i	stoichiometric number of reactants
φ	Galvani potential
φ_S	Galvani solution potential
φ_M	Galvani metal potential
φ_{OHP}	potential of the outer Helmholtz plane
φ_{mix}	mixed potential
$\Delta\varphi$	Galvani potential difference
$\Delta\varphi_{diff}$	diffusion potential
$\Delta\varphi_H$	potential drop in the Helmholtz layer
$\Delta\varphi_S$	liquid-junction potential
$\Delta\varphi_{SC}$	internal potential drop in semiconductor in reverse bias
γ	surface tension
γ_i	activity coefficient
η	overpotential
η_r	reaction overpotential
η	viscosity
ρ	resistivity
ρ	density
θ	coverage of an adsorbate

Units

Force = Mass × Acceleration
$1\ N = 1\,kg\ m/s^2$

Energy = Force x Distance = Work
$1\ VAs = 1\ Ws = 1\ J = 1\ Nm = 10^7 erg$
$1\ eV = 1.602\times10^{-19}\ J$
$1\ cal = 4.1868\ J$

Pressure
$1\ Pa = 1\ N/m^2$
$1\ bar = 10\ N/cm^2 = 10^5\ N/m^2 = 10^5\ Pa$
1 atm = 1013mb = 760 torr

Electrical Charge
1 C = 1 As

Electrical Capacity
1 F = As/V

Fundamental Constants

Speed of light (in vacuum)	$c = 2.99792 \times 10^8\,\mathrm{ms}^{-1}$
Elementary charge	$e_0 = 1.60218 \times 10^{-19}$ C
Avogadro Number	$N_A = 6.022 \times 10^{23}$ mol^{-1}
Faraday Constant	$F = N_A\, e = 0.964853 \times 10^5$ C mol^{-1}
Boltzmann Constant	$k_B = 1.38065 \times 10^{-23}$ JK^{-1}
Gas Constant	$R = N_A\, k_B = 8.31447$ JK^{-1} mol^{-1}
Permittivity of free space	$\varepsilon_0 = 8.85419 \times 10^{-12}$ AsV^{-1}m^{-1}

At 25^{0}C (298.15 K):

$RT/F = k_B T/e = 25.693\,\mathrm{mV}$

$2.303\ RT/F = 59.16\,\mathrm{mV}$

$k_B T = 4.116 \times 10^{21}$ J

$RT = N_A\, k_B T = 2.478$ kJ mol^{-1} = 592 cal mol^{-1}

Monolayer

Au(111) = 1.5×10^{15} Atome cm^{-2}

Mathematical relationships

e = 2.71828

$e^{ix} = \cos x + i \sin x (i = \sqrt{-1})$

$e^x = 1 + x + x^2/2! + x^3/3! \ldots$

$\pi = 3.14159$

$$\frac{dx^n}{dx} = nx^{n-1};\ \frac{d\ln x}{dx} = \frac{1}{x};\ \frac{d(e^{ax})}{dx} = ae^{ax}$$

1
Foundations, Definitions and Concepts

1.1
Ions, Electrolytes and the Quantisation of Electrical Charge

In solid crystals such as NaCl, electrical charges are localised on those sites that form the lattice. These lattice sites are occupied not by (neutral) atoms but by negatively charged chlorine or positively charged sodium ions and the crystal is held together by the coulombic forces that exist between all electrically oppositely charged species. The energy of interaction between two particles of charge q_1 and q_2 at a distance r from one another is given by

$$U_{12} = \frac{q_1 \cdot q_2}{4\pi\varepsilon_r\varepsilon_0 r} \tag{1.1}$$

where ε_r is the relative permittivity of the medium and ε_0 the permittivity of free space. The energy is positive (ie unfavourable) when q_1 and q_2 have the same sign, but is negative (implying an attractive force) when q_1 and q_2 have opposite signs. The corresponding force between the two particles is a *vector* quantity, directed along the interparticle axis, **r**; if the two charges have the same sign, the force **F** is repulsive:

$$\mathbf{F}_{12} = -\frac{\partial U_{12}}{\partial \mathrm{r}} = \frac{q_1 q_2}{4\pi\varepsilon_r\varepsilon_0 r^2}\frac{\mathbf{r}}{r} \tag{1.1a}$$

These coulombic forces are very powerful, and ions are drawn close together before short-range repulsive forces come into play and an equilibrium interionic distance is established. The result that very large amounts of energy are needed to break down the lattice, leading, for example, to the familiar observation of high melting points in ionic crystals. The actual calculation of the energy required completely to break up the crystal lattice can be carried out provided some analytical form for the ionic repulsion can be written down. Commonly this repulsion is deemed to take the form $R_{12} = B/r^n$, where B depends on the relative extension of valence and core-electron clouds. The total interaction energy between pairs of ions can then be written $E_{12} = U_{12} + R_{12}$, and by summing over all pairs of ionic sites in the lattice, the total crystal can be calculated. In general this is difficult to do mathematically, since the sum will obviously consists of a very large number of positive and negative terms that will nearly cancel. However, the final result for a simple cubic lattice such as NaCl is

Electrochemistry. Carl H. Hamann, Andrew Hamnett, Wolf Vielstich

ISBN: 978-3-527-31069-2

$$E = \frac{MN_A|q|^2}{4\pi\varepsilon_0 r}\left(1 - \frac{1}{n}\right) \tag{1.2}$$

where N_A is the Avagadro constant (6.203×10^{23} mol^{-1}), and M is a number (which has the value 1.7476 for the NaCl structure) termed the Madelung constant. We note that the value of ε_r in eqn. (1.1) has been put equal to unity since the ions in the lattice are formally separated by vacuum.

If the NaCl crystal be now introduced into a solvent such as water, the attractive forces between the ions will be much reduced since the relative permittivity for water, ε_w, is 78.3 (at 25°C). The result is that ionic attraction between Na^+ and Cl^- ions dissolved in water is much weaker, sufficiently so that NaCl will readily dissolve in water to give freely moving Na^+ and Cl^- ions (a word derived from the Greek word for wanderer); the salt is said to dissociate into *ions*.

Actually, the reduction of attractive energy in the aqueous phase would not be sufficient by itself to dissolve NaCl. What is decisive is the fact that water has a strong dipole which preferentially orients itself around each ion in a process termed *solvation*, or, for water, *hydration*. Each positive or negative ion in aqueous solution is surrounded by a sheath of water molecules, and it is the energy gained by the ions from this solvation process that usually tips the balance in favour of dissolution, or at least reduces the dissolution enthalpy to the point where the *entropy* of dissolution (which is always positive) can tip the balance. The process is shown schematically in Fig. 1.1.

Chemical compounds that are dissociated into ions in solid, liquid or dissolved forms are termed electrolytes. The example of NaCl above is termed a 1—1 electrolyte, since two ions are formed for every formula unit of the material, and each carries one unit of elementary charge, e_0, of magnitude 1.602×10^{-19} C. From the dissolution

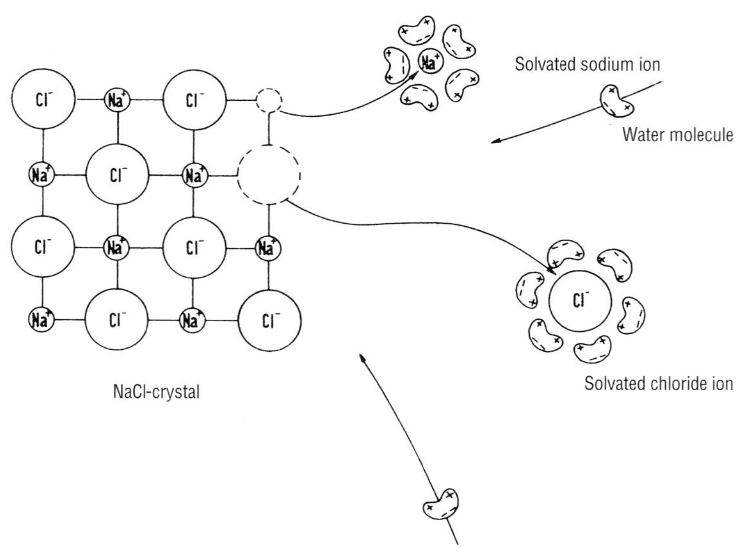

Fig. 1.1 Ionic solvation through dissolution of a NaCl crystal in water (two dimensional representation).

of multivalent electrolytes, more than two ions per formula unit may be obtained and such ions may carry a multiple of the elementary charge, $\pm ze_0$, where z is *the charge number of the ion*. As an example, the dissolution of Na_2SO_4 will give two Na^+ ions and one SO_4^{2-} ion.

It can be seen that for every electrolyte of the general form $A_{\nu_+}B_{\nu_-}$ which dissolves to yield A^{z_+} and B^{z_-} ions, then electroneutrality requires that $z_+\nu_+ = z_-\nu_- = z_\pm\nu_\pm$, where $z_\pm\nu_\pm$ is termed the equivalent number of the electrolyte. It is clear that for Na_2SO_4, $z_\pm\nu_\pm = 2$.

1.2 Transition from Electronic to Ionic Conductivity in an Electrochemical Cell

If the ions in an electrolyte solution are subjected to an electric field, $\boldsymbol{E}$, one has with the definition for the field $\boldsymbol{E} = \boldsymbol{F}/q$ as expression for the force on the ions

$$\boldsymbol{F} = ze_0\boldsymbol{E} \tag{1.3}$$

which will induce motion in or against the direction of the field depending on the sign of the charge on the ion, i.e. whether z is positive or negative. This ion motion leads to the transport of charge and hence to the flow of electrical current through the electrolyte solution.

An electric field can be applied across an electrolyte solution quite straightforwardly by introducing two electronic conductors (solids or liquids containing free electrons, such as metals, carbon, semiconductors etc), and applying a *dc* potential difference. These electronic conductors are termed *electrodes.*

The actual arrangement of the electrodes is shown in Fig. 1.2. The electrical circuit between the electrodes is completed by a resistor, an ammeter and a *dc* voltage source connected by external wiring from one electrode to the other. The electrolyte solution in Fig. 1.2 is formed from the dissolution of $CuCl_2$ in water, which leads to one Cu^{2+} and two Cl^- ions per formula unit. The electrodes themselves are formed from a suitable inert metal such as platinum.

When current flows in the cell, the negatively charged chloride ions migrate to the positive electrode and the positively charged ions to the negative electrode. At the

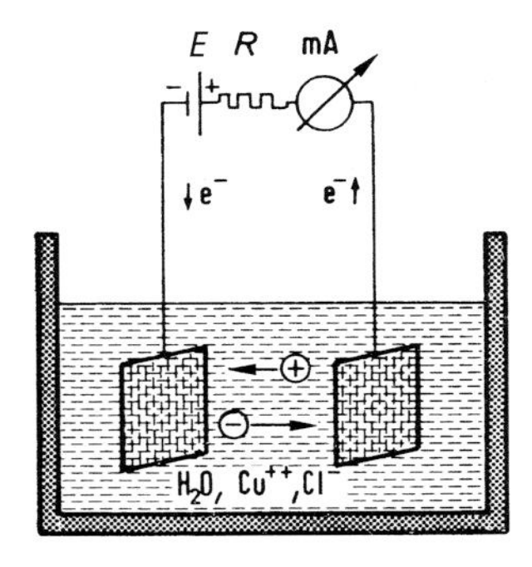

Fig. 1.2 Electrochemical cell for the electrolysis of aqueous $CuCl_2$ solution - *E*: *dc* voltage; *R*: resistance; mA: galvanometer for current measurement.

phase boundary between ionic and electronic conductors, the ions arriving are transformed by capture or release of electrons. At the negative electrode, the Cu^{2+} ions are plated out as copper metal:

$$Cu^{2+} + 2e^- \rightarrow Cu^0 \tag{1.4}$$

and at the positive electrode the chloride ions release electrons to form chlorine gas:

$$2Cl^- \rightarrow Cl_2 + 2e^- \tag{1.5}$$

It can be seen that the fundamental difference between charge transport through electrolyte solution by ion migration and through electronic conductors by electron migration is that the latter leaves the conductor essentially unaltered whereas the former leads to changes in the electrolyte. In the case above, the flow of current leads to the appearance of concentration differences, since the Cu^{2+} ions move from right to left and the chloride ions from left to right, and it also leads to changes in the total electrolyte concentration as both copper and chlorine are lost from the solution. From the addition of the two electrode reactions, (1.4) and (1.5) above, the overall cell reaction follows

$$\begin{array}{ll} \text{Electrode Reactions} & \left\{ \begin{array}{l} Cu^{++} + 2e^- \rightarrow Cu^0 \\ 2Cl^- \rightarrow Cl_2 + 2e^- \end{array} \right. \\ \text{Cell Reactions} & Cu^{++} + 2Cl^- \rightarrow Cu^0 + Cl_2 \end{array} \tag{1.6}$$

It should be emphasised at this point that a constant *direct current* through an ionic conductor is only possible if electrode reactions take place at the phase boundaries between the electronic and ionic components of the circuit. These reactions must permit electrons to be exchanged between the two phases, in the manner indicated in the fundamental considerations given above. However, a constant *alternating* current can flow through an ionic conductor without *any* electrode reaction taking place, as will be seen in section 2.1.2.

1.3
Electrolysis Cells and Galvanic Cells: The Decomposition Potential and the Concept of emf.

If the aqueous $CuCl_2$ solution in the electrochemical cell of Fig. 1.2 is replaced by an aqueous solution of HCl, which dissociates into aquated hydrogen ions (protons) of approximate formula H_3O^+ and chloride ions, a *dc* current flow will again lead to the loss of Cl^- at the positive electrode. However, at the negative electrode, the H_3O^+ ions are reduced to hydrogen. Thus, the flow of charge is accompanied by the electrochemical decomposition of HCl to its component parts

$$\begin{array}{ll} 2Cl^- & \rightarrow Cl_2 + 2e^- \\ 2H_3O^+ + 2e^- & \rightarrow H_2 + 2H_2O \\ \hline 2HCl & \rightarrow H_2 + Cl_2 \end{array} \tag{1.7}$$

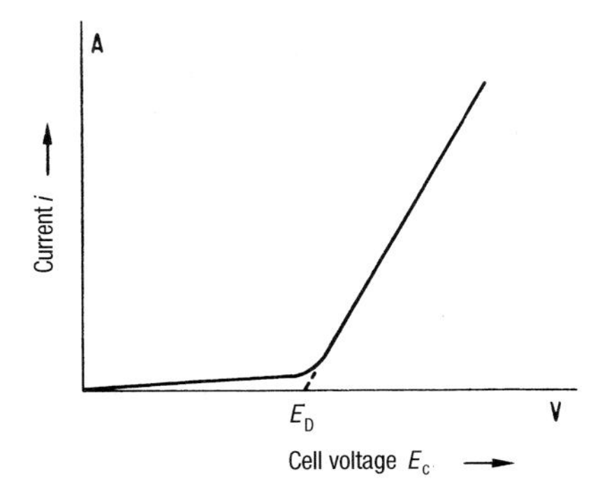

Fig. 1.3 Electrolysis current as a function of the cell voltage E. E_D is the decomposition voltage.

The electrochemical decomposition of a substance through the passage of an electrical current is termed *electrolysis*, and corresponds to the conversion of electrical to chemical energy.

For a significant rise in the current through an electrolysis cell, the potential difference between the electrodes, or *cell voltage, E*, must exceed a certain value, the *decomposition potential difference*, E_D as shown schematically in Fig. 1.3. The decomposition potential difference for HCl at a concentration of 1.2M has the value 1.37V at 25°C, and the value usually lies between one and four volts for other electrolytes.

If the electrolysis process is suddenly brought to a halt by removing the voltage source in the external circuit of Fig. 1.2, and swiftly reconnecting the electrodes through a voltmeter, a voltage of about 1V is observed. If the electrodes are reconnected by a resistor and ammeter, a current can also be observed. This current has its origin in the reversal of the two electrode reactions of equation 1.7; hydrogen is oxidised to hydrated protons, the electrons released travel round the external circuit and reduce chlorine gas to chloride ions. Actually, the current in this case would decay rapidly since the gases hydrogen and chlorine are only slightly soluble in water, and will, for the most part, have escaped from solution. However, if their concentrations can be maintained by bubbling the two gases over the electrodes, as shown in Fig. 1.4, then the current will remain fairly constant with time; the cell can continually provide electrical energy from chemical energy.

Electrochemical cells in which the electrode reactions take place spontaneously, giving rise to an electrical current, are termed *galvanic* cells, and are capable of the

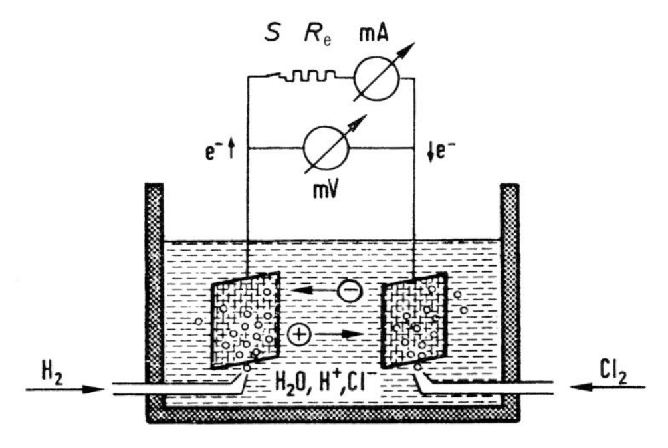

Fig. 1.4 Galvanic cell based on the H_2/Cl_2 reaction - mV: voltage measurement; mA: current measurement; R_e: external resistance; S: switch.

direct conversion of chemical to electrical energy. If the external resistance between the electrodes is made very high, so that very little current flows, the observed potential difference of the galvanic cell becomes the rest potential, E_r, which, under equilibrium conditions, is equal to the *electro-motive force or EMF* of the cell, E_0, related, in turn, to the free energy of the overall cell reaction (for a thermodynamic treatment, see chapter 3).

Irrespective of whether electrolysis is taking place or the cell is behaving in a galvanic mode, the electrode at which negative charge *enters* the electrolyte solution is termed the ***cathode***. Equivalently, it is at this electrode that positive charge may be said to leave the solution. Thus, typical cathode reactions are:

$$Cl_2 + 2e^- \rightarrow 2Cl^-$$

$$Cu^{2+} + 2e^- \rightarrow Cu^0$$

In these cases, the reactant, Cl_2 or Cu^{2+} is said to be *reduced*. In a similar fashion, at the ***anode*** negative charge leaves the electrolyte solution, or, equivalently, positive charge enters the solution, and typical anode reactions include:

$$2Cl^- \rightarrow Cl_2 + 2e^-$$

$$2H_2O + H_2 \rightarrow 2H_3O^+ + 2e^-$$

In these cases, the reactant is said to be *oxidised*.

It should now be clear that for the electrolysis cell above, the cathode is that electrode at which hydrogen is evolved, whereas in the case of the galvanic cell, the cathode is that electrode at which chlorine gas is reduced to chloride. Since, in electrolysis, the positively charged ions migrate towards the cathode, they are termed *cations*, and the negative ions are termed *anions*. That part of the solution near the cathode is termed the *catholyte*, and that part near the anode is termed the *anolyte*.

If a galvanic cell is set up as shown in Fig. 1.4, and both current and potential difference are monitored simultaneously, it is found that as the current increases, the measured external potential difference between the electrodes, E, decreases, as shown schematically in Fig. 1.5. In fact, the potential drop is partitioned between internal (R_i) and external (R_e) resistances as:

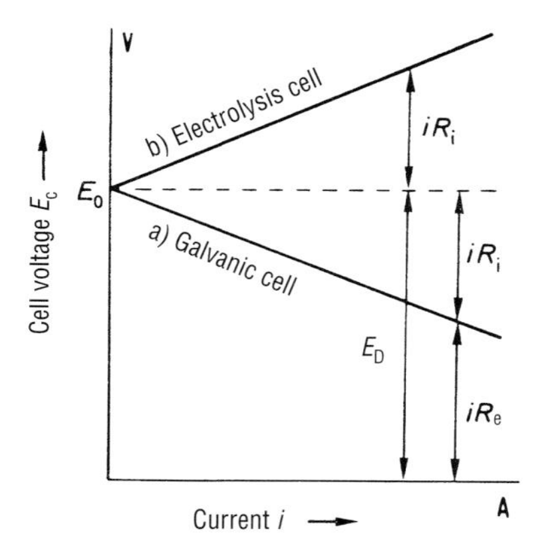

Fig. 1.5 Schematic variation of cell voltage, E_C, against load current i for (a) a galvanic cell; (b) an electrolysis cell. E_0 is the voltage without current passing, also termed the *electro-motive force* or EMF. E_D is the decomposition voltage, R_i the internal resistance of the cell and R_e the external resistance.

$$\begin{aligned} E_0 &= iR_i + iR_e \\ E &= E_0 - iR_i \end{aligned} \tag{1.8}$$

The power output, P, of the cell is clearly the product of potential and current:

$$P = iE = i(E_0 - iR_i) \tag{1.9}$$

and this is a maximum for $i = E_0/2R_i$, and $E = E_0/2$.

In the case of an electrolysis cell, when the current is forced through the cell, the potential drop across the inner resistance of the cell must be added to the decomposition potential, E_D, so we have

$$E = E_D + iR_i \tag{1.10}$$

1.4 Faraday's Laws

If each ion is assigned a charge of $\pm ze_0$, and the electron current in the external circuit has the value i_e (which will evidently be equal to the current flow associated with the ion flux of both positive and negative ions, i_I), then, at either electrode, the mass of material converted by the total charge passing between ionic and electronic conductors must be proportional to this charge. The charge, Q, itself will just be the product of current and time, i.e. $Q = i_e \cdot t$, so the mass, m, of material that reacts is given by:

$$m = \text{const.}Q = \text{const.}i_e \cdot t \tag{1.11}$$

For an ion with unit elementary charge, e_0, the total charge passed in oxidising or reducing one mole, Q_M, is evidently equal to $N_A e_0$, where N_A is the Avagadro constant, and for the conversion of one mole of z-valent ion a charge of $zN_A e_0$ will be necessary. The value of the product $N_A e_0$ is numerically equal to 96485 Coulomb mol^{-1}, and this quantity is termed the Faraday, and denoted by the symbol F. It follows that the passage of one Coulomb of charge will lead to the conversion of $M/(96485z)$ g of material, where M is the molar mass. As an example, silver metal can be deposited at the cathode from solutions of $AgNO_3$; the passage of one Coulomb will lead to the deposition of $107.88/96485 \equiv 1.118$ mg of silver metal.

It follows from this that the ratio of the masses of material converted at the two electrodes will have the form

$$m_1/m_2 = (M_1/z_1)/(M_2/z_2) \tag{1.12}$$

The ratio M/z is termed the molar mass of an ion-equivalent, and equation (1.12) tells us that for the passage of equal amounts of charge, the ratio of the masses of material converted at the two electrodes is equal to the ratio of the molar masses of the ion equivalents.

The relationships (1.11) and (1.12) were first expressed as laws by Faraday in 1833 and are known as Faraday's first and second laws. Faraday established his laws purely experimentally and Helmholtz went on to deduce from this that there must be an

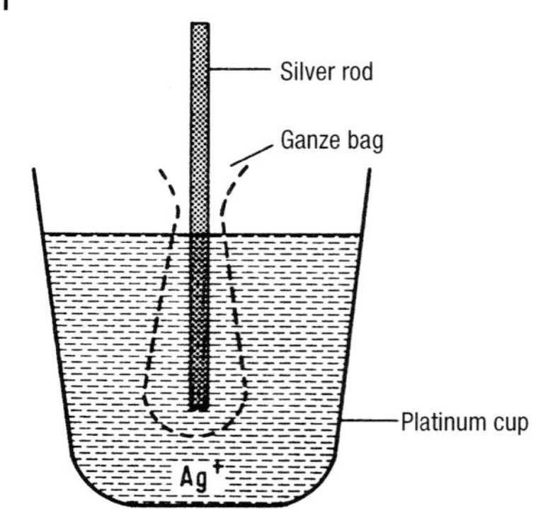

Fig. 1.6 Schematic construction of a silver coulometer for determination of the total quantity of electricity passed.

elementary unit of charge, a fact we have used effectively to derive his laws from a theoretical standpoint.

The quantitative laws found by Faraday can be used in order to determine the total amount of charge passed in a circuit if we insert a suitable electrolysis cell in the circuit and measure the amount of material formed or deposited at one of the electrodes. Commonly the measurement is carried out by measuring the amount of metal deposited as a coating on an inert electrode, such as platinum, or in the form of an amalgam on a mercury cathode. From this mass, m, and from the electrochemical ion-equivalent of the material deposited, the total charge passed is given by

$$Q = \frac{m}{M/zF} \tag{1.13}$$

In practice, the so-called silver coulometer is normally used for the measurement of charge. A silver coulometer consists, in principle, of a platinum crucible which is filled with ca 30% $AgNO_3$ solution and into which is inserted a silver rod, as shown in Fig. 1.6. The crucible is connected to the negative pole of the circuit and the silver rod to the positive pole, so that passage of charge results in a build up of a silver coating on the inside of the platinum crucible whilst at the same time, silver dissolves as Ag^+ from the rod. The electrode reactions are:

$$\begin{aligned} &\text{Platinum Crucible:} \quad && Ag^+ + e^- \rightarrow Ag^0 \\ &\text{Silver Rod:} && Ag^0 \rightarrow Ag^+ + e^- \end{aligned} \tag{1.14}$$

Around the silver rod is arranged a porous bag, whose main purpose is to prevent any particles that may fall off the silver rod due to mechanical damage from collecting in the crucible and being weighed as well. By carefully weighing the mass of the crucible before and after passage of charge, the total amount of charge passed can be determined by the expression $Q(/\text{Coulomb}) = m(/\text{mg})/1.118$.

Another type of coulometer, used particularly for the measurement of small amounts of charge, is the so-call *combustion-gas coulometer*. In this device, water is made ionically conducting by addition of Na_2CO_3, and electrolysed between two platinum electrodes. The electrode reactions are:

$$\begin{aligned}
&\text{Cathode:} && 2H_2O + 2e^- \rightarrow H_2 + 2OH^- \\
&\text{Anode:} && 2OH^- \rightarrow H_2O + {}^1/_2O_2 + 2e^- \qquad (1.15)\\
&\text{Overall cell reaction:} && H_2O \rightarrow {}^1/_2O_2 + H_2
\end{aligned}$$

The mixture of O_2 and H_2 produced is termed 'combustion gas', and is determined volumetrically.

It is evident that the current passed during the measurement need not be constant with time, since the amount of material deposited or converted at the electrode depends solely on the integral of the current with respect to time. In measurements using modern electronic equipment, the charge $Q = \int i dt$ is determined by integration of the accurately measured current and displayed directly.

Accurate measurement of charge is necessary above all in the analytical technique of *coulometry*, in which either the material to be analysed is converted completely into a form suitable for weighing, or is converted into a species that can be determined volumetrically. Further information can be found in Chapter 10 and in the references cited there on analytical chemistry.

1.5 Systems of Units

The fundamental quantities measured in electrochemical studies are current, charge and voltage, the latter being the potential difference between two defined points in a circuit. The current itself is often replaced by the *current density*, defined as the current per unit area of electrode surface.

From the discussion above, it is obviously possible to use Faraday's laws to relate current and charge to mass and time. This gives rise to the 'practical international electromagnetic' system of measurement, in which the ampère is *defined* as that current which, flowing for one second, deposits 1.118 mg of silver in a silver coulometer. In this system, the unit of potential difference, the volt, is fixed by reference to the potential difference of the 'Weston' cell, an easily prepared galvanic cell which is described in detail in section 4.11. The potential difference of the Weston cell is defined to be 1.01830V at 20°C. The unit of electric field strength is the volt per centimetre, and is defined as the *gradient* of the potential at any given point. It is clearly a *vector* quantity, in contrast to potential, which is a *scalar*.

The connection between the electrical current, I flowing through a simple electronic conductor, and the potential difference, V, applied between its ends is given by Ohm's law:

$$V = IR \qquad (1.16)$$

where R is termed the resistance and is strongly dependent on the material from which the conductor is made. Its unit is the ohm (Ω) $\equiv$ V A^{-1}, and the value is defined by the resistance of mercury. In fact, a column of mercury of 1mm^2 cross-section and length 106.300 cm at 0°C and 1 atm pressure ($\equiv$ 101325 Pa) is defined as having a resistance of one ohm. The necessity of specifying cross-section and length arises because the resistance of any material can be factored into two components, one dependent on the intrinsic properties of that material and one dependent on the geometry of the particular sample. For samples of uniform cross-sectional area A, and length l, it is easily shown that

$$R = \rho(l/A) \tag{1.17}$$

where ρ is termed the resistivity of the material and has the units ohm-cm.

Although this system of units is quite internally consistent, there was considerable pressure from professional organisations and learned journals for the universal use of a system based on the ampère as the basic electrical unit, with mass, length and time being in units of kg, metre and second. The basic units of this internationally agreed system, the 'Système International d'Unités' or SI system, are defined in Table 1.1. This set of units is the preferred set for scientific publishing, but the reality is that the older units, particularly those based on the unit of length one centimetre and unit of weight one gram are exceedingly well established, especially in the secondary literature of compilations of physical quantities; both sets will be used throughout this book, and the inter-relationships described where appropriate.

In the S.I. system of units, the unit of potential difference is still the volt, but it is now defined in terms of the units of energy, the joule or kg m^2 s^{-2} and current: in fact, 1V $\equiv$ 1 joule/(1 ampere second). The unit of resistance is similarly defined as 1 ohm $\equiv$ 1 volt/1 ampere. In this system, the unit of resistivity should now be the ohm-metre. However, it should be again emphasised that the overwhelming number of values quoted in the older literature are in the units of ohm-cm, and caution should be used in importing literature values into the formulae quoted.

Table 1.1 The International System of Units.

Quantity	Unit	Symbol	Definition (abridged)
Length	metre	m	The length of path travelled by light in vacuum during a time interval of 1/299 792 458 of a second
Mass	kilogram	kg	The mass of a piece of platinum-iridium kept in Sèvres.
Time	second	s	9 192 631 770 periods of the radiation corresponding to the transition between the hyper-fine levels of the ground state of the ^{133}Cs atom.
Electrical Current	ampere	A	The current that produces a force of 2.10^{-7} newton per metre of length between two infinitely long negligibly thick parallel conductors 1 metre apart in vacuo.
Thermo-dynamic temperature	kelvin	K	The fraction 1/273.16 of the thermodynamic temperature of the triple point of water.
Amount of substance	mole	mol	The amount of substance of a system containing as many specified elementary entities (atoms, molecules, ions etc) as there are atoms in 12g of ^{12}C.

Two units of concentration are used in the rigorous SI system. The concentration per unit volume has units mol m^{-3} (mole per cubic metre), and denoted by *c*. However, the subsidiary unit of dm^3 (known familiarly as the litre) is permitted, and concentrations of solutions are often quoted in units of mol dm^{-3}, and denoted by the symbol M the molarity of the solution. Whilst it is correct to refer to a solution as having concentration *x*M, the concentration units used in rational S.I. formulae must be the (rational) mol m^{-3}.

The second unit of concentration has units of mol kg^{-1}, i.e. moles per kg of *solvent*. This is the more useful measure of concentration when considering the thermodynamics of solutions, as will become apparent below, and it has the further advantage of being independent of temperature, but it is less useful from the practical point of view. For dilute solutions in water at 25°C, to a high degree of accuracy, a solution of x mol kg^{-1} is very close to xM or to $1000x$ mol m^{-3}. For dilute solutions in other solvents, a molality of x mol kg^{-1} will correspond to a concentration of $\rho_S x$ mol m^{-3}, provided the density of the solvent, ρ_S, is expressed in kg m^{-3}.

In order to ensure that equilibrium constants involving concentrations of reactant and product do not possess units themselves, all concentration terms in such formulae are expressed as ratios of the molarity or molality to a standard value, usually written c^0 or m^0. The choice of m^0 is straightforward, and is invariably taken as one mol kg^{-1}. The choice of c^0 is less obvious, since values of one mol m^{-3} or one mol dm^{-3} might be chosen. Conventionally, the latter has been used, leading to some confusion in formulae. To avoid this, we shall use *molality* as the main unit of concentration in this book, save in those areas, such as electrolyte conductivity, where volume-based units are normally used.

Although this discussion of units may seem quite dry, it actually has considerable practical importance. It should become second-nature to check any formula for consistency of units, a process that is immeasurably helped by having a rational system of units in the first place. Unfortunately, the unit of concentration remains a problem; the unit of molarity is deeply ingrained in chemical practice, and cannot be erased easily. This is a particular problem in the development of solution thermodynamics, and the choice of standard states remains a real difficulty, for which the only remedy is constant caution. Let the reader be warned!

References for Chapter One

The most useful background reading is in elementary physics and physical chemistry, and the textbooks listed below are intended to provide some further material on electricity and electrical measurements.

A good background text in physics is:

R. Muncaster: "'A'-level Physics", 3rd. Edition, Stanley Thornes (Publishers) Ltd, 1989.

More advanced but with a wealth of illustrative material is:

D. Halliday, R. Resnick and K.S. Krane: "Physics", Volume 2, 4th Edition, John Wiley and Sons Inc., 1992.

The best introductory text in physical chemistry at this level is

P.W. Atkins: "Introduction to Physical Chemistry", Oxford University Press, 1998.

2
Electrical Conductivity and Interionic Interactions

2.1
Fundamentals

2.1.1
The Concept of Electrolytic Conductivity

The ability of an electrolyte solution to sustain the passage of electrical current depends on the mobility of its constituent charged ions in the electric field between electrodes immersed in the solution. Ions of charge ze_0, accelerated by the electric field strength, $\boldsymbol{E}$, are subject to a frictional force, $\boldsymbol{F}$, that increases with velocity, v, and is given, for simple spherical ions of radius r_1, by the Stokes formula, $\boldsymbol{F} = 6\pi\eta r_1 \mathrm{v}$, where η is the viscosity of the medium. The result is that after a short induction period, the velocity attains a limiting value, $\mathrm{v}_{\mathrm{max}}$, corresponding to the exact balance between the electrical and frictional forces:

$$ze_0|\boldsymbol{E}| = 6\pi\eta r_1 \mathrm{v}_{\mathrm{max}} \tag{2.1}$$

and the terminal velocity is given by

$$\mathrm{v}_{\mathrm{max}} = \frac{ze_0|\boldsymbol{E}|}{6\pi\eta r_1} \tag{2.2}$$

It follows that for given values of η and $|\boldsymbol{E}|$, each type of ion will have a transport velocity dependent on the charge and the radius of the solvated ion, and a direction of migration dependent on the sign of the charge.

A note on units may be helpful at this point: In equation (2.1), the units of $\boldsymbol{E}$ are V m^{-1}, and of e_0 are C; the product is thus C V $\mathrm{m}^{-1} \equiv$ J $\mathrm{m}^{-1} \equiv$ kg m $\mathrm{s}^{-2} \equiv$ N, where N is the newton, the unit of force. On the right-hand side of equation (2.1), given that the product $r\mathrm{v}$ has units $\mathrm{m}^2\ \mathrm{s}^{-1}$, it follows that the viscosity has the units kg $\mathrm{m}^{-1}\mathrm{s}^{-1}$. This compares to the c.g.s. unit of viscosity, the poise, expressed as g $\mathrm{cm}^{-1}\ \mathrm{s}^{-1}$, and it can be seen that the SI unit is a factor of ten larger. Thus, the viscosity of liquid water is close to one centipoise $\equiv 10^{-3}$ kg $\mathrm{m}^{-1}\mathrm{s}^{-1}$.

Electrochemistry. Carl H. Hamann, Andrew Hamnett, Wolf Vielstich

ISBN: 978-3-527-31069-2

We consider in the following discussion an electrolyte solution containing both anions and cations. If the magnitude of the terminal velocity of the cations is v^+_{max}, and the number of ions of charge z^+e_0 per unit volume is n^+, the product $An^+v^+_{max}$ corresponds just to that quantity of positive ions that passes per unit time through a surface of area A normal to the direction of flow. The product $An^-v^-_{max}$ can be defined analogously, and assuming v^+_{max} and v^-_{max} are in opposite directions, the amount of charge carried through this surface per unit time, or the current per area A, is given by

$$i = i^+ + i^- = \frac{dQ^+}{dt} + \frac{dQ^-}{dt} = e_0(n^+z^+v^+_{max} + n^-z^-v^-_{max}) \tag{2.3}$$

If there are more than two types of ion present, then the summation in equation (2.3) must be extended to incorporate all charged species.

From (2.2), the velocities v^+_{max} and v^-_{max} are dependent on the magnitude of the electric field, $|\boldsymbol{E}|$, and hence, for a fixed electrode geometry, on the potential difference, ΔV, between the electrodes. The *scalar* quantity, u, the mobility, is defined as:

$$u = v_{max}/|\boldsymbol{E}| \tag{2.4}$$

and if the expression (2.2) for the terminal velocity is valid, $u = ze_0/6\pi\eta r$; clearly u has the units $m^2V^{-1}s^{-1}$. Substituting for the mobilities in equation 2.3, we find that the current *per unit area* is

$$i = Ae_0(n^+z^+u^+ + n^-z^-u^-) \cdot |\boldsymbol{E}| \tag{2.5}$$

If the potential difference between the electrodes is ΔV, and the distance apart of the electrodes is l, then the magnitude of the electric field $|\boldsymbol{E}| = \Delta V/l$. It follows that we can write

$$i = L\Delta V \tag{2.6}$$

where L is termed the conductance, and is given, generally, for area A by

$$L = (A/l)e_0(n^+z^+u^+ + n^-z^-u^-) \tag{2.7}$$

In accord with equation (2.7), L can be split into two components, L^+ and L^-. It can be seen that the conductance depends on the nature of the ions dissolved (through z and r), the concentrations of these ions (through n^+ and n^-), the viscosity, η, and hence the temperature T, and the dimensions of the conductivity cell, expressed through the area A of the electrodes and their separation l. A comparison with equation (1.16) shows that the units of conductance are reciprocal ohms or siemens, Ω^{-1}, and clearly the conductance of the solution is the reciprocal of its resistance.

2.1.2 The Measurement of Electrolyte Conductance

From equation (2.6), we would expect that the current flowing through a conductivity cell would depend linearly on the potential difference applied across the electrodes. In other words, we would expect that electrolyte solutions would obey Ohm's law. However, as we saw in Fig. 1.3, such a linear relationship is not, in general, found for electrolytes. Only if the potential exceeds the decomposition potential, E_D does the resistance of the solution fall rapidly to a relatively constant value and only then does the current show a rise with increasing potential. This behaviour is associated with the problems discussed above of the transport of charge across the electrode-electrolyte interface.

Even if the conductance of the electrolyte were substantially increased by, for example, the addition of large quantities of NaCl to the solution, the conductance of the *cell* would hardly change in the low potential region, since in this region, the overall resistance of the cell is determined primarily by the large non-ohmic resistances of the two electrode-electrolyte interfaces, which are in series with the electrolyte itself. It is clear that these large non-ohmic contributions must be eliminated before practical measurements can be carried out.

This can be achieved by measurements with *alternating current* (*ac*) techniques. If a potential difference ΔV is applied between the electrodes of an electrochemical cell, a double layer forms by migration of ions of opposite charge to the charge on the each of the two electrodes, as shown in Fig. 2.1. If an *ac* potential is now applied, this double layer will charge and discharge in the same manner as a capacitor, as shown in Fig. 2.2. Thus, the circuit formed of voltage source, wires, interfacial electrode impedance and electrolyte resistance can allow an alternating current to flow without any necessity for the actual transfer of charge across the electrode-electrolyte interface.

In describing the behaviour of an electrolysis cell when an alternating potential is applied, the cell can be replaced by an equivalent electrical circuit, as shown in Fig. 2.3. Each electrode can be represented by a parallel combination of a capacitance, corresponding to the capacitance of the double layer, C_D, and a non-linear resistance, represented by the symbol —▭— which models charge transfer across the interface. The ohmic resistance of the electrolyte is represented by the symbol ⊓⊔⊓⊔⊓.

If the alternating potential has the form $A_S \cdot \sin \omega t$, where $\omega \equiv 2\pi f$, and f the frequency, and where the signal amplitude A_s is such that $A_S \ll E_D$, where E_D is the decomposition potential defined in Fig. 1.3, then the non-linear interfacial resistances

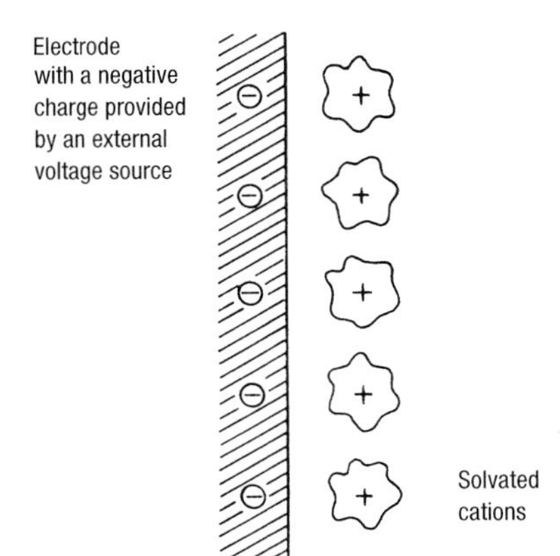

Fig. 2.1 Schematic representation of the electrolytic double layer at the metal/electrolyte interface.

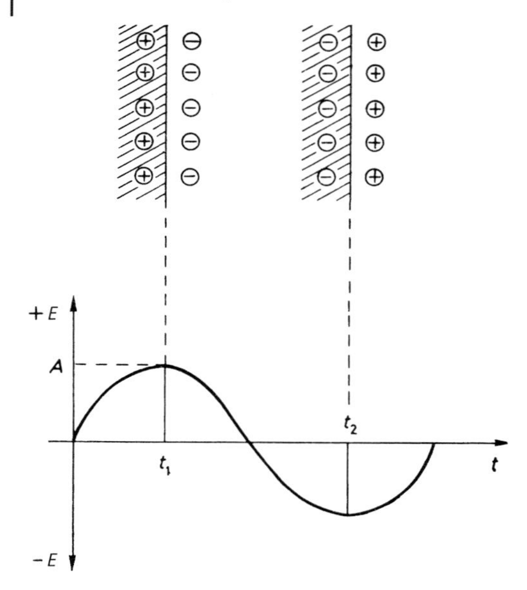

Fig. 2.2 Charging and discharging of the electrolytic double layer on imposition of an *ac* voltage $E_c = A \sin \omega t$; $\omega = 2\pi f$, f is the frequency of the *ac* signal. The potential drop across the interface, E, is shown as a function of time t.

R^+ and R^- will both be large and little faradaic current will flow. It can be shown that provided the angular frequency, ω, is such that $\omega \gg (1/R^+C_D^+)$, $(1/R^-C_D^-)$, then the impedances of the two interfaces will become very small compared to the resistance of the electrolyte solution. The equivalent circuit under these circumstances will then collapse to a simple resistor, and the relationship between potential and current will be linear, as shown schematically in Fig. 2.4.

The inequalities for ω can be satisfied either by working at high frequencies or by ensuring high values for the double-layer capacitancies. However, if measuring frequencies in excess of 50 kHz are used, considerable care must be taken to eliminate stray inductances in the circuit, since the impedance associated with an inductance rises linearly with frequency. Provided the frequency is not so high that this is a problem, then it is possible to carry out high precision measurements by extrapolating to infinite frequency as shown in Fig. 2.5. High double-layer capacitances can be

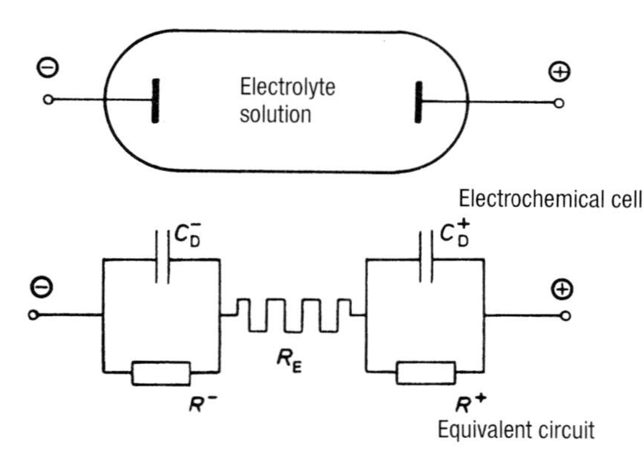

Fig. 2.3 Schematic diagram of an electrolysis cell and the equivalent *ac* circuit. R_E is the electrolytic resistance, R^-, C_D^- and R^+ and C_D^+ are the interfacial resistance and double layer capacity of the cathode and anode respectively.

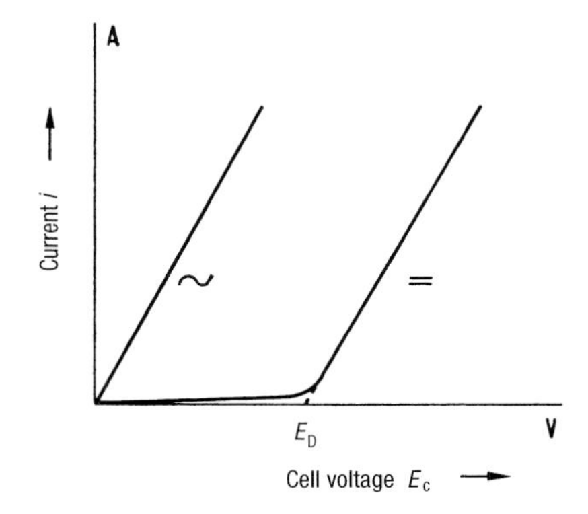

Fig. 2.4 Electrolytic current, i, as a function of the cell voltage, E_C, for $ac(\sim)$ and $dc(=)$ voltages. The decomposition voltage is E_D.

achieved by increasing the real surface area of the electrode. One way that this can be done is by platinisation of the electrode surface, a process in which a layer of platinum in finely divided form is deposited electrolytically from solution under carefully controlled conditions of potential and current.

As is generally the case in physics, resistance measurements on electrolyte solutions may be carried out with a Wheatstone Bridge using the arrangement shown in Fig. 2.6. In order to balance the double-layer capacitance, a variable capacitance is necessary, at least in principle, in the balance arm, as shown. Once a true balance is achieved, the ac potential between the points A and B will be zero, and this condition may be accurately detected by the use of an amplifier and oscilloscope.

Measurements which use both a variable capacitor and resistor to obtain a true balance are time-consuming, and only necessary if an accuracy of better than 0.1 % is sought. For routine measurements, it is usual to use just a resistor in the balance arm, and to adjust this until a minimum potential is found between A and B, or, more simply, until there is a change of sign in the potential between A and B.

In all measurements of impedance, careful temperature control is essential, since the viscosity of water, for example, changes in the region near room temperature by about 3 % for each degree rise. In addition, if highly accurate measurements are

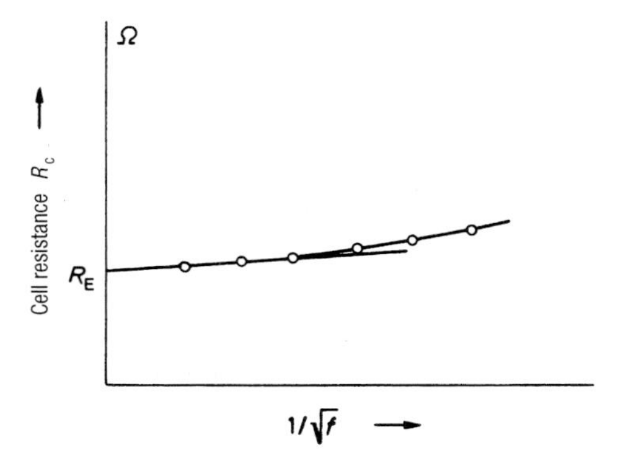

Fig. 2.5 Frequency dependence of the cell resistance, R_C. Extrapolation to $f \rightarrow \infty$ in the plot of R_C vs $1/\sqrt{f}$ allows the determination of the true electrolyte resistance.

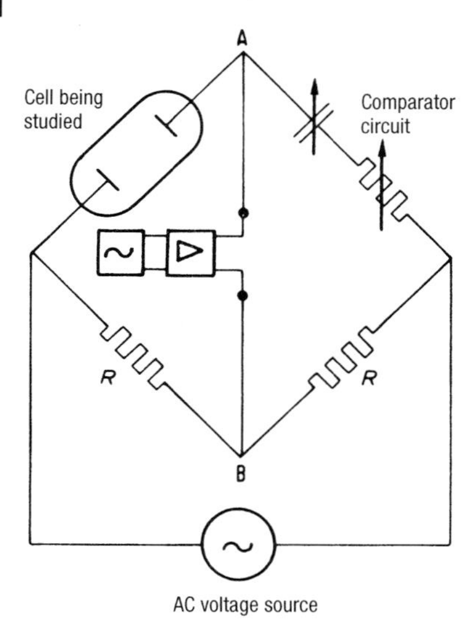

Fig. 2.6 Wheatstone bridge circuit for the measurement of electrolyte resistance; null detection is by an amplifier and oscilloscope.

needed, a whole series of further factors needs to be considered. Thus, for the determination of the conductance of dilute solutions, a correction must be made for the conductivity of the solvent, which may be significant, particularly if there are small concentrations of residual electrolyte present as impurities as well. Correction must also be made for the adsorption of ions from solution onto the electrode surface, an effect that may be reduced in importance by careful attention to the conditions of platinisation. One aspect of this problem that is universal in aqueous electrolyte is the presence of dissolved CO_2; accurate measurements should be carried out under nitrogen or argon on solutions freed of CO_2. Further details of the procedure are given, for example, in the text by Robinson and Stokes.

Determination of the *ac* impedance can also be carried out with an automatic bridge that employs a frequency generator and gives a direct read out of the in-phase and quadrature values of the impedance for each frequency. By exploring the impedance as a function of frequency, a complete correction may be made for the capacitance of the double layer.

2.1.3
The Conductivity

Just as the resistivity of a material is a characteristic property of that material, rather than the resistance itself, which contains contributions from purely geometrical factors, so we seek to define the conductivity for an electrolyte solution, which will be characteristic of that solution and independent of geometrical factors. For a cell of uniform cross-section, A, with electrodes at either end separated by a distance l, the ionic conductivity κ_I is related to the conductance L by the expression

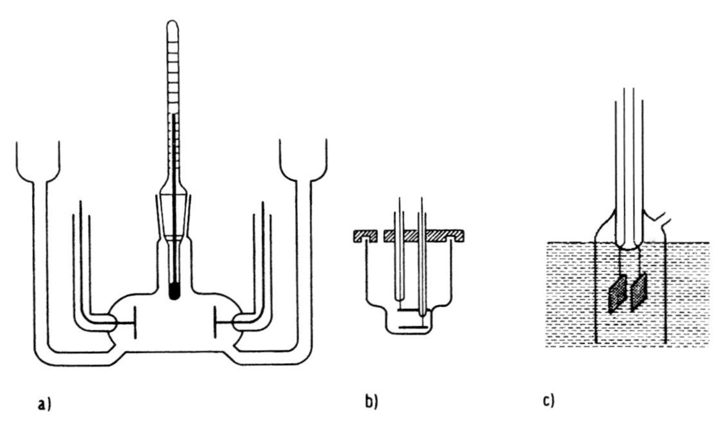

Fig. 2.7 Different types of cells for measurement of conductivity of electrolyte solutions (a) conductivity cell for precision measurements with inbuilt thermometer; (b) cell with variable inter-electrode separation; (c) immersion cell for rapid laboratory measurements.

$$\kappa_{\mathrm{I}} = \frac{lL}{A} \tag{2.8}$$

and evidently has the units $\Omega^{-1}\mathrm{m}^{-1}$ or Sm^{-1}. From equation (2.7) we see that κ_I takes the value:

$$\kappa_{\mathrm{I}} = e_0(n^+z^+u^+ + n^-z^-u^-) \tag{2.9}$$

In principle, the measurement of conductivity could be carried out in a cell, with rectangular electrodes of known area A m^2 positioned l m apart. However, in practice a number of complicated corrections would have to be made for edge effects. Rather than do this for all measurements, use is now made of definitive conductivity measurements for certain standard solutions carried out under very carefully controlled conditions in specially designed cells. By making a single measurement of one of these standards in a simple conductivity cell, such as one of those shown in Fig. 2.7, the so-called cell constant, $l/A = K_{\mathrm{cell}}$, can be obtained. Correction of all other measurements in this cell to absolute conductivities can then be made by multiplying the measured conductance by K_{cell}. The commonest standard used is aqueous KCl at 25°C, whose conductivity at various concentrations can be found in a number of standard compilations: for 0.01M ($\equiv$ 10 mol m^{-3}) KCl, the conductivity takes the value 0.1413 $\Omega^{-1}\mathrm{m}^{-1}$, so that if the actual measured conductance of the cell with 0.01M KCl was $L\ \Omega^{-1}$, $K_{\mathrm{cell}} = 0.1413/L$.

2.1.4 Numerical Values of Conductivity

As we might expect from equation (2.9), the values of the electrolyte conductivity obtained from experiment show a strong dependence on the concentration of the solution, and high conductivities are expected for high concentrations of fully dissociated electrolyte. If the degree of dissociation of the electrolyte is small, however, the conductivity will be considerably smaller. Most pure solvents themselves possess very low conductivities, which may be either intrinsic, arising from slight dissociation, or may be extrinsic, and due to irremovable quantities of more conducting impurities. Some representative examples of both are given in Table 2.1.

Table 2.1 Conductivities of different solvents and electrolytes, and some comparative values for molten salts, metals and solid electrolytes.

System	$T°$ C	κ_1 Ω^{-1}m^{-1}	Origin of conductivity
Pure benzene C_6H_6	20	$5 \cdot 10^{-12}$	Dissociation of traces of water to protons and OH^- ions
Pure methanol CH_3OH	25	$2 - 7.10^{-7}$	Slight dissociation to CH_3O^- and $CH_3OH_2^+$
Pure Acetic Acid	25	$\sim 4 \cdot 10^{-7}$	Slight dissociation to CH_3COO^- and $CH_3COOH_2^+$
Millipore® Water	25	$\sim 5.5 \cdot 10^{-6}$	Slight dissociation to OH^- and H_3O^+
Distilled Water	20	$10^{-3} - 10^{-4}$	Dissociation of CO_2 (aq.)
Saturated aq. AgCl	25	$1.73 \cdot 10^{-4}$	Ag^+ and Cl^- ions that form as slightly soluble AgCl completely dissociates
1.0M Aqueous Acetic Acid	25	0.13	Partial dissociation of the acid to CH_3COO^- and H_3O^+
1.0M LiCl in Methanol	20	1.83	Dissociation of LiCl to solvated Li^+ and Cl^- in methanol
1.0M LiCl in Water	18	6.34	Dissociation of LiCl to hydrated Li^+ and Cl^- in water
1.0M NaCl in Water	18	7.44	Dissociation to Na^+ and Cl^-
1.0M MgSO4 in Water	18	4.28	Near complete dissociation to Mg^{2+} and SO_4^{2-}
Saturated aq. NaCl ($\sim$ 5M)	18	21.4	Near complete dissociation to Na^+ and Cl^-
1.0M KOH in Water	18	18.4	Dissociation to K^+ and OH^-
1.0M H_2SO_4 in Water	18	36.6	Dissociation to H_3O^+ and SO_4^{2-}
3.5M H_2SO_4 in Water	18	73.9	Dissociation to H_3O^+, HSO_4^- and SO_4^{2-}
Stabilised Zirconia (85 % ZrO_2;15 % Y_2O_3)	1000	5.0	Oxide-ions migrating through the oxide-deficient lattice
Molten NaCl	1000	417	Completely dissociated Na^+ and Cl^- ions
Mercury	0	$1.063 \cdot 10^6$	Electronic conductor
Copper	0	$6.452 \cdot 10^7$	Electronic conductor

(Data from Landolt-Bornstein; Numerical Values and Functions, II 7, Springer-Verlag, 1960)

According to equation (2.9), increasing the charge on the ions should cause an increase in the conductivity, but this is clearly not the case if the conductivities of 1.0M (10^3 mol m^{-3}) NaCl and $MgSO_4$ are compared. The reason for this is that increasing the charge on the ion will increase the extent of interaction with the surrounding water dipoles, and hence the extent of hydration of the ion. Since the ionic radius in solution is effectively determined by the radius of the solvation sheath, then we might expect ions of higher charge to have larger radii and, hence, from equation (2.2), smaller mobilities. The balance between these two effects makes it very difficult to predict with confidence the comparative conductivities of strong 1–1 and 2–2 electrolytes. A discussion will be given below of the detailed variation of conductivity with concentration.

2.2 Empirical Laws of Electrolyte Conductivity

2.2.1 The Concentration Dependence of the Conductivity

Electrolytes that are completely dissociated irrespective of their concentration, the so-called *strong* electrolytes, should show a linear dependence of the conductivity on concentration from equation (2.9). However, this relationship is only approximately true even for dilute solutions, as shown in Fig. 2.8. At higher concentrations, the conductivity rises less rapidly than expected from extrapolation of the results in dilute solutions, owing to the increasing inter-ionic interactions as the mean distance between ions decreases. Given that electrostatic forces increase with the inverse square of the separation between ions, as shown in equation (1.1), and that such forces between oppositely charged ions would be expected to hinder the mobility, this behaviour

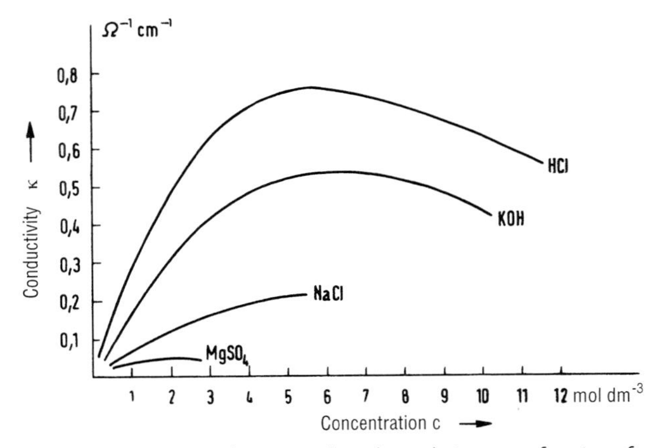

Fig. 2.8 Conductivity of aqueous electrolyte solutions as a function of concentration at 18°C (from Landolt Bornstein).

at higher concentrations can be rationalised. On this basis, we can also explain the difference between NaCl and $MgSO_4$ solutions: in the latter case, the Coulomb forces are expected, from equation (1.1) to be four times larger, and the deviation from linearity would be expected to be more marked, as indeed is observed in Fig. 2.8.

In very concentrated solutions, the conductivity actually begins to fall as the concentration increases. This is the result of ion-pair formation: as the mean interionic distance decreases, there comes a point at which the attractive forces between oppositely charged ions are sufficient to cause association, leading to neutral particles that cannot contribute to the overall conductivity.

2.2.2 Molar and Equivalent Conductivities

From equation (2.9), we recall that the conductivity of an electrolyte solution is given by:

$$\kappa_I = e_0(n^+z^+u^+ + n^-z^-u^-) \tag{2.9}$$

where n has the rational units m^{-3} and u the units $m^2V^{-1}s^{-1}$. If the concentration of the electrolyte is expressed as c mol m^{-3}, and the formula of the electrolyte is $A_{\nu+}B_{\nu-}$, which dissociates into A^{z^+} and B^{z^-}, then from the requirement for each formula unit to be neutral, $\nu^+z^+ = \nu^-z^- \equiv \nu_\pm z_\pm$, and given that the solution overall must also be neutral, it follows that

$$n^+z^+ = n^-z^- = \nu_\pm z_\pm cN_A \tag{2.10}$$

where N_A is Avogadro's constant. Substituting equation (2.10) into equation (2.9) we obtain

$$\kappa_I = \nu_\pm z_\pm cN_Ae_0(u^+ + u^-) \tag{2.11}$$

and the simple linear relationship between κ and c is evident. Now, if we wish to explore the relationship between concentration and mobility in more detail, then we should *divide* the conductivity by the *concentration* to obtain the *molar conductivity*, Λ; the rational units of Λ will be $\Omega^{-1}mol^{-1}m^2$, and clearly

$$\Lambda = \nu_\pm z_\pm N_Ae_0(u^+ + u^-) \tag{2.12}$$

In earlier texts, it was normal further to divide Λ by the equivalence number $\nu_\pm z_\pm$; the result is the equivalent conductivity, Λ_{eq}, which has the value

$$\Lambda_{eq} = N_Ae_0(u^+ + u^-) = F(u^+ + u^-) \tag{2.13}$$

Whilst the equivalent conductivity is very common in older textbooks and in major compilations of conductivity, its use is no longer encouraged as it is so closely related to molar conductivity.

2.2.3 Kohlrausch's Law and the Determination of the Limiting Conductivities of Strong Electrolytes

From equation (2.12), it is evident that the molar conductivity for a strong electrolyte should be independent of concentration provided that we can neglect interionic interactions. However, as we saw above, the effect of the electrostatic interactions between ions gives rise to a strongly non-linear variation of κ_I with c, and this implies that Λ is *not* independent of concentration.

In Table 2.2 and Fig. 2.9 the experimental values of κ_I and Λ for aqueous NaCl at 25°C are given as a function of concentration. If the molar conductivity is plotted against the square root of the concentration, $\sqrt{c/c^0}$, where c^0 is a standard concentration, taken here to be 1 mol dm^{-3}, a *linear* relationship is found at low concentrations, as can be seen from Fig. 2.9. This relationship is found to be universal for strong electrolytes, and was first enunciated by Kohlrausch in 1900 in the form:

$$\Lambda = \Lambda_0 - k\sqrt{c/c^0} \tag{2.14}$$

where Λ_0 is termed the molar conductivity at infinite dilution. The significance of Λ_0 is considerable: since it refers to infinite dilution, there will be no effects from interionic interactions. Its measurement is also straightforward: obviously direct measurement is impossible, since an infinitely dilute solution will have no conductivity contribution from the movement of the ions, but the use of Kohlrausch's law allows us to make a simple extrapolation from experimentally attainable values of Λ.

Table 2.3 gives a series of values of equivalent conductivity at different concentrations, including those at infinite dilution, and the graphical representation of these data as a function of $\sqrt{c/\nu_\pm z_\pm}$ is given in Fig. 2.10. It can clearly be seen that the value of k in equation (2.14) is larger for those electrolytes that dissociate into more highly charged ions, a reasonable result since interionic interactions will be inherently larger in such systems.

Table 2.2 Conductivity, κ_I, and molar conductivity, Λ for aqueous solutions of NaCl at 25°C.

c/mol dm^{-3}	$\kappa_I/\Omega^{-1}m^{-1}$	$\Lambda/10^{-4}\ \Omega^{-1}mol^{-1}m^2$
0	-	126.45
0.0005	6.2250×10^{-3}	124.5
0.001	1.2374×10^{-2}	123.74
0.005	6.0325×10^{-2}	120.65
0.01	1.1851×10^{-1}	118.51
0.02	2.3152×10^{-1}	115.76
0.05	5.5530×10^{-1}	111.06
0.1	1.0674	106.74

Table 2.3 $\Lambda/\nu_\pm z_\pm$ for some aqueous electrolytes at 25°C at different concentrations.

$(c/\nu_\pm z_\pm)$/mol dm^{-3}				$(\Lambda/\nu_\pm z_\pm)/10^{-4}\ \Omega^{-1}$ mol^{-1} m^2				
$(\nu_\pm z_\pm)$	HCl (1)	H_2SO_4 (2)	NaOH (1)	KCl (1)	NaCl (1)	$CuSO_4$ (2)	CH_3COONa (1)	CH_3COOH (1)
0	426.16	429.8	247.8	149.86	126.45	133.6	91.0	390.57
0.0001114								127.71
0.0005	422.74	413.7	245.6	147.81	124.5	121.6	89.2	
0.001	421.36		244.7	146.95	123.74	115.26	88.5	
0.001028								48.13
0.005	415.8	390.8	240.8	143.35	120.65	94.07	85.72	
0.005912								20.96
0.01	412.0		238.0	141.27	118.51	83.12	83.76	
0.0125		327.5						
0.01283								14.37
0.02	407.24			138.34	115.51	72.20	81.24	11.56
0.05	399.09	273.0		133.37	111.06	59.05	76.92	7.36
0.1	391.32	251.2		128.96	106.74	50.58	72.80	5.20

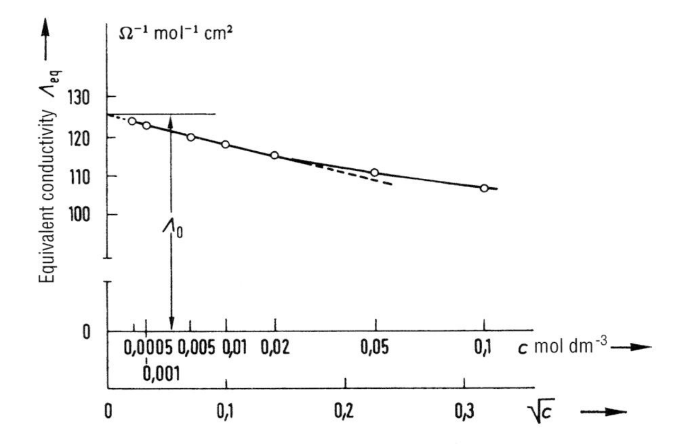

Fig. 2.9 Equivalent conductivity of aqueous NaCl electrolyte solutions as a function $\sqrt{c}$ of (from Table 2.2).

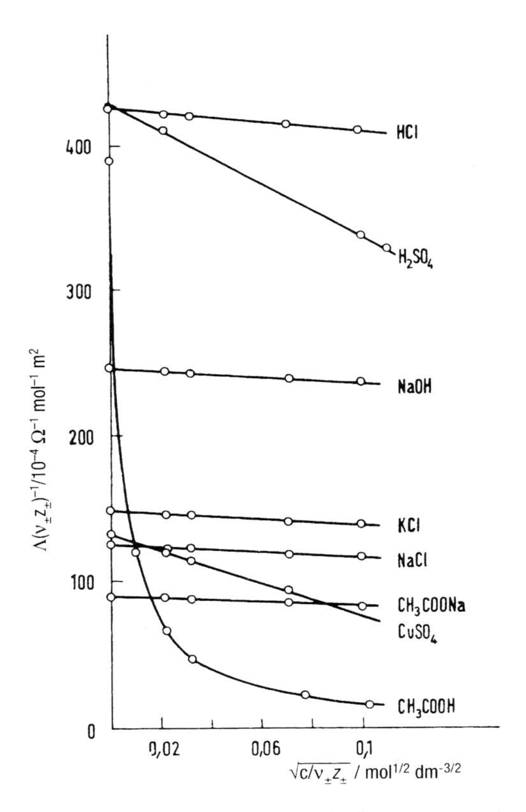

Fig. 2.10 Equivalent conductivity of different aqueous electrolyte solutions at 25°C as a function of $\sqrt{c/v_\pm z_\pm}$ (from Table 2.3).

2.2.4 The Law of Independent Migration of Ions and the Determination of the Molar Conductivity of Weak Electrolytes

The variation of the molar (= equivalent) conductivity of acetic acid is given in Table 2.3 and in Fig. 2.10. Acetic acid is a representative *weak electrolyte*, a family comprising, in aqueous solution, primarily organic acids and bases. In contrast to strong electrolytes, weak electrolytes show a concentration dependence of the *degree of dissociation*, which last can be defined as the ratio of the amount of dissociated species to the total amount of species present. In the case of acetic acid, the extent of dissociation remains comparatively small even at quite high dilutions, and dissociation is only complete at infinite dilution. It follows that the variation of equivalent conductivity with concentration is determined primarily by the change in dissociation, and as a result, the plot of Λ vs. $\sqrt{c}$ is non-linear, as shown in Fig. 2.10.

For weak electrolytes, the limiting equivalent conductivity can, as a result, be determined directly only with great difficulty, since accurate extrapolation to infinite dilution using equation (2.14) is not possible. However, an indirect method is possible from a second law discovered by Kohlrausch: in ideal dilute solutions, in which interactions between the ions can be neglected completely, the individual ions migrate in the electric field essentially independently of each other. As a consequence, we can associate with each ion an individual limiting molar ionic conductivity, $\lambda_0^{\pm}$. The limiting conductivity of the solution, Λ_0, can simply be written as the stoichiometrically weighted sum of the limiting molar conductivities of the anion and cation:

$$\Lambda_0 = \nu_+ \lambda_0^+ + \nu_- \lambda_0^- \tag{2.15}$$

The essential difference between (2.15) and equation (2.12) is that the mobilities u^+ and u^- are *not*, in general, independent either of concentration or of the nature of the two ions. Thus, the mobility of the K^+ ion in 0.1M (100 mol m^{-3}) KCl is not the same as the mobility of the same ion in 0.1M KNO_3. However, at infinite dilution, when the ions move entirely independently of each other, the mobilities, and hence the limiting molar or equivalent ionic conductivities, *are* independent of each other and can, in principle, be tabulated without reference to the identity of the counterion.

The additivity of mobilities encapsulated in the *law of independent migration of ions at infinite dilution* can be verified immediately from a comparison between the differences in limiting conductivities of pairs of electrolytes each of which has one type of ion in common, as can be seen from Table 2.4.

With the help of the law of independent migration of ions, we can now calculate the limiting conductivities of weak electrolytes from those of strong electrolytes. Thus, for acetic acid, we can write:

$$\begin{aligned}
\Lambda_0^{CH_3COOH} &= \lambda_0^{H_3O^+} + \lambda_0^{CH_3COO^-} \\
&\equiv \lambda_0^{H_3O^+} + \lambda_0^{CH_3COO^-} + \lambda_0^{Na^+} - \lambda_0^{Na^+} + \lambda_0^{Cl^-} + \lambda_0^{Na^+} \\
&\equiv \Lambda_0^{HCl} + \Lambda_0^{CH_3COONa} - \Lambda_0^{NaCl}
\end{aligned} \tag{2.16}$$

where the last line is derived by rearranging the terms.

Table 2.4 Values of the limiting molar conductivities, Λ_0, of aqueous electrolytes at 25°C to illustrate the law of independent migration of ions in ideal dilute solutions.

Electrolyte	$\Lambda_0/10^{-4}\Omega^{-1}$ $m^2 mol^{-1}$	$\Delta\Lambda_0 = \lambda_0^{K^+} - \lambda_0^{Na^+}$	Electrolyte	$\Lambda_0/10^{-4}\Omega^{-1}$ $m^2 mol^{-1}$	$\Delta\Lambda_0 = \lambda_0^{Cl^-} - \lambda_0^{NO_3^-}$
KCl	149.86	23.4	KCl	149.86	4.9
NaCl	126.46		KNO_3	144.96	
KI	150.4	23.4	LiCl	115.25	4.3
NaI	127.0		$LiNO_3$	111.0	
KOH	271.0	23.2	NH_4Cl	149.9	4.6
NaOH	247.8		NH_4NO_3	145.3	

After putting in the values for the limiting equivalent conductivities of the strong electrolytes: hydrochloric acid, sodium acetate and sodium chloride, from Table 2.3, we find

$$\Lambda_0^{CH_3COOH} = (426.16 + 91.0 - 126.45) \cdot 10^{-4} = 390.71 \cdot 10^{-4}\ \Omega^{-1}m^2mol^{-1} \tag{2.17}$$

As a footnote to this section, it is worth pointing out that we can, of course, split up the molar conductivity of any solution into anion and cation contributions, even where the concentration is finite and the solution non-ideal. Thus, we can always write, by analogy with equation (2.12)

$$\Lambda = (\nu_+\lambda^+ + \nu_-\lambda^-) \tag{2.18}$$

However, it has to be emphasised that since the mobilities u^+ and u^- are not independent of each other or of the overall concentration, the values of λ^+ and λ^- are also no longer independent of the identity of their counterions, since incorporated into them are the important interionic interactions of the types discussed above. For example, the ionic conductivity of a cation will be a function both of the concentration of the electrolyte, and the identity of the counter-anion. It will also depend on the concentrations and identities of any other ions present in the same solution.

2.3 Ionic Mobility and Hittorf Transport

We have seen that only at infinite dilution can the molar conductivity be split into two limiting molar conductivities associated with individual ions, which are independent of each other. This is because only at infinite dilution can we completely neglect interionic interactions. However, in order to determine the values of the individual ionic conductivities, we must carry out an additional measurement, since it is obvious from

Table 2.4 above that only differences are available from conductivity measurements themselves. In order to partition Λ_0 into λ_0^+ and λ_0^- we must determine the so-called *transport numbers* of the individual ions.

2.3.1
Transport Numbers and the Determination of Limiting Ionic Conductivities

We recall, from equation (2.3), that the total current i, can be written as the sum of partial currents i^+ and i^-, corresponding to the currents carried by the cations and anions. We define the transport number of the cations, t^+, as:

$$t^+ = \frac{i^+}{i^+ + i^-} \tag{2.19}$$

and similarly

$$t^- = \frac{i^-}{i^+ + i^-} \tag{2.20}$$

and from these definitions it can clearly be seen that $t^+ + t^- = 1$. From this it can also be seen that the transport numbers of cation and anion are *not* independent of concentration; in fact, from equations (2.3) and (2.12)

$$t^+ = \frac{u^+}{u^+ + u^-} \tag{2.21}$$

$$= \frac{\nu_+ \lambda^+}{(\nu_+ \lambda^+ + \nu_- \lambda^-)} \tag{2.22}$$

and

$$t^- = \frac{(\nu_- \lambda^-)}{(\nu_+ \lambda^+ + \nu_- \lambda^-)} \tag{2.23}$$

so we can see that the larger the transport number of an ion, the larger is its contribution to the overall conductivity. From (2.22) and (2.23) we see that

$$\lambda^+ = \frac{t^+ \Lambda}{\nu_+} \tag{2.24}$$

$$\lambda^- = \frac{t^- \Lambda}{\nu_-} \tag{2.25}$$

and at infinite dilution

$$\lambda_0^+ = \frac{t_0^+ \Lambda_0}{\nu_+} \tag{2.26}$$

$$\lambda_0^- = \frac{t_0^- \Lambda_0}{\nu_-} \tag{2.27}$$

It follows that the values of the limiting ionic conductivities λ_0^+, λ_0^- are accessible provided that we can measure Λ_0 and we can determine the transport numbers of anion and cation at infinite dilution.

The experimental determination of transport numbers is described in the next section, 2.3.2., but it is, of course, the case that such measurements can only be carried out in solutions of finite concentration. However, since the values of λ^+ and λ^- vary in the same sense with increasing concentration, as we saw above, the transport number itself varies relatively little with concentration. Indeed, in the so-called Debye-Hückel region, with the concentration less than ca. 0.001 mol dm^{-3}, there is very little error is involved in simply equating t_0^+, t_0^- with t^+ and t^-.

If there are more than two sorts of ion present in the electrolyte solution, the transport number of the ith ion can be defined as:

$$t_i = \frac{i_i}{\left(\sum_i i_i\right)} \tag{2.28}$$

and from (2.5) above, we see, by extension, that

$$t_i = \frac{|z_i| c_i u_i}{\left(\sum_i |z_i| c_i u_i\right)} \tag{2.29}$$

a formula that we shall need in Section 3.2.4.

2.3.2
Experimental Determination of Transport Numbers

Experimentally, the partition of the charge transport through an electrolyte solution into contributions from anion and cation was first investigated by Hittorf in the years 1853–1859. For this purpose, Hittorf used, inter alia, the arrangement shown in Fig. 2.11. As an example, we could have a cell consisting of an anode and cathode formed from platinum and an electrolyte of hydrochloric acid. On current flow, hydrogen is evolved at the cathode and chlorine gas at the anode, and on the passage of one Faraday of charge through the cell, the following changes in composition will be observed in the various parts of the cell, as given in Table 2.5.

In total, t^+ mol HCl have disappeared from the anode compartment and t^- mol HCl from the cathode compartment. By quantitatively analysing the solutions around the anode and cathode before and after the passage of a known amount of charge, it is, therefore, possible to determine the transport numbers. It is also clear that the principle of Hittorf's method, sketched above, can be extended without difficulty to other systems whose ionic composition and electrode reactions are known. However, in any measurement, care must be always be taken to ensure that the concentration differences generated by the passage of current are not reduced either by back-diffusion or by stirring of the solution associated, for example by the evolution of gas bubbles at either of the two electrodes. The first problem can be tackled by using a long diffusion path length between the anode and cathode compartments and short measuring times, and the second by placing frits between the two compartments. In more precise mea-

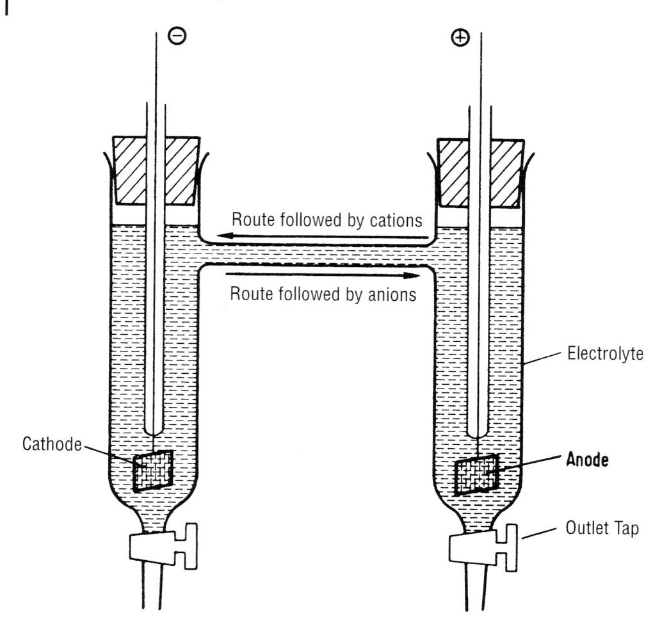

Fig. 2.11 Hittorf transport cell.

surements, correction must also be made for concentration changes arising due from the unequal transport of water molecules in the hydration sheaths of the two ions. It is also possible to carry out such measurements in non-aqueous solvents and in molten

Table 2.5 Changes in composition on passage of one Faraday of charge through hydrochloric acid.

Anode Space	Cathode Space	Processt
$-t^{+}$ mol H^{+}	$+t^{+}$ mol H^{+}	The cation contribution to the charge passed is $t^{+}F$, and this is the charge transported by the cations as they move from the anode compartment to the cathode compartment.
$+t^{-}$ mol Cl^{-}	$-t^{-}$ mol Cl^{-}	This corresponds to the transport of charge $t^{-}F$ by the anions which migrate from the cathode compartment towards the anode compartment.
-1 mol Cl^{-}	-1 mol H^{+}	Evolution of $^{1}/_{2}$ H_2 from the cathode compartment and $^{1}/_{2}$ Cl_2 from the anode compartment.
$(t^{-}-1)$ mol Cl^{-}	$(t^{+}-1)$ mol H^{+}	
$-t^{+}$ mol H^{+}	$-t^{-}$ mol Cl^{-}	Sum of processes
$= -t^{+}$ mol HCl	$-t^{-}$ mol HCl	

and solid electrolytes. It is also possible to measure transport numbers in non-aqueous solutions and in molten and solid electrolytes both by Hittorf and by other methods described below.

2.3.3 Magnitudes of Transport Numbers and Limiting Ionic Conductivities

Transport numbers of anions and cations will be about equal if the contributions of the two ions to the overall charge transport are about equal, but will differ appreciably if charge transport is predominantly through one of the ions. Some examples of both types of situation are given in Table 2.6

It can be seen that, in accord with the definition, the transport number of an ion depends on the identity of the counter-ion.

From the transport number on the one hand and the total limiting conductivity Λ_0 on the other, we can, from equations (2.26) and (2.27) immediately obtain the limiting ionic conductivities of all the ions separately. A representative number of values are given in Table 2.7, which has been compiled from a number of different sources.

The data above show that the extraordinarily high ionic conductivity found for the mineral acids can clearly be traced to the high conductivity of the proton, and in a similar way the high conductivity of the hydroxide ion is found to lead to high conductivities for the simple inorganic bases. In contrast, the limiting ionic conductivities of other ions are appreciably smaller, and there is a general trend, shown in the column on the left, for the mobilities of divalent cations to be smaller than monovalent cations when the molar ionic conductivities are corrected for charge. As is apparent from the rest of the right-hand column, anions show the opposite trends. Finally, it should be pointed out that from the additivity law of equation (2.15), it is, of course, possible to calculate the limiting conductivities of weak electrolytes, which are, of course, not experimentally accessible.

Table 2.6 Transport numbers t of anions and cations in selected aqueous electrolytes at 25°C, extrapolated to infinite dilution.

Electrolyte	t_0^+	$t_0^+ (= 1 - t^+)$
KCl	0.4906	0.5094
NH_4Cl	0.4909	0.5091
HCl	0.821	0.179
KOH	0.274	0.726
NaCl	0.3962	0.6038
$NaOOCCH_3$	0.5507	0.4493
$KOOCCH_3$	0.6427	0.3573
$CuSO_4$	0.375	0.625

Table 2.7 Limiting ionic conductivities, λ_0, in aqueous solution at 25°C.

Ion	λ_0^+, λ_0^-/$10^{-4}\Omega^{-1}mol^{-1}m^2$	Ion	λ_0^+, λ_0^-/$10^{-4}\Omega^{-1}mol^{-1}m^2$
		Ag^+	62.2
H^+	349.8	Na^+	50.11
OH^-	197	Li^+	38.68
K^+	73.5	$[Fe(CN)_6]^{4-}$	440
NH_4^+	73.7	$[Fe(CN)_6]^{3-}$	303
Rb^+	77.5	$[CrO_4]^{2-}$	166
Cs^+	77	$[SO_4]^{2-}$	161.6
		I^-	76.5
Ba^{2+}	126.4	Cl^-	76.4
Ca^{2+}	119.6	NO_3^-	71.5
Mg^{2+}	106	CH_3COO^-	40.9
		$C_6H_5COO^-$	32.4

2.3.4 Hydration of Ions

We know from equations (2.12) and (2.15) that the limiting ionic conductivities are directly proportional to the limiting ionic mobilities: in fact

$$\lambda_0^+ = z^+ F u_0^+ \tag{2.30}$$

$$\lambda_0^- = z^- F u_0^- \tag{2.31}$$

and we recall from equations (2.1) and (2.4) that if Stokes law is valid, we can write, approximately,

$$u_0 = \frac{z e_0}{6\pi\eta r_1} \tag{2.32}$$

We can now compare the limiting mobilities of ions having different charges, as, for example, K^+ and Ca^{2+}, which are nearest neighbours in the periodic table. For this purpose, we calculate from Table 2.7 the respective equivalent conductivities. These are $73.5 \cdot 10^{-4}$ for K^+ and $(119.6/2) \cdot 10^{-4} = 59.8 \cdot 10^{-4}\ \Omega^{-1}\ \text{equiv}^{-1}\ m^2$ for Ca^{2+}. We conclude that the Ca^{2+} species has the lower mobility (see equation 2.30). However, according to equation 2.32, we would expect a higher mobility for Ca^{2+} since it has the smaller ionic radius. This apparent contradiction can be explained by taking into account the fact that ions in solution possess a solvation sheath, which moves with the ion. Because the Ca^{2+} ion has a larger interaction with the water dipoles than K^+, owing to its larger charge, it has a larger solvation sheath and hence a larger effective radius. The observed mobilities are, therefore, in agreement with equation (2.32) provided that we remember to include the solvation sheath in the overall ionic radius.

It is also possible to explain, from hydration models, the differences between equally-charged cations, such as the alkali metals ($\lambda_0^{K^+} = 73.5$, $\lambda_0^{Na^+} = 50.11$ and $\lambda_0^{Li^+} = 38.68$ all in units of $10^{-4}\ \Omega^{-1}mol^{-1}m^2$). From atomic physics, it is known that the radii of the bare ions is in the order $Li^+ < Na^+ < K^+$. The attraction of the water dipoles to the cation increases strongly as the distance between the charge centres of cation and water molecule decreases, with the result that the total radius of ion and bound water molecules actually increases in the order $K^+ < Na^+ < Li^+$, and this accounts for the otherwise rather strange order of mobilities.

The differing extent of hydration shown by different types of ion can be determined experimentally from the amount of water carried over with each type of ion. A simple possible measurement can be carried out by adding an electrolyte such as LiCl to an aqueous solution of sucrose in a Hittorf cell. On passage of charge, the strongly hydrated Li^+ ions will migrate from the anode to the cathode compartment, whilst the more weakly hydrated Cl^- ions migrate towards the anode compartment; the result is a slight *increase* in concentration of sucrose in the anode compartment, since the sucrose itself is essentially electrically neutral and does not migrate in the electric field. The change in concentration of the sucrose can either be determined analytically or by measuring the change in rotation of plane polarised light transmitted through the compartment. Measurements carried out in this way lead to hydration *numbers* for ions, these being the number of water molecules that migrate with each cation or anion. Values of 10-12 for Mg^{2+}, 5.4 for K^+, 8.4 for Na^+ and 14 for Li^+ are clearly in reasonable qualitative agreement with the values inferred from the Stokes law arguments above. They are also in agreement with measurements carried out using large organic cations to calibrate the experiment, since these are assumed not to be hydrated at all.

By contrast, anions are usually only weakly hydrated, and from equation (2.32), this would suggest that increasing the charge on the anion should lead unequivocally to an increase in mobility and hence to an increase in limiting ionic conductivity. The rather low conductivities exhibited by organic anions is a result of their considerably larger size: even taking hydration into account, their total diameter normally exceeds that of the simple anions.

It must be emphasised that the dependence on radius discussed here is only a rule of thumb. The reason is that the use of equation (2.32) presupposes that Stokes' law for the frictional drag of a sphere through a liquid medium is valid even at molecular dimensions. This is most unlikely to be rigorously true, since the use of a macroscopically averaged quantity such as viscosity to describe the essentially microscopic process of molecular interaction is at best an approximation.

2.3.5
The Enhanced Conductivity of the Proton, the Structure of the H_3O^+ Ion and the Hydration Number of the Proton

Protons and hydroxide ions are hydrated in a similar manner to metal ions, and have comparable radii; they should, therefore, possess similar mobilities and limiting equivalent conductivities. However, it is clear from Table 2.7 that their mobilities are far higher than might be expected, and we must seek a special mechanism to account for this. To understand this mechanism, it is first helpful to describe in more detail the structure of the hydrated proton in aqueous solution.

The water molecule possesses a large dipole moment, which can be traced to the fact that the molecule itself is *non-linear*, with an H-O-H angle of 104.45°, as shown in Fig. 2.12. This non-linearity ensures that the centres of positive and negative charge cannot be coincident, and hence there must be a dipole moment. Free protons cannot exist as such in aqueous solution: the very small radius of the proton would lead to an enormous electric field that would polarise any molecule, and in aqueous solution, the proton immediately attaches itself to the oxygen atom of a water molecule giving rise to an H_3O^+ ion. In this ion, however, the positive charge does not simply reside on a single hydrogen atom; all three hydrogen atoms are equivalent, giving a structure similar to that of the NH_3 molecule (as confirmed by NMR, and shown in Fig. 2.13), and the positive charge is shared equally amongst them.

Let us now consider the approach of a proton to the water molecule of Fig. 2.12 from the left hand side. With the formation of the H_3O^+ ion, the charge is shared equally among the three hydrogen atoms of the ion, and if the hydrogen on the opposite side now splits off from the water molecule as a proton, the effect will have been the transport of a proton through a distance equal to the diameter of the water molecule without this transport having been accompanied by the physical movement through this distance of the incoming proton. In fact, this apparent transport has its origin in the rearrangement of the bonding electrons, which have in effect moved from one side of the molecule to the other.

To see how this mechanism applies to the movement of protons in water, we consider now the scheme shown in Fig. 2.14. The splitting off of a proton from an H_3O^+ molecule can take place provided there is a water molecule nearby that can act as an

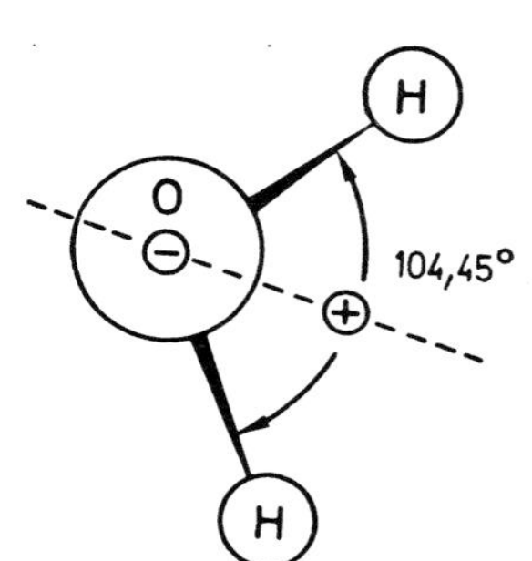

Fig. 2.12 Schematic representation of an H_2O molecule with centers of positive and negative charges.

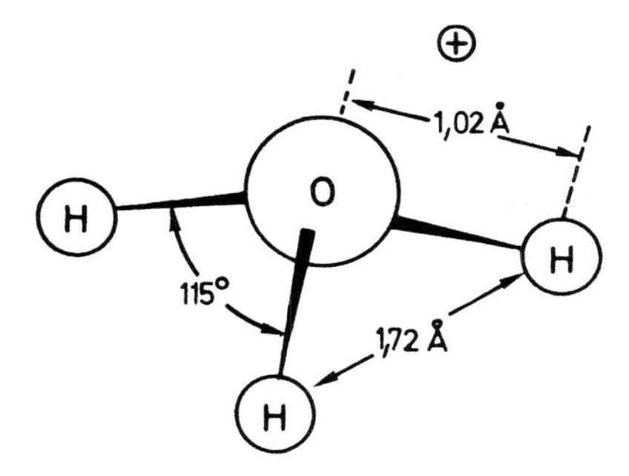

Fig. 2.13 Structure of the hydronium ion, H_3O^+.

acceptor. It is evident from Fig. 2.14 that this acceptor molecule must be in the correct orientation with respect to the H_3O^+ ion before transfer of the proton takes place, but that once the orientation is favourable, the proton can be transferred rapidly by a process known as tunnelling. This is a quantum mechanical effect: the very low mass of the proton facilitates a non-classical penetration of the energy barrier that in purely classical mechanics would have to be surmounted if the transfer is to be effected. This penetration is an extremely sensitive function not only of the mass of the particle but of the distance over which tunnelling must take place. It follows that the relative orientation of the H_3O^+ ion and H_2O molecule is critical, and in proton transport it is this re-orientation that determines the frequency with which proton movements can take place.

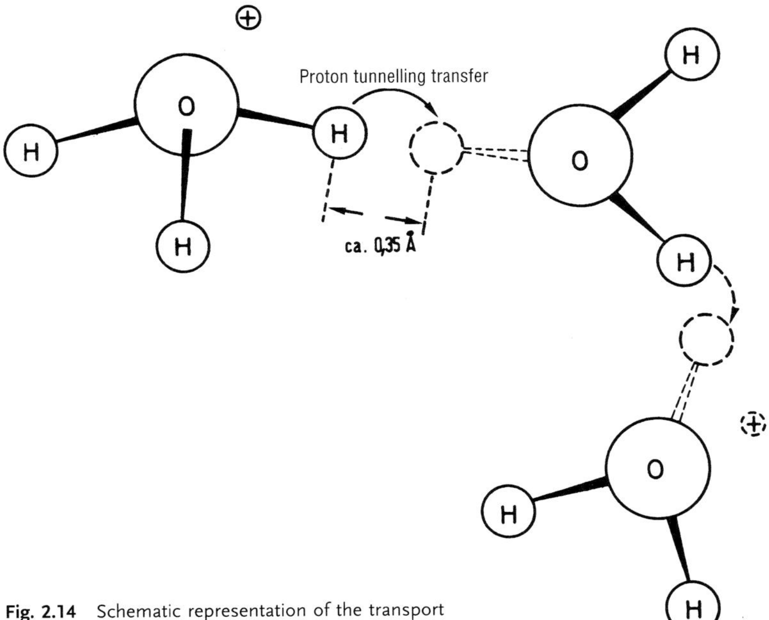

Fig. 2.14 Schematic representation of the transport mechanism of protons through water.

As Fig. 2.14 shows, proton transfer by this mechanism would not lead to any nett charge transfer to take place in any particular direction. However, if an electric field be applied, transfer will be favoured in the direction of the field, leading to the passage of a net current, whose origin, as we have seen, is the rearrangement of bonding electron density and the tunnelling processes themselves. We also saw that the reorientation of the water molecules is the rate limiting step. The need for reorientation arises because the local ordering of water molecules in the liquid phase is dominated by hydrogen bonding: the acceptor water molecule must first be decoupled from this local ordering before proton transfer from the neighbouring H_3O^+ can be effected. Although hydrogen bonding is comparatively weak (usually in the order 10–40 kJ mol^{-1}), the cumulative effect is substantial since several hydrogen bonds must be broken before re-orientation takes place. This obviously disfavours the tunnelling process, but the re-orientation is actually favoured by the electrostatic effect of the H_3O^+ field on the H_2O molecule. Thus, we can see that the enhanced conductivity of the proton is intimately bound up to the structure of the H_3O^+ ion and water molecule, and we expect, and find, that in other solvents the H_3O^+ ion has a similar mobility to that of other cations of comparable size.

The mechanism described in the previous paragraph is supported by the observation of the effect of temperature. It is found that the mobility of the proton goes through a maximum at a temperature of 150°C (where, of course, the measurements are carried out under pressure). This arises because as the temperature is increased from ambient, the main initial effect is to loosen the hydrogen bonded local structure that inhibits re-orientation. However, at higher temperatures, the thermal motion of the water molecules becomes so marked that the tunnelling probability itself begins to decline.

It is clear that this model can be applied to the anomalous conductivity of the hydroxide ion without any further modification. Hydrogen-atom tunnelling from a water molecule to a OH^- ion will leave behind a OH^- ion, and the migration of OH^- ions is, in fact, traceable to the migration of H^+ in the opposite direction.

The complete hydration shell of the proton consists of both the central H_3O^+ unit and further associated water molecules; mass spectrometric evidence would suggest that a total of four water molecules form the actual $H_9O_4^+$ unit, giving a hydration number of four for the proton. Of course, the measurement of this number by the Hittorf method is not possible since the transport of protons takes place by a mechanism that does not involve the actual movement of this unit.

2.3.6
The Determination of Ionic Mobilities and Ionic Radii: Walden's Rule

From equations (2.30) and (2.31), we can calculate the magnitudes of the mobilities for cations and anions. As an example, from Table 2.7, the limiting ionic conductivity for the Na^+ ion is $50.11 \times 10^{-4}\ \Omega^{-1}\ m^2\ mol^{-1}$. From this we obtain a value of $u_0^+ = \lambda_0^+/F \equiv 5.19 \times 10^{-8}\ m^2\ V^{-1}\ s^{-1}$, which implies that in a field of 100 V m^{-1}, the sodium ion would move a distance of ca. 2 cm in one hour. The mobilities of other ions have about the same magnitude (4–8 × 10^{-8} m^2 V^{-1} s^{-1}), with the marked exception of the proton. This has an *apparent* mobility of 3.63 × 10^{-7} m^2 V^{-1} s^{-1}, almost an order of magnitude higher, reflecting the different conduction mechanism described above.

Ionic mobilities decrease with increasing concentration, as interionic interactions become more important. From Tables 2.2 and 2.6, for example, the contribution of the Na^+ ion to the conductivity of 0.1M (100 mol m^{-3}) NaCl is clearly $106.7 \times 10^{-4} \times 0.396 \equiv 42.2 \times 10^{-4}$ $\Omega^{-1}m^2mol^{-1}$, which corresponds to a mobility of 4.37 × 10^{-8} m^2 V^{-1} s^{-1}.

The mobilities of ions can be directly measured if two electrolyte solutions, each of which contains one ion in common but having a different counter ion, can be arranged so that a sharp boundary between the two solutions can be maintained. On application of an electric field, and subsequent flow of charge, the movement of this boundary can be followed, usually by virtue of the different refractive indices of the two solutions, the technique being known as the "moving boundary method". The experimental setup is shown in Fig. 2.15.

With the knowledge now of the magnitude of the mobility, we can use equation (2.32) to calculate the radii of the ion; thus for lithium, using the value of 0.00089 kg m^{-1} s^{-1} for the viscosity of pure water (since we are using the conductivity at infinite dilution), the radius is calculated to be 2.38 × 10^{-10} m (≡ 2.38 Å). This can be contrasted with the crystalline ionic radius of Li^+, which has the value 0.78Å. The difference between these values reflects the presence of the hydration sheath of water

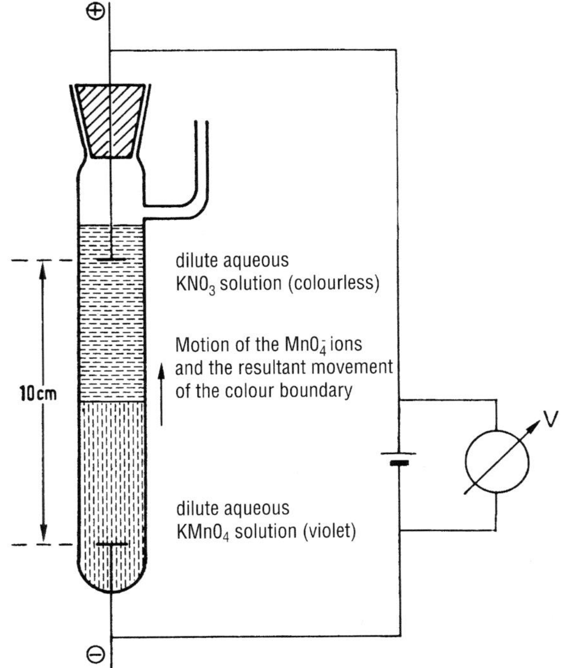

Fig. 2.15 Construction of a cell for observation of the "moving boundary" between two electrolytes. For accurate work, the boundary must not be disturbed by gas evolution from the cathode.

molecules; as we showed above, transport measurements suggest that Li^+ has a hydration number of 14.

From equation (2.32) we can, finally, deduce Walden's rule, which states that the product of the ionic mobility at infinite dilution and the viscosity of the pure solvent is a constant. In fact:

$$u_0 \eta = \frac{z e_0}{6 \pi r_I} = \text{const.} \tag{2.33}$$

whereby, $\lambda_0 \eta =$ constant and $\Lambda_0 \eta =$ constant. This rule permits us to make an estimate of the change of u_0 and λ_0 with change in temperature and alteration of the solvent, simply by incorporating the changes in viscosity.

It must be emphasised that Walden's rule is only valid so long as Stokes' law can be used to describe the frictional drag on the ion. It is thus clearly not valid for the H_3O^+ or OH^- ion, and it is also not valid if the ionic radius itself changes with temperature or solvent. This latter problem will be minimised if we use large molecules that are only slightly solvated, and Walden's rule is expected to work best for large organic ions.

2.4 The Theory of Electrolyte Conductivity: The Debye-Hückel-Onsager Theory of Dilute Electrolytes

2.4.1 Introduction to the Model: Ionic Cloud, Relaxation and Electrophoretic Effects

In 1923, Debye and Hückel developed a simplified model to account for the observed deviations from ideal behaviour found for electrolyte solutions.

Furthermore, they applied their theory to calculate the electrical mobility of ions in solution, a treatment that was improved by Onsager. The result is termed the Debye-Hückel-Onsager theory of conductivity, and we begin by describing the basic ideas of this theory.

Solvated ions dissolved in an appropriate solvent arrange themselves at equilibrium as a result of two opposing effects. The electrostatic forces between two ions will tend to ensure that any ion attracts ions of opposite charge and repels ions of the same charge, leading to the establishment of local ordering in which an ion will be surrounded by a centrosymmetric cloud of oppositely charged ions. Opposing this local ordering is the thermal motion of the ions, which tends to smear out this simple picture by introducing random motion of the two types of ion, and the net picture is shown in Fig. 2.16, in which the attempts of each ion to surround itself with a cloud of oppositely charged ions is illustrated.

On application of an electric field, positive and negative ions are accelerated in opposite directions, and the charge distribution shown in Fig. 2.16 becomes severely distorted. Each migrating ion will attempt to rebuild its atmosphere during its motion, but this rebuilding process will require a certain time, termed the *relaxation time*, so that the central ion, on its progress through the solution, will always be a

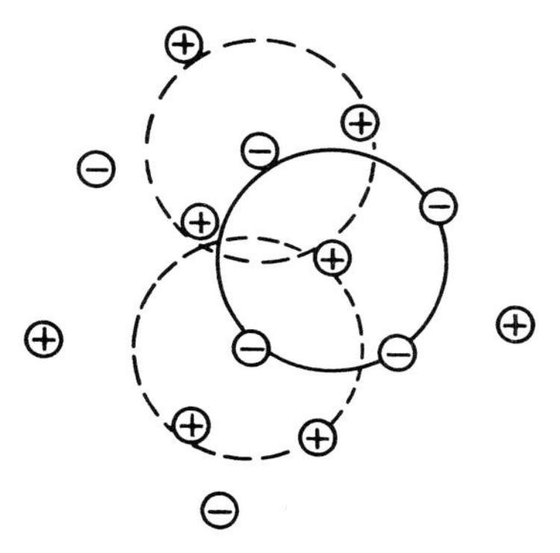

Fig. 2.16 Schematic representation of the ionic atmosphere in solution.

little displaced from the centre of charge of its ionic cloud. The result of this is that each central ion will experience a retarding force arising from its associated ionic cloud, which is migrating in the opposite direction, an effect termed the *relaxation or asymmetry effect*. Obviously this effect will be larger the nearer, on average, the ions are in solution; in other words, the effect will increase at higher ionic concentrations.

In addition to the relaxation effect, the theory of Debye, Hückel and Onsager also takes into account a second effect, which arises from the Stokes law discussed above. We saw that each ion, travelling through the solution, will experience a frictional effect owing to the viscosity of the liquid. However, this frictional effect itself depends on concentration, since, with increasing concentration, encounters between the solvent sheaths of oppositely charged ions will become more frequent. The solvent molecules in the solvation sheaths are moving with the ions, and therefore an individual ion will experience an additional drag associated with the solvent molecules in the solvation sheaths of oppositely charged ions; this is termed the *electrophoretic effect*.

2.4.2 The Calculation of the Potential due to the Central Ion and its Ionic Cloud: Ionic Strength and Radius of the Ionic Cloud

As a first step in the calculation of the electrostatic interaction between ions in an electrolyte, we shall consider how the potential ϕ resulting from both the central ion and its associated charge cloud varies with position. Initially, we will consider the potential for this system at *equilibrium* without the presence of an external electrical field, and incorporate this latter later in the development of the theory.

The fundamental equation relating the charge density, ρ (in units of Coulomb m^{-3}), to the electric field strength, $\boldsymbol{E}$ (units of V m^{-1}), is:

$$\operatorname{div} \boldsymbol{E} = \frac{\rho}{\varepsilon_0 \varepsilon_r} \tag{2.34}$$

where $\boldsymbol{E}$ is a position-dependent *vector* quantity, ε_r is the relative permittivity of the medium (a pure number), ε_0 is the absolute permittivity of free space, and has the value 8.854×10^{-12} F m^{-1} and

$$\text{div } \boldsymbol{E} = \frac{\partial \boldsymbol{E}}{\partial \mathrm{x}} + \frac{\partial \boldsymbol{E}}{\partial \mathrm{y}} + \frac{\partial \boldsymbol{E}}{\partial \mathrm{z}} \tag{2.35}$$

In turn, the electric field is related to the potential through the expression:

$$\boldsymbol{E} = -\boldsymbol{grad}\varphi \equiv -\frac{\boldsymbol{i}\partial\varphi}{\partial x} - \frac{\boldsymbol{j}\partial\varphi}{\partial y} - \frac{\boldsymbol{k}\partial\varphi}{\partial z} \tag{2.36}$$

where **i,j,k** are unit vectors in the directions x,y,z.

Combining (2.34) and (2.36) we obtain

$$\text{div } \mathbf{grad}\varphi \equiv \frac{\partial^2\varphi}{\partial x^2} + \frac{\partial^2\varphi}{\partial y^2} + \frac{\partial^2\varphi}{\partial z^2} = -\frac{\rho}{\varepsilon_r\varepsilon_0} \tag{2.37}$$

Although we have written equation (2.37) in Cartesian coordinates, it is more sensible to take advantage of the obvious *spherical* symmetry of the central ion and ionic cloud. Rewriting (2.37) in spherical coordinates, and taking advantage of the fact that we need only consider the *radial* variation of φ, and not the angular variation, (2.37) reduces to the simple differential equation

$$\frac{1}{r^2}\frac{d}{dr}\left(r^2\frac{d\varphi}{dr}\right) = -\frac{\rho(r)}{\varepsilon_r\varepsilon_0} \tag{2.38}$$

where we have explicitly written $\rho(r)$ to remind ourselves that the charge density of the ionic cloud surrounding the central ion will itself be a function of the distance from that ion.

In equation (2.38), we have two unknowns, $\varphi(r)$ and $\rho(r)$, and we must find a second relationship between these variables if (2.38) is to be solved; this we can do by considering the energy associated with an ion of charge z_ie_0 in the potential field. If the potential at any point is $\varphi(r)$, then the energy required to bring an ion of charge z_ie_0 from some remote point in the solution to the neighbourhood of our central ion is, of course, $z_ie_0\phi(r)$. This will clearly be favourable if, for example, $z_i < 0$ and $\phi(r) > 0$: that is, a positive central ion will tend to attract ions of negative charge. Opposing this attraction are thermal effects, that can be taken into account with the Boltzmann equation, giving us the expression

$$n_i(r) = n_i^0 \exp\left(-\frac{z_ie_0\varphi(r)}{k_BT}\right) \tag{2.39}$$

where $n_i(r)$ is the number per unit volume of ions of charge z_ie_0 at a distance r from the central ion, and n_i^0 the concentration of such ions in the bulk of the solution. The density of ions at any point r can now be found by multiplying the number density of ions at that point by the charges on those ions:

$$\rho(r) = \sum_i z_ie_0n_i(r) = \sum_i z_ie_0n_i^0 \exp\left(-\frac{z_ie_0\varphi(r)}{k_BT}\right) \tag{2.40}$$

where the summation includes the central ion. We note that if $T \to \infty$, then the expression: $\exp(-z_ie_0\varphi(r)/k_BT) \to 1$ and $\rho(r) \to \sum_i z_ie_0n_i^0 \equiv 0$; the summation here is zero as the total charge concentration in the solution itself must be zero from the principle of electroneutrality.

By putting equation (2.40) into equation (2.38) we obtain the Poisson- Boltzmann equation:

$$\frac{1}{r^2}\frac{d}{dr}\left(r^2\frac{d\varphi}{dr}\right) = -\sum_i\left(\frac{z_i e_0 n_i^0}{\varepsilon_r\varepsilon_0}\right)\exp\left(-\frac{z_i e_0\varphi(r)}{k_B T}\right) \tag{2.41}$$

This is a non-linear differential equation, and is not only very difficult to solve but cannot physically be correct, since we would expect, if all the charges were to be doubled in the system, that the potential would double too. This principle, known as the principle of linear superposition, is quite fundamental in electrostatics, and the fact that equation (2.41) does not obey this principle is, in itself, a serious problem. Originally, Debye and Hückel circumvented the problem by considering only dilute solutions, for which the interionic interactions were small compared to $k_B T$. Under these circumstances, the exponential can be expanded: quite generally, if $x << 1$, $e^{-x} \sim (1-x)$, so

$$\sum_i\left(\frac{z_i e_0 n_i^0}{\varepsilon_r\varepsilon_0}\right)\exp\left(-\frac{z_i e_0\varphi(r)}{k_B T}\right) \approx \sum_i\left(\frac{z_i e_0 n_i^0}{\varepsilon_r\varepsilon_0}\right)\left(1-\frac{z_i e_0\varphi(r)}{k_B T}\right) =$$

$$\sum_i\left(\frac{z_i e_0 n_i^0}{\varepsilon_r\varepsilon_0}\right) - \sum_i\left(\frac{z_i e_0 n_i^0}{\varepsilon_r\varepsilon_0}\right)\left(\frac{z_i e_0\varphi(r)}{k_B T}\right) \tag{2.42}$$

As before, from electroneutrality, the first summation in (2.43) is zero, and we finally end up with the linearised Poisson-Boltzmann equation

$$\frac{1}{r^2}\frac{d}{dr}\left(r^2\frac{d\varphi}{dr}\right) = \frac{e_0^2}{\varepsilon_r\varepsilon_0 k_B T}\sum_i z_i^2 n_i^0\varphi(r) \tag{2.43}$$

It may be helpful at this juncture to consider the units of equation (2. 43). The units of $(e_0^2/\varepsilon_r\varepsilon_0 k_B T)$ are m (metre), and n has units m^{-3}, so the units of the RHS of (2.43) are $V\ m^{-2}$, as we might expect from the LHS. If the concentration is expressed as a molality, m (units of mol kg^{-1}), and if the bulk molality of the ith ion is m_i, then $n_i^0 = m_i\rho_s N_A$, where N_A is Avogadro's constant as before and ρ_s is the solvent density in units of kg m^{-3}. We thus have, for (2.43)

$$\frac{1}{r^2}\frac{d}{dr}\left(r^2\frac{d\varphi}{dr}\right) = \frac{e_0^2\rho_s N_A}{\varepsilon_r\varepsilon_0 k_B T}\sum_i z_i^2 m_i\varphi(r) \tag{2.44}$$

The Ionic Strength, **I**, is defined as

$$\mathbf{I} = \frac{1}{2}\sum_i z_i^2\left(\frac{m_i}{m^0}\right) \tag{2.45}$$

where we divide the molality m by the standard molality m^0 in order to ensure that **I** is dimensionless. In our system of units, $m^0 = 1$ mol kg^{-1}, defining

$$\kappa^2 = \frac{2e_0^2 N_A \rho_s m^0}{\varepsilon_r \varepsilon_0 k_B T} \mathbf{I} \tag{2.46}$$

we can write (2.44) as:

$$\frac{1}{r^2}\frac{d}{dr}\left(r^2 \frac{d\varphi}{dr}\right) = \kappa^2 \varphi(r) \tag{2.47}$$

where it is clear that κ has units of inverse length: m^{-1}. Equation (2.47) is now linear and can be solved straightforwardly. Details of the solution may be found in standard mathematics texts, and the boundary conditions for the equation are:

(1) $\varphi(r) \rightarrow 0$ as $r \rightarrow \infty$

(2) we presume the ion to have a finite radius, a_0: for $r < a_0$, the equation to be solved must have the form

$$\frac{1}{r^2}\frac{d}{dr}\left(r^2 \frac{d\varphi}{dr}\right) = 0 \tag{2.48}$$

since we assume that there is no charge density *inside* the ion; all the charge of the ion is treated as being on the surface.

By matching the solutions to (2.47) and (2.48) at the surface of the ion (i.e. at the radius a_0) the potential at $r = a_0$ is found to be

$$\varphi(a_0) = \frac{z_0 e_0}{4\pi\varepsilon_r \varepsilon_0 a_0 (1 + \kappa a_0)} \tag{2.49}$$

where $z_0 e_0$ is the charge *on the central ion*. The complete solution for $r \geq a_0$ then takes the form:

$$\varphi(r) = \frac{z_0 e_0}{4\pi\varepsilon_r \varepsilon_0 r}\left(\frac{e^{\kappa a_0}}{1 + \kappa a_0}\right) e^{-\kappa r} \tag{2.50}$$

and we have implicitly assumed that the radii of *all* the ions are the same.

It is important, at this stage, to be reminded of the limits of validity of equation (2.50). The most important limitation arises from the need to linearise equation (2.41); in general, the requirement that $z_i e_0 \varphi(r) \ll k_B T$ limits total ionic molalities, in principle, to below 0.01 mol kg^{-1}. In aqueous solutions this corresponds to less than ca. 0.01 M ($\equiv$10 mol m^{-3}). However, the approximations made in neglecting higher terms in the expansion are fortuitously much better for symmetrical electrolytes (that is, those that dissociate into two ions of equal and opposite charge). This is because all the terms in the expansion (2.43) containing *odd* powers of z_i will vanish identically, and this includes the next term in the expansion leading to (2.43). Thus, for symmetrical electrolytes, the expansion is actually correct to fourth power in z_i. It is easy to see that this cancellation does *not* take place for unsymmetrical electrolytes, and the concentration range over which (2.50) is valid is correspondingly much more restricted in the latter case.

The potential of equation (2.50) can be divided into that due to the central ion, and that due to the surrounding ionic cloud, again invoking the superposition principle. The former can be written from elementary electrostatics as:

$$\varphi_0(r) = \frac{z_0 e_0}{4\pi\varepsilon_r\varepsilon_0 r} \tag{2.51}$$

and it follows that the potential due to the cloud has the form

$$\varphi_c(r) = \frac{z_0 e_0}{4\pi\varepsilon_r\varepsilon_0 r}\left\{\left(\frac{e^{\kappa a_0}}{1+\kappa a_0}\right)e^{-\kappa r} - 1\right\} \tag{2.52}$$

so that the potential (2.50) can be written as the sum of the potentials (2.51) and (2.52). It is clear that equation (2.52) is a measure of the deviation of the electrolyte solution from ideality. The potential around the central ion of an *ideal* solution must have the form of (2.51), and this is confirmed by the observation that as the concentration, and hence the ionic strength, becomes very small, $\kappa \to 0$ and, from (2.52), $\varphi_c(r) \to 0$.

The distance $r_c = 1/\kappa$ is usually regarded as an approximate measure of the radius of the ionic cloud. This is because at this value of r, the charge density of the ionic cloud surrounding the central ion is a maximum. This can easily be shown by considering expression (2.40) for the charge density. Remembering that we are working under the assumption that $z_i e_0 \varphi(r) << k_B T$, then we can linearise (2.40) to obtain

$$\rho(r) \approx -\sum_i \frac{z_i^2 e_0^2 n_i^0 \varphi(r)}{k_B T} \equiv -\sum_i z_i^2 e_0^2 n_i^0 \frac{z_0 e_0}{4\pi\varepsilon_r\varepsilon_0 r k_B T}\left(\frac{e^{\kappa a_0}}{1+\kappa a_0}\right)e^{-\kappa r} \tag{2.53}$$

and the total charge $q(r)$ in a spherical shell, distance r from the central ion and thickness dr is clearly $4\pi r^2 \rho(r) dr \equiv \text{const.} \cdot re^{-\kappa r} \cdot dr$, where the constant term includes all charge and concentration factors. Simple differentiation of the expression for $q(r)$ shows that it has its maximum when $r = r_c = 1/\kappa$.

The value of κ is, therefore, of considerable importance in electrolyte solution theory. Its inverse, κ^{-1}, is usually referred to as the Debye screening length, since at distances greater than κ^{-1} from the central ion, the potential (2.51) that would be expected for that ion alone is reduced by a factor of order e. In other words, we can say, at least approximately, that if we are further away from a particular ion than the distance κ^{-1}, the electrostatic effect due to that ion becomes very small.

Returning to the actual definition of κ, its value in water at 25°C (for which $\varepsilon_r = 78.54$ and $\rho_S = 997$ kg m^{-3}) is given by

$$\kappa^{-1}/\text{metre} = \frac{3.046 \times 10^{-10}}{(\mathbf{I})^{1/2}} \tag{2.54}$$

where, as above, **I** is referred to a standard solution of unit molality. It follows that the smaller the concentration of ions, and hence the ionic strength, the larger the size of the ionic cloud and the corresponding Debye screening length. Table 2.7 gives some idea of magnitudes:

Table 2.8 Radius of the ionic cloud in aqueous solution at 25°C for different electrolyte types and concentrations. All radii in units of 10^{-10} m ($\equiv 1\text{Å}$).

Molality/mol kg^{-1}	Electrolyte type		
	1 – 1	1 – 2 or 2 – 1	2 – 2
	Ionic Cloud Radius/10^{-10} m		
10^{-4}	304	176	152
10^{-3}	96	55.5	48.1
10^{-2}	30.4	17.6	15.2
10^{-1}	9.6	5.5	4.8

2.4.3 The Debye-Onsager Equation for the Conductivity of Dilute Electrolyte Solutions

The quantitative calculation of the dependence of electrolyte conductivity on concentration begins from the expression (2.50) for the potential exerted by a central ion and its associated ionic cloud. The relaxation effect, introduced in Section 2.4.1, arises as a result of the asymmetry in the ionic cloud, which itself is caused by the movement of the central ion away from the centre of the cloud. As soon as this ionic motion begins, the ion will experience an effective electric field $\boldsymbol{E}_{\text{rel}}$ in a direction opposite to that of the applied electric field, whose magnitude will depend on the ionic mobility. The second effect identified by Onsager concerns the additional force experienced by an ion due to the movement of solvent sheaths associated with oppositely charged ions encountered during its own migration through the solution. This second term, the electrophoresis term will depend on the viscosity of the liquid, and, combining this with the reduction in conductivity due to relaxation terms, we finally emerge with the Debye-Hückel-Onsager equation:

$$\Lambda = \Lambda_0 - \Lambda_0\left(\frac{z^+z^-e_0^2}{24\pi\varepsilon_r\varepsilon_0 k_B T}\right)\left(\frac{2q\kappa}{1+\sqrt{q}}\right) - \frac{N_A e_0^2(z^+ + z^-)\kappa}{6\pi\eta} \tag{2.55}$$

where

$$q = \frac{z^+z^-}{z^+ + z^-}\left(\frac{\lambda_0^+ + \lambda_0^-}{z^-\lambda_0^+ + z^+\lambda_0^+}\right) \tag{2.56}$$

For a completely dissociated 1-1 electrolyte $q = 0.5$, and expression (2.55) can be rewritten in terms of the molarity, c/c^0, remembering that κ can be expressed either in terms of molalities or molarities; in the latter case, from (2.46) we have

$$\kappa^2 = \left(\frac{e_0^2 N_A c^0}{\varepsilon_r\varepsilon_0 k_B T}\right)\sum_i z_i^2(c_i/c^0) = \left(\frac{e_0^2 N_A c^0}{\varepsilon_r\varepsilon_0 k_B T}\right)\sum_i z_i^2 v_i(c/c^0), \text{ and we can write}$$

$$\Lambda = \Lambda_0 - (B_1\Lambda_0 + B_2)\sqrt{c/c^0} \tag{2.57}$$

Table 2.9 Experimental and theoretical values of $B_1\Lambda_0 + B_2$ for various salts in aqueous solution at 291 K.

Salt	Observed value of $(B_1\Lambda_0 + B_2)/m^2\Omega^{-1}\ mol^{-1}$	Calculated value of $(B_1\Lambda_0 + B_2)/m^2\Omega^{-1}\ mol^{-1}$
LiCl	7.422×10^{-3}	7.343×10^{-3}
NaCl	7.459×10^{-3}	7.569×10^{-3}
KCl	8.054×10^{-3}	8.045×10^{-3}
$LiNO_3$	7.236×10^{-3}	7.258×10^{-3}

in which B_1 and B_2 are independent of concentration. This is evidently identical in form to Kohlrausch's empirical law (equation 2.14) already discussed in section 2.2.3. Given that (2.55) derives from (2.50), the same assumptions underlying the latter must underlie the former as well. In particular, the ionic concentration must be low enough for the expansion of (2.42) to be valid. As indicated above, this will restrict the molarity to below 0.01M for symmetrical electrolytes and for unsymmetrical electrolytes to even lower values. In fact, for symmetrical singly-charged 1-1 electrolytes, useful estimations of the behaviour can be obtained, with a few % error, for up to 0.1M concentrations, but symmetrical multi-charged electrolytes ($z^+ = z^- \neq 1$) usually show deviations, even at 0.01M owing to the formation of ion pairs.

In aqueous solution, the values of B_1 and B_2 can be calculated straightforwardly. If Λ is expressed in $m^2\ \Omega^{-1}\ mol^{-1}$ and c^0 is taken as 1 mol dm^{-3}, then for water at 298 K, and $z^+ = |z^-| = 1$, $B_1 = 0.229$ and $B_2 = 6.027 \cdot 10^{-3}\ m^2\ \Omega^{-1}\ mol^{-1}$. At 291 K, the corresponding values are 0.229 and $5.15 \cdot 10^{-3}\ m^2\ \Omega^{-1}\ mol^{-1}$ respectively. Some data for selected 1:1 electrolytes is given in Table 2.9; it can be seen that Onsager's formula is very well obeyed.

For electrolytes that are not completely dissociated (weak electrolytes), with degree of dissociation α, where

$$\alpha = \frac{\text{Amount of material dissociated}}{\text{Total amount of material dissolved}} \tag{2.58}$$

Onsager showed that the conductivity was given by

$$\Lambda = \alpha[\Lambda_0 - (B_1\Lambda_0 + B_2)\sqrt{\alpha c/c^0}] \tag{2.59}$$

Evidently, in order to use this expression, the variation of α with c must be known, and this is further discussed in Section 2.6 below.

2.4.4
The Influence of Alternating Electric Fields and Strong Electric Fields on the Electrolyte Conductivity

Accepting that the conductivities of electrolyte solutions are given to a reasonable degree of accuracy by (2.57) for molarities below 0.01M and, at least for 1-1 electrolytes, usable values are obtained for concentrations as high as 0.1M it is clear that the underlying model leading to (2.59) is highly likely to be correct. However, there are two further indications from experiment that strongly support the model.

Debye and Falkenhagen were the first to suggest, on the basis of the Onsager model, that if the conductivity were measured by an alternating electric field of sufficiently high frequency, f, such that the asymmetry effect could not develop (that is, the inverse relaxation time of the ionic atmosphere, $\tau^{-1} < f$), the conductivity would rise. This is termed the Debye-Falkenhagen effect, and is observed experimentally for frequencies above 10^7–10^8 Hz. On that basis, we can estimate the relaxation time of the ionic atmosphere about a central ion to be ca. 10^{-8} s.

The second effect arises from the fact that an ion accelerated in a sufficiently strong electric field will attain a speed at which it will traverse the diameter of its ionic cloud in a time shorter than the relaxation time of the cloud. Under these circumstances, the cloud cannot form, and its inhibiting effects on the mobility of the central ion will be nullified.

Taking some typical values, if $\kappa^{-1} \approx 10^{-8}$ m and $\tau \approx 10^{-8}$ s, then a velocity of 1 m s^{-1} is necessary. Given a typical ionic mobility of the order $5 \cdot 10^{-8}$ $m^2V^{-1}s^{-1}$, a field strength of $2 \cdot 10^7$ Vm^{-1} would be required to ensure that the condition above can be satisfied.

This is indeed observed: above a critical field strength of ca. 10^7 Vm^{-1}, the apparent ionic mobility does rise, and the critical field strength does decrease with increasing concentration, as predicted by the model. This rise in mobility at very high electric fields is termed the Wien Effect.

2.5
The Concept of Activity from the Electrochemical Viewpoint

2.5.1
The Activity Coefficient

We have seen that an ion in a solution of finite concentration will accumulate around itself a cloud formed from ions of opposite charge. Before the central ion can itself react, either in solution or at an electrode surface, it must divest itself of this cloud, a process requiring the input of a certain energy. This energy represents a loss to the system, and leads to the ion having a lower free energy than expected, as well as being less reactive than it would have been if isolated in an ideal solution. This lowering of energy and loss of reactivity is, in turn, expected to be the more marked, the greater the concentration, since at higher concentration, the density of the ionic cloud rises.

It follows that for an accurate description of the electrochemical or, indeed, thermodynamic properties of dissolved ions, a simple measure of concentration such as the molality, m, is not adequate, and instead an *effective* molality, or *activity*, a_i, for ion i is employed, where

$$a_i = \gamma_i \frac{m_i}{m^0} \tag{2.60}$$

and γ_i, the activity coefficient, represents the deviation from ideality, this latter being defined as the situation when $\gamma_i = 1$ and $a_i = (m_i/m^0)$, and as in equation (2.46), the standard molality, m^0 is taken here to be 1 mol kg^{-1}. As above, this is very close, in aqueous solution, to a *concentration* standard of 1 mol dm^{-3}, which is commonly adopted in the older literature. From the discussion above, clearly γ_i is a function of the total ionic strength, **I**, defined in equation (2.45):

$$\gamma_i \equiv \gamma_i(\mathbf{I}) \tag{2.61}$$

It is clear that at infinite dilution, where there are no interionic interactions,

$$\lim_{\mathbf{I}\to 0} \gamma_i = \lim_{\mathbf{I}\to 0} \left\{ \frac{a_i}{m_i/m^0} \right\} = 1 \tag{2.62}$$

In non-ideal solutions, we might expect that the activity coefficient would decrease with increasing concentration from the argument above, so that $\gamma_i < 1$.

Since electrolyte solutions containing only a single type of ion are not capable of separate existence, single-ion activities and activity coefficients are not accessible. Instead, only the properties of the complete solution can be determined, and we define for the anions and cations of the solution together a mean activity, $a_\pm(\mathbf{I})$, and a mean activity coefficient, $\gamma_\pm(\mathbf{I})$. For a 1-1 electrolyte, the mean activity and activity coefficient are defined as the geometric means of the individual ion values:

$$a_\pm = (a_+ a_-)^{\frac{1}{2}} = \gamma_\pm \left(\frac{m_i}{m^0}\right) = \left(\gamma_+ \gamma_- \left[\frac{m}{m^0}\right]^2\right)^{\frac{1}{2}} \tag{2.63}$$

and for an electrolyte that dissociates into ν_+ cations and ν_- anions, the resultant equations are:

$$a_\pm = (a_+^{\nu_+} a_-^{\nu_-})^{\frac{1}{(\nu_+ + \nu_-)}}; \ \gamma_\pm = (\gamma_+^{\nu_+} \gamma_-^{\nu_-})^{\frac{1}{(\nu_+ + \nu_-)}} \tag{2.64}$$

As an example, if we require the activity of a solution of K_2SO_4, then, we obtain:

$$a_\pm = \left(a_{K^+}^2 a_{SO_4^{2-}}\right)^{\frac{1}{3}} = \gamma_\pm \left(m_{K^+}^2 m_{SO_4^{2-}}\right)^{\frac{1}{3}} / m^0 \tag{2.65}$$

2.5.2 Calculation of the Concentration Dependence of the Activity Coefficient

As the ionic cloud forms around a particular central ion, this latter will suffer a loss of potential energy as compared to its original state. This energy is liberated on formation of the cloud, and must be supplied to the ion again to free it from its cloud.

If we write the chemical potential of the ion, i, with its associated cloud, μ_i^{real}, as equal to the sum of the chemical potential in the absence of the cloud, μ_i^{ideal}, and the potential energy associated with the ionic cloud, U, then a calculation of the later will permit us, in turn, to calculate the difference between the real and ideal chemical potentials, assuming that this difference arises solely through ion-ion interactions.

However, there are subtleties in this calculation: for an ideal solution, we would like to write

$$\mu_i = \mu_i^0 + RT \ln x_i \tag{2.66}$$

where x_i is the mole fraction of the ions of type i in the solution, and is defined as

$$x_i = n_i/(n_i + n_s) \tag{2.67}$$

where n_i is the number of moles of i and n_S the number of moles of solvent. However, this formulation leads to difficulties in the definition of the quantity μ_i^0; in principle, it would correspond to the free energy of i in a hypothetical material formed solely of ions i, with $n_S = 0$: hardly a useful standard for a dilute ionic solution! Even if we modify equation (2.66) to take into account the fact that we cannot write down the chemical potential of a single ion, the standard state remains remote from the physical state we are attempting to describe.

In order to circumvent this problem, we seek to write down the chemical potential of an ideal solution of ions i as:

$$\mu_i \sim \mu^{0\dagger} + RT \ln(m_i/m^0) \tag{2.68}$$

where m^0 is the standard molality introduced earlier, which we can take as 1 mol kg^{-1}, and $\mu^{0\dagger}$ is the chemical potential of ions i in a *hypothetical solution of molality m^0 in which the ion-ion interactions are assumed to be insignificant.* Equation (2.68) would certainly be expected to be the limiting form as $m_i \to 0$, since only ion-solvent interactions are present at high dilution, and the central advantage of the way in which equation (2.68) is written is that for dilute solutions, $\mu^{0\dagger}$ contains all the ion-solvent interactions. Provided that the solutions remain sufficiently dilute that the individual ion-solvent interactions do not alter with ion concentration, then the main reason for deviations from (2.68) at higher concentrations must be ion-ion interactions only. From above, we can formally express deviations from (2.68) by writing

$$\mu_i^{\text{real}} \equiv \mu_i^{0\dagger} + RT \ln a_i \tag{2.69}$$

$$\equiv \mu_i^{0\dagger} + RT \ln(m_i/m^0) + RT \ln \gamma_i \tag{2.70}$$

$$\equiv \mu_i^{\text{ideal}} + RT \ln \gamma_i \tag{2.71}$$

$$\equiv \mu_i^{\text{ideal}} + N_A U \tag{2.72}$$

where we have multiplied the potential energy U associated with the ionic cloud by the Avogadro Number N_A. If our ion has radius a_0 and charge $z_i e_0$, then the associated potential energy needed to charge up the ion to the potential $\phi(a_0)$ is just $\frac{1}{2} z_i e_0 \phi(a_0)$. From (2.52), we have

$$U = \frac{1}{2} z_i e_0 \varphi(a_0) = z_i e_0 \left(\frac{z_i e_0}{8\pi\varepsilon_r \varepsilon_0 a_0} \right) \left\{ \left(\frac{e^{\kappa a_0}}{1 + \kappa a_0} \right) e^{-\kappa a_0} - 1 \right\} = \frac{-z_i^2 e_0^2 \kappa}{8\pi\varepsilon_r \varepsilon_0 (1 + \kappa a_0)} \tag{2.73}$$

where it can be seen that the expression (2.73) is *always negative*, a result that derives essentially from the fact that the ion in solution is stabilised by its ionic cloud. By comparing (2.71) and (2.73) we obtain, finally,

$$-RT \ln \gamma_i = \frac{N_A z_i^2 e_0^2 \kappa}{8\pi\varepsilon_r \varepsilon_0 (1 + \kappa a_0)} \tag{2.74}$$

For very dilute solutions, the radius of the ionic cloud, κ^{-1}, is very much larger than a_0, the ionic diameter, and $a_0 \kappa \ll 1$. By substituting expression (2.47) for κ, we obtain an expression:

$$\ln \gamma_i = -A z_i^2 \sqrt{\mathbf{I}} \tag{2.75}$$

where **I** is the ionic strength referred to a standard concentration of 1 mol kg^{-1} and A is a constant that depends solely on the properties of the solvent. For water at 25°C, we find $A = 1.172$ and

$$\ln \gamma_i = -1.172 z_i^2 \sqrt{\mathbf{I}} \tag{2.76}$$

Assuming a mean ionic diameter of ca. 2Å, it can easily be seen that the molality range of validity of (2.76) is up to ca. 10^{-2} mol kg^{-1} for 1-1 valued electrolytes and ca. 10^{-3} mol kg^{-1} for multicharged electrolytes.

The expressions for ionic activity coefficient in equations (2.74) to (2.76), are given for a single ion; in fact, the only experimentally accessible activity coefficient is the mean, defined in equations (2.63) et seq. Straightforward algebra shows that

$$\ln \gamma_{\pm} = -1.172 |z^+ z^-| \sqrt{\mathbf{I}} \tag{2.77}$$

This is the standard form of the *Debye-Hückel limiting law*. From this, values for the mean activity coefficients in aqueous solution may easily be calculated, at least over the range of validity of the formula.

For different electrolytes of the same charge, expression (2.77) predicts the same values for $\gamma_{\pm}$; in other words, the limiting law does not make allowance for any differences in size or other ionic property. For 1-1 electrolytes, this is experimentally found to be the case within the expected concentration range of validity, though for multicharged electrolytes, the agreement is less good, even for 10^{-3} mol kg^{-1}. In Table 2.10 some values of $\gamma_{\pm}$ calculated from (2.77) are collected and compared to some measured activity coefficients for a few simple electrolytes. Fig. 2.17 shows these properties graphed for 1-1 electrolytes to emphasise the nature of the deviations from the limiting law.

Table 2.10 Calculated (from 2-77) and experimental values of the mean ionic activity coefficient for different concentrations and electrolytes at 25°C.

1-1 Electrolytes

			$\gamma_{\pm}$		
$m/\text{mol kg}^{-1}$	I	(Eq. 2-77)	HCl	KNO_3	LiF
0.001	0.001	0.9636	0.9656	0.9649	0.965
0.002	0.002	0.9489	0.9521	0.9514	0.951
0.005	0.005	0.9205	0.9285	0.9256	0.922
0.010	0.010	0.8894	0.9043	0.8982	0.889
0.020	0.020	0.8472	0.8755	0.8623	0.850
0.050	0.050		0.8304	0.7991	
0.100	0.100		0.7964	0.7380	

1-2 or 2-1 Electrolytes

			$\gamma_{\pm}$	
$m/\text{mol kg}^{-1}$	I	(Eq. 2-77)	H_2SO_4	Na_2SO_4
0.001	0.003	0.8795	0.837	0.887
0.002	0.006	0.8339	0.767	0.847
0.005	0.015	0.7504	0.646	0.778
0.010	0.030	0.6662	0.543	0.714
0.020	0.060		0.444	0.641
0.050	0.150			0.536
0.100	0.300		0.379	0.453

2-2 Electrolytes

			$\gamma_{\pm}$	
$m/\text{mol kg}^{-1}$	I	(Eq. 2-77)	$CdSO_4$	$CuSO_4$
0.001	0.004	0.7433	0.754	0.74
0.002	0.008	0.6574	0.671	
0.005	0.020	0.5152	0.540	0.53
0.010	0.040		0.432	0.41
0.020	0.080		0.336	0.315
0.050	0.200		0.277	0.209
0.100	0.400		0.166	0.149

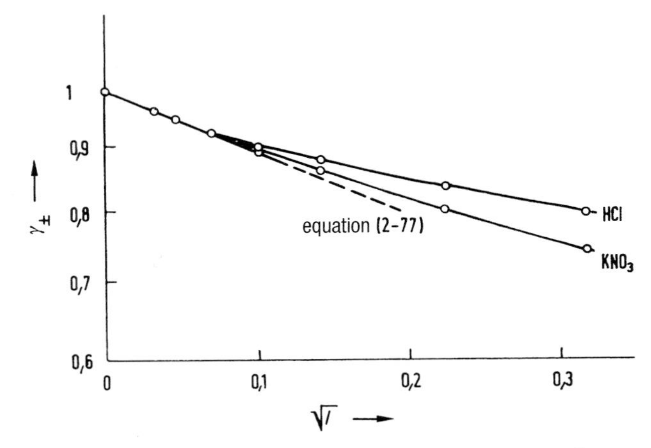

Fig. 2.17 Theoretical variation of the activity coefficient γ with $\sqrt{I}$ from equation 2-77 and experimental results for 1-1 electrolytes at 25°C (from table 2.10).

Experimentally, the most important consequence of (2.77) is that the activity coefficient depends solely on the ionic strength. If two electrolytes are present, with the first in excess, then regardless of the concentration of the other, its activity coefficient will remain effectively constant provided that the concentration of the first is constant. Thus, by using an excess of some supporting electrolyte such that we work effectively at constant ionic strength, the dilute electrolyte will appear to behave in a quasi-ideal manner, a result that greatly simplifies a large number of analytical procedures.

2.5.3 Extensions to More Concentrated Electrolytes

Modern-day approaches to ionic solutions need to be able to contend with the following problems:

1. The nature of the solvent itself, and the interactions taking place in that solvent;
2. The changes taking place on dissolution of ionic electrolyte in the solvent;
3. Macroscopic and microscopic studies of the properties of electrolyte solutions.

2.5.3.1 Calculations on the Solvent Alone

Even the description of the solvent itself presents major theoretical problems. In the gas phase, the mean distance apart of the molecules means that we can consider intermolecular interactions from the point of view of perturbation theory. In the case of a solid, we can treat the molecular motion in terms of coupled vibrations, using the translational symmetry of the solid to simplify the analysis. In the case of liquids, neither of these simplifications is possible, we have to consider all possible arrangements of the molecules of the liquid and calculate, for each arrangement, the associated energy. By weighting each arrangement by the appropriate Boltzmann factor, and carrying out a complete summation, we obtain the partition function of the liquid and hence its thermodynamic properties. Such a procedure is clearly infeasible for macroscopic samples of liquid, and methods of simplifying the problem need to be sought. There are, broadly, two approaches: to find analytical approximations that permit at least some comparative statements to be made, and computational approximations that usually amount to considering a relatively small number of molecules in the hope that the behaviour of a small ensemble will reflect the macroscopic behaviour.

Both analytical and computational approximations for liquids centre on the introduction of a pairwise density function, $g(r)$, defined in such as way that the probability of a second molecule being found at a distance r from a central molecule is $\rho g(r) r dr \sin\theta d\theta d\phi$, where ρ is the density of the liquid. Integration over the angular variables gives the number of molecules found in a spherical shell at a distance r from a central molecule as:

$$N(r)dr = 4\pi r^2 \rho g(r) dr \tag{2.78}$$

The function $g(r)$ is central to the modern theory of liquids, since it can be measured experimentally using neutron or X-ray diffraction and can be related to the inter-particle potential energy. Experimental data for liquids show that $g(r)$ goes through a series of maxima and minima. For $r < \sigma$, the distance of closest approach of two molecules of the liquid, $g(r)$ falls to zero, and the first maximum for most liquids is found at $r = \sigma$. For water, as shown in Fig. 2.18a, the first maximum is at $r = \sigma = 2.82\,\text{Å}$, very close to the intermolecular distance in the normal tetrahedrally bonded form of ice, and for argon (iso-electronic with water), the first maximum is found close to the expected inter-nuclear separation assuming close packed spheres of $r = \sigma = 3.4\,\text{Å}$. The second peak for argon is at roughly double this value, as expected for a spherical molecular system consisting roughly of concentric spheres. However, for water, the second peak in $g(r^*)$ is found at only $r^* = 1.6$, which corresponds closely to the second-nearest-neighbour distance in ice ($\sim 4.5\,\text{Å}$), strongly supporting a model for the structure of water that is ice-like over short distances as described in more detail below.

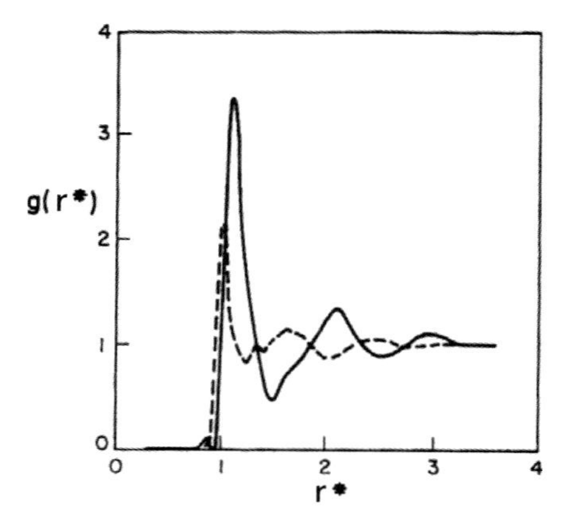

Fig 2.18a Radial distribution function $g(r^*)$ for water (dashed curve) at 4°C and 1 atm. and for liquid argon (full curve) at 84.25 K and 0.71 atm. as functions of the reduced distance $r^* = r/\sigma$, where σ is the atomic/molecular diameter and r the separation of the atomic or molecular centres [from A. Ben-Naim, "Statistical Thermodynamics for Chemists and Biologists", Plenum Press, London, 1992].

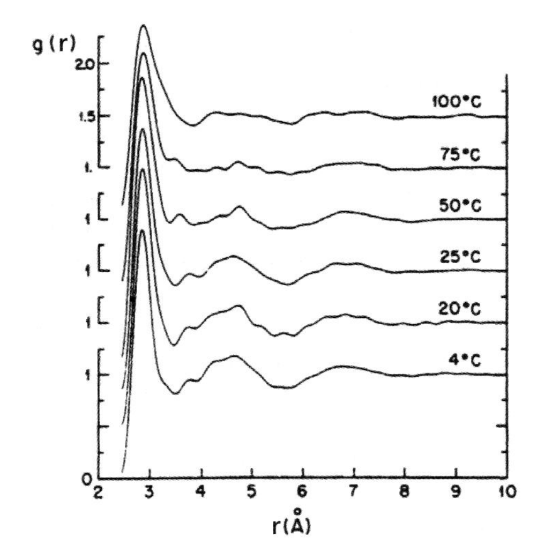

Fig. 2.18b The temperature dependence of $g(r)$ of water [from A. Ben-Naim, loc. cit.].

This strongly structured model for water in fact dictates many of its anomalous properties.

From knowledge of $g(r)$, it is possible to calculate all the required thermodynamic properties of the liquid, at least to a first approximation. In analytical approximations, $g(r)$ is a function both of the *direct* interaction between the two molecules, obtained from the intermolecular potential $u(r)$, and *indirect* interactions arising by averaging over the presence of all the other molecules in the system. It is clear from Fig. 2.18 that $g(r)$ extends over a long range, whereas the intermolecular potential $u(r)$ is frequently of shorter range, and the usual development of the theory seeks to partition $g(r)$ into a short-range direct correlation function $c(r)$ and a second term containing the indirect contributions. This leads to equations of some complexity, even for spherically symmetrical molecules, and a variety of simplifications have been proposed, details of which can be found in the references. For water, the situation is even more complex since the intermolecular potential includes not only short-range repulsive contributions, but electric multipole interactions at long range, and hydrogen-bonded interactions of uncertain analytical form at intermediate ($2.4 \leq r \leq 4\,\text{Å}$) distances. These forms can be further approximated by representing water molecules by a system of three point charges, which may or may not reside on the atomic sites and which are chosen to reproduce the dipole and quadrupolar structure of the molecule. Such models may also be extended to include polarisability to allow the description of the response of intra-molecular geometries to strong inter-molecular forces.

The form of the hydrogen-bonded interactions leads to a strongly structured model for water. In principle, this structure can be defined in terms of the average number of hydrogen bonds formed by a single water molecule with its neighbours: in normal ice, this is four, and we expect a value close to this in water close to the freezing point. We also intuitively expect that this number will decrease with increasing temperature, an expectation confirmed by the temperature dependence of $g(r)$ for water as shown in

Fig. 2.18b; it can clearly be seen that the second maximum in $g(r)$ disappears at higher temperatures. The picture should be seen as highly dynamic, with these hydrogen bonds forming and breaking continuously, with the result that the clusters of water molecules characterising this picture are themselves in a continuous state of flux.

The alternative to quasi-analytical techniques, which has attracted substantial attention in recent years, is to use computer simulation methods for calculating $g(r)$, and these have come into their own in the past twenty years as the cost of computing has fallen. The Monte-Carlo method is perhaps the commonest in use: this depends essentially on identifying a Monte-Carlo approach to the calculation of the partition function. As with all Monte-Carlo integrations, a series of random values of the molecular co-ordinates of all N molecules (where N may lie in the range of several hundred to many thousands) is generated, and the partition function evaluated. The essential art in the technique is to pick predominantly configurations of high probability, or at the least to eliminate wasteful complete evaluation of the components of the summation that have high energy. This is achieved by moving one particle randomly from the previous configuration, i, and checking the energy difference $\Delta U = U_{i+1} - U_i$. If $\Delta U < 0$ the configuration is accepted, and if $\Delta U > 0$, then the value of $\exp[-\beta \Delta U]$ is compared to a second random number ξ, where $0 < \xi < 1$. If $\exp[-\beta \Delta U] > \xi$ the configuration is accepted, otherwise it is rejected and a new single particle movement generated. A second difficulty is that the total number of particles that can be treated by Monte-Carlo techniques is relatively small unless huge computing resource is available: given this, boundary effects would be expected to be dominant, and so periodic boundary conditions are imposed, in which any particle leaving through one surface re-enters the system through the opposite surface. Detailed treatments of the Monte-Carlo technique were first described by Metropolis and co-workers; the method has proved valuable not only in the simulation of realistic inter-particle potentials, but also in the simulation of model potentials for comparison with the integral equation approaches above.

The alternative simulation approaches are based on molecular dynamics calculations. This is conceptually simpler that the Monte Carlo method: the equations of motion are solved for a system of N molecules, and for the kth molecule, we have:

$$m_k \frac{d^2 \boldsymbol{r}_k}{dt^2} = \sum_{j=1, j \neq k}^{j=N} \boldsymbol{F}(\boldsymbol{r}_k - \boldsymbol{r}_j) = - \sum_{j=1, j \neq k}^{j=N} \nabla_k U(\boldsymbol{r}_k - \boldsymbol{r}_j) \tag{2.79}$$

and periodic boundary conditions are again imposed, permitting both the equilibrium and transport properties of the system to be evaluated numerically, by integrating over discrete time intervals δt. Details are given in the references cited at the end of the chapter, but clearly the method requires that the differing timescales associated with internal and external molecular motions are somehow incorporated into the model. It will be clear that for water this is highly problematic, and small time steps must be used unless rigid molecular geometries are assumed, in which case high frequency vibrations can be ignored. As with the Monte Carlo method, there is a strong limit imposed on the number of molecules in the simulation, essentially determined by the power of the computer available. This has the consequence that simulation methods for electrolyte solution are restricted to relatively high concentrations of ions, since at low concentrations there may be insufficient ions to probe details of ion-ion interactions.

2.5.3.2 **Studies of Ionic Solvation**

There is, in essence, no limitation other than computing time to the accuracy and predictive capacity of molecular dynamic and Monte Carlo methods, and although the derivation of realistic potentials for water is a formidable task in its own right, accurate simulations of water are now becoming available. However, there remain major theoretical problems in deriving any analytical theory for water, and indeed any other highly polar solvent of the sort encountered in normal electrochemistry. It might be felt, therefore, that extension of the theory to analytical descriptions of ionic solutions was a well-nigh hopeless task. However, a major simplification of our problem is allowed by the possibility, at least in more dilute solutions, of smoothing out the influence of the solvent molecules, reducing their influence to such average quantities as the dielectric permittivity, ε_s, of the solvent. Such a viewpoint is developed within the McMillan-Mayer theory of solutions, which essentially seeks to partition the interaction potential into three parts: that due to interaction between the solvent molecules themselves, that due to interaction between solvent and solute, and that due to interaction between the solute molecules dispersed within the solvent. It is this approach that we used in deriving equation (2.69) above.

In formulating equation (2.69), we assumed that it was possible, at least for more dilute solutions, to partition the energetics of the solution in such a way that the ion-solvent interactions could be subsumed into the standard free energy of an ideal (if imaginary) solution of unit molality containing *non-interacting* ions; only the inter-ionic interactions were then considered as leading to non-ideal behaviour. However, we must understand these ion-solvent interactions if we are to have a complete picture, either for water or any other solvent.

The basic model for ionic hydration is shown in Fig. 2.19: there is an inner hydration sheath of water molecules whose orientation is determined essentially entirely by the field due to the central ion. The number of water molecules in this inner sheath depends on the size and chemistry of the central ion, being, for example, four for Be^{2+} but six for Mg^{2+}, Al^{3+} and most of the first-row transition ions. Outside this primary shell, there is a secondary sheath of more loosely bonded water molecules oriented essentially by hydrogen bonding, the evidence for which was initially indirect and derived from ion mobility measurements. More recent evidence for this secondary shell has now come from X-ray diffraction and scattering studies and IR measurements. A further highly diffuse region, the tertiary region, is probably present, marking a transition to the hydrogen-bonded structure of water described above. As we discussed above when considering the migration of such ions in an electric field, the ion, as it moves, will drag at least part of this solvation sheath with it, but the picture should be seen as essentially dynamic, with the well-defined and essentially completely bound inner sheath structure of Fig. 2.19 being mainly found in highly charged ions of high electronic stability, such as Cr^{3+}.

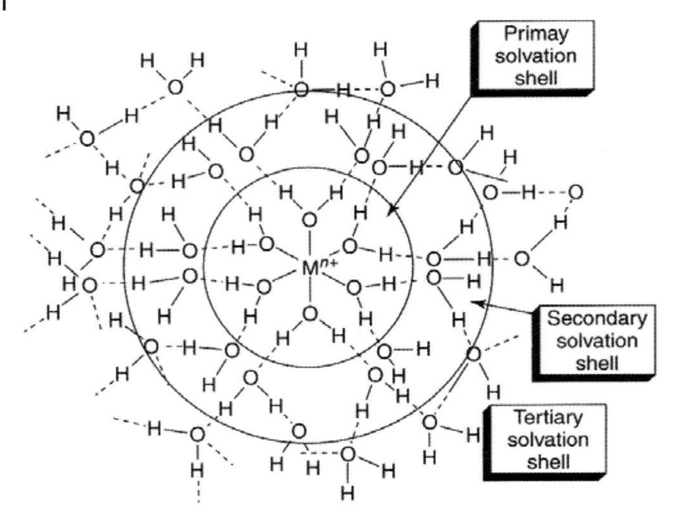

Fig. 2.19 The localised structure of a hydrated metal cation in aqueous solution (the metal ion being assumed to have a primary hydration number of six) [from D.T. Richens: "The Chemistry of Aqua-ions"; Wiley, Chichester, 1997].

The enthalpy of solvation of *cations* primarily depends on the charge on the central ion and the effective ionic radius, the latter being the sum of the normal Pauling ionic radius and the radius of the oxygen atom in water (0.85 Å). A reasonable approximate formula has the form

$$\Delta_{\text{hyd}}H^0 = -695z^2/(r_+ + 0.85)\text{kJ mol}^{-1} \tag{2.80}$$

In general, anions are less strongly hydrated than cations, but recent neutron diffraction data have indicated that even around the halide ions there is a well-defined primary hydration shell of water molecules, which, in the case of Cl^- varies from four to six in constitution, the exact number being a sensitive function of concentration and the nature of the accompanying cation.

Structural investigations of metal-ion hydration have been carried out by spectroscopic, scattering and diffraction techniques, but these techniques do not always give identical results, since they measure in different timescales. There are three distinct types of measurement:

1. Those giving an average structure, such as neutron and X-ray scattering and diffraction studies;
2. Those revealing dynamic properties of co-ordinated water molecules, such as NMR and quasi-elastic scattering methods;
3. Those based on energetic discrimination between water in the different hydration sheaths and bulk water, such as IR, Raman and thermodynamic studies.

First-order difference neutron scattering methods for the analysis of concentrated solutions of anions and cations show sharp M-O pair correlations typical of long-lived inner hydration sheaths, with broader structure, showing the second hydration sheath, being clearly present but more diffuse. Their results also show that the inner solvating water molecule is tilted so for Ni^{2+}, for example, the H-O-H..Ni^{2+} species are not co-planar. This tilt appears to be concentration dependent, and decreases to zero below 0.1M $NiCl_2$. It is almost certainly caused by interaction between the hydrogen bonded secondary sheaths around the cations. The secondary hydration sheaths have been studied by large-angle X-ray scattering (LAXS): for Cr^{3+}, a well-defined secondary sheath containing 13 ± 1 molecules of water could be identified some 4.02 ± 0.2Å distance from the central ion. The EXAFS technique has also been used to study the local environment around anions and cations: in principle the technique is ideally suited for this, since it has high selectivity for the central ion and can be used in solutions more dilute than those accessible to neutron or X-ray scattering. However, the technique also depends on the capacity of the data to resolve different structural models that may actually give rise to rather similar EXAFS spectra. The sensitivity of the technique also falls away for internuclear distances in excess of 2.5Å.

The secondary hydration sheath has also been studied using vibrational spectroscopy: in the presence of highly charged cations, such as Al^{3+}, Cr^{3+} and Rh^{3+}, frequency shifts can be seen due to the entire primary and secondary hydration structure, though the *number* of water molecules hydrating the cation is rather lower than that expected on the basis of neutron data or LAXS data. By contrast, comparison of the Raman and neutron diffraction data for Sc^{3+} indicates the presence of $[Sc(H_2O)_7]^{3+}$ in solution, a result supported by the observation of pentagonal bipyramidal co-ordination in the X-ray structure of the aqua di-μ-hydroxo dimer $[Sc_2\ (OH)_2]^{4+}$.

The hydration of more inert ions has been studied by ^{18}O labelling mass spectrometry. ^{18}O-enriched water is used, and equilibrium between the solvent and the hydration around the central ion is first attained, after which the cation is extracted rapidly and analysed. The method essentially reveals the number of oxygen atoms that exchange slowly on the timescale of the extraction, and has been used to establish the existence of the stable $[Mo_3O_4]^{4+}$ cluster in aqueous solution.

One of the most powerful methods for the investigation of hydration is NMR, and both ^1H and ^{17}O nuclei have been used. By using paramagnetic chemical shift reagents such as Co^{2+} and Dy^{3+}, which essentially shift the peak position of bulk water, hydration measurements have been carried out using ^1H NMR on a number of tri-positive ions. ^{17}O measurements have also been carried out, and by varying the temperature the dynamics of water exchange can also be studied. The hydration numbers measured by this technique are those for the inner hydration sheath, and again values of 4 are found for Be^{2+} and six for many other di- and tri-positive cations. The hydration numbers for the alkali metals singly-positive cations have also been determined by this method, with values of ~ 3 being found.

2.5.3.3 Extensions to the Debye-Hückel Treatment

We have seen that in the molality range >0.001 mol kg^{-1}, equation (2.77) becomes progressively less and less accurate, particularly for unsymmetrical electrolytes. It is also clear, from Table 2.10, that even the properties of electrolytes of the same charge type are no longer independent of the chemical identity of the electrolyte itself, and our neglect of the factor κa_0 in the derivation of (2.77) is also not valid. As indicated above, a partial improvement in the DH theory may be made by including the effect of finite size of the central ion alone. This leads to the expression:

$$\ln \gamma_{\pm} = -\frac{|z_+ z_-| e_0^2 \kappa}{8\pi\varepsilon\varepsilon_0 k_B T(1 + a_0\kappa)} = -\frac{A|z_+ z_-|\sqrt{\mathbf{I}}}{(1 + Ba_0\sqrt{\mathbf{I}})} \tag{2.81}$$

where the parameter B also depends only on the properties of the solvent, and has the value 3.28×10^9 m^{-1} for water at 25°C. The parameter a_0 is adjustable in the theory, and usually the product Ba_0 is close to unity. The problem with equation (2.81) is that whilst it affords a finite size to the central ion, it does not afford this dignity to the surrounding ions, and we must expect this approach also to fail at higher concentrations. Details of this type of methodology can be found in the cited review article by Outhwaite.

Technically more sophisticated statistical thermodynamic calculations have been carried out for electrolyte concentrations up to 2M using interionic potentials of the form $u_{ij}(r) = u_{ij}{}^*(r) + \dfrac{q_i q_j e_0^2}{4\pi\varepsilon_0\varepsilon_r}$ where $u_{ij}{}^*$ is the short range inter-ionic potential. Perhaps surprisingly, the this type of approach even gives reasonable results for molten salts away from the critical point, but fails for very concentrated aqueous solutions.

The failure of both the Debye-Hückel and more advanced statistical thermodynamic methods using the type of potential given above particularly for higher valence electrolyte solutions points to neglect of another effect, first identified by Bjerrum: that of ion pairing. Bjerrum's original theory assumed that all oppositely charged ions within a distance $q_+ q_- e_0^2/8\pi\varepsilon_r\varepsilon_0 k_B T$ became paired whilst the rest remained free and could be treated within the Debye-Hückel approximation. More complete theories have been derived; Mayer was the first to show that the terms in the virial expansion for electrolyte solutions could be re-summed to cancel out the divergences arising from the long-range nature of the Coulomb potential. Whilst Mayer's theory is formally exact within the radius of convergence of the virial series, however, the difficulties involved in calculating the higher-order coefficients have restricted its use; nonetheless, even the second-order theory reproduces qualitatively both the effects of ion-pairing and excluded volume effects.

In spite of the limitations of the various approaches used, and the sometimes very drastic approximations employed, the results of calculations are often surprisingly good. Provided good quality potentials are used for both solvent and ions, then both analytical and Monte Carlo type methods usually converge over a wide concentration range, and show the right limiting behaviour at low concentrations. Furthermore, both approaches also show the expected rise in activity coefficients at higher concentrations owing to the excluded volume effects.

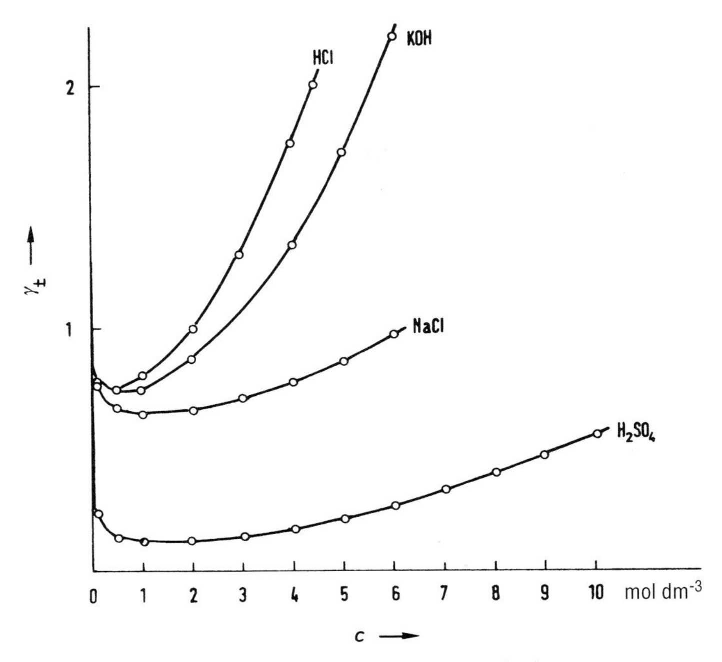

Fig. 2.20 Variation of activity coefficients in concentrated solutions.

However, encouraging though such agreement is, it has been obtained at the expense of any simplicity. Equation (2.81) fails at concentrations above *c.* 0.1M, and the mean activity coefficient for NaCl and other electrolytes shown in Fig. 2.20 demonstrates that in more concentrated solutions, the activity coefficients of all electrolytes begin to rise, often exceeding the value of unity. In addition to excluded volume effects and ion pairing, other contributions to this rise come from alterations in ion-solvent interactions: at higher concentrations, the number of solvent molecules essentially immobilised in the solvent sheath about each ion becomes a significant fraction of the total amount of solvent present. This can be exemplified by the case of sulphuric acid: given that each proton requires four water molecules for solvation, and the sulphate ion can be estimated to require one, each mole of H_2SO_4 will require 9 moles of water. One kg of water contains ca. 55 moles, so that a 1 mol kg^{-1} solution of H_2SO_4 will only leave 46 moles of 'free' water. The effective concentration of an electrolyte will, therefore, be appreciably higher than its analytical value, and this effect becomes more marked the higher the concentration. A further effect also becomes important at higher concentrations: implicit in our whole approach is the assumption that the free energy of solvation of the ions is independent of concentration. However, if we look again at our example of sulphuric acid, it is clear that for $m > 6$ mol kg^{-1}, apparently all the water is present in the solvation sheaths of the ions! Of course what actually occurs is that the extent of solvation of the ions changes, in effect decreasing the stability of the ions, but this process essentially invalidates the McMillan-Mayer approach, or at the least requires the potential of mean force to be chosen in such a way as to reproduce the change in solvation energy.

Empirically, equation (2.81) may be further modified by adding a linear term, as suggested by Hitchcock:

$$\ln \gamma_{\pm} = -\frac{A|z_+ z_-|\sqrt{\mathbf{I}}}{(1 + Ba_0\sqrt{\mathbf{I}})} + b\mathbf{I} \tag{2.82}$$

where b is a parameter to be fitted to the data. This does account for the rise in $\gamma_{\pm}$ quite well, but the empirical nature of the parameter b and the lack of agreement on its interpretation do mean that (2.82) can only be used phenomenologically.

A quite different approach was adopted by Robinson and Stokes [R.A. Robinson and R.H. Stokes; "Electrolyte Solutions", Butterworths, London, 1959] who emphasised, as above, that if the solute dissociated into ν ions, and a total of h molecules of water are required to solvate these ions, then the real concentration of the ions should be corrected to reflect only the bulk solvent. Robinson and Stokes derive, with these ideas, the following expression for the activity coefficient:

$$\ln \gamma_{\pm} = \frac{A|z_+ z_-|\sqrt{\mathbf{I}}}{(1 + Ba_0\sqrt{\mathbf{I}})} - \frac{h}{\nu}\ln a_A - \ln[1 + 0.001 M_A(\nu - h)m] \tag{2.83}$$

where a_A is the activity of the solvent, M_A its molar mass and m the molality of the solution. The equation (2.83) has been extensively tested for electrolytes and provided h is treated as a parameter, fits remarkably well for a large range of electrolytes up to molalities in excess of 2. Unfortunately, the values of h so derived, whilst showing some sensible trends, also show some rather counter-intuitive effects, such as an increase from Cl^- to I^-. Furthermore, the values of h are not additive between cations and anions in solution, leading to significant doubts about the interpretation of the equation. Although considerable effort has gone into finding alternative more accurate expressions, there remains considerable doubt about the overall physical framework.

The activity coefficients of neutral molecules – undissociated dissolved species or solvent molecules – can, in principle, be calculated provided the nature of the interactions of these species with each other and, if applicable, with the surrounding solvent are known. These interactions are usually of the Van der Waals or dipole-dipole type and are generally substantially weaker than the interionic interactions that we have been concerned with above. The result is that the activity coefficients for neutral solutes are usually in the neighbourhood of unity. This is valid for the solvent itself only as long as the latter is in excess, and the amount of solvent bound in the solvation sheaths can be neglected in comparison to the free solvent. Formally, this condition is equivalent to the demand that the mole fraction of solvent, $x_S \gg \sum_j x_j$ where x_j refers to the mole fraction of the jth dissolved species.

2.6 The Properties of Weak Electrolytes

2.6.1 The Ostwald Dilution Law

Weak electrolytes can be described by the introduction of an equilibrium constant that relates the splitting of the parent molecule into its constituent ions. As an example, we can consider the ionisation of a weak organic acid:

$$\mathrm{R{-}COOH + H_2O \rightleftharpoons R{-}COO^- + H_3O^+} \tag{2.84}$$

or, in short form

$$\mathrm{HA + H_2O} \underset{k_r}{\overset{k_d}{\rightleftharpoons}} \mathrm{A^- + H_3O^+} \tag{2.85}$$

where k_d and k_r are the rate constants for the dissociation and recombination reactions.

In equilibrium, we have

$$k_d c_{\mathrm{HA}} c_{\mathrm{H_2O}} - k_r c_{\mathrm{A^-}} c_{\mathrm{H_3O^+}} = 0 \tag{2.86}$$

where the rate at which the molecules of HA dissociate is exactly balanced by the rate at which they reform. If we write molalities in place of concentrations and recall that the molality of the solvent water is effectively a constant, provided that the solution is reasonably dilute, we can incorporate the molality of water into k_d. Rearranging (2.86) then leads to the equation

$$\frac{m_{\mathrm{A^-}} m_{\mathrm{H_3O^+}}}{m_{\mathrm{HA}}} = \frac{k_d}{k_r} = K'_m \tag{2.87}$$

where the expression K'_m is referred to as the 'dissociation constant' of the acid, or, classically, as the 'acidity constant'. In fact, the 'rate constants' appearing in (2.87) are not independent of concentration, and, as a result, neither is K'_m. In addition, the molalities are not referred to any standard, and formally K'_m would have units. The corresponding thermodynamic acid dissociation constant, K_a, defined as

$$K_a = \frac{a_{\mathrm{A^-}} a_{\mathrm{H_3O^+}}}{a_{\mathrm{HA}}} = \frac{(m_{\mathrm{A^-}}/m^0)(m_{\mathrm{H_3O^+}}/m^0)}{(m_{\mathrm{HA}}/m^0)} \frac{\gamma_\pm}{\gamma_{\mathrm{HA}}} \approx K_m \gamma_\pm^2 \tag{2.88}$$

where we have defined the molality form of the dissociation constant as:

$$K_m = \frac{(m_{\mathrm{A^-}}/m^0)(m_{\mathrm{H_3O^+}}/m^0)}{(m_{\mathrm{HA}}/m^0)} \tag{2.89}$$

and made the assumption that $\gamma_{\mathrm{HA}} \approx 1$, as justified above, so that K_m is now defined in terms of dimensionless molality ratios, and is itself dimensionless. Although K_a and K_m are dimensionless, nevertheless their numerical values depend on the choice for m^0. It is also true that K_m will retain a concentration dependency, through the concentration dependence of the $\gamma_\pm$ term, though at low concentrations, K_a and K_m can be assumed to be equal without great error. However, at molalities greater than 0.01 mol kg^{-1} errors in excess of 20 % will be encountered, and the use of activity coefficients becomes mandatory. In the calculation of those activity coefficients, expressions (2.77) or (2.81) to (2.83) can be used and m^0 can be taken as one mol kg^{-1}.

Ostwald was the first to clarify the connection between the equilibrium properties of weak electrolytes and the electrical conductivities of their solutions. If m is the stoichiometric molality of a 1-1 weak electrolyte, HA, and α is the degree of dissociation, then

$$m_{A^-} = m_{H_3O^+} = \alpha m \text{ and } m_{HA} = (1 - \alpha)m \tag{2.90}$$

Substituting these values into (2.89) gives

$$K_m = \frac{\alpha^2 (m/m^0)^2}{(1-\alpha)(m/m^0)} = \left(\frac{\alpha^2}{(1-\alpha)}\right)(m/m^0) \tag{2.91}$$

and, if $\alpha << 1$, we see

$$\alpha\sqrt{\frac{m}{m^0}} \approx \sqrt{K_m} \tag{2.92}$$

It follows from (2.91) that at infinite dilution, the weak electrolyte will be completely dissociated, a result that can be understood if we remember that the recombination reaction is effectively prevented at infinite dilution by the large distances between the ions. It also follows that the equivalent conductivity at infinite dilution will correspond to that of the completely dissociated electrolyte, and the rapid fall in the equivalent conductivity of a weak electrolyte with increasing concentration is due to the decrease in α. In fact, it can easily be seen that

$$\alpha = \Lambda/\Lambda_0 \tag{2.93}$$

a result that follows immediately from the definition of Λ in equation (2.12), with the proviso that the measured conductivity is to be divided by the stoichiometric concentration c_{HA} of the weak acid, *expressed in mol m*$^{-3}$. Implicit in (2.93) is the assumption that the mobilities of anion and cation are independent of concentration; this is only approximate, though the actual concentration of ions will, of course, be much smaller than the total analytical concentration of acid.

If (2.93) be inserted into (2.91), then an expression for K_m can be obtained. If all the assumptions are valid, K_m should be a constant, and Table 2.11 shows that this is indeed the case to considerable accuracy over a wide concentration range.

In fact, the consistency is remarkable, given that we have neglected the effects of both activity coefficient and variation of mobility with concentration. It is true that the ionic strength remains low: even at an acetic acid molality of 0.1 mol kg^{-1} the ionic strength is only ca. 0.003. However, this should lead to an estimate of K_m some 10 % low, and the constancy of the last column in Table 2.11 is certainly better than this. In fact, fortuitously, the neglect of the change in mobility with ionic strength leads to α

Table 2.11 Conductivity, Λ, degree of dissociation, α and acid dissociation constant, K_m (referred to $m_0 = 1$ mol kg^{-1}) for aqueous acetic acid at 25°C.

m/m^0	$\Lambda/10^{-4}\Omega^{-1}mol^{-1}m1^2$	$\alpha = \Lambda/\Lambda_0$	$K_m = \alpha^2(m/m^0)/1-\alpha)$
0	390.59	1	
0.0001114	127.71	0.327	1.77 4 10^{-5}
0.001028	48.13	0.123	1.77 4 10^{-5}
0.005912	20.96	0.0537	1.80 4 10^{-5}
0.01283	14.37	0.0368	1.80 4 10^{-5}
0.02000	11.56	0.0296	1.81 4 10^{-5}
0.05000	7.36	0.0188	1.80 4 10^{-5}
0.1000	5.20	0.0133	1.79 4 10^{-5}

being overestimated, an error that almost exactly cancels that caused by neglect of activity coefficients, at least in the case of acetic acid.

2.6.2 The Dissociation Field Effect

The equivalent conductivity of weak acids shows a marked increase at high values of the applied electric field, an increase above that expected from the counterbalancing of the relaxation and electrophoretic effects described above. This increase can, in fact, be traced to an increase in the degree of dissociation, α.

In fact, the bond between the acid anion and the proton is stretched in the presence of an electric field and this weakens it, leading to an increase in the rate constant for dissociation, k_d. By contrast, the recombination velocity between anion and proton is not affected by the field, and the net effect is, therefore, to increase α. Onsager has shown that the value of k_d^E in the presence of an electric field $\boldsymbol{E}$ is given by

$$k_d^E = k_d^0\left(1 + \frac{z_+^2 z_-^2(\lambda_0^+ + \lambda_0^-)}{z_+\lambda_0^+ + z_-\lambda_0^-}\frac{1.211|\boldsymbol{E}|}{4\pi\varepsilon_r T^2}\right) \tag{2.94}$$

where $\boldsymbol{E}$ is given in units of Vm^{-1}, and ε_r is the relative permittivity of the solvent.

This effect is also known as the *Second Wien Effect*, and is of particular electrochemical significance in that the electric fields encountered in the electrical double layer close to the electrode surface are of the order of $10^8 - 10^9$ Vm^{-1}, sufficient, according to (2.94), to raise the value of k_d by more than an order of magnitude. This has been confirmed by rotating-disc studies (see Section 4.3.5).

2.7
The Concept of pH and the Idea of Buffer Solutions

Three steps are necessary in order to approach the modern concept of pH. Sørensen was the first to suggest that, given the huge range of proton concentrations encountered in solution (which, for example, extends from 10M to 10^{-15}M in aqueous solution), it would be sensible to introduce a logarithmic scale of concentration. He defined the pH as the negative of the decadic logarithm of the proton concentration, the latter *being expressed relative to a standard of 1 mol dm*$^{-3}$ *or 1000 mol dm*$^{-3}$:

$$pH_c \equiv -\log_{10}\left(\frac{c_{H_3O^+}}{c^0}\right) \tag{2.95}$$

Evidently, for dilute aqueous solutions, this is closely approximated by $pH_m \equiv -\log_{10}\left(\frac{m_{H_3O^+}}{m^0}\right)$ where m^0 is taken as 1 mol kg^{-1}. However, for very many solution properties it is not the concentration but the activity that is significant. This led to the definition of an activity-based pH as:

$$pH_a \equiv -\log_{10} a_{H_3O^+} \equiv -\log_{10} \gamma_{H_3O^+}\left(\frac{H_3O^+}{m^0}\right) \tag{2.96}$$

Unfortunately, the activity of a single ion is not measurable, even in principle, as discussed above, and in order to find a definition of pH that allowed comparison with experiment, the *conventional pH scale* has been introduced, and is described in detail in Section 3.6.6. Individual values of this scale are fixed by the use of standard buffer solutions, which are made up and their pH measured using agreed procedures.

The definition of pH given above is a useful heuristic device, and interpolation between fixed values uses techniques essentially dependent on the formula (2.96). It should be emphasised that (2.96) is by no means limited to aqueous solution, though in practice the great majority of measurements are carried out in water. For both water and other solvents there are now available agreed standard buffer solutions for which the pH is *defined*. In water the following equilibrium takes place:

$$2H_2O \rightleftharpoons H_3O^+ + OH^- \tag{2.97}$$

and, bearing in mind the essential constancy of the total concentration of undissociated water (since the equilibrium constant of 2.97 is very small), we have

$$a_{H_3O^+}a_{OH^-} = K_w^{H_2O} \approx \left(\frac{m_{H_3O^+}}{m^0}\right)\cdot\left(\frac{m_{OH^-}}{m^0}\right) \tag{2.98}$$

where $K_w^{H_2O}$ is termed the ionic product of water and takes the value 1.0084×10^{-14} (referred to a standard concentration of 1 mol kg^{-1}) at 25°C. It follows that in water

$$pH = -\log_{10} K_w^{H_2O} - \log_{10} a_{OH^-} \equiv 13.9965 - pOH \tag{2.99}$$

an expression that will be true regardless of other ionic equilibria or species present. At the neutral point (pH = 7.00 at 25°C), the values of pH and pOH are equal, and in aqueous solutions, we have, in practice, a pH-scale that extends from near −1 to +15.

Buffer solutions are electrolyte solutions whose pH is very stable to either dilution or to addition of small quantities of acid or base. This stability is conferred by having simultaneously in solution relatively high concentrations of both a weak acid (or base) together with a completely dissociated salt of the acid (or base). Then the equilibrium for the reaction of (for example) a weak acid with water:

$$\mathrm{HA} + \mathrm{H_2O} \rightleftharpoons \mathrm{A^-} + \mathrm{H_3O^+} \tag{2.100}$$

can be shifted to the left or right as we add or take away protons. If the molality of HA is close to the analytical concentration of the added weak acid, m_{acid}, and the molality of A^- is similarly close to that of the added salt, m_{salt}, then we have, from the equilibrium (2.100)

$$a_{\mathrm{H_3O^+}} = K_a\left(\frac{a_{\mathrm{HA}}}{a_{\mathrm{A^-}}}\right) \approx K_a\left(\frac{m_{\mathrm{HA}}}{m_{\mathrm{A^-}}}\right)\cdot\left(\frac{1}{\gamma_{\mathrm{A^-}}}\right) \approx K_a\left(\frac{m_{\mathrm{acid}}}{m_{\mathrm{salt}}}\right)\cdot\left(\frac{1}{\gamma_{\mathrm{A^-}}}\right) \tag{2.101}$$

where we have assumed in (2.101) that the activity coefficient of the undissociated acid is unity and that we can replace the molalities of HA and A^- by the added molalities of weak acid and completely dissociated salt as discussed.

If we write $\mathrm{p}K_a \equiv -\log_{10} K_a$, then taking logarithms of (2.101) gives

$$\mathrm{pH} = \mathrm{p}K_a + \log_{10}(m_{\mathrm{salt}}/m_{\mathrm{acid}}) + \log_{10}\gamma_{\mathrm{A^-}} \tag{2.102}$$

It can be seen that the pH of a buffer solution depends, to a good approximation, only on the equilibrium constant for (2.100) and on the relative concentrations of weak acid and salt, and does not change on dilution, provided we can neglect the change in the activity coefficient. If protons are added or taken away, we expect, from (2.100), that there will be an (almost) quantitative recombination or dissociation of the weak acid to maintain equilibrium, provided that the concentrations of HA and A^- greatly exceed the proton concentration. It follows that the proton concentration itself will remain essentially unaltered. This will be valid just as long as the concentration of added proton is small compared to the acid and its salt, i.e. the *buffer capacity* is not exceeded.

Very similar considerations apply in the case of a weak base, the equation corresponding to (2.100) being

$$\mathrm{B} + \mathrm{H_2O} \rightleftharpoons \mathrm{BH^+} + \mathrm{OH^-} \tag{2.103}$$

and corresponding to (2.102) we obtain

$$\mathrm{pOH} = \mathrm{p}K_b + \log_{10}(m_{\mathrm{BH^+}}/m_{\mathrm{base}}) + \log_{10}\gamma_{\mathrm{BH^+}} \tag{2.104}$$

In Table 2.12 we present some standard buffer solutions, whose concentrations are referred to the molarity scale, and the corresponding pH values in the temperature range 0–95°C.

The temperature variation arises since both K_a and γ are functions of temperature. In order to make up these solutions, the purest grade of chemical should be used and CO_2-free water that has been double-distilled through quartz condensers. Storage of buffer solutions should be in polythene flasks with tight-fitting screw-top lids to prevent any CO_2 being absorbed from the atmosphere. All solutions should be renewed at least every two months.

Table 2.12 Standard Buffer solutions in which the molalities of acid and anion are equal unless otherwise stated.

Temp/°C	K-oxalate ($c = 0.05$ mol dm^{-3})	KH-tartrate (saturated at 25°C)	KH-phthalate ($c = 0.05$ mol dm^{-3})	KH_2PO_4 ($c = 0.025$ mol dm^{-3}) Na_2HPO_4 ($c = 0.025$ mol dm^{-3})	Borax ($c = 0.01$ mol dm^{-3})
0	1.67		4.01	6.98	9.46
5	1.67		4.01	6.95	9.39
10	1.67		4.00	6.92	9.33
15	1.67		4.00	6.90	9.27
20	1.68		4.00	6.88	9.22
25	1.68	3.56	4.01	6.86	9.18
30	1.69	3.55	4.01	6.85	9.14
35	1.69	3.55	4.02	6.84	9.10
40	1.70	3.54	4.03	6.84	9.07
45	1.70	3.55	4.04	6.83	9.04
50	1.71	3.55	4.06	6.83	9.01
55	1.72	3.56	4.08	6.84	8.99
60	1.73	3.57	4.10	6.84	8.96
70		3.59	4.12	6.85	8.92
80		3.61	4.16	6.86	8.88
90		3.64	4.20	6.86	8.85
95		3.65	4.22	6.87	8.83

2.8 Non-aqueous Solutions

Non-aqueous liquids are often better solvents than water, particularly for organic and organo-metallic compounds. Materials that are inert in water can react completely in other solvents, and, by way of contrast, solutions of alkali metals can be stabilised in non-aqueous solvents. Moreover, many non-aqueous solvents exhibit higher decomposition potentials than water, and provide a broad potential region of several volts for electrochemical investigation. Recently, non-aqueous solvents have been the focus of additional scientific and technical interest, particularly in the area of electro-organic synthesis and lithium-based batteries.

2.8.1
Ion Solvation in Non-aqueous Solvents

The calculation of the interaction of two ions within a solvent of relative permittivity ε_r can be carried out by assuming following Bjerrum and Fuoss, that the ions can be treated as charged hard spheres in a continuum. As above, the force between two such charged ions can be written:

$$F = \frac{q_1 \cdot q_2}{4\pi\varepsilon_r\varepsilon_0 r^2}\frac{\boldsymbol{r}}{r} \quad (2.105)$$

Where r is the distance apart of the ions, ε_r the relative permittivity (dielectric constant) of the medium, and ε_0 the permittivity of free space, which has the approximate value 8.854×10^{-12} $CV^{-1}m^{-1}$. The dielectric constant, or relative permittivity of a medium is, as shown in Table 2.13, related to the dipole moment, μ, of its constituent molecules, the latter being the measure of charge separation in the neutral species.

The dipole moment is of importance since, in an electric field, the molecules of solvent will tend to be oriented with their dipoles along the field direction. Thus, the polarity of the solvent has two important roles to play: it weakens ion-ion interactions and determines the direction and extent of solvent ordering in the presence of the ions. The values of ε_r for a series of solvents are given in Table 2.14. It will be noted that, in general, the relative permittivities of non-aqueous solvents are smaller than that of water itself, and this limits greatly the choice of conducting salts that can be dissolved in such solvents to give conducting solutions. Salts that do show good dissociative properties in non-aqueous solutions include the alkali-metal perchlorates, $NaBF_4$, LiCl, $LiAlCl_4$ and $LiAlH_4$, as well as some organic electrolytes such as quaternary ammonium salts.

Table 2.13 Dipole moment, μ, and relative permittivity, ε_r, for different solvents.

Solvent	$\mu/10^{-30}$ Cm	ε_r	Temperature/°C
CO_2	0	1.6	– 5
Benzene	0	2.24	20
Ammonia	4.90	14.9	24
Methanol	5.70	31.2	20
Water	6.17	81.1	18

Table 2.14 Relative permittivities of important solvents.

Solvent	Abbreviation	Temp./°C	ε_r
Water		25	78.3
Acetone		25	20.7
Acetonitrile	ACN	25	36.0
Ammonia		− 34	22.0
1,2-Dimethoxyethane	DME	25	7.2
N,N-Dimethylformamide	DMF	25	36.7
Dimethylsulphoxide	DMSO	25	46.6
Dioxane		25	2.2
Ethanol		25	24.3
Ethylene Carbonate	EC	40	89.0
Methyl Formate	M	20	8.5
Nitromethane	NM	25	35
Proplylene Carbonate	PC	25	64
Pyridine		25	12.0
$POCl_3$		25	13.7
$SOCl_2$		22	9.1
SO_2Cl_2		22	9.2
H_2SO_4		25	101

2.8.2
Electrolytic Conductivity in Non-aqueous Solutions

The Debye-Hückel-Onsager theory covered in section 2.4.3 is also valid for dilute non-aqueous solutions of electrolytes, Fig. 2.21 shows the linearity of the molar conductivity Λ *vs.* $\sqrt{c/c^0}$ plot for solutions of alkali-metal thiocycanates in methanol as solvent. The important variables affecting the magnitude of the slope are, as before, the relative permittivity and the viscosity, and both occur in the denominator of equation (2.55). This leads to a much stronger dependence of molar conductivity on concentration than is normally found in water, with Λ falling much more rapidly with concentration.

This effect is shown clearly in figure 2.22, where the variation of Λ with concentration for water-dioxane mixtures of different relative permittivity is shown. The stronger influence of the concentration in solvents of smaller ε_r has the consequence that the conductivity of solutions of electrolytes in non-aqueous solvents is significantly

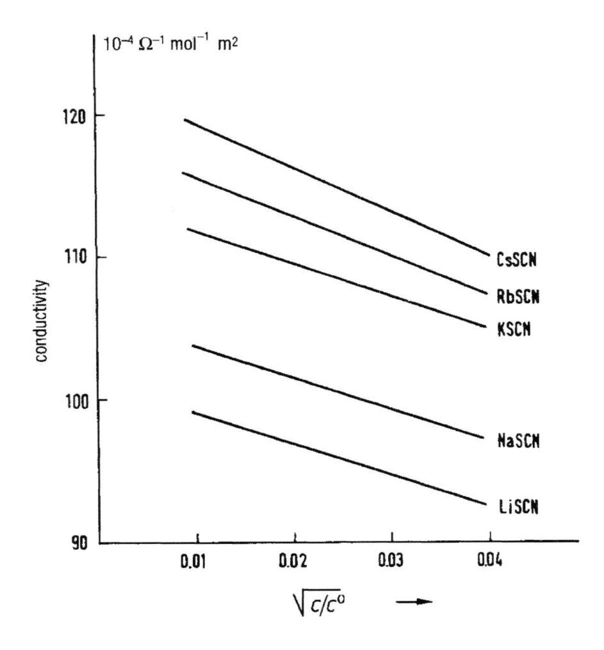

Fig. 2.21 Molar conductivities of some alkali-metal thiocyanates in methanol as a function of concentration. The decrease on going from Cs^+ to Li^+ thiocyanates corresponds to an increase in ionic solvation, as found in aqueous solutions (see section 2.3.4).

smaller than that of the corresponding aqueous solutions, at least for practical concentrations. However, the molar conductivities at infinite dilution for individual ions are rather similar in magnitude in most solvents.

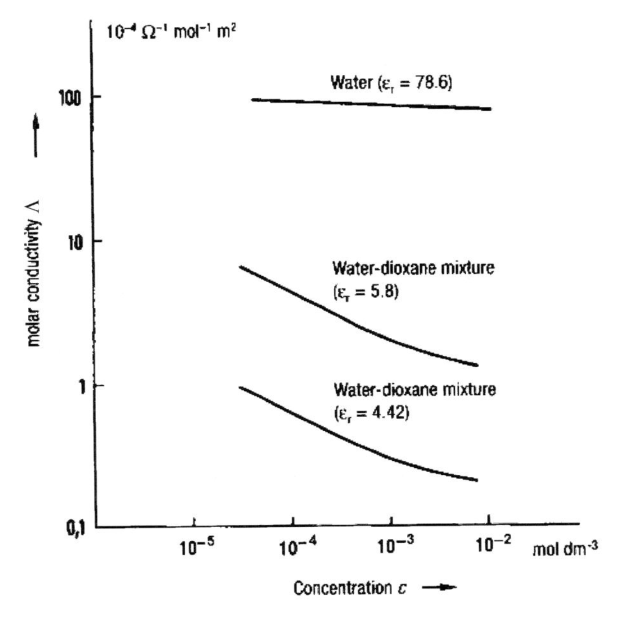

Fig. 2.22 Concentration dependence of the molar conductivity of tetraisoamylammonium nitrate in solvents of different dielectric constant.

2.8.3
The pH-Scale in Protonic Non-aqueous Solvents

For water-free *protonic* solvents of the form HL, self ionisation equilibria of the form:

$$2HL \rightleftharpoons L^- + LH_2^+ \tag{2.106}$$

will always exist, at least formally. One well-known example is liquid ammonia, in which the equilibrium

$$2NH_3 \rightleftharpoons NH_4^+ + NH_2^- \tag{2.107}$$

is found. In such solvents, the definition of pH established for aqueous solutions;

$$pH_{aq} = -\log_{10} a_{H_3O^+} \tag{2.108}$$

can be broadened to

$$pH_p = -\log_{10} a_{LH_2^+} \tag{2.109}$$

where the subscript p stands for protonic.

A protonic solvent is said to be at neutral pH_p when the two ions generated by the equilibrium (2.96) are present at equal activities. One example is formic acid, for which the equilibrium dissociation constant, K_f, as given by

$$K_f = a_{HCOO^-} \cdot a_{HCOOH_2^+} \tag{2.110}$$

has the value $10^{-6.2}$. Thus, at the neutral point, the activities of $HCOO^-$ and $HCOOH_2{}^+$ will both be equal to $(K_f)^{1/2}$, giving a pH_p value of 3.1. The neutral points

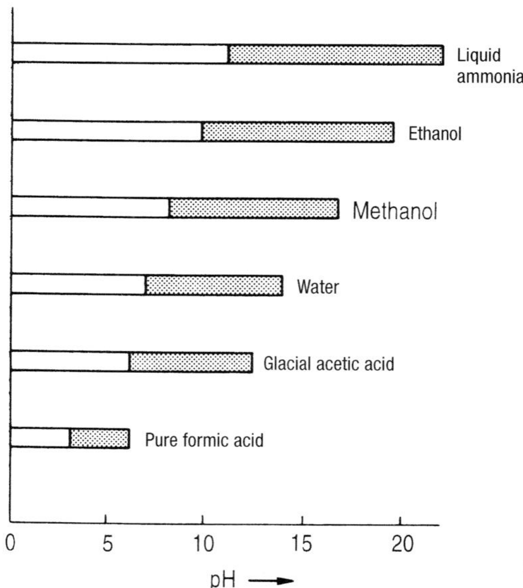

Fig. 2.23 Position of the neutral point in different protonic solvents.

of the pH_p-scales of different water-free protonic solvents do differ, though different pHp-values in one and the same solvent are comparable. In Fig. 2.23, the positions of the neutral points of a number of different protonic solvents are shown for comparison.

Measurement with glass electrodes (see section 3.6.7) have shown that these electrodes are also sensitive to the activity of protonated species even in non-aqueous solvents, a result important, for example, in following titrations in glacial acetic acid.

2.9 Simple Applications of Conductivity Measurements

2.9.1 The Determination of the Ionic Product of Water

Water dissociates to a small extent according to the equation

$$2H_2O \rightleftharpoons H_3O^+ + OH^- \tag{2.111}$$

If the molalities of H_3O^+ and OH^- are both equal to m mol kg^{-1}, then the molar conductivity is given approximately, from equations (2.11) and (2.12), by

$$\Lambda = \frac{\kappa_{H_2O}}{\rho_S m} \tag{2.112}$$

where ρ_s is the density of water in kgm^{-3} and κ_{H_2O} is the ionic conductivity of pure water. Whence

$$m = \frac{\kappa_{H_2O}}{\rho_S \Lambda} \tag{2.113}$$

Now, since the extent of self-ionisation of water is extremely small, we can, in place of the molar conductivity, Λ, write the limiting ionic conductivity $\Lambda_0 = \lambda_0^{H_3O^+} + \lambda_0^{OH^-}$ and

$$m = \frac{\kappa_{H_2O}}{\{\rho_S(\lambda_0^{H_3O^+} + \lambda_0^{OH^-})\}} \tag{2.114}$$

From the experimental values given in Tables 2.1 and 2.7 we find:

$$m = \frac{5.5 \cdot 10^{-6}}{\{\rho_S(349.8 + 197) \cdot 10^{-4}\}} = 1.00 \times 10^{-7} \text{ molkg}^{-1} \tag{2.115}$$

The ionic product of water can now be obtained immediately since

$$K_w = \left(\frac{m_{H_3O^+}}{m^0}\right)\left(\frac{m_{OH^-}}{m^0}\right) = 1.00 \times 10^{-14} \tag{2.116}$$

where m^0 is taken as 1 mol kg^{-1}. This value should also be the activity product or thermodynamic product since $\gamma_\pm = 1$ in this concentration region. This value is close to that found from thermodynamic considerations, and will be discussed further in section 3.6.

2.9.2
The Determination of the Solubility Product of a Slightly Soluble Salt

The addition to water of a slightly soluble salt such as AgCl leads to only a small amount of the material actually going into solution as completely dissociated ion, with the bulk of the substance remaining as a solid suspension. The contribution to the measured specific conductivity, κ_M, of the resultant dilute solution from the dissolved salt, $\kappa_S \equiv \kappa_M - \kappa_{H_2O}$, leads to a measure of the saturated molality of the dissolved salt, m_S, using exactly the same argument as in the previous section:

$$m_S = \frac{\kappa_S}{\{\rho_S(\nu_+\lambda_0^+ + \nu_-\lambda_0^-)\}} \tag{2.117}$$

From Tables 2.1 and 2.7, for saturated AgCl at 25°C, $\kappa_S = 1.73 \times 10^{-4} - 5.47 \times 10^{-6} = 1.675 \times 10^{-4}\ \Omega^{-1}\mathrm{m}^{-1}$, $\nu_+ = \nu_- = 1$, and

$$m_S = \frac{1.675 \times 10^{-4}}{\{\rho_S(62.2 + 76.4) \times 10^{-4}\}} = 1.21 \times 10^{-5}\mathrm{mol\ kg}^{-1} \tag{2.118}$$

From the definition of the solubility product, we find

$$K_S^{AgCl} = \left(\frac{m_{AG^+}}{m^0}\right)\left(\frac{m_{Cl^-}}{m^0}\right) = 1.46 \times 10^{-10} \tag{2.119}$$

Given the very low concentration of the ions, this may be considered equal to the thermodynamic product of activities, which is close to the thermodynamic value of 1.78×10^{-10}.

2.9.3
The Determination of the Heat of Solution of a Slightly Soluble Salt

The molar heat of solution of a slightly soluble salt, $\Delta_{sol}H_m$, is very difficult to determine calorimetrically owing to the very small amount of dissolved material. Using straightforward thermodynamic arguments, it can be shown that the temperature variation of the activity of the ion in a saturated solution will be of the form:

$$a_S = \mathrm{const.exp}\left(\frac{\Delta_{\mathrm{sol}}H_m}{RT}\right) \tag{2.120}$$

and for low concentrations, we may replace a_S by (m_S/m^0). The measurement of the temperature dependence of the conductivity of the saturated solution will then allow us to determine the temperature variation of m_S, provided that at each temperature, the values of λ_0 *appropriate to that temperature are employed.* The value of $\Delta_{sol}H_m$ can then be obtained from the slope of the plot of ln m_S vs. $1/T$.

2.9.4 The Determination of the Thermodynamic Dissociation Constant of a Weak Electrolyte

We have already discussed in section 2.6.1 the determination of the classical equilibrium constant from the expression Λ/Λ_0 for the degree of dissociation α. To obtain the thermodynamic equilibrium constant it is necessary to correct K_m for $\gamma_\pm$ using the expression $K_a = K_m\,\gamma_\pm^2$. From this we obtain:

$$\log_{10} K_m = \log_{10} K_a - 2\log_{10}\gamma_\pm \tag{2.121}$$

and using the Debye-Hückel limiting law expression for $\gamma_\pm$, we find

$$\log_{10} K_m = \log_{10} K_a + 1.018|z_+ z_-|\sqrt{\mathbf{I}} \tag{2.122}$$

For a 1-1 electrolyte, $\mathbf{I} = \alpha(m/m^0)$, and we find

$$\log_{10} K_m = \log_{10} K_a + 1.018\sqrt{\alpha(m/m^0)} \tag{2.123}$$

It follows that a plot of $\log_{10} K_m$ vs. $\sqrt{\alpha(m/m^0)}$ should be linear with an intercept corresponding to $\log_{10} K_a$. The problem thus reduces to finding a reasonable manner of determining α. From the measured conductivity, we expect that

$$\Lambda = \alpha\left(\Lambda_0 - (B_1\Lambda_0 + B_2)\sqrt{\alpha(m/m^0)}\right) \tag{2.124}$$

as discussed in section 2.4.3, where we have replaced concentration by molality. For water at 25°C, this becomes

$$\Lambda = \alpha\left(\Lambda_0 - (0.229\Lambda_0 + 60.20 \times 10^{-4})\sqrt{\alpha(m/m^0)}\right) \tag{2.125}$$

and from this expression, values of α and $\alpha(m/m^0)$ may be obtained. By substituting these into equation (2.123), a value of K_a can be found.

2.9.5 The Principle of Conductivity Titrations

The measured conductivity of an electrolyte solution depends primarily on the concentration and types of the ions. A conductivity measurement can thus provide a sensitive measure of the changes taking place in ionic concentration in the course of a chemical reaction in the solution phase.

An example of such a reaction is the precipitation of magnesium hydroxide and barium sulphate from a aqueous solution of $MgSO_4$ when a solution of $Ba(OH)_2$ is added; the more $Ba(OH)_2$ solution is added, the lower the conductivity will fall as all the ions present are removed.

Even if the total ionic concentration of a solution does not alter during the reaction, the course of the reaction may be followed by conductivity measurements provided one sort of ion is exchanged for another with a lower or higher equivalent ionic conductivity. One example is the neutralisation of a strong acid, such as HCl, with a strong

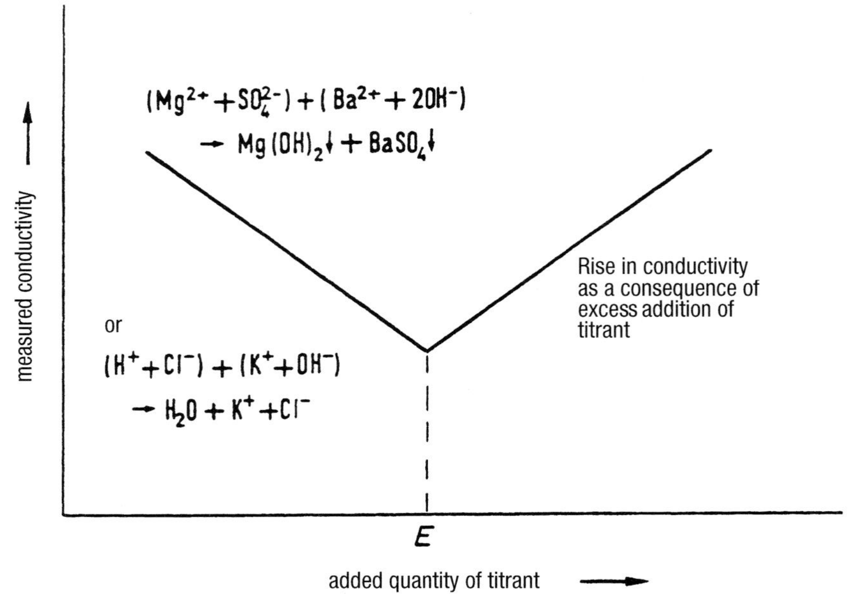

Fig. 2.24 Schematic representation of conductimetric precipitation and acid-base titrations. *E*: end- or equivalence point. In order to obtain *E* as the clear intersection of two lines of different slope, the volume of solution added must be as small a fraction of the total volume possible, otherwise curves are obtained.

base, such as KOH. In the course of the reaction, H_3O^+ ions are replaced by K^+ ions, whose mobility is considerably smaller. The measured conductivity thus declines as the base is added.

In both the above examples, the ions originally present in solution in the conductivity cell are consumed completely at the end point, and further addition of titrant, whether $Ba(OH)_2$ or KOH, will result in a rise ion conductivity. The change in direction of the conductivity with volume of added titrant allows us unambiguously to fix the end or equivalence point, as shown in Figure 2.24; this plot is applicable to both precipitation and acid-base reactions.

Conductivity titrations can also be carried out on redox reactions, such as the oxidation of Fe^{2+} with Ce^{4+}. The great advantage of conductimetric titrations lies in the fact that the end-point determination is not interfered with by either highly coloured reagents already present in solution or turbidity, and that the process is applicable to very dilute solutions.

Experimentally, it is sufficient in most conductivity titrations to use *ac* bridges and simple conductivity cells as shown in Figs. 2.6 and 2.7c. With modern equipment, it is even possible to automate the process. A detailed description is beyond the scope of this book, but details may be found in standard texts. The detection of the end-point in some cases is far from easy: for example, in the titration of a weak acid by a strong base, the end point is just a more or less prominent kink on the titration curve.

References for Chapter 2

Detailed discussions of the problems of electrolyte resistance measurements, as well as a very detailed discussion of activity coefficients and their determination are given in:

Robinson R. A., Stockes R. H., *Electrolyte Solutions*, Butterworth, London, 1959.

Bockris J. O'M., Reddy A. K. N., *Modern Electrochemistry*, Plenum Press, New York, 1970.

Levine I. N., *Physical Chemistry*. Mc.Graw Hill, New York, 1978.

Summaries of electrolyte conductivities are given in:

Landolt-Bornstein, *Zahlenwerte und Funktionen*, II. Band, 7. Teil. Springer-Verlag, Berlin 1960.

Conway B. E., *Electrochemical Data*. Elsevier, Amsterdam, 1952.

Background readings in the problems of consistency in the thermodynamic treatment of activity coefficients can be found in:

Denbigh K. G., *The principles of Chemical Equilibrium*. Cambridge University Press, 1981.

Introduction to the theory of liquids and electrolyte solutions:

A. Ben-Naim, *Statistical Thermodynamics for Chemists and Biologists*, Plenum Press, London, 1992

J.C. Rasaiah, *Statistical mechanics of strongly interacting systems: liquids and solids*, in J.H. Moore and N.D. Spencer (eds.), *Encyclopaedia of Chemical Physics and Physical Chemistry, vol. I: Fundamentals*, Institute of Physics Publ., Bristol, 2001.

L.L. Lee, *Molecular Thermodynamics of Non-ideal Fluids*, Butterworths, Boston, 1988.

C.W. Outhwaite: "Equilibrium theories of electrolyte solutions"; *Specialist Periodical Reports Chem. Soc. Vol.* **1**, 1974.

J. Mayer *Theory of Ionic Solutions* J. Chem. Phys. **18** (1950) 1426.

F. Hirata (ed.), *Molecular Theory of Solvation*, Kluwer Academic Publishers, Dordrecht, 2003.

Introduction to Simulation Methods

N.A. Metropolis, A.W. Rosenbluth, M.N. Rosenbluth, A.H. Teller and E. Teller: J. Chem. Phys. **21** (1953) 1087.

K. Binder and D.W. Heermann, *Monte Carlo Simulation in Statistical Physics, An Introduction*, 3rd Edition, Springer, Berlin, 1997.

D.W. Heermann, *Computer Simulation Methods in Theoretical Physics*, 2nd Edition, Springer-Verlag, Berlin, 1990.

D. Fraankel and B. Smit, *Understanding Molecular Simulation*, Academic Press 2002.

3
Electrode Potentials and Double-Layer Structure at Phase Boundaries

3.1
Electrode Potentials and their Dependence on Concentration, Gas Pressure and Temperature

3.1.1
The EMF of Galvanic Cells and the Maximum Useful Energy from Chemical Reactions

The discharge of a galvanic cell always corresponds to a chemical process that can be divided into two spatially separated electrochemical half-reactions or electrode reactions. These electrochemical half-reactions can then be combined to form an electronically neutral overall transformation termed the cell reaction.

In order for the overall reaction to be stoichiometrically balanced, the same number of electrons, n, must be exchanged in both half-reactions, with the stoichiometric factors, ν, chosen appropriately. Thus, we have in general (for a simple example see eq. (1.7))

$$\begin{array}{c} \nu_1 S_1 + \ldots + \nu_i S_i + ne^- \rightleftharpoons \nu_j S_j + \ldots + \nu_k S_k \\ \nu_1 S_1 + \ldots + \nu_m S_m \rightleftharpoons \nu_n S_n + \ldots + \nu_p S_p + ne^- \\ \hline \nu_1 S_1 + \ldots + \nu_m S_m \rightleftharpoons \nu_j S_j + \ldots + \nu_p S_p \end{array} \qquad (3.1)$$

If now we allow just one formula unit of reactants to be transformed to products in a galvanic cell, for example one mole of chlorine and one mole of hydrogen in the example in section 1.3, then, at maximum, the amount of electrical work equal to $nFE_{c,0}$ can be performed in the external electrical circuit, where $E_{c,0}$ is the potential difference between the electrodes at open circuit. This electrical work corresponds to the maximum cell voltage for reaction (3.1), which will be the zero-current voltage $E_{c,0}$ as illustrated in Fig. 1.5 and will then be obtained only if we allow the reaction to proceed essentially infinitely slowly (i.e., $i \rightarrow 0$, $E \rightarrow E_{c,0}$ as in Fig. 1.5).

The maximum electrical work must correspond to the maximum useful work obtainable from a chemical reaction. In the example of section 1.3 this would correspond to the maximum work obtainable when gaseous Cl_2 and H_2 are combined and the resultant hydrogen chloride dissolved in water to form hydrochloric acid, Eq. (1.7). For the transformation of one mole of reactants, this is the molar free energy of

Electrochemistry. Carl H. Hamann, Andrew Hamnett, Wolf Vielstich

ISBN: 978-3-527-31069-2

the reaction and is designated $\Delta_r G$; it can be found from thermodynamic tables. The electrical work obtained is a positive number, but it is conventional in thermodynamics to designate work done by the system on the surroundings as negative. We thus have the fundamental relationship:

$$\Delta_r G = -nFE_0 \tag{3.2}$$

We saw, in section 1.3, that if the current flows as a result of a chemical process (i.e. the cell is discharging or behaving as a galvanic cell), then $E_c^{\text{galv}} < E_0$, and it follows that the inequality

$$\Delta_r G + nFE_c^{\text{galv}} < 0 \tag{3.3}$$

must hold for any finite current flow in the system. In the same way, if the current flows in the reverse direction, and electrolysis takes place to convert hydrochloric acid to H_2 and Cl_2, then the potential required to drive the cell, $E_c^{\text{elec}} > E_0$ and

$$\Delta_r G + nFE_c^{\text{elec}} > 0 \tag{3.4}$$

is the inequality satisfied.

If the reactants and products of the cell reaction are in their *standard states* at the temperature of the transformation, i.e. the gases are at unit atmospheric pressure and the soluble species at unit mean activity, then the free energy change is denoted by the standard value, $\Delta_r G^0$, and we have the familiar expression

$$\Delta_r G^0 = -nFE^0 \tag{3.5}$$

3.1.2
The Origin of Electrode Potentials, Galvani Potential Difference and the Electrochemical Potential

In chapter 2, we defined the chemical potential for the ith component of a mixture through the equation

$$\mu_i = \mu_i^{0\dagger} + RT \ln a_i \tag{3.6}$$

where a_i is the activity of the ith component and $\mu_i^{0\dagger}$ the corresponding chemical potential at unit activity, the asterisk reminding us that this standard state does not refer to the pure substance. The chemical potential can be regarded either as the change in free energy when one mole of i is added to an infinite amount of the mixture, so that the mole fractions of all the components are unaltered, or it can be treated as a differential quantity

$$\mu_i = \left(\frac{\partial G}{\partial n_i}\right)_{n_j \neq n_i, p, T} \tag{3.7}$$

where the partial derivative is evaluated with the amounts of all other components j being maintained constant. The two definitions are equivalent, and it is apparent that $\mu_i^{0\dagger}$ will be a function of the temperature and pressure of the mixture.

The total free energy of a mixture may be written

$$G = \sum_i n_i \mu_i \tag{3.8}$$

and it follows that if a reaction occurs in the mixture such as that shown in equation (3.1), in which stoichiometric amounts of $S_1..S_m$ are consumed and $S_j..S_p$ generated, then the free energy change, assuming that the reaction is taking place in a large excess of the mixture, such that the mole fractions of both reactants and products is not sensibly affected is given by

$$\Delta_r G = \sum_i v_i \mu_i \tag{3.9}$$

and we assign *negative* values to those stoichiometric numbers v_i of the *reactants*. At equilibrium, $\Delta_r G$ must be zero, since otherwise, it will be possible to lower the free energy of the mixture by allowing reaction (3.1) to proceed either from left to right or right to left. We then reach the fundamental expression for a chemical reaction to be at equilibrium:

$$\sum_i v_i \mu_i = 0 \tag{3.10}$$

If two mixtures or solutions are in contact with each other such that chemical equilibrium is established between the phases, then for each component *i* present in the two phases

$$\mu_i(\text{I}) = \mu_i(\text{II}) \tag{3.11}$$

If this condition in not fulfilled, then reaction will take place spontaneously until it is. The reason for this is that until (3.11) is satisfied, it will be possible to lower the free energy by transfer of *i* from one phase to the other.

If a metal, such as copper, is placed in contact with a solution containing the ions of that metal, such as aqueous copper sulphate, then we expect an equilibrium to be set up of the following form:

$$\text{Cu}^0 \rightleftharpoons \text{Cu}^{2+}(aq.) + 2e^-(\text{M}) \tag{3.12}$$

where the label (M) refers to the metal. At the moment at which the metal is placed in contact with the solution, condition (3.11) will not be fulfilled. In order to satisfy the condition, some chemical reaction, either deposition of copper ions onto the metal or dissolution of copper ions into solution according to the dominant energetic relationships, must take place. However, it can be seen from (3.12) that copper deposition or dissolution cannot be treated as a simple chemical reaction in a single phase since the process must generate or consume electrical charge, and this will always be accompanied by the formation of a potential difference between the two phases. If the chemical potential of the copper in the metal exceeds that of the copper ions in solution and electrons in the metal, then metal dissolution will take place, and the solution next to the metal will become positively charged with respect to the metal itself. A so-called double layer is set up, the details of which we will treat later in this chapter, but the

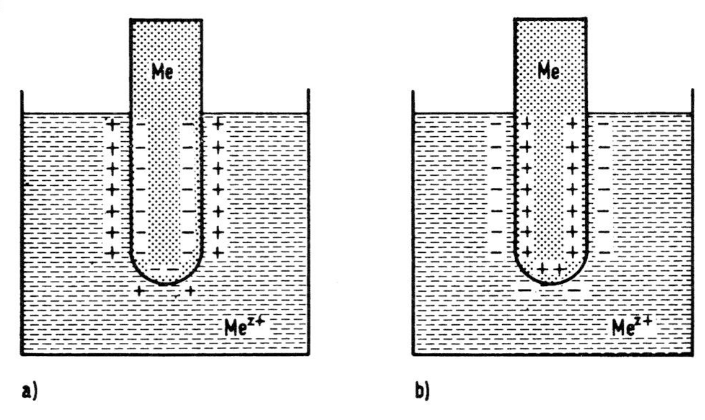

Fig. 3.1 Origin of the potential difference between electrode and electrolyte as a consequence of the formation of an electrolytic double layer at the phase boundary between metal and electrolyte: (a) $\mu_{Me^{2+}}$ (Metal) $\geq \mu_{Me^{2+}}$ (Solution); (b) $\mu_{Me^{2+}}$ (Metal) $\leq \mu_{Me^{2+}}$ (Solution).

important point here is that the potential difference established will inhibit further dissolution. In a similar way, if the chemical potential of Cu^{2+} ions in solution and electrons in the metal exceeds copper metal itself, then there will be a tendency for the metal ions to deposit on the copper, with the result that a potential difference of opposite polarity is established, which again acts to inhibit further deposition; both situations are shown in Fig. 3.1.

It is clear that we must incorporate the effects of this potential difference into our thermodynamic equations. If one mole of z-valent ions is brought from a remote position to the interior of the solution, in which there exists a potential φ, then the work done will be $zF\varphi$; this work term must be added to or subtracted from the free energy μ depending on the sign of z, and the conditions for equilibrium for component i partitioned between two phases with potentials φ(I) and φ(II) is

$$\mu_i(\mathrm{I}) + zF\varphi(\mathrm{I}) = \mu_i(\mathrm{II}) + z_i F\varphi(\mathrm{II}) \tag{3.13}$$

where φ(I) and φ(II) are the electrical potentials *in the interior* of phases (I) and (II), and are referred to as the Galvani potentials. The expression $\mu_i + z_i F\varphi$ is referred to as the *electrochemical potential*, $\tilde{\mu}_i$. We have:

$$\tilde{\mu}_i = \mu_i + z_i F\varphi = \mu_i^{0\dagger} + RT \ln a_i + z_i F\varphi \tag{3.14}$$

and the condition for electrochemical equilibrium is now:

$$\sum_i \nu_i \tilde{\mu}_i = 0 \tag{3.15}$$

where the ν_i are taken as positive numbers on the right-hand side of the equilibrium and negative numbers on the left-hand side. It should be emphasised at this stage that the condition for electrochemical equilibrium is essentially a *dynamic* one. In the example given above of copper immersed in copper sulphate, equilibrium is reached when the dissolution of copper is taking place as rapidly as its deposition. At this point, an external probe would reveal no *nett* current passing between electrode and solution, a result that can be used as the basis of a quantitative treatment of electrode kinetics.

3.1.3
Calculation of the Electrode Potential and the Equilibrium Galvani Potential Difference between a Metal and a Solution of its Ions – The Nernst Equation

The difference in inner potentials between two phases is referred to as the Galvani potential difference, and it can be calculated for the equilibrium case when no nett current flows. We represent by φ_S the Galvani potential of the solution and by φ_M the Galvani potential of the metal. According to equation 3.15, the equilibrium condition for reaction (3.12) can be written

$$\tilde{\mu}_{Cu}(M) = \tilde{\mu}_{Cu^{2+}}(aq.) + 2\tilde{\mu}_{e^-}(M) \tag{3.16}$$

Assuming the copper atoms in the metal to be neutral, so that $\tilde{\mu} = \mu$, then we have

$$\mu^0_{Cu}(M) + RT \ln a_{Cu}(M) = \mu^0_{Cu^{2+}} + RT \ln a_{Cu^{2+}} + 2F\varphi_S + 2\mu^0_{e^-}(M) + 2RT \ln a_{e^-} - 2F\varphi_M \tag{3.17}$$

The concentration of both copper atoms and electrons in copper metal will be effectively constant, so two of the activity terms can be neglected, and we finally have, on rearranging (3.17),

$$\Delta\varphi \equiv \varphi_M - \varphi_S = \frac{\mu^0_{Cu^{2+}} + 2\mu^0_{e^-} - \mu^0_{Cu}(M)}{2F} + \frac{RT}{2F} \ln a_{Cu^{2+}} \equiv \Delta\varphi^0 + \left(\frac{RT}{2F}\right) \ln a_{Cu^{2+}} \tag{3.18}$$

where $\Delta\varphi^0$ is the Galvani potential difference between electrode and solution in the case where $a_{Cu^{2+}} = 1$, and is referred to as the *standard Galvani Potential Difference.* It can be seen that the Galvani potential difference will alter, in general, by a factor of $(RT/zF) \ln\ 10 \equiv (0.059/z)\ V$ at 298K, for every order of magnitude change in activity of the metal ion, where z is the valence of the metal ion in solution.

Although neither $\Delta\varphi$ nor $\Delta\varphi^0$ are experimentally accessible, provided a *second electrode can be introduced into the solution, whose Galvani Potential Difference, $\Delta\varphi'$, is constant,* then the potential of our working electrode with respect to this second electrode, E, is measurable: $E = \Delta\varphi - \Delta\varphi'$. Furthermore, the standard electrode potential at unit activity, $E^0 = \Delta\varphi^0 - \Delta\varphi'$, and so $E - E^0 = \Delta\varphi - \Delta\varphi^0$. The practical problems involved in setting up a suitable reference electrode system are further discussed in sections 3.1.6 and 3.1.10 below; assuming this can be done, then we can write, for the potential of a metal electrode, M, in contact with a solution of its ions, M^{z+}:

$$E = E^0 + \left(\frac{RT}{zF}\right) \ln a_{M^{2+}} \tag{3.19}$$

This dependence of the equilibrium potential on the concentration of ions in solution is termed the Nernst equation.

3.1.4
The Nernst Equation for Redox Electrodes

In addition to the case of a metal in contact with its ions in solution, there are other cases in which a Galvani potential difference between two phases may be found. One such is the immersion of an inert electrode, such as platinum metal (usually in the form of a foil attached to a platinum rod fused into a glass tube), into an electrolyte solution containing a substance S that can exist in oxidised or reduced form through the loss or gain of electrons from the electrode. In the simplest case, we have

$$S_{ox} + ne^- \rightleftharpoons S_{red} \tag{3.20}$$

and S_{ox}, S_{red} are termed the oxidised and reduced components respectively of a redox pair. A simple example of this is the redox reaction

$$\mathrm{Fe}^{3+} + e^- \rightleftharpoons \mathrm{Fe}^{2+} \tag{3.21}$$

and for all these types of reaction, the important point to emphasise is that it is *electrons* that are exchanged between the solution phase and the electrode and not ions, as in the case of the metal-ion electrodes discussed above. In equations (3.20) and (3.21), it is envisaged that the electrons are transferred directly from the redox species to the electrode; at no point is the electron conceived as being free in the solution. It is true that there are certain solvents, such as liquid ammonia, in which the solvated electron has a long lifetime, but although it is possible, by radiolytic means, to generate solvated electrons, their lifetime is very small (in the region of 10^{-3} sec), and they are very powerful reducing agents.

The equilibrium properties of the redox reaction (3.20) can, in principle, be treated in the same way as above. At equilibrium, once a double layer has formed and a Galvani potential difference set up, we can write

$$\tilde{\mu}_{ox} + n\tilde{\mu}_{e^-}(\mathrm{M}) = \tilde{\mu}_{red} \tag{3.22}$$

and, bearing in mind that the positive charge on '*ox*' must exceed '*red*' by $|ne^-|$ if we are to have electroneutrality, then (3.22) becomes

$$\mu_{ox}^{0\dagger} + RT\ln a_{ox} + nF\varphi_S + n\mu_{e^-}^0 - nF\varphi_\mathrm{M} = \mu_{red}^{0\dagger} + RT\ln a_{red} \tag{3.23}$$

whence

$$\Delta\varphi = \varphi_M - \varphi_S = \frac{\mu_{ox}^{0\dagger} + n\mu_{e^-} - \mu_{red}^0}{nF} + \frac{RT}{nF}\ln\left(\frac{a_{ox}}{a_{red}}\right) = \Delta\varphi^0 + \frac{RT}{nF}\ln\left(\frac{a_{ox}}{a_{red}}\right) \tag{3.24}$$

where the standard Galvani potential difference is now defined as that for which the activities of Ox and Red are equal. As for the case of the metal-ion electrode, the Galvani potential difference must be measured experimentally with respect to a reference electrode. Assuming this can be done, we can write, for the electrode potential of the metal electrode itself

$$E = E^0 + (RT/nF)\ln(a_{ox}/a_{red}) \tag{3.25}$$

and once again it can be seen that the electrode potential can be altered by altering the ratio of the activities of oxidised and reduced states. In fact, as before, an alteration of this ratio by a factor of ten leads to a change of 0.059/n volt in E. It can also be seen that E will be independent of the magnitudes of a_{ox} and a_{red} provided their ratio is a constant.

For more complicated redox reactions, a general form of the Nernst equation may be derived by analogy with (3.25). If we consider a stoichiometric reaction of the following type

$$\nu_1 S_1 + \ldots + \nu_i S_i + ne^- \rightleftharpoons \nu_j S_j + \ldots \nu_k S_k \tag{3.26}$$

which can be written in the abbreviated form

$$\sum_{ox} \nu_{ox} S_{ox} + ne^- \rightleftharpoons \sum_{red} \nu_{red} S_{red} \tag{3.27}$$

then straightforward manipulation leads to the generalised Nernst equation

$$E = E^0 + \frac{RT}{nF} \ln \left(\frac{\Pi_{ox} a_{ox}^{\nu_{ox}}}{\Pi_{red} a_{red}^{\nu_{red}}} \right) \tag{3.28}$$

where we are using the notation

$$\prod_t a_t^{\nu_t} \equiv a_{S_1}^{\nu_1} \cdot a_{S_2}^{\nu_2} \ldots a_{S_i}^{\nu_i} \tag{3.29}$$

As an example, the reduction of permanganate in acid solution follows the equation

$$MnO_4^- + 8H_3O^+ + 5e^- \rightleftharpoons Mn^{2+} + 12H_2O \tag{3.30}$$

and the potential of a platinum electrode immersed in a solution containing both permanganate and Mn^{2+} is given by

$$E = E^0 + \left(\frac{RT}{5F} \right) \ln \left(\frac{a_{MnO_4^-} a_{H_3O^+}^8}{a_{Mn^{2+}}} \right) \tag{3.31}$$

assuming that the activity of neutral H_2O can be put equal to unity.

3.1.5
The Nernst Equation for Gas-electrodes

The Nernst equation for the dependence of the equilibrium potential of redox electrodes on the activity of solution species is also valid for uncharged species in the gas phase that take part in electron exchange reactions at the electrode-electrolyte interface. For the specific equilibrium process involved in the reduction of chlorine:

$$Cl_2 + 2e^- \rightleftharpoons 2Cl^- \tag{3.32}$$

the corresponding Nernst equation can easily be shown to be

$$E = E^0 + \left(\frac{RT}{2F}\right)\ln\left(\frac{a_{Cl_2}(aq)}{a^2_{Cl^-}}\right) \tag{3.33}$$

where $a_{Cl_2}(aq)$ is the activity of the chlorine gas dissolved in water. If the Cl_2 solution is in equilibrium with chlorine at pressure p_{CL_2} in the gas phase, then

$$\mu_{Cl_2}(gas) = \mu_{Cl_2}(aq) \tag{3.34}$$

Given that

$$\mu_{Cl_2}(gas) = \mu^0_{Cl_2}(gas) + RT\ln\left(\frac{p_{Cl_2}}{p^0}\right) \tag{3.35}$$

and

$$\mu_{Cl_2}(aq) = \mu^0_{Cl_2}(aq) + RT\ln a_{Cl_2}(aq) \tag{3.36}$$

where p^0 is the standard pressure of 1 atmosphere ($\equiv$ 101325 Pa), then it is clear that

$$a_{Cl_2}(aq) = \left(\frac{p_{Cl_2}}{p^0}\right)\cdot\exp\left(\frac{\mu^0_{Cl_2}(gas) - \mu^0_{Cl_2}(aq)}{RT}\right) \tag{3.37}$$

and we can write

$$E = E^{0\dagger} + \left(\frac{RT}{2F}\right)\ln\left(\frac{p_{Cl_2}}{p^0 a^2_{Cl^-}}\right) \tag{3.38}$$

where $E^{0\dagger}$ is the Galvani potential difference under the standard conditions of $p_{Cl_2} = p^0$ and $a_{Cl^-} = 1$.

For the hydrogen electrode

$$2H^+(aq.) + 2e^- \rightleftharpoons H_2 + 2H_2O \tag{3.39}$$

we have similarly

$$E = E^{0\dagger} + \left(\frac{RT}{2F}\right)\ln\left(\frac{a^2_{H_3O^+}}{p_{H_2}/p^0}\right) \tag{3.40}$$

3.1.6
The Measurement of Electrode Potentials and Cell Voltages

A direct measurement of the Galvani potential difference between electrode and electrolyte is not possible, since any device to measure such a voltage would have to be simultaneously in contact with both phases. Any contact of a measurement probe with the solution phase will have to involve a second phase boundary between metal and electrolyte somewhere; at this boundary an electrochemical equilibrium will be set up and with it a second equilibrium Galvani potential difference, and the overall potential difference measured by this instrument will in fact be the difference of two Galvani potential differences $\Delta\varphi(\mathrm{I})$ and $\Delta\varphi(\mathrm{II})$ for the two interfaces, as shown in Fig. 3.2.

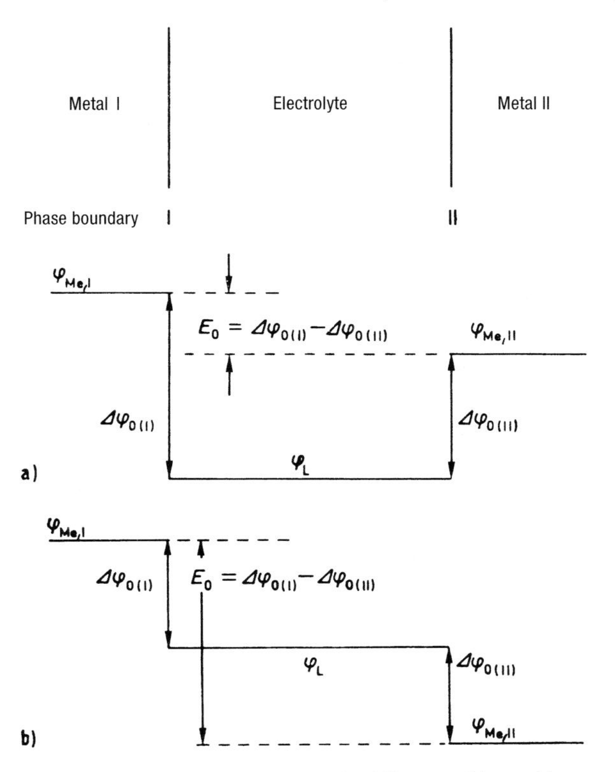

Fig. 3.2 The EMF of a galvanic cell as the difference of the equilibrium Galvani potentials at the two electrodes: (a) $\Delta\varphi(\mathrm{I}) > 0$; $\Delta\phi(\mathrm{II}) > 0$; (b) $\Delta\varphi(\mathrm{I}) > 0$; $\Delta\varphi(\mathrm{II}) < 0$.

Fig 3-2 shows the two possibilities that can exist, in which the Galvani potential of the solution, φ_S, lies between $\varphi(\mathrm{I})$ and $\varphi(\mathrm{II})$ and in which it lies below (or, equivalently, above) the Galvani potentials of the metals. It should be emphasised that Fig. 3.2 is highly schematic: in reality the potential near the phase boundary in the solution changes initially linearly and then exponentially with distance away from the electrode surface, as we shall see later in this chapter. The other point is that we have assumed that φ_S is a constant in the region between the two electrodes; this will only be true provided the two electrodes are immersed in the same solution and that no current is passing.

It is clear from Fig. 3.2 that the potential difference, E, between the two metals, which is termed the electromotive force (EMF), is given by

$$E = \Delta\varphi(\mathrm{II}) - \Delta\varphi(\mathrm{I}) = \varphi(\mathrm{II}) - \varphi(\mathrm{I}) \tag{3.41}$$

We adopt here the convention, discussed in detail below, that the EMF is always equal to the potential on the metal on the right of the figure less the potential of the metal on the left. It follows that once the Galvani potential of any one electrode is known, it should be possible, at least in principle, to determine the potentials for all other electrodes. In practice, since the Galvani potential of no single electrode is known, the method adopted is to arbitrarily choose one electrode and *assign* a value for its Galvani potential. The choice made will be further discussed in section 3.1.10.

The practical measurement of EMF must be carried out, as far as possible, under conditions approaching those for which no current flows in the cell, in order that the electrochemical equilibria be not disturbed. In general, it is sufficient for exploratory measurements to be carried out with a simple electronic voltmeter, but for accurate investigations, a high impedance electronic voltmeter must be used. The latter has an input impedance of extremely high value ($> 10^{12}\Omega$), with the result that measurements are carried out at very close to the ideal conditions. The great advantage of such voltmeters is that the behaviour of the cell can be measured over a period of hours with very little risk of error.

3.1.7
Schematic Representation of Galvanic Cells

Let us consider the cell in Fig. 3.3, in which the electrodes are made from the same metal, but in contact with two solutions of differing ionic activities separated from each other by a glass frit that permits contact but impedes diffusion. The EMF of such a cell, termed *a concentration cell*, is given by

$$E = E^0 + \left(\frac{RT}{zF}\right) \ln a_{M^{2+}}(II) - E^0 - \left(\frac{RT}{zF}\right) \ln a_{M^{2+}}(I) \tag{3.42}$$

$$= \left(\frac{RT}{zF}\right) \ln\left(\frac{a_{M^{2+}}(II)}{a_{M^{2+}}(I)}\right) \tag{3.43}$$

Equation (3.43) shows that the EMF increases by $0.059/z$ V per decade change in the activity ratio in the two solutions.

In order to proceed further, we must establish a convention for writing down electrochemical cells, such as the concentration cell we have described. This convention should establish clearly where the boundaries between different phases exist and also what the overall cell reaction is. It is now standard to use vertical lines, |, to delineate phase boundaries, such as those between a solid and a liquid or between two immiscible liquids. The junction between two miscible liquids, which might be maintained by the use of a porous glass frit, is represented by a single vertical dashed line, ¦, and two dashed lines, ¦¦, are used to indicate two liquid phases joined by an appropriate electrolyte bridge for which the junction potential (see section 3.2) has been eliminated.

The cell is written such that the *cathode* is to the right when the cell is acting in galvanic mode, and electrical energy is being generated from electrochemical reactions at the two electrodes. From the point of view of external connections, the cathode will appear to be the *positive* terminal, since electrons will travel in the external circuit from the *anode,* where they pass from electrolyte to metal, to the *cathode,* where they pass back into the electrolyte (for details see part 4.1). The EMF of such a cell will then be the difference in Galvani potentials of the metal electrode on the right-hand side

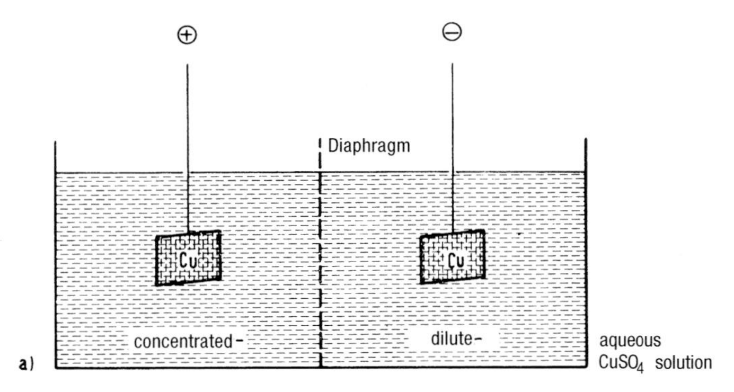

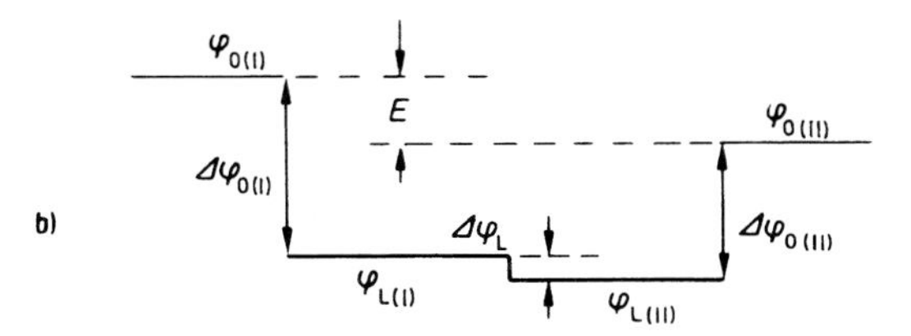

Fig. 3.3 Concentration cell: (a) Construction; (b) Galvani potentials.

and the left-hand side. Thus, the concentration cell of Fig. 3.3 would be represented by $M|M^{z+}(\mathrm{I}) \,\vdots\, M^{z+}(\mathrm{II})|M$.

Electrodes involving a redox pair in the solution such as $Fe^{3+}(aq)/Fe^{2+}(aq)$ or gases, as for example Cl_2/Cl^- or H_2/H^+, require an inert metal, such as platinum, for the transport of electrons. This metal must be included in the schematic representation of the cell. Thus, the hydrogen/chlorine cell is written $Pt|H_2(g)|H_3O^+(aq)$, $Cl(aq)|Cl_2(g)|Pt$. Note that the species in the same solution are separated by commas. For this cell, the reactions are:

$$\begin{array}{ll} H_2 + 2H_2O \rightarrow 2H_3O^+ + 2e^- & (\mathit{anode}) \\ Cl_2 + 2e^- \rightarrow 2Cl^- & (\mathit{cathode}) \\ \hline H_2 + 2H_2O + Cl_2 \rightarrow 2H_3O^+ + 2Cl^- & (\mathit{overall\ reaction}) \end{array}$$

And the Nernst equation is written as the difference in potentials of the cathode and anode (cf. eqs. 3.38 and 3.40):

$$\begin{aligned} E_c &= E(Cl_2/Cl^-) - E(H_2/H^+) \\ &= \left(E^{0\dagger}_{Cl_2/Cl^-} + \frac{RT}{2F}\ln\left(\frac{p_{Cl_2}}{p^0 a^2_{Cl^-}}\right)\right) - \left(E^{0\dagger}_{H_2/H^+} + \frac{RT}{2F}\ln\left(\frac{a^2_{H_3O^+}}{p_{H_2}/p^0}\right)\right) \\ &= E^{0\dagger} + \frac{RT}{F}\ln a_{H_3O^+}a_{Cl^-} - \frac{RT}{2F}\ln\left((p_{Cl_2}/p^0)(p_{H_2}/p^0)\right) \end{aligned} \tag{3.44}$$

where $E^{0\dagger}$ is the standard EMF of the cell, or the EMF at which the activities of H_3O^+ and Cl^- are unity and the pressures of H_2 and Cl_2 are both equal to the standard pressure, p^0. By analogy with this, it is clearly possible to write the EMF between two arbitrary electrodes as the difference of two standard EMFs and terms dependent on activities and gas-pressures.

3.1.8
Calculation of Cell emf's from Thermodynamic Data

Calculation of cell voltages can be carried out using the thermodynamic data for the *overall cell reaction*. For the standard cell voltage we have from equation (3.5)

$$E_c^0 = -\Delta_r G^0/nF \tag{3.45}$$

where $\Delta_r G^0$ is evaluated under the standard conditions of unit activities of all ions in solution and atmospheric pressure for all gases. The EMF of a cell for arbitrary ionic activities and gas pressures is then given by

$$E_c = -\Delta_r G/nF \tag{3.46}$$

where

$$\Delta_r G = \sum_i \nu_i \mu_i = \sum_i \nu_i \mu_i^{0\dagger} + \sum_i \nu_i RT \ln a_i = \Delta_r G^0 + \sum_i RT \ln a_i \tag{3.47}$$

From which it follows that

$$E = E^0 - \left(\frac{RT}{nF}\right) \cdot \sum_i \nu_i \ln a_i \tag{3.48}$$

and we recall that the summation in (3.48) employs the convention that ν_i is positive for products and negative for reactants. If a reactant or product is present in the electrolyte as a dissolved gas in equilibrium with the corresponding gas in the vapour phase with partial pressure p_i, then the activity a_i must be replaced by the term (p_i/p^0).

If we take, as a specific example, the chlorine-hydrogen cell described above, then the overall cell reaction is

$$H_2 + Cl_2 \rightleftharpoons 2H^+(aq) + 2Cl^-(aq) \tag{3.49}$$

then equation (3.48) becomes

$$E = E^0 + \left(\frac{RT}{2F}\right) \ln\{(p_{H_2}/p^0) \cdot (p_{Cl_2}/p^0)\} - \left(\frac{RT}{F}\right) \ln(a_{H^+} \cdot a_{Cl^-}) \tag{3.50}$$

which is identical to equation (3.44).

It follows that the theoretical calculation of E essentially reduces to the calculation of E_c^0 from equation (3.45), which can be effected from thermodynamic tables. From the expression

$$\Delta_r G^0 = \Delta_r H^0 - T\Delta_r S^0 \tag{3.51}$$

then a value of $\Delta_r G^0$ can be obtained from the standard enthalpies of formation and the standard entropies of the reactants and products. We have, quite generally, for a reaction of the form (3.1)

$$\Delta_r H^0 = \sum_{R,P} (v_p \Delta_f H_P^0 - v_R \Delta_f H_R^0) \tag{3.52}$$

$$\Delta_r S^0 = \sum_{R,P} (v_P S_P^0 - v_R S_R^0) \tag{3.53}$$

where the subscript P refers to products and R to reactants. Some standard enthalpies of formation and standard entropies are shown in Table 3.1 for a temperature of 298K. It will be observed that the standard enthalpies of formation of the elements in their standard states are set at zero by definition, since the enthalpy of formation is *defined* as the enthalpy change when a compound is formed from its constitutive elements in their standard state at one atmosphere pressure and at some pre-determined temperature. Note also in Table 3.1 that the enthalpy of formation of the solvated proton at 298K and its standard entropy are both put equal to zero; all other ions are then referred to this standard.

For the chlorine-hydrogen cell, equation 3.49, we can now calculate the free energy of the cell reaction straightforwardly from data in the table. We have

$$\Delta_r H^0 = (2\Delta_f H_{H^+}^0 + 2\Delta_f H_{Cl^-}^0 - \Delta_f H_{H_2}^0 - \Delta_f H_{Cl_2}^0) = -335.08 \text{ kJ mol}^{-1}$$

$$\Delta_r S^0 = (2S_{H^+}^0 + 2S_{Cl^-}^0 - S_{H^2}^0 - S_{CL_2}^0) = -243.57 \text{ JK}^{-1} \text{ mol}^{-1}$$

whence

$$\Delta_r G^0 = \Delta_r H^0 - T\Delta_r S^0 = -262.27 \text{ kJ mol}^{-1} \tag{3.54}$$

Table 3.1 Enthalpies and entropies of formation for selected elements and compounds at 298 K. Abbreviations: g – gas phase; l – liquid phase; s – solid phase; aq – in aqueous solution with unit activity.

Substance	State	$\Delta_f H_{298}^0/\text{kJ mol}^{-1}$	$S_{298}^0/\text{JK}^{-1}\text{mol}^{-1}$
H_2	g	0	130.74
Cl_2	g	0	223.09
H^+	aq	0	0
Cl^-	aq	– 167.54	55.13
O_2	g	0	205.25
H_2O	l	– 285.25	70.12
Zn	s	0	41.65
Zn^{2+}	aq	– 152.51	– 106.54
HCl	g	– 92.35	186.79
C (graphite)	s	0	5.69
CO	g	– 110.5	198.0

It should be emphasised that the molar free energy is a state function, i.e. its value depends only on the identity of the reactants and products themselves and not on the actual sequence of chemical reactions that take place to convert the former into the latter. Thus, we would obtain the same value for $\Delta_r G^0$ if the free energy is first calculated for the process $H_2 + Cl_2 \rightarrow 2HCl(g)$ followed by $2HCl(g) \rightarrow 2H^+(aq.) + 2Cl^-(aq.)$, as can be verified by direct calculation using the data in Table 3.1.

By substitution of the value obtained in equation (3.54) into (3.45) we obtain

$$E_c^0 = -(-262.270/(2 \times 96.485)) = 1.36 \text{ V} \tag{3.55}$$

since $n = 2$ for this reaction.

3.1.9
The Temperature Dependence of the Cell Voltage

From the expression $E = -\Delta_r G/nF$ relating the cell voltage to the free energy of the reaction, we can calculate the temperature dependence of the EMF by direct differentiation

$$\left(\frac{\partial E}{\partial T}\right)_p = -\frac{1}{nF} \cdot \left(\frac{\partial \Delta_r G}{\partial T}\right)_p = \frac{\Delta_r S}{nF} \tag{3.56}$$

using a fundamental relationship in thermodynamics. It can be seen from (3.56) that if the entropy *decreases* during the cell reaction, i.e. the state of order of the system *increases* in the course of the reaction, then the cell voltage will *decrease* with increasing temperature. From the values in Table 3.1, we can easily calculate the temperature dependence of the standard EMF as:

$$\left(\frac{\partial E^0}{\partial T}\right)_p = \frac{2S^0_{H^+} + 2S^0_{Cl^-} - S^0_{H_2} - S^0_{Cl_2}}{2 \times 96485} = -1.2 \text{ mV K}^{-1} \tag{63.57}$$

If the number of molecules in the gas phase decreases during the cell reaction, as in the case above, then the entropy change will generally be negative, and $\left(\frac{\partial E^0}{\partial T}\right)_p < 0$.

One further example is of great importance technically; the hydrogen-oxygen fuel cell has the cell reaction

$$H_2 + \frac{1}{2}O_2 \rightarrow H_2O \tag{3.58}$$

and for this, it is easily verified from Table 3.1 that $E_c^0 = 1.23\text{V}$ and $(\partial E^0/\partial T)_p = -0.85 \text{ mV K}^{-1}$.

An example for which $\Delta_r S > 0$ is the partial combustion of carbon to form CO

$$C + \frac{1}{2}O_2 \rightarrow CO \tag{3.59}$$

which can be verified from Table 3.1. The temperature dependence of the cell potential differences for the three cell reactions considered above are given in Fig. 3.4; for the

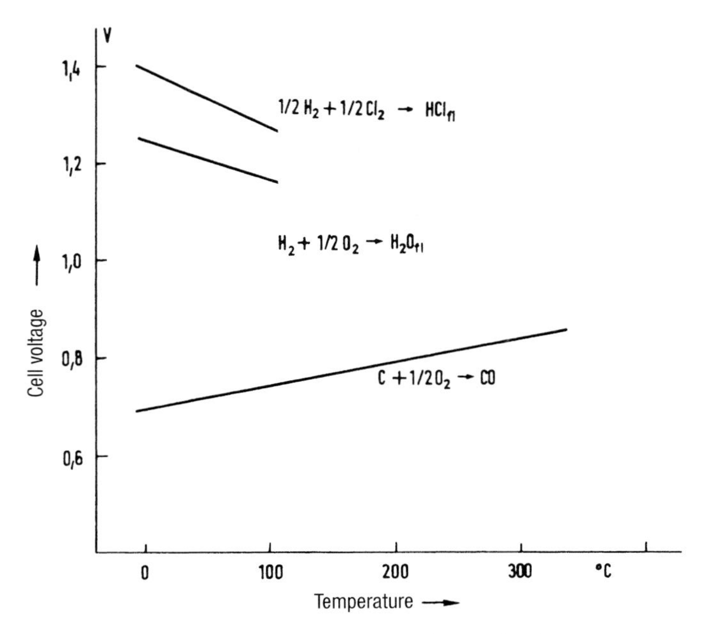

Fig. 3.4 Temperature dependence of the cell voltages of the hydrogen/chlorine, hydrogen/oxygen and carbon/carbon monoxide cells.

first two the plot is only extended as far as the boiling point of water. It can be seen that if the change in entropy is positive, then not only can the reaction enthalpy be converted into electrical energy, but in addition the amount of heat $T\Delta_r S$ can be extracted from the environment and converted to electrical energy (see Chapter 9). It is also evident from the discussion above that the reverse procedure, i.e. the determination of thermodynamic data ($\Delta_r G$, $\Delta_r H$ and $\Delta_r S$) from experimental EMF measurements, is also possible (see section 3.2.5).

3.1.10 The Pressure Dependence of the Cell Voltage – Residual Current for the Electrolysis of Aqueous Solutions

For the change of *EMF* with pressure we have

$$\left(\frac{\partial E}{\partial p}\right)_{\mathrm{T}} = -\frac{1}{nF} \cdot \left(\frac{\partial \Delta_r G}{\partial p}\right)_{\mathrm{T}} = -\frac{\Delta_r \bar{V}}{nF} \tag{3.60}$$

where $\Delta_r \bar{V}$ is the molar volume change associated with the reaction. Owing to the very small dependence of volume on pressure for liquids and solids, it is normally only necessary to consider the gaseous components of the cell reaction.

The change in volume, $\Delta_r \bar{V}$ can then be written

$$\Delta_r \bar{V} = \sum_j v_j \frac{RT}{p_j} \tag{3.61}$$

where p_j is the partial pressure of the jth component, and we assume that the ideal gas-law is valid. It follows that

$$\left(\frac{\partial E}{\partial p}\right)_T = -\sum_j v_j \frac{RT}{nFp_j} \equiv \sum_j \left(\frac{\partial E}{\partial p_j}\right)_T \tag{3.62}$$

and this may be integrated to yield

$$E = E^0 - \sum_j v_j \left(\frac{RT}{nF}\right) \ln\left(\frac{p_j}{p^0}\right) \tag{3.63}$$

where E^0 is defined as the EMF when all partial pressures are one atmosphere, and is identical to the E^0 defined above. At 298K, (3.64) reduces to

$$E = E^0 - \sum_j v_j \left(\frac{0.059}{n}\right) \log_{10}\left(\frac{p_j}{p^0}\right) \tag{3.64}$$

and, as a specific example, for the chlorine/hydrogen cell, with the cell reaction (3.49), we have, for unit activities of H^+(aq.) and Cl^-(aq.),

$$E = E^0 + \frac{1}{2} \times 0.059 \log_{10}\left\{\left(\frac{p_{H_2}}{p^0}\right)\left(\frac{p_{Cl_2}}{p^0}\right)\right\} \tag{3.65}$$

A simultaneous increase of the pressure of both working gases, Cl_2 and H_2, from 1 to 10 bar leads, from (3.65) and (3.55) to $E^0 = 1.42$ V, whereas a simultaneous decrease from 1 bar to 1 mbar in the gas pressures at both electrodes leads to a reduction of the cell potential difference to 1.18 V.

Equation (3.65), as well as the fundamental equation (3.46) applies provided that the electrode reactions are at equilibrium. For a galvanic cell, this is achieved by maintaining the partial pressures of the two working gases at the desired level, usually by bubbling the gases, if necessary diluting with an inert gas, through the electrolyte. If one goes from a galvanic cell to an electrolysis cell to which is applied an external potential, then, provided that the two working gases are not continuously supplied to the electrodes by bubbling, the partial pressures of the gases appropriate to that potential will be established through the cell reactions themselves. Thus, if a potential of 1.18 V is applied to a cell consisting of anode and cathode immersed in HCl(aq, $a = 1$), then gas pressures of 1 mbar will be formed at both electrodes.

In order to form these pressures, some electrolysis must occur. Thus, if the electrodes are immersed in aqueous HCl of unit activity and then both are connected to a potential of 1.18 V, an initial current flow is observed that swiftly decays to a low approximately stationary value as the equilibrium gas pressures are established. This low current is termed the *residual* current and it flows primarily to maintain the partial pressure of the gas at 1 mbar in the neighbourhood of the electrode as the H_2 or

Cl_2 formed initially begins to diffuse into the bulk of the electrolyte. Its order of magnitude, for the example considered above, is $\sim 10^{-6}$ A cm^{-2}.

As the applied potential in our electrolysis cell approaches the decomposition potential, E_D, defined in section 1.3, it is clear that the gas pressure required to maintain equilibrium increases as does the residual current necessary to maintain the steady state. Once the equilibrium gas pressure reaches one atmosphere, which corresponds to $E_c = E_c^0 = E_D^0$ under the conditions of unit activity in solution, then a vigorous evolution of gas, accompanied by a marked rise in current, begins at the electrodes. Both current and gas evolution increase rapidly as E rises above E_D, as is shown in Fig. 1.3.

For the cell described here, the slope of the current-voltage line in Fig. 1.3 for $E > E_D$ corresponds to the cell resistance, and this line can be extended back to cut the potential axis at the decomposition potential, E_D. The value of E_D determined by this means is, however, only in rare cases such as the present example equal to the thermodynamically calculated value. For the most part, the decomposition potentials lie appreciably higher than E_0 owing to kinetic limitations at the electrodes themselves; the difference between the value of E_0 and the observed decomposition potential is termed the activation overpotential, and is discussed in detail in chapters 4 and 8.

The magnitude of the residual current depends on both the area of the electrode and on the rate at which material is transported from the electrode. Clearly, if the electrolyte is strongly stirred, a higher current will be needed to maintain the equilibrium gas pressure at the electrode surface.

3.1.11
Reference Electrodes and the Electrochemical Series

As has already been discussed in section 3.1.6, it is not possible to determine experimentally the Galvani potential difference between electrode and electrolyte interior. We cannot, therefore, use the solution potential, φ_S, as a reference point for the tabulation of the values of standard potential for all the large number of possible potential determining systems.

On this basis, the electrode potential differences as measured with a high impedance voltmeter (which will always yield the potential difference between two electrodes immersed either in the same electrolyte or in electrolytes in contact with each other) may most sensibly be quoted with *reference* to an arbitrarily chosen reference electrode, whose Galvani potential difference is set equal to *zero*. This reference electrode is then made the common electrode for a series of Galvanic cells formed between it and all other electrodes.

The reference electrode chosen is the Standard Hydrogen Electrode, which consists of a platinised platinum sheet immersed in an aqueous solution of unit activity of H^+(aq) in contact with hydrogen gas at a pressure of one atmosphere. The potential-determining reaction has the form

$$2H^+(aq.) + 2e^- \rightleftharpoons H_2 \tag{3.66}$$

and the Galvani potential difference, or single electrode potential, will take the form

$$E^{\mathrm{HE}} = E^{0,\mathrm{SHE}} + \frac{RT}{F}\ln\left\{a_{\mathrm{H}^+}/\sqrt{p_{\mathrm{H}_2}}\right\} \tag{3.67}$$

The standard potential, $E^{0,\mathrm{SHE}}$ of (3.67) is, from our definition, put equal to zero. The Standard Hydrogen Electrode (SHE) possesses the advantage, compared to many other electrode systems, of achieving its equilibrium potential quickly and reproducibly, and of maintaining its potential well with time, making it particularly well suited to act as a comparison. Fig. 3.5 shows a scheme of a SHE and a simple arrangement to measure the potential of a Cu electrode.

Once a definite comparison has been established, all other electrode systems can be tabulated against this, and the potential values obtained are termed standard electrode potentials.

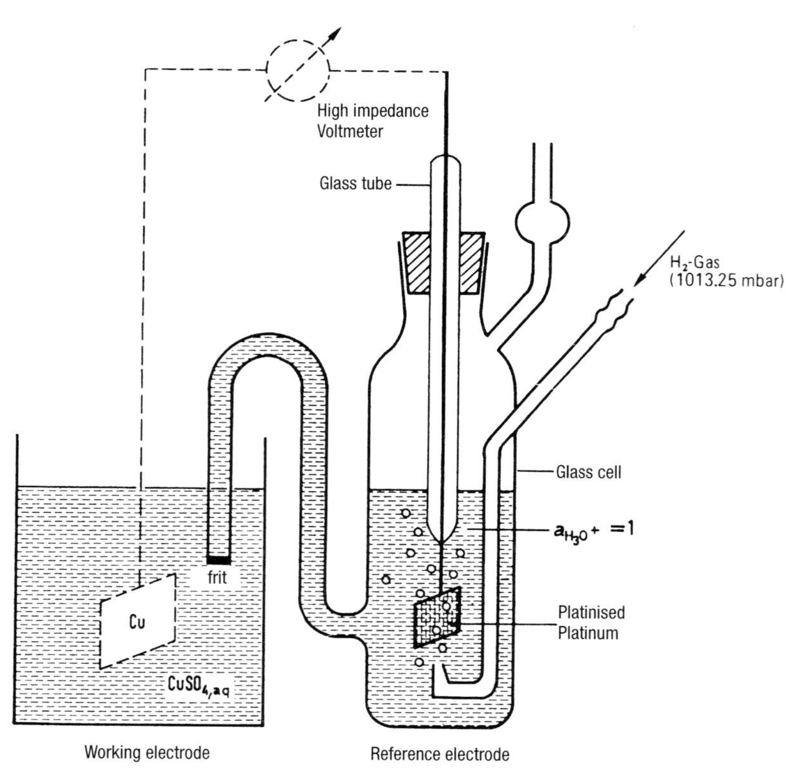

Fig. 3.5 Construction of a simple Normal Hydrogen Electrode. The electrode is robust and stable, but is not suited to precision measurements owing to the presence of a small potential jump at the junction between the frit and the electrolyte solution.

We can further illustrate this convention by considering two cells

$$Pt|H_2|H_3O^+,\ SO_4^{2-}||Cu^{2+},\ SO_4^{2-}|Cu \tag{3.68}$$

$$Pt|H_2|H_3O^+,\ SO_4^{2-}||Zn^{2+},\ SO_4^{2-}|Zn \tag{3.69}$$

For (3.68) the cell reactions are

$$H_2 + 2H_2O \rightarrow 2H_3O^+ + 2e^-; \quad \textit{anode} \tag{3.70}$$

$$Cu^{2+} + 2e^- \rightarrow Cu; \quad \textit{cathode} \tag{3.71}$$

and the EMF is given by

$$E^0 = E^{0,\text{copper}} - E^{0,\text{SHE}} \equiv E^{0,\text{copper}} \tag{3.72}$$

Provided all ions are present in unit activity and the hydrogen pressure is one atmosphere.

In fact, if the cell (3.68) is connected up in the way shown, then experimentally the copper electrode would indeed be found to be positive, as in Table 3.2, and $E^0 = +0.3402$ V. The overall cell reaction is

$$H_2 + 2H_2O + Cu^{2+} \rightarrow 2H_3O^+ + Cu \tag{3.73}$$

and the standard free energy of this reaction is

$$\Delta_r G^0 = -nFE^0 = -2F.(0.3402) = -65.66 \text{ kJ } mol^{-1} \tag{3.74}$$

Since $\Delta_r G < 0$, the reaction will take place spontaneously as written, so that we can immediately see that a *positive EMF* implies that the cell reaction is thermodynamically favourable.

For cell (3.69), the cell reactions are

$$H_2 + 2H_2O \rightarrow 2H_3O^+ + 2e^-; \quad \textit{anode} \tag{3.75}$$

$$Zn^{2+} + 2e^- \rightarrow Zn; \quad \textit{cathode} \tag{3.76}$$

and the EMF is given by

$$E^0 = E^{0,\text{zinc}} - E^{0,\text{SHE}} \equiv E^{0,\text{zinc}} \tag{3.77}$$

If Table 3.2 is consulted, it will be found that the standard electrode potential for zinc is negative, − 0.7628 V. Thus, for the overall cell reaction for the cell *as written*

$$H_2 + 2H_2O + Zn^{2+} \rightarrow 2H_3O^+ + Zn \tag{3.78}$$

the free energy is

$$\Delta_r G^0 = -2F(-0.7628) = +147.21 \text{ kJ mol}^{-1} \tag{3.79}$$

Thus, the EMF actually found for the cell as written would be negative. In turn this implies that the cell reaction as written is thermodynamically unfavourable. If we reverse the cell, writing it as

Table 3.2 Standard Potentials for different metal-ion, gas or redox electrodes vs. SHE at 25°C. In the column ‚Half cell' the construction of the electrode is given schematically in accord with the convention established above.

Half Cell	Electrode Process	Potential/V
$Li^+ \vert Li$	$Li^+ + e^- \rightleftharpoons Li$	– 3.045
$Rb^+ \vert Rb$	$Rb^+ + e^- \rightleftharpoons Rb$	– 2.925
$K^+ \vert K$	$K^+ + e^- \rightleftharpoons K$	– 2.924
$Ca^{2+} \vert Ca$	$Ca^{2+} + 2e^- \rightleftharpoons Ca$	– 2.76
$Na^+ \vert Na$	$Na^+ + e^- \rightleftharpoons Na$	– 2.7109
$Mg^{2+} \vert Mg$	$Mg^{2+} + 2e^- \rightleftharpoons Mg$	– 2.375
$Al^{3+} \vert Al$	$Al^{3+} + 3e^- \rightleftharpoons Al$	– 1.706
$Zn^{2+} \vert Zn$	$Zn^{2+} + 2e^- \rightleftharpoons Zn$	– 0.7628
$Fe^{2+} \vert Fe$	$Fe^{2+} + 2e^- \rightleftharpoons Fe$	– 0.409
$Cd^{2+} \vert Cd$	$Cd^{2+} + 2e^- \rightleftharpoons Cd$	– 0.4026
$Ni^{2+} \vert Ni$	$Ni^{2+} + 2e^- \rightleftharpoons Ni$	– 0.23
$Pb^{2+} \vert Pb$	$Pb^{2+} + 2e^- \rightleftharpoons Pb$	– 0.1263
$Cu^{2+} \vert Cu$	$Cu^{2+} + 2e^- \rightleftharpoons Cu$	+ 0.3402
$Ag^+ \vert Ag$	$Ag^+ + e^- \rightleftharpoons Ag$	+ 0.7996
$Hg_2^{2+} \vert 2Hg$	$Hg_2^{2+} + 2e^- \rightleftharpoons 2Hg$	+ 0.7961
$Au^+ \vert Au$	$Au^+ + e^- \rightleftharpoons Au$	+ 1.42
$Pt \vert H_2 \vert H_3O^+$	$2H_3O^+ + 2e^- \rightleftharpoons H_2 + 2H_2O$	0.0000
$Pt \vert H_2 \vert OH^-$	$2H_2O + 2e^- \rightleftharpoons H_2 + 2OH^-$	– 0.8277
$Pt \vert Cl_2 \vert Cl^-$	$Cl_2 + 2e^- \rightleftharpoons 2Cl^-$	+ 1.37
$Pt \vert O_2 \vert H_3O^+$	$1/2O_2 + 2H_3O^+ + 2e^- \rightleftharpoons 3H_2O$	+ 1.229
$Pt \vert O_2 \vert OH^-$	$1/2O_2 + H_2O + 2e^- \rightleftharpoons 2OH^-$	+ 0.401
$Pt \vert F_2 \vert F^-$	$F_2 + 2e^- \rightleftharpoons 2F^-$	+ 2.85
$Pt \vert Co(CN)_6^{3-}, Co(CN)_5^{3-}$	$Co(CN)_6^{3-} + e^- \rightleftharpoons Co(CN)_5^{3-} + CN^-$	– 0.83
$Pt \vert Cr^{3+}, Cr^{2+}$	$Cr^3 + e^- \rightleftharpoons Cr^{2+}$	– 0.41
$Pt \vert Cu^{2+}, Cu^+$	$Cu^{2+} + e^- \rightleftharpoons Cu^+$	+ 0.167
$Pt \vert Fe(CN)_6^{3-}, Fe(CN)_6^{4-}$	$Fe(CN)_6^{3-} + e^- \rightleftharpoons Fe(CN)_6^{4-}$	+ 0.356
$Pt \vert Fe^{3+}, Fe^{2+}$	$Fe^{3+} + e^- \rightleftharpoons Fe^{2+}$	+ 0.771
$Pt \vert Au^{3+}, Au^+$	$Au^{3+} + 2e^- \rightleftharpoons Au^+$	+ 1.29
$Pt \vert Mn^{3+}, Mn^{2+}$	$Mn^{3+} + e^- \rightleftharpoons Mn^{2+}$	+ 1.51
$Pt \vert PbO_2, Pb^{2+}, H_3O^+$	$PbO_2 + 4H_3O^+ + 2e^- \rightleftharpoons Pb^{2+} + 6H_2O$	+ 1.69
$Pt \vert MnO_4^-, Mn^{2+}, H_3O^+$	$MnO_4^- + 8H_3O^+ + 5e^- \rightleftharpoons Mn^{2+} + 12H_2O$	+ 1.491
$Pt \vert Cr_2O_7^{2-}, Cr^{3+}, H_3O^+$	$Cr_2O_7^{2-} + 14H_3O^+ + 6e^- \rightleftharpoons 2Cr^{3+} + 21H_2O$	+ 1.36
$Pt \vert ClO_3^-, Cl^-, H_3O^+$	$ClO_3^- + 6H_3O^+ + 6e^- \rightleftharpoons Cl^- + 9H_2O$	+ 1.45

$$Zn|Zn^{2+},\ SO_4^{2-}|H_3O^+,\ SO_4^{2-}|H_2|Pt \tag{3.80}$$

then the cell reaction will be

$$Zn + 2H_3O^+ \rightarrow H_2 + 2H_2O + Zn^{2+} \tag{3.81}$$

The EMF will be

$$E^0 = E^{0,SHE} - E^{0,zinc} \equiv -E^{0,zinc} = +0.7628\ V \tag{3.82}$$

and the change in free energy is now $-$ 147.21 kJ mol^{-1}, corresponding to a thermodynamically favoured process. Table 3.2 gathers together a series of standard electrode potentials for a variety of half cells. The number given in each case is the EMF of the total cell

$$H_2|Pt|H_3O^+(a_\pm = 1)||M^{z+}(a_\pm = 1)|M \tag{3.83}$$

or more generally

$$H_2|Pt|H_3O^+(a_\pm = 1)||Ox(a_\pm = 1),\ Red(a_\pm = 1)|Pt \tag{3.84}$$

It should be pointed out that the values calculated from thermodynamic data cannot always be verified experimentally in cells of the type shown above. Owing either to sluggish electrochemical kinetics or to undesired side reactions taking place in the cell (such as solvent decomposition) the apparent value obtained experimentally is often different from the thermodynamically calculated value. In addition, for many organic redox processes, particularly those taking place in non-aqueous solvents, thermodynamic data may not be available, preventing any tabulation of E^0 values and only permitting tabulation of observed decomposition potentials.

The difference of any two values from Table 3.2 allows us to obtain the EMF and polarity of any cell combined as indicated above either in galvanic mode or, in terms of the decomposition potential, as an electrolysis cell. Thus, for the Daniell cell, $Zn|Zn^{2+}$, $SO_4^{2-}||Cu^{2+}$, $SO_4^{2-}|Cu$, the calculated EMF is $+0.3402 - (-0.7628) = +1.1030$ V, assuming all electrolytes are present at unit mean activity.

For the technically important hydrogen-oxygen fuel cell, which can be written, $Pt|H_2|H_3O^+\ (a_\pm = 1)|O_2|Pt$ with half-cell reactions

$$H_2 + 2H_2O \rightarrow 2H_3O^+ + 2e^- \tag{3.85}$$

$$\frac{1}{2}O_2 + 2H_3O^+ + 2e^- \rightarrow 3H_2O \tag{3.86}$$

and with an overall cell reaction

$$H_2 + \frac{1}{2}O_2 \rightarrow H_2O \tag{3.87}$$

the standard EMF for the cell is $+1.229 - 0.0 = 1.229$ V. It is evident from Table 3.2 that the same value results for the cell if it is constructed in an alkaline medium, $Pt|H_2|OH^-(a\pm = 1)$ since the EMF now is $+0.401 - (-0.8277) = 1.229$ V. The reason for this is that the cell reaction (3.87) is actually the same whether the electrolyte is acid or alkaline, and the corresponding free energy must, therefore, be the same. Fuel cells based on reaction (3.87) have assumed considerable significance recently as energy sources for space-missions and as environmentally attractive energy sources for the future. Unfortunately, the oxygen half-cell is an example of a kinetically sluggish electrode: instead of the expected 1.229 V for the acid cell of (3.86), only a value of ca. 1.1 V is normally found on platinum, and this value is extremely sensitive to impurities in the electrolyte. A similar problem exists in alkaline solution, where the measured cell potential is also 0.1 – 0.2 V lower then expected.

3.1.12
Reference Electrodes of the Second Kind

The normal hydrogen electrode as shown in Fig. 3.5 is well suited to carrying out potential measurements, since it establishes its equilibrium potential rapidly and reproducibly. It does, however, possess certain disadvantages: the electrolyte must be prepared with an accurately known H_3O^+ activity, the hydrogen gas used must be purified, particularly from oxygen, and the platinum electrode employed must be frequently re-platinised, since the adsorption of impurities present in the solution prevents the establishment of a reproducible potential, and the electrode is said to become 'poisoned'.

Less sensitive, and simpler to handle is an electrode formed from a platinised platinum gauze placed half-way inside a test-tube that is filled with electrolyte and then inverted into the cell such that the sealed end of the tube is uppermost. A negative potential is then applied to the platinum such that evolution of hydrogen takes place until the gauze is half in the hydrogen gas that now fills the upper part of the tube and half is still immersed in the electrolyte. The rest potential of the gauze is then close to that of the normal hydrogen electrode.

In spite of these different designs, the disadvantages of the hydrogen electrode have led electrochemists to search for secondary reference electrodes that are easier to construct and whose potentials are established rapidly and reproducibly. The most important types of secondary reference electrode are metal-ion electrodes, where the activity of the potential-determining metal, $a_{M^{2+}}$ in the solution phase is itself controlled by ensuring that the solution is in contact with a second solid phase composed of a sparingly soluble salt of M^{z+}. Under these circumstances, $a_{M^{2+}}$ is determined by the solubility product of salt. Such electrodes are termed reference electrodes of the second kind.

If we take the silver ion as an example, then for the silver electrode we have

$$E_{Ag|Ag^+} = E^0_{Ag|Ag^+} + \frac{RT}{F} \ln a_{Ag^+} \tag{3.88}$$

Now, if the sparingly soluble salt AgCl is present, then we must have the equilibrium

$$AgCl \rightleftharpoons Ag^+ + Cl^- \tag{3.89}$$

from which we find

$$a_{Ag^+} \cdot a_{Cl^-} = K_S^{AgCl} \quad (3.90)$$

and for the equilibrium potential of the silver-silver chloride half-cell, consisting of $Ag|AgCl|Cl^-$ we have

$$E_{Ag|AgCl|Cl^-} = E^0_{Ag|Ag^+} + \frac{RT}{F} \ln K_S^{AgCl} - \frac{RT}{F} \ln a_{Cl^-} \quad (3.91)$$

We can obviously write $E_{Ag|Ag^+} + \frac{RT}{F} \ln K_g^{AgCl}$ as the standard potential of the silver-silver chloride electrode, and we obtain, finally

$$E_{Ag|AgCl|Cl^-} = E^0_{Ag|AgCl|Cl^-} - \frac{RT}{F} \ln a_{Cl^-} \quad (3.92)$$

The value of $E^0_{Ag|AgCl|Cl^-}$ can be obtained from the known value of $E^0_{Ag|Ag^+}$ (from Table 3.2, this is + 0.7996 V) and the solubility product of AgCl, which has the value $1.784 \cdot 10^{-10}$; its value is +0.2224 V vs. the SHE. Clearly the actual potential of this particular secondary reference electrode still depends on the solution activity of the anion that forms the sparingly soluble salt, and this is controlled by the addition of an easily soluble salt of the same anion, such as KCl. If a KCl solution of concentration 1.000 mol dm^{-3} is employed (which has $a_{Cl^-} < 1$) then the equilibrium potential of the Ag|AgCl electrode is 0.2368 V at 25°C.

Fig. 3.6 shows a frequently used construction for the silver-silver chloride electrode. The electrode shown is easily put together, usually ready for use immediately after assembly and can be placed directly into the electrochemical cell. This convenience of use has led to the frequent employment of this or other secondary electrodes by

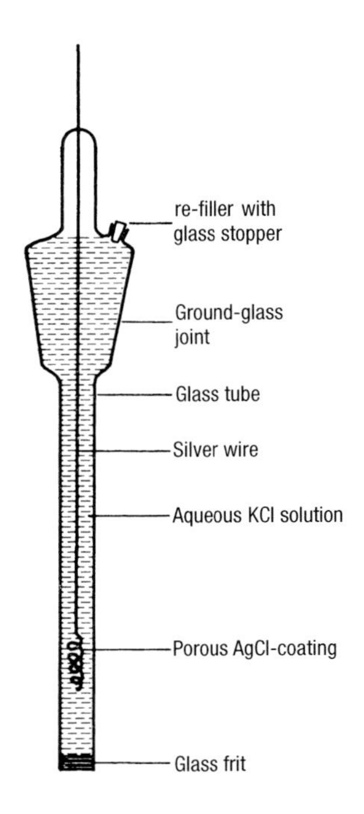

Fig. 3.6 Construction of a silver-silver chloride electrode. To obtain a good junction between the silver wire and the AgCl, the wire is first electrolysed in a chloride-containing solution to coat it with a red-violet AgCl deposit.

electrochemists instead of the hydrogen electrode itself. Obviously, the proliferation of measurements carried out with these different reference electrodes necessitates that every potential measurement must always be accompanied by a clear statement as to the reference electrode used.

Another common reference electrode is the so-called calomel electrode, for which the potential-determining reaction is $Hg_2^{2+} + 2e^- \rightarrow 2Hg$. The sparingly soluble salt in this case is calomel itself, Hg_2Cl_2, which floats as a paste on the top of a liquid mercury drop and which dissociates slightly into Hg_2^{2+} and Cl^- ions in solution.

Following the same approach as for the silver-silver chloride electrode, we find

$$E_{Hg|Hg_2Cl_2|Cl^-} = E^0_{Hg|Hg_2Cl_2|Cl^-} - \frac{RT}{F} \ln a_{Cl^-} \tag{3.93}$$

with a standard reference potential of + 0.2682 V vs. SHE.

Usually calomel electrodes are prepared in KCl solutions of strength 0.1 mol dm^{-3}, 1.0 mol dm^{-3} (NCE) or in saturated KCl solution (SCE). From the different chloride activities in these solutions, the potentials of the corresponding reference electrodes at 25°C are + 0.3337 V, + 0.2807 V and + 0.2415 V vs. SHE respectively. The preparation of the saturated calomel electrode is especially straightforward, since there is no need to regulate the gas pressure, weigh out any materials or titrate any solutions. A disadvantage of the SCE is, though, that the solubility of KCl shows a strong temperature dependence, and this leads, in turn, to a strong temperature dependence of the electrode potential itself (ca. 1 mV/°C as compared to 0.1 mV/°C for the NCE).

The practical construction of calomel electrodes is illustrated in Fig. 3.7, and Table 3.3 summarises the potentials obtained for a number of reference electrodes of the second kind, including those dealt with above.

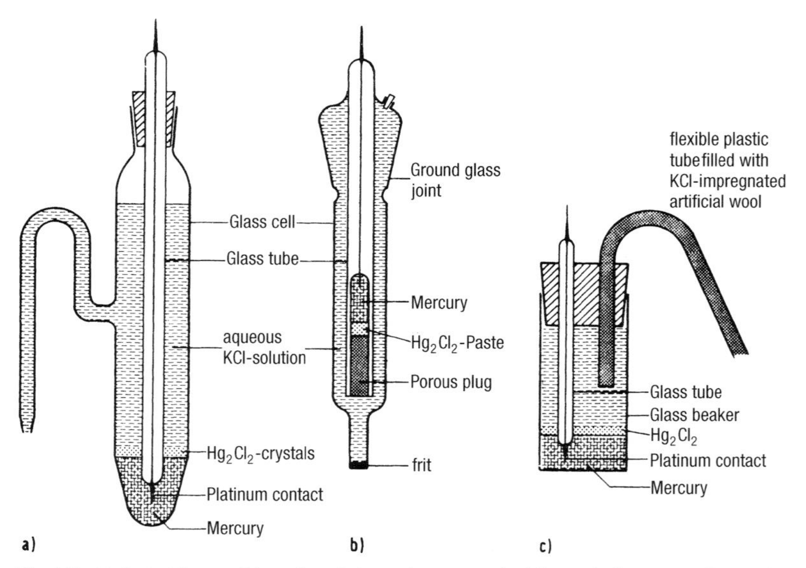

Fig. 3.7 Technical forms of the calomel electrode: (a) standard form; (b) form manufactured as a miniature; (c) form suitable for technical measurements.

Table 3.3 Standard potentials for some reference electrodes of the second kind vs. SHE at 25°C. The electrode processes in each case incorporate the dissociation reaction of the sparingly soluble salt. The preparation of the mercury-sulphate and -oxide electrodes is closely analogous to Fig. 3.7 and that of the lead-sulphate electrode to Fig. 3.6. Further information on reference electrodes can be found in: D.J.G. Ives and J. Janz; Reference Electrodes, Academic Press, New York, 1961.

Half Cell	Conditions	Electrode Process	Potential/V
$Ag\|AgCl\|Cl^-$	$a_{Cl^-} = 1$	$AgCl + e^- \rightarrow Ag + Cl^-$	0.2224
(silver-silver chloride electrode)	saturated KCl		0.1976
	KCl ($c = 1M$)		0.2368
	KCl ($c = 0.1M$)		0.2894
$Hg\|Hg_2Cl_2\|Cl^-$	$a_{Cl^-} = 1$	$Hg_2Cl_2 + e^- \rightarrow 2Hg + 2Cl^-$	0.2682
(calomel-electrode)	saturated KCl		0.2415
	KCl ($c = 1.0M$)		– 0.2807
	KCl ($c = 0.1M$)		0.3337
$Pb\|PbSO_4\|SO_4^{2-}$ (Lead sulphate electrode)	$a_{SO_4^{2-}} = 1$	$PbSO_4 + 2e^- \rightarrow Pb + SO_4^{2-}$	–0.276
$Hg\|Hg_2SO_4\|SO_4^{2-}$	$a_{SO_4^{2-}} = 1$	$Hg_2SO_4 + 2e^- \rightarrow 2Hg + SO_4^{2-}$	0.6158
(Mercury-sulphate electrode)	H_2SO_4 ($c = 0.5M$)		0.682
	saturated K_2SO_4		0.650
$Hg\|HgO\|OH^-$	$a_{OH^-} = 1$	$HgO + H_2O + 2e^- \rightarrow 2Hg + 2OH^-$	0.097
(Mercury-oxide electrode)	NaOH ($c = 1.0M$)		0.140
	NaOH ($c = 0.1M$)		0.165

If a galvanic cell is formed from two reference electrodes of the second kind, whose potentials are controlled by the same anion, then the effect of altering the activity of the anion on the overall EMF of such a cell becomes extremely small. Even the use of a saturated solution has little effect on the temperature dependence of the EMF, leading to a very low temperature dependence of the overall cell voltage.

An example of such a cell is the Weston cell, introduced in section 1.5 as a secondary potential standard. This cell is a combination of a cadmium amalgam with a saturated cadmium sulphate solution, $Cd(Hg)|CdSO_4 \cdot \frac{8}{3}H_2O$, which is an electrode of the second kind, together with a similar mercury-mercury sulphate electrode also operating in saturated $CdSO_4$. The *EMF* of the cell

$$Cd(Hg,\ \omega = 12.5)|CdSO_4 \cdot \frac{8}{3}H_2O|Cd^{2+}(sat.), Hg_2^{2+}(sat.), SO_4^{2-}|Hg_2SO_4|Hg \tag{3.94}$$

where ω is the weight percent of Cd in the amalgam, is given by

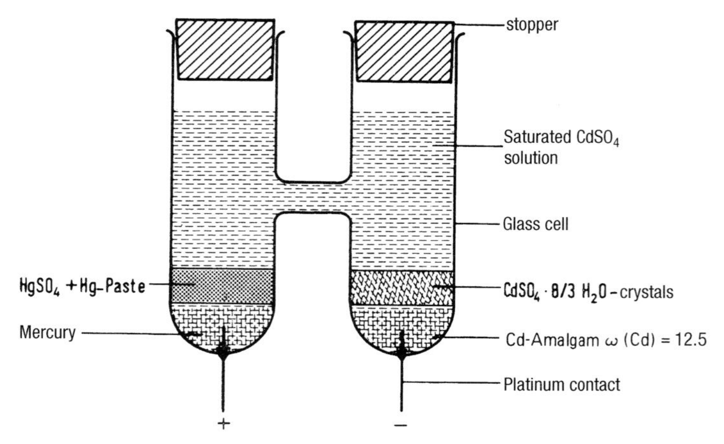

Fig. 3.8 The Weston Cell.

$$E = E^0_{Hg|Hg_2SO_4|SO_4^{2-}} - \left(\frac{RT}{2F}\right) \ln a_{SO_4^{2-}} - E^0_{Cd(Hg)|CdSO_4 \cdot \frac{8}{3}H_2O|SO_4^{2-}} + \left(\frac{RT}{2F}\right) \ln a_{SO_4^{2-}} \tag{3.95}$$

$$\equiv E^0_{Hg|Hg_2SO_4|SO_4^{2-}} - E^0_{Cd(Hg)|CdSO_4 \cdot \frac{8}{3}H_2O|SO_4^{2-}} \tag{3.96}$$

and is clearly independent of $a_{SO_4^{2-}}$. Its construction is shown in detail in Fig. 3.8.

3.1.13
The Electrochemical Series in Non-aqueous Solvents

In the same way as in aqueous solutions, half-cells can be set up and compared with each other in non-aqueous solution, and for each solvent an electrochemical series can be established. However, we can also ask how the potentials of such a series would compare with the corresponding potentials in water. It is not possible simply to carry out a direct potential difference measurement between half cells in two different solvents owing to the problem of the diffusion potential (see below), but if the free energy of transfer of even one single electroactive species between the solvent of interest and water were known, then the two electrochemical series could be connected together and the potentials of half-cells in solvent S could all be related to the standard aqueous reference $E_{H_2|H^+} = 0$. In addition, the free energies for transfer of all other electroactive species could be immediately obtained from a comparison of the half-cell potentials in water and solvent S.

Unfortunately, calculation of the free energy of transfer for a single electroactive species in a half-cell is not possible, since we would have to know the free energy of solvation of individual ions, whereas thermodynamically we can only know the corresponding values for the sum of both the anions and cations.

A way out of this difficulty was first suggested by Pleskov, who proposed, as a first approximation, that the free energy of solvation of the rubidium ion be taken as the same in all solvents. The justification for this assumption is that the rubidium ion, owing to its very large radius, should only be slightly solvated in any solvent. This should lead to very little inaccuracy in the basic assumption that the free energy of transfer, $\Delta_t G(\mathrm{Rb/Rb^+})$, from water to any solvent is zero, or, equivalently, $E^0_{\mathrm{Rb|Rb^+}}(S) = E^0_{\mathrm{Rb|Rb^+}}(\mathrm{H_2O})$.

In Fig. 3.9, the electrochemical series in seven different solvents are compared directly to each other assuming that the potential $E^0_{\mathrm{Rb|Rb^+}} = -2.92$ V in all the solvents. The gross effects in the shift of standard potentials from solvent to solvent are clearly correctly reproduced.

One disadvantage of the use of the rubidium electrode (in the form of a rubidium *amalgam* in contact with Rb^+) is encountered experimentally, in that the overpotential for hydrogen evolution on a mercury surface in non-aqueous solvents is frequently smaller than that found for water, and so corrosion effects can disturb the potential measurements (see sections 4.7 and 8.1.4).

A second suggestion to establish a connection between potentials in different solvents was made by Strehlow, who assumed that the free energy of transfer of both the cation and the neutral molecule in redox systems of the form A/A^+ would be the same from solvent to solvent, provided that relatively large, approximately spherical, molecules were used, such as the ferrocene/ferricinium system, $(C_5H_5)_2Fe/(C_5H_5)_2Fe^+$,

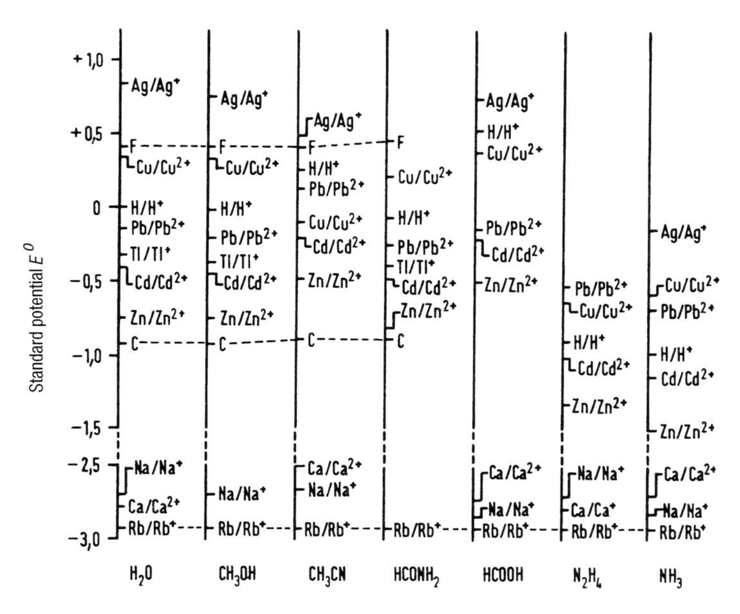

Fig. 3.9 The Electrochemical Series in different solvents, assuming the electrode potential for the Rb^+/Rb couple is constant at − 2.92 V vs. NHE in all solvents. The ferrocene/ferrocinium (F) and cobaltocene/cobalticinium (C) couples are also shown in some of these solvents, and both can also be seen to be close to constant.

or its cobalt analogue, $(C_5H_5)_2Co/(C_5H_5)_2Co^+$. In this case, as in the one above, standard electrode potentials in all solvents can be compared. It can be seen from Fig. 3.9 that this assumption and that of the constancy of the rubidium electrode are in reasonable agreement; the dashed lines in Fig. 3.9 show the ferrocene and cobaltocene couples and in both cases the lines are closely parallel to the rubidium line.

3.1.14
Reference Electrodes in Non-aqueous Systems and Usable Potential Ranges

Even in non-aqueous solvents, the most frequently used reference electrode is the aqueous saturated calomel electrode (SCE) connected to the measurement solution by a salt-bridge. If no water can be permitted in the solvent, then the connecting salt-bridge must contain a suitable non-aqueous electrolyte.

In order to minimise diffusion potentials in this type of measurement, it would clearly be preferable to use the same solvent in the reference electrode. However, some new difficulties may be encountered, such as the fact that Hg_2Cl_2 tends to disproportionate in acetonitrile, and in this solvent a completely different reference system, such as Ag/Ag^+ must be used. For measurements in those solvents that can sustain the cathodic evolution of hydrogen, the use of a hydrogen electrode as described above offers a perfectly practicable alternative.

As already mentioned in the introduction, the much wider usable potential ranges encountered in non-aqueous solvents are of great interest. In Table 3.4, some data are shown for two technically important solvents, acetonitrile and dimethyl sulphoxide with a variety of dissolved electrolyte salts. As for aqueous systems, the potential range depends strongly on conditions, and particularly on the electrode material used.

Table 3.4 Usable potential ranges in acetonitrile and dimethyl sulphoxide.

Solvent	Electrolyte	Electrode	Potential Range
		Material	(V vs. SCE)
ACN	$Na^+ClO_4^-$	Hg	– 1.7 to + 0.6
ACN	$Na^+ClO_4^-$	Pt	– 1.5 to + 1.8
ACN	$Et_4N^+ClO_4^-$	Hg	– 2.8 to + 0.6
ACN	$But_4N^+I^-$	Hg	– 2.8 to – 0.6
DMSO	$Na^+ClO_4^-$	Hg	– 1.90 to + 0.25
DMSO	$Et_4N^+ClO_4^-$	Hg	– 2.80 to + 0.25
DMSO	$Na^+ClO_4^-$	Pt	– 1.85 to + 0.70
DMSO	$Et_4N^+ClO_4^-$	Pt	– 1.85 to + 0.70
DMSO	$But_4N^+ClO_4^-$	Hg	– 3.00 to + 0.30
DMSO	$But_4N^+I^-$	Hg	– 2.85 to – 0.41

3.2 Liquid-junction Potentials

3.2.1 The Origin of Liquid-junction Potentials

In addition to phase boundaries between electrode and electrolyte solutions in electrolysis and galvanic cells, we saw in section 3.1 that there can arise additional phase boundaries between different electrolytes, as, for example, between anolyte and catholyte. The anolyte and catholyte may consist of different electrolytes present in different concentrations, as, for example, the Daniell cell

$$\mathrm{Zn}|\mathrm{ZnSO_4(dilute)} \,\vdots\, \mathrm{CuSO_4(conc.)}|\mathrm{Cu} \tag{3.97}$$

A simpler case is that in which both sides of the liquid-liquid junction are the same electrolyte, but present in different concentrations. If the electrodes are also the same, this is termed a concentration cell, and a simple example would be

$$\mathrm{Cu}|\mathrm{CuSO_4(dilute)} \,\vdots\vdots\, \mathrm{CuSO_4(conc.)}|\mathrm{Cu} \tag{3.98}$$

in which the double dashed bars indicate that a bridge is present between the two solutions and transport between them does not take place. Before we discuss the practicality of devising such a bridge, we need to consider a cell with a simple liquid-liquid interface across which transport may take place. In order to prevent the physical mixing of anolyte and catholyte, the phase boundary may be formed by careful addition of the two solutions in a capillary tube or by stabilisation with a membrane or diaphragm, and the resultant cell is denoted by the expression:

$$\mathrm{Cu}|\mathrm{CuSO_4(dilute)} \,\vdots\, \mathrm{CuSO_4(conc.)}|\mathrm{Cu} \tag{3.99}$$

These concentration cells should be distinguished from electrode concentration cells, in which two electrodes of similar type but with differing 'electrode activities' are immersed in the same electrolyte. One example of such a cell would one in which two hydrogen electrodes with differing partial pressures of hydrogen were both immersed in the same acid. Such a cell would be represented as $\mathrm{Pt}|\mathrm{H_2}(p_2)|\mathrm{HCl}(a_\pm)|\mathrm{H_2}(p_1)|\mathrm{Pt}$, and its EMF is given by $E = (RT/2F)\ln\{p_2/p_1\}$. A second type of electrode concentration cell could be constructed by using two zinc amalgam electrodes of different composition again immersed in the same solution, forming the cell $(\mathrm{Hg})\mathrm{Zn}(a_2)|\mathrm{ZnSO_4}(a_\pm)|\mathrm{Zn}(a_1)(\mathrm{Hg})$ whose EMF is similarly given by $E = (RT/2F)\ln\{a_2/a_1\}$.

In order to bring out the origin of the potential change at a liquid-liquid interface, we consider now in more detail the boundary between a dilute and concentrated solution in an electrolyte concentration cell. On both sides of this boundary, the dissolved ions have differing chemical potentials $\mu_i = \mu_i^{0\dagger} + RT\ln(m_i/m^0) + RT\ln\gamma_i$. Under these circumstances, the ions will experience a diffusional driving force in the direction of the more dilute solution, giving rise to an ion flux

$$J_i = -\frac{D_i c_i}{RT}\,\mathbf{grad}\,\mu_i \tag{3.100}$$

where J_i is the amount of material transported through unit area per second, D_i is the diffusion coefficient of ion i and **grad** is defined in equation (2.36). Equation (3.100) reduces to the familiar Fick's first law of diffusion under the assumption that γ_i is the same on both sides of the boundary. Under these circumstances, approximating the molality m_i to the concentration c_i,

$$J_i = -\frac{D_i c_i}{RT}\,\mathbf{grad}\left(\mu_i^{0\dagger} + RT\ln(m_i/m^0) + RT\ln\gamma_i\right) = -D_i\,\mathbf{grad}\,c_i \tag{3.101}$$

which, for one-dimensional diffusion through a planar membrane, takes the form

$$J_i = -D_i\left(\frac{dc_i}{dx}\right) \tag{3.102}$$

Anions and cations of the dissolved electrolyte at first diffuse independently of each other, and, since in general the diffusion coefficients of anion and cation will be different, they will diffuse at different speeds. As might be expected, there is a relationship between the diffusion coefficient and the ionic mobility, u_i, defined in chapter 2. This relationship is termed the Nernst-Einstein equation, which states that

$$D_i = u_i \cdot k_B T \tag{3.103}$$

where k_B is Boltzmann's constant, which has the value R/N_A, where N_A is the Avogadro constant. In the example shown in Fig. 3.3, the sulphate ions, with their higher mobilities, will outpace the copper ions, and this leads to an excess of negatively charged ions in the direction of diffusion on the dilute electrolyte side of the phase boundary, and, conversely, an excess of positively charged ions on the other side of the membrane. This is the origin of the potential jump, $\Delta\varphi_{diff}$, at the membrane itself. The resultant electric field, $\boldsymbol{E} \equiv -\mathbf{grad}\,\varphi$, operates in such a way as to *accelerate* the cations that lag behind and to *decelerate* the anions that have moved ahead. In the steady state, the value of $\Delta\varphi_{diff}$, the junction or diffusion potential, is just such that both types of ions are transported at the same rate. This diffusion potential inside the cell will contribute to the overall measured EMF with a sign that depends on the relative mobilities of the two ions, and diffusion potentials will be the larger, the greater the difference between the mobilities of the corresponding charge carriers.

3.2.2
The Calculation of Diffusion Potentials

We have seen that the origin of the diffusion potential, $\Delta\varphi_{diff}$ is the difference in mobility of the various charge carriers in the two phases. The calculation of $\Delta\varphi_{diff}$ starts from a consideration of the fluxes of the positive and negative ions as a function of **grad** μ and **grad** φ, and their opposing effects on the potential. A complete treatment is, however, rather demanding; it starts from the phenomenological reciprocity relationships derived originally by Onsager, and the individual ion properties are introduced through the transport numbers t_i. Following a lengthy calculation, we emerge with the formula

$$\Delta\varphi_{diff} = \int \sum_i \frac{t_i}{z_i F} \cdot \frac{\partial \mu_i}{\partial x} \cdot dx \tag{3.104}$$

Rather than follow this derivation in detail, we present a considerably simpler treatment using thermodynamic relationships. Consider a galvanic cell operating close to the reversible thermodynamic limit, such that $i \to 0$ with the passage of n Faradays. Then, from (3.2) we have

$$E = -\frac{\Delta_r G}{nF} \tag{3.105}$$

where $\Delta_r G$ is the free energy change associated with the cell reaction. If this cell reaction consists, in a concentration cell, of the transfer of ions i through a phase boundary from a region of chemical potential $\mu_i(\mathrm{I})$ to one with chemical potential $\mu_i(\mathrm{II})$, we can imagine the phase boundary to be a region of continuous change in μ_i. If $\mathrm{d}\mu_i$ is the change in chemical potential of the ith type of ion on moving an infinitesimal distance dx through the boundary region, then the associated change in free energy is

$$d\Delta_r G = \sum_i v_i d\mu_i \tag{3.106}$$

and integration over the whole boundary layer gives

$$\Delta_r G_{diff} = \sum_i \int_{\mu_i(\mathrm{I})}^{\mu_i(\mathrm{II})} v_i d\mu_i \tag{3.107}$$

It may be helpful at this stage to reconsider the concentration cell of (3.99), which we write as $Cu|CuSO_4$(dilute) ¦ $CuSO_4$(conc.)$|Cu$. The passage of two faradays of charge in such a cell will lead to the following changes, *provided the transport numbers are independent of* μ_i:

- One mole of Cu^2+ will be electro-deposited as Cu^0 at the cathode (right-hand electrode)
- One mole of Cu metal will be dissolved as Cu^{2+} at the anode (left-hand electrode)
- t^+ moles of Cu^{2+} will be transported from anolyte to catholyte (i.e., from I (dilute) to II (concentrated) copper sulphate)
- t^- moles of SO_4^{2-} will be transported from catholyte to anolyte (giving a nett *gain* in concentration in the anolyte of t^- moles of $CuSO_4$ and loss in the catholyte of the same)

Thus, in equation (3.107), $v^+ = +t^+$ as given and $v^- = -t^-$ for the direction of integration specified. In fact, in turns out that for any electrolyte concentration cell, we can write $v_i = t_i n/z_i$, where z_i is the algebraic charge on the ion (in units of electron charge) and is therefore positive for positive ions and negative for negative ions. We can, therefore, write (3.107) as

$$\Delta_r G_{diff} = \sum_i \int_{\mu_i(\mathrm{I})}^{\mu_i(\mathrm{II})} n(t_i/z_i)d\mu_i \tag{3.108}$$

from which, the diffusion potential is

$$\Delta\varphi_{diff} = -\frac{\Delta_r G_{diff}}{nF} = -\sum_i \int_{\mu_i(\mathrm{I})}^{\mu_i(\mathrm{II})} \left(\frac{t_i}{z_i F}\right) d\mu_i \equiv -\frac{RT}{F}\sum_i \int_{a_i(\mathrm{I})}^{a_i(\mathrm{II})} \left(\frac{t_i}{z_i}\right) d\ln a_i \tag{3.109}$$

since $\mu_i = \mu_i^{0\dagger} + RT\ln a_i$. The overall change in free energy for the cell, $\Delta G'$, is then the algebraic sum of the free energy in (3.109) and the free energy change for the two electrode reactions.

$$\Delta G' = \Delta G_{\mathrm{electrodes}} + \Delta_r G_{diff} = -nFE - nF\Delta\varphi_{diff} \tag{3.110}$$

where E is given by equation (3.43). Given that we have assumed the t_i to be independent of the μ_i, (3.109) may be integrated immediately to yield

$$\Delta_r G_{diff} = t^+(\mu^+(\mathrm{II}) - \mu^+(\mathrm{I})) - t^-(\mu^-(\mathrm{II}) - \mu^-(\mathrm{I})) \tag{3.111}$$

where (II) refers to the catholyte (more concentrated) and (I) to the anolyte. The electrode reactions as given lead to

$$\Delta G_{\mathrm{electrodes}} = (\mu^+(\mathrm{I}) - \mu^+(\mathrm{II})) \tag{3.112}$$

since one mole of Cu^{2+} is deposited from (II) and gained by dissolution in (I), whence

$$\Delta G' = t^-(\mu^+(\mathrm{I}) - \mu^+(\mathrm{II})) + t^+(\mu^-(\mathrm{I}) - \mu^-(\mathrm{II})) \tag{3.113}$$

which does indeed correspond to the transfer of t^- moles of Cu^{2+} ions and SO_4^{2-} ions from region (II), the more concentrated copper sulphate solution, to region (I), the less concentrated one.

3.2.3
Concentration Cells with and without Transference

Provided that the concentration differences between the electrolyte solutions on both sides of the phase boundary are not too large, the transport numbers will be independent of a_i in the first approximation, since we recall from section 2.3.1 that transport numbers are only weakly dependent on concentration in any case. This allows us to integrate (3.109), and for a zz electrolyte, where $z = z^+ = |z^-|$, we obtain, assuming mean activities $a_\pm(\mathrm{II})$ and $a_\pm(\mathrm{I})$ and transfer of cation from I to II

$$\Delta\varphi_{diff} = -\left(\frac{RT}{zF}\right)\cdot(t^+ - t^-)\cdot\ln\{a_\pm(\mathrm{II})/a_\pm(\mathrm{I})\} \tag{3.114}$$

The diffusion potential will thus be the larger, the more different the properties of the individual cation and anion. Thus, for HCl, with $t^+ = 0.821$ and $t^- = 0.179$, and

$z = 1$, a diffusion potential of $-$ 37.9 mV per decade difference in activity between anolyte and catholyte is expected. Conversely, particularly low values of the diffusion potential are expected if $t^+ \approx t^- \approx 0.5$. An example would be KCl, where $t^+ = 0.4906$ and $t^- = 0.5094$, giving a diffusion potential of + 1.1 mV per decade difference in activity. In the particular case of the copper sulphate concentration cell considered above, $t^+ = 0.375$ and $t^- = 0.625$, so the diffusion potential is 7.4 mV per decade activity difference.

Obviously, the diffusion potential can be measured in this case since, from (3.110), it is part of the measured emf of the concentration cell. For the general concentration cell, we have

$$E = (RT/zF) \cdot \ln\{a_\pm(\mathrm{II})/a_\pm(\mathrm{I})\} \tag{3.115}$$

and the measured EMF, E', is given by

$$E' = E + \Delta\varphi_{diff} = \left(\frac{RT}{zF}\right) \cdot (1 - t^+ + t^-) \cdot \ln\{a_\pm(\mathrm{II})/a_\pm(\mathrm{I})\} =$$

$$2t^- \left(\frac{RT}{zF}\right) \cdot \ln\{a_\pm(\mathrm{II})/a_\pm(\mathrm{I})\} \tag{3.116}$$

since $t^+ + t^- = 1$. It is clear from this expression that is the transport number of the anions is greater than 0.5, then the EMF of a concentration cell with transport is greater than the corresponding cell without transport, for which, as we saw in equation (3.43), $E = (RT/zF) \ln\{a(\mathrm{II})/a(\mathrm{I})\}$. The converse is true if $t^- < 0.5$.

3.2.4 Henderson's Equation

We now turn to the more general case of the diffusion potential between two solutions that have differing ionic compositions. Clearly, if the concentration of some species i changes across the boundary from an essentially constant value to zero, then the assumption of constant transport number must be incorrect, and we must seek a more realistic equation.

In order to integrate the expression (3.110), we must be able to express the variation of t_i with distance x across the interface. Replacing the concentration terms in (2.29) by activities, which is an approximation, then

$$t_i \approx \frac{a_i u_i |z_i|}{\sum_i a_i u_i |z_i|} \tag{3.117}$$

We will assume that the ionic mobilities are independent of the activities (an assumption that is certainly not fulfilled with any accuracy), and we will also assume that the activities of the ions will alter linearly with fractional distance x through the membrane

$$a_i(x) = a_i(\mathrm{I}) + [a_i(\mathrm{II}) - a_i(\mathrm{I})]x \tag{3.118}$$

From which we obtain

$$t_i \approx \frac{a_i(\mathrm{I})u_i|z_i| + [a_i(\mathrm{II}) - a_i(\mathrm{I})]u_i|z_i|x}{\sum_i a_i(\mathrm{I})u_i|z_i| + [a_i(\mathrm{II}) - a_i(\mathrm{I})]u_i|z_i|x} \tag{3.119}$$

and substituting this into equation (3.109) leads to

$$\Delta\varphi_{diff} = -\frac{RT}{F}\int_0^1 \sum_i \frac{[a_i(\mathrm{II}) - a_i(\mathrm{I})]u_i|z_i|}{z_i} \cdot \frac{dx}{\sum_i a_i(\mathrm{I})u_i|z_i| + [a_i(\mathrm{II}) - a_i(\mathrm{I})]u_i|z_i|x} \tag{3.120}$$

Carrying out the integration leads to the Henderson equation

$$\Delta\varphi_{diff} = -\frac{RT}{F} \cdot \frac{\sum_i \frac{[a_i(\mathrm{II}) - a_i(\mathrm{I})]u_i|z_i|}{z_i}}{\sum_i [a_i(\mathrm{II}) - a_i(\mathrm{I})]u_i|z_i|} \cdot \ln\left\{\frac{\sum_i a_i(\mathrm{II})u_i|z_i|}{\sum_i a_i(\mathrm{I})u_i|z_i|}\right\} \tag{3.121}$$

With equation (3.121) we can, in principle, calculate the diffusion potential for any chosen liquid-liquid interface. It can also be seen that (3.121) will reduce to (3.114) under the conditions where we have the same pair of ions in both phases with charge $z = z^+ = |z^-|$, with $a_+(\mathrm{I}) = a_-(\mathrm{I}) = a_\pm(\mathrm{I})$ and similarly for II, and we recall that $(u^+ - u^-)/(u^+ + u^-) = t^+ - t^-$.

The assumptions underlying (3.121) are that activities and concentrations are interchangeable in the equation for the transport numbers, that the activities themselves vary linearly with distance across the interface, and that the mobilities are not functions of activity. Although these are approximations, the Henderson equation does, in fact, in certain simple cases, give rather good agreement with experiment. Thus, for two liquid phases containing respectively KCl and HCl solutions both with mean ionic activities of 0.1, we have, for the Henderson equation

$$\Delta\varphi_{diff} = -\frac{RT}{F}\frac{0.1(u^{K^+} - u^{Cl^-}) - 0.1(u^{H^+} - u^{Cl^-})}{0.1(u^{K^+} + u^{Cl^-}) - 0.1(u^{H^+} + u^{Cl^-})} \cdot \ln\frac{0.1(u^{K^+} + u^{Cl^-})}{0.1(u^{H^+} + u^{Cl^-})} \tag{3.122}$$

where we have designated phase (I) as HCl and phase (II) as KCl. Using the limiting mobility values of $u^{K^+} = 7.61 \times 10^{-8}\ \mathrm{m^2V^{-1}s^{-1}}$, $u^{Cl^-} = 7.91 \times 10^{-8}\ \mathrm{m^2V^{-1}s^{-1}}$ and $u^{H^+} = 3.63 \times 10^{-7}\ \mathrm{m^2V^{-1}s^{-1}}$ a diffusion potential of 26.85 mV is calculated, which compares well to an experimental value of 28 mV, as shown in Table 3.5 for a concentration of 0.1M (corresponding closely to a molality of 0.1 mol kg^{-1}). In fact, given all the other approximations in the Henderson equation, it is sufficient, at least for 1:1 electrolytes, to replace activities by concentrations or molalities, at least when considering aqueous solutions, and Table 3.5 reports experimental values of the diffusion potential as a function of concentration, all of which are close to the predictions of the Henderson equation with activities replaced by concentrations.

Table 3.5 Numerical experimental values of the diffusion potential.

Solution I	Solution II	$\Delta\varphi_{diff}$/V
HCl (0.1M)	KCl (0.1M)	0.028
HCl (0.1M)	KCl (0.05M)	0.053
HCl (0.1M)	LiCl (0.1M)	0.035
HCl (0.1M)	LiCl (0.05M)	0.058
HCl (0.1M)	LiCl (0.01M)	0.091
HCl (0.01M)	KCl (0.1M)	0.010
NaCl (0.1M)	KCl (0.1M)	0.005
KCl (0.2M)	KBr (0.2M)	0.004
NaCl (0.2M)	NaOH (0.2M)	0.019

3.2.5
The Elimination of Diffusion Potentials

The magnitude of diffusion potentials (normally in the order of a few tens of millivolts) are at least an order of magnitude smaller than the EMF values of galvanic cells, but they nevertheless have the ability to disturb measurements carried out at high precision, such as those aimed at determining the thermodynamic properties reviewed in the previous chapter.

In principle, the effect of diffusion potential could be incorporated by direct calculation, following the equations given above, but these equations are not completely accurate, and it is better experimentally to eliminate the diffusion potential between two solutions if at all possible. One technique for achieving this can be deduced from equation (3.121): if a solution of some electrolyte, such as HCl, is brought into contact with a solution of an electrolyte whose ions have mobilities as close as possible to each other, such as KCl, the pre-logarithmic term will be reduced in value. It can be further minimised if the concentration of the KCl is much larger than that of the HCl, since although this will increase the size of the logarithmic term, the overall magnitude of $\Delta\varphi_{diff}$ will fall further as the pre-logarithmic term is decreased faster than the logarithmic term increases. Thus, the diffusion potential between 0.1M HCl and saturated (3.3M) KCl is only a few mV.

We can make use of this effect by replacing the direct contact between two solutions with contact through a so-called 'salt-bridge' that is filled with a concentrated electrolyte whose ions have equal mobilities, as shown in Fig. 3.10. In this case, we have replaced a single liquid-liquid junction with a high diffusion potential with two junctions, both of which have low junction potentials. Furthermore, these junction potentials have opposite signs in the example of Fig. 3.10, leading to an overall diffusion potential of less than 1 mV.

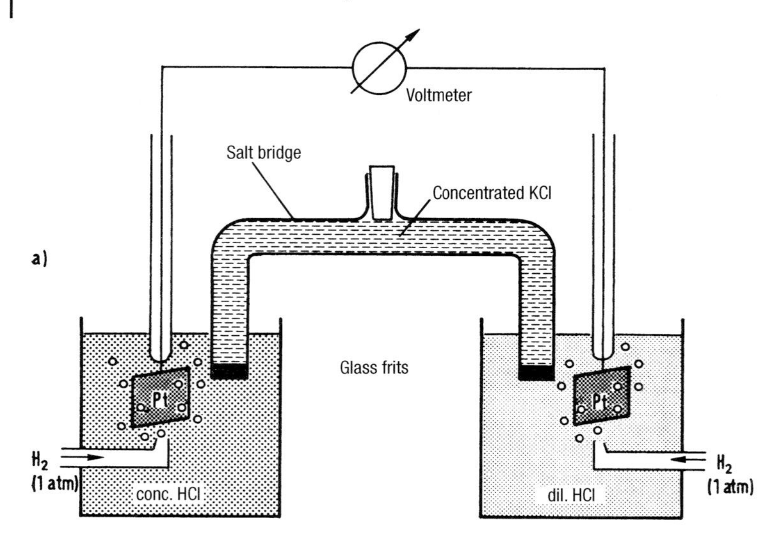

Fig. 3.10 The elimination of diffusion potentials in a practical concentration cell with a salt-bridge.

The cell in Fig. 3.10 can be represented by the scheme

$$\mathrm{Pt}|\mathrm{H_2}(p = 1\ \mathrm{atm.})|\mathrm{HCl_{conc}}\mathbin{\vdots}\mathrm{KCl_{conc}}\mathbin{\vdots}\mathrm{HCl_{dil}}|\mathrm{H_2}(p = 1\ \mathrm{atm.})|\mathrm{Pt}$$

The salt-bridge $\vdots\mathrm{KCl_{conc}}\vdots$ is conventionally replaced by the symbol $||$, which implies that the diffusion potential has been eliminated, or at least reduced to a very low level. Thus we can write our concentration cell as

$$\mathrm{Pt}|\mathrm{H_2}(p = 1\ \mathrm{atm.})|\mathrm{HCl_{conc}} \;||\; \mathrm{HCl_{dil}}|\mathrm{H_2}(p = 1\ \mathrm{atm.})|\mathrm{Pt}$$

A further possibility for reducing diffusion potential is to add an excess of an indifferent electrolyte to both solutions, where the added electrolyte again contains ions of near equal mobilities. Under these circumstances, most of the charge is carried by the ions of the added electrolyte, and the transport numbers of the other ions will be small. It is evident from equation (3.122) that under these circumstances, the diffusion potential must inevitably be very small, since the only terms of significance are those for which the difference in activity is large, and it is these that are multiplied by small transport numbers.

3.3
Membrane Potentials

If two solutions of ions are separated by a semi-permeable membrane such that the free diffusion of some subset of the ions across the membrane is prevented, an electrochemical osmotic equilibrium will be set up leading to a potential drop across the

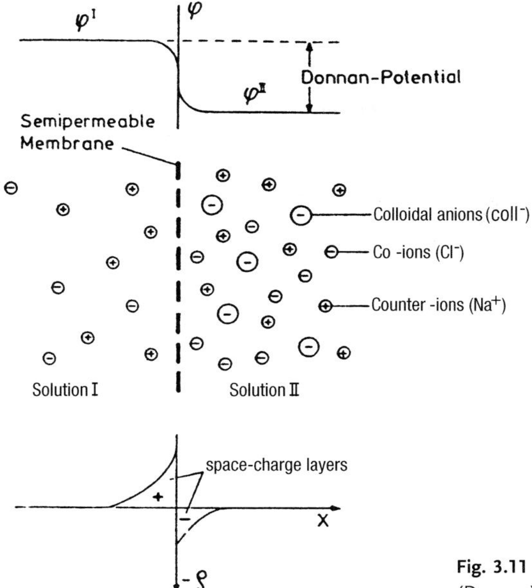

Fig. 3.11 Schematic origin of a membrane (Donnan) potential.

membrane. In a similar way, if ions in solution are in equilibrium with those present in a membrane in contact with that solution, so that ion exchange can occur, a potential drop at the phase boundary between solution and membrane may also be produced.

If we consider the first case, it is possible to draw some quantitative conclusions about the magnitude of the membrane potential relatively straightforwardly. Let the membrane separate a purely inorganic solution of NaCl on the left hand side from a solution of an anionic colloidal sol with Na+ ions on the right hand side, as shown in Fig. 3.11. Once equilibrium is established, we expect that some Cl^- ions will have diffused through to the right and some Na^+ ions will have diffused in whatever direction is dictated by the concentration gradient at the membrane. The colloidal anions, represented by $coll^-$, cannot diffuse through the membrane. At equilibrium, assuming the potentials φ^{I} and φ^{II} refer to left and right hand sides:

$$\text{For } Na^+ : \mu^{\mathrm{I}}_{Na^+} + F\varphi^{\mathrm{I}} = \mu^{\mathrm{II}}_{Na^+} + F\varphi^{\mathrm{II}}$$

$$\text{For } Cl^- : \mu^{\mathrm{I}}_{Cl^-} - F\varphi^{\mathrm{I}} = \mu^{\mathrm{II}}_{Cl^-} - F\varphi^{\mathrm{II}}$$

whence

$$\mu^{\mathrm{I}}_{Na^+} + \mu^{\mathrm{I}}_{Cl^-} = \mu^{\mathrm{II}}_{Na^+} + \mu^{\mathrm{II}}_{Cl^-} \tag{3.123}$$

Now, in general,

$$\mu_i = \mu_i^0 + RT \ln a_i \tag{2.69}$$

from which we see that

$$a^{\mathrm{I}}_{\mathrm{Na^+}} \cdot a^{\mathrm{I}}_{\mathrm{Cl^-}} = a^{\mathrm{II}}_{\mathrm{Na^+}} \cdot a^{\mathrm{II}}_{\mathrm{Cl^-}} \tag{3.124}$$

(the so-called Donnan distribution). The boundary conditions in the bulk of the solution are:

$$c^{\mathrm{I}}_{\mathrm{Na^+}} = c^{\mathrm{I}}_{\mathrm{Cl^-}} \text{ and } c^{\mathrm{II}}_{\mathrm{Na^+}} = c^{\mathrm{II}}_{\mathrm{Cl^-}} + c_{\mathrm{coll^-}} \tag{3.125}$$

derived from electroneutrality requirements. In essence, the diffusion of the anions and cations gives rise to a *space charge* layer on either side of the membrane: the chloride ions diffuse into the right-hand side, giving rise to a slight excess of negative charge near the membrane surface, and leaving a slight excess of positive charge on the left-hand side, as shown in Fig. 3.11. The overall potential drop across the membrane, $\Delta\varphi$, is given by

$$\Delta\varphi = \varphi^{\mathrm{II}} - \varphi^{\mathrm{I}} = \frac{RT}{F}\ln\left(\frac{a^{\mathrm{I}}_{\mathrm{Na^+}}}{a^{\mathrm{II}}_{\mathrm{Na^+}}}\right) = \frac{RT}{F}\ln\left(\frac{a^{\mathrm{II}}_{\mathrm{Cl^-}}}{a^{\mathrm{I}}_{\mathrm{Cl^-}}}\right) \tag{3.126}$$

and is termed the Donnan potential. If the colloidal particles are *positively* charged, the sign of the potential changes.

It is also possible to extend the considerations above to ion-exchange membranes, such as those based on polymeric sulphonic acids, of the form R-$SO_3^-Na^+$, where the sulphonic acid groups are immobilised in the membrane and the Na^+ can diffuse at least a small distance. This situation is shown in Fig. 3.12, in which the dashed line

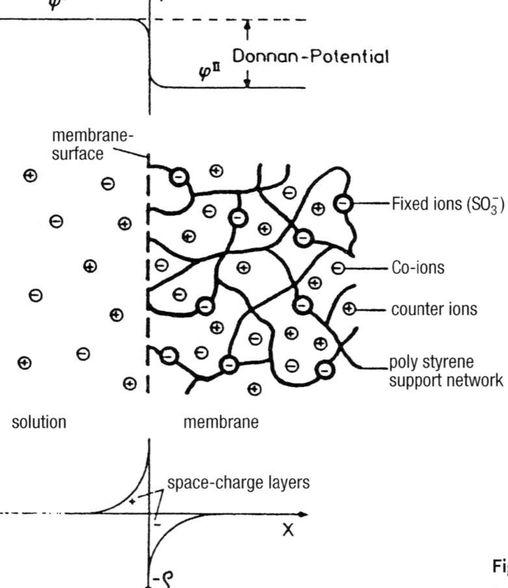

Fig. 3.12 Donnan potential at the surface of a cation-exchange membrane.

represents the boundary between the membrane and the solution. Once again, a space-charge region is set up both in the solution and within the membrane, giving rise to a Donnan potential at the interface whose calculation follows essentially the discussion above, though with the simplification that only one equilibrium, that of the cations, is necessary. If the ion-exchange membrane separates two solutions of differing activity, $a^{\mathrm{I}}_{\mathrm{Na}^+}$ and $a^{\mathrm{II}}_{\mathrm{Na}^+}$, then the potential across the first face has the form

$$\Delta\varphi^{(\mathrm{I})} = \frac{RT}{F}\ln\left(\frac{a^{\mathrm{I}}_{\mathrm{Na}^+}}{a^{\mathrm{M}}_{\mathrm{Na}^+}}\right) \quad (3.127)$$

where $a^{\mathrm{M}}_{\mathrm{Na}^+}$ is the activity of Na^+ ions in the membrane, and the potential across the entire membrane, termed the *dialysis* potential, is given by:

$$\Delta\varphi_{dial} = \Delta\varphi^{(2)} - \Delta\varphi^{(1)} = \frac{RT}{F}\ln\left(\frac{a^{\mathrm{II}}_{\mathrm{Na}^+}}{a^{\mathrm{I}}_{\mathrm{Na}^+}}\right) \quad (3.128)$$

similar to the expression obtained for the semi-permeable membrane case.

3.4 The Electrolyte Double-Layer and Electrokinetic Effects

When a metal electrode is brought into contact with a solution containing the corresponding metal ion M^{z+}, then the reaction

$$\mathrm{M}^{z+} + ze^- \rightleftharpoons \mathrm{M} \quad (3.129)$$

can take place at the surface, with either the forward or reverse reactions being favoured. If the forward reaction dominates, then the electrode will become denuded of electrons and acquire a positive charge. This positive charge must reside on the surface of the metal, since the high conductivity inside the metal does not permit the development of an extended internal space-charge region, and it will attract anions towards the neighbourhood of the electrode, setting up a charged double layer of the type already discussed earlier in this chapter. As a result, the interior of the metal and the solution phase find themselves at different potentials, $\varphi_M \neq \varphi_S$. A similar analysis for redox electrodes or gas-electrodes leads to similar conclusions.

This equilibrium potential at the electrode can be disturbed if we apply an external potential difference between the working electrode and a counter electrode immersed in the same solution. Under these circumstances, the charge at the double layer may be increased or decreased, or may even change sign altogether. In the case of electrochemically active electrodes, in addition to the charging of the double layer, the passage of electrical current will also produce an electrochemical reaction at the electrode (as in Eq. 3.129). However, if an electrode is immersed in a solution containing only "inert" ions, i.e., ions having a large decomposition potential (see Section 2.1.2), then alteration of the potential will cause only an alteration in the charge associated with the double layer, as long as the decomposition potential is not reached. Within this potential range, the electrode is said to be *polarisable*.

3.4.1
Helmholtz and Diffuse Double Layer: the Zeta-Potential

If an electrolytic double layer is formed at an electrode, an excess of ions of charge opposite to that on the electrode will be found on the solution side of the phase boundary. The simplest view of the phase boundary is to imagine that these ions approach the surface of the electrode as closely as possible, and the double layer then consists of two parallel layers of charge, one on the metal surface and one comprising anions at the distance of closest approach, as shown in Fig. 3.13. If we identify the electrode side of the double layer with the metal surface itself, and the solvent side with the centre of charge of the (possibly solvated) ions present in excess, then the separation of the charged layers may be taken as half the diameter, a, of the (possibly solvated) ions themselves. This simplest possible model was termed by Helmholtz the compact double layer: it is now known as the Helmholtz-layer model. It is clear that the Helmholtz layer will behave as a capacitor with a plate separation of $a/2$, and the plane passing through the centre of charge of the ions is termed the Helmholtz plane.

The connection between the space-charge density and the potential is given by the Poisson equation (2.37); for a one-dimensional problem we have

$$\frac{d^2\varphi}{dx^2} = -\frac{\rho}{\varepsilon_r \varepsilon_0} \tag{3.130}$$

where x is the direction perpendicular to the electrode surface. A further approximation is made that the ions are treated as point charges: this allows us to assume that the charge density between the electrode surface and the Helmholtz plane is *zero*.

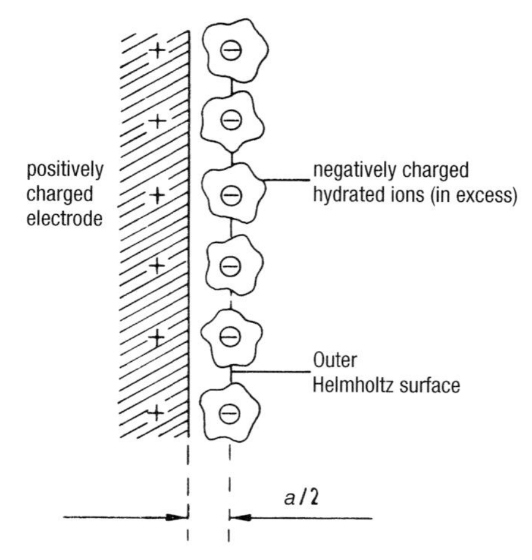

Fig. 3.13 Helmholtz double layer of width $a/2$, where a is the diameter of the solvated counter ion.

Hence, within the Helmholtz plane,

$$\frac{d^2\varphi}{dx^2} = 0 \tag{3.131}$$

which integrates to

$$\frac{d\varphi}{dx} = \text{constant}$$

Further integration leads to

$$\varphi = \varphi_M - 2(\varphi_M - \varphi_S)x/a \tag{3.132}$$

for the region $0 \leq x \leq a/2$, as shown in Fig. 3.14.

The Helmholtz model is undoubtedly incomplete, since it takes no account of the thermal motion of the ions, which will tend to loosen ions from the compact layer. Goüy and Chapman were the first to consider the effect of thermal motion on the ions near an electrode surface. Their model did not consider the possibility of an inner Helmholtz layer at all, and led to the picture of a *diffuse* double layer, consisting of positive and negative ions, in a spatially quite extended region near the electrode surface. Ions of charge opposite to that on the electrode surface were found to be present in excess, compared to the bulk of the electrolyte, and ions of the same charge as the electrode surface were found to be present in lower concentration compared to the bulk. It was left to Stern to point out that the most realistic model was a combination of the Helmholtz layer and diffuse layer models, as shown in Fig. 3.15. An important point, brought out by Stern, is that the position of the Helmholtz plane will vary with the type of ion attracted to the surface: some ions may be able to lose their solvation sheaths and approach the surface very closely, whereas others may only approach to within a distance determined by the extent of their solvation sheaths. The two planes so defined are termed the *inner* and *outer* Helmholtz planes, and we will use a to denote

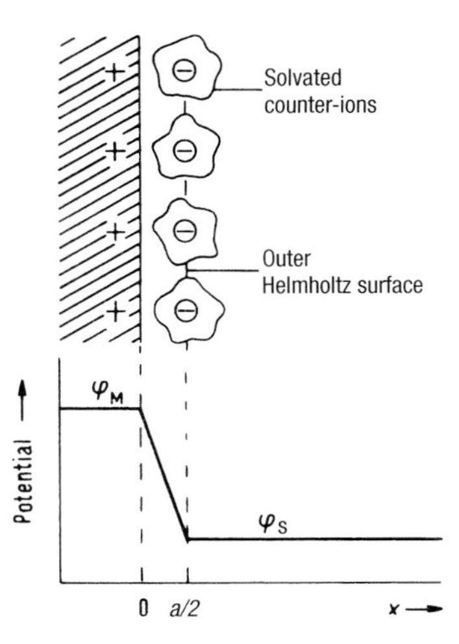

Fig. 3.14 Potential distribution between electrode and outer Helmholtz surface: φ_M is the potential in the metal and φ_S the potential of the electrolyte at the outer Helmholtz layer.

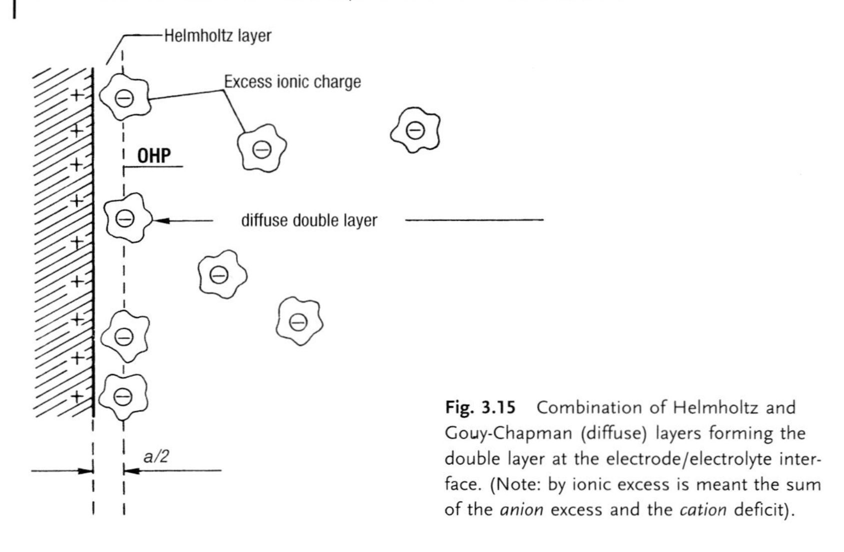

Fig. 3.15 Combination of Helmholtz and Gouy-Chapman (diffuse) layers forming the double layer at the electrode/electrolyte interface. (Note: by ionic excess is meant the sum of the *anion* excess and the *cation* deficit).

the diameter of the largest solvated ions, whose centres constitute the outer Helmholtz plane.

We can quantify the Goüy-Chapman or diffuse-layer model straightforwardly. If n_i^0 is the number of ions of type i per unit volume present in the bulk of the electrolyte at equilibrium, then

$$n_i = n_i^0 \exp\left[\frac{-z_i e_0 \varphi(x)}{k_B T}\right] \tag{3.133}$$

where we anticipate that equation (3.133) will be valid only for $x \geq a/2$, since this is the distance of closest approach of the ions to the surface. The discussion can be simplified by the replacement of x by the variable ξ, where $\xi = x - a/2$. Also, we anticipate that far out in the electrolyte, where $\varphi(x) \to \varphi_S$, we will find that $n_i \to n_i^0$, so (3.133) can be written

$$n_i = n_i^0 \exp\left[\frac{-z_i e_0 (\varphi(\xi) - \varphi_S)}{k_B T}\right] \tag{3.134}$$

Further treatment of equation (3.134) follows the course of Section 2.4.2. By considering all the ions present in the solution, and defining the ionic strength I as

$$\mathbf{I} = \frac{1}{2}\sum_i z_i^2 \frac{m_i}{m^0} \tag{3.135}$$

where the molality is expressed in mol kg^{-1}, and

$$\kappa^2 = \frac{2e_0^2 \rho_S N_A m^0}{\varepsilon_r \varepsilon_0 k_B T}\mathbf{I} \tag{3.136}$$

where ρ_S is the density of the solvent. We can write the Poisson equation as

$$\frac{d^2\varphi}{d\xi^2} = \kappa^2(\varphi(\xi) - \varphi_S) \tag{3.137}$$

which solves to give

$$\varphi(\xi) - \varphi_S = const.e^{-\kappa\xi} \tag{3.138}$$

At $\xi = 0$, the potential will be that present at the outer Helmholtz plane, φ_{OHP}, and we may evaluate the integration constant to give

$$\varphi(\xi) - \varphi_S = (\varphi_{OHP} - \varphi_S)e^{-\kappa\xi} \tag{3.139}$$

It can be seen that according to equation 3.139, the potential will fall (or rise) exponentially from the Helmholtz plane to the interior of the electrolyte. The total Galvani potential between the interior of the electrode and the interior of the solution can now be divided into two contributions:

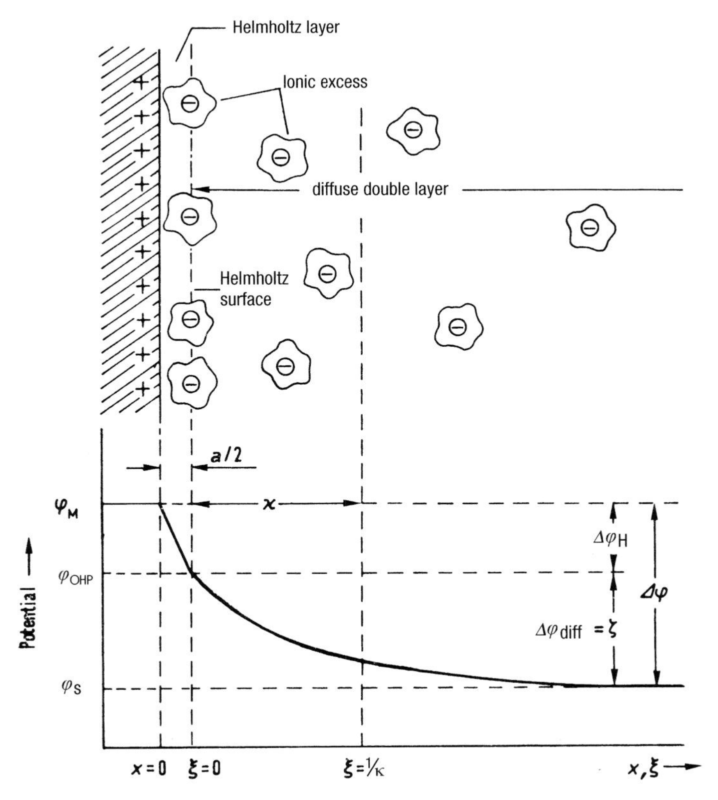

Fig. 3.16 Potential distribution at the general double layer: $\Delta\varphi_H$ is the potential dropped in the Helholtz layer and $\Delta\varphi_{diff}$ that dropped in the diffuse layer. The zeta potential is numerically the same as $\Delta\varphi_{diff}$.

$$\Delta\varphi = (\varphi_M - \varphi_{\text{OHP}}) + (\varphi_{\text{OHP}} - \varphi_S) = \Delta\varphi_H + \Delta\varphi_{\text{diff}} \tag{3.140}$$

where $\Delta\varphi_H$ is the potential drop across the Helmholtz layer and $\Delta\varphi_{\text{diff}}$ the potential drop across the diffuse double layer; this latter potential drop is known as the zeta-potential, ζ, and $\Delta\varphi = \Delta\varphi_H + \zeta$. The distance $\xi = 1/\kappa$ from the outer Hemholtz plane to the point at which the diffuse-layer potential has dropped to a value 1/e of its total change from OHP to bulk solution is a measure of the thickness of the double layer. Figure 3-16 shows the relationship between this and the other potentials and distances characterising the total double-layer structure.

The thickness of the double layer depends primarily on the ionic strength, **I**, of the solution (s. Eq 2.136). From the data in Table 2.8, it can be seen that the diffuse double layer may, in dilute solutions, extend more than 10 nm; but already for a concentration of 0.1 M, the thickness of the double layer decreases to values not much greater than that of the Helmholtz layer itself. (From Table 2.8, for $c = 0.1\text{M}$, $\kappa^{-1} < 1\text{nm}$, and $a/2 \approx 0.1\text{nm}$, as in Section 2.3.6). It follows that for higher ionic strengths, the diffuse double layer will become sufficiently small that it can be neglected, and the entire potential drop is accommodated across the Helmholtz layer; under these circumstances, ζ can be neglected.

3.4.2
Adsorption of Ions, Dipoles and Neutral Molecules – the Potential of Zero Charge

Ions, solvent dipoles and neutral molecules with or without dipole character may be adsorbed onto a metal electrode surface, either through van der Waals or Coulombic interactions with the charged surface, or even forming stronger bonds, as in the case of chemisorption. In general, the adsorption may be weakened or strengthened by potential changes on the electrode, but since anions, in particular, have a tendency to undergo *specific* adsorption through van der Waals interactions, they can be adsorbed on an electrode surface even if it is negatively charged. For such anions, given that the solvent sheath must be stripped off, at least on the metal side of the anions, before specific adsorption can take place, it is a general rule that the weaker the solvation of the anion, the stronger will be its specific adsorption. The fact that anions may adsorb more closely than cations gives rise, as stated above, to the existence of an *inner* Helmholtz plane. The final picture of the electrified double layer is given in Fig. 3.17, where the inner and outer Helmholtz planes are specifically indicated for a negatively charged electrode surface.

The inner potential of an electrode on whose surface there are no free excess charges, either specifically adsorbed ions or excess ions of either charge in the diffuse double layer, is termed its *potential of zero charge*, φ_{pzc}. In general, the value of φ_{pzc} is not identical to φ_S, the potential in the interior of the electrolyte, owing to the presence of solvent dipoles at the surface, which give rise to an unknown additional potential drop. The determination of E_{pzc}, the electrode potential at the pzc with respect to some standard reference electrode, is discussed in Section 3.4.5.

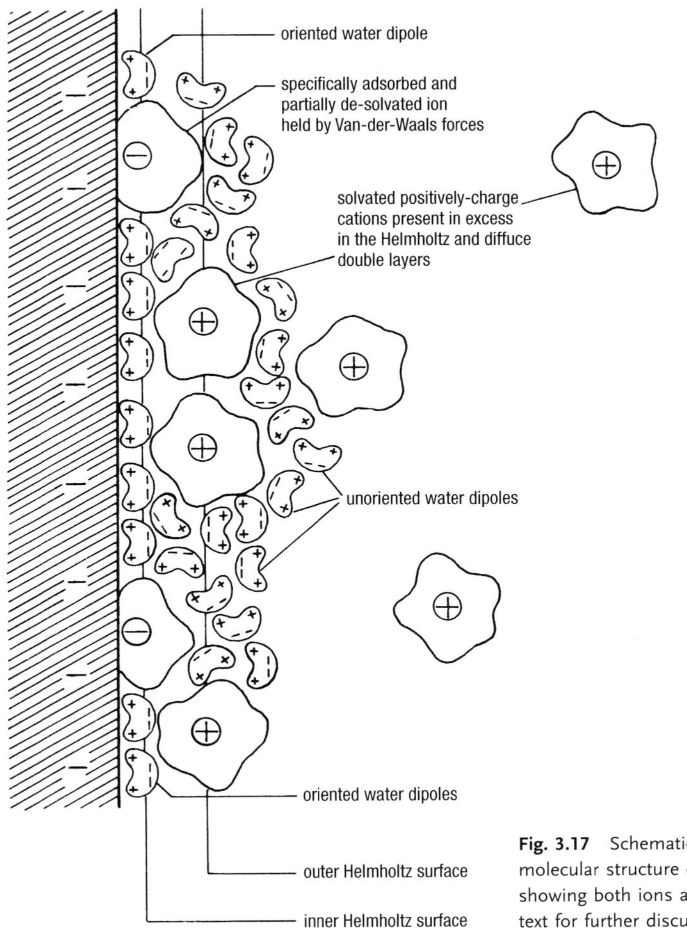

Fig. 3.17 Schematic representation of the molecular structure of the double layer showing both ions and water dipoles. See text for further discussion.

3.4.3
The Double-Layer Capacity

The electrolytic double layer can be considered most simply as a parallel plate capacitor. For such a capacitor, there will be a linear relationship between the charge on the plates, Q, and the potential difference between the plates, $\Delta\varphi$, the constant of proportionality being the capacitance, C

$$Q = C\Delta\varphi \tag{3.141}$$

In the case of the double-layer, Q corresponds to the charge excess on the solution side of the interface, and $\Delta\varphi$ to the Galvani potential difference between metal and solution interiors. However, there is one subtlety: we have already indicated that even when $Q = 0$, there will still be a potential difference between metal and solution owing to the solvent dipole layer at the electrode surface. This is not included in (3.141) explicitly, but if we assume that this dipole layer does not change greatly with potential, we can retain the form of (3.141) by writing the potential difference $\Delta\varphi$ as $E - E_{pzc}$. Since $\Delta\varphi - \Delta\varphi_{pzc} = E - E_{pzc}$, where E is measured with respect to a standard reference electrode,

$$Q = C(E - E_{pzc}) \tag{3.142}$$

which ensures that the charge Q will be zero at the point of zero charge.

Experimentally, the value of C might be determined by measuring the total charge passed when the potential is stepped from E_{pzc} to some convenient potential E. However, in this type of experiment, great care is necessary to ensure that there is no contribution from any ongoing electrochemical reaction; in other words, it is essential to work only in that region of the potential for which the electrode is *polarisable*. A second experimental method, which is described in more detail elsewhere, involves the use of *impedance* methods.

The comparison of the electrolytic double layer with a parallel-plate capacitor will be most apt where the double layer is most closely described by the Helmholtz model, i.e. at higher concentrations of electrolyte. In general, however, we do not expect an exact proportionality between the excess charge Q and the potential $E - E_{pzc}$, since alteration of the potential may cause the re-orientation of the solvent dipole layer or produce specific adsorption or desorption of ions, which will act to shield or de-shield the diffuse double layer from the electrode surface. In fact what is found is that the alteration in charge δQ that accompanies a minute change δE in electrode potential is itself in general a function of E, and it follows that the derivative, dQ/dE is a better description of the process of double layer charging than the value of C itself. Since dQ/dE also has the dimensions of capacitance, it is usually referred to as the *differential* capacitance, C_d, to distinguish it from the *integral* capacitance, C calculated above.

$$C_d = dQ/dE \tag{3.143}$$

The measurement of C_d can be carried out very straightforwardly. Given that $i = dQ/dt$, we have, for the capacitance charging current, i_C

$$i_C = dQ(E)/dt = (dQ/dE) \cdot (dE/dt) = C_d \cdot dE/dt \tag{3.144}$$

It is, therefore, sufficient to impress upon the electrode a linear rise in potential with time in order to measure C_d. In general values of C_d for flat metal surfaces lie in the range $0.05 - 0.5\ \mathrm{Fm}^{-2}$ depending on the type of metal, composition of electrolyte, ionic strength, temperature and electrode potential. However, on rough metal electrodes, such as platinised platinum, values a thousand times larger may be obtained.

The variation of C_d with potential has been the subject of an immense amount of study, since it offers a very direct test of the structure of the double layer. Some experimental curves for the variation of differential capacitance with electrode potential for mercury in various concentrations of aqueous NaF are shown in Fig. 3.18. In the absence of specifically adsorbed ions, the minimum is observed at the potential of zero charge.

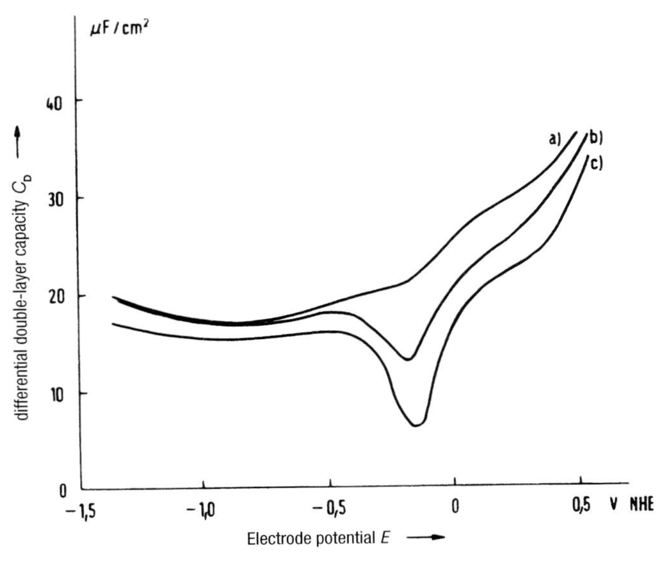

Fig. 3.18 Differential double-layer capacity C_D at mercury at 25 °C in NaF solution of concentrations: (a) 0.1 M; (b) 0.01 M; (c) 0.001 M.

3.4.4
Some Data for Electrolytic Double Layers

We have seen that in concentrated electrolyte solutions, the double layer will closely resemble that predicted by Helmholtz and therefore, the integral capacitance of this layer should be easily calculated from the expression

$$C = \frac{\varepsilon_r \varepsilon_0}{l} \tag{3.145}$$

where l is the distance separating the plates. If we put $l = a/2 \approx 0.2$ nm (which is about half the diameter of a typical hydrated cation) and $\varepsilon_r = 80$, the approximate value for water at 25°C, then substitution into (3.145) gives us a value for the integral capacitance C of ca. 3.5 Fm^{-2} or ca. 350 $\mu\text{F cm}^{-2}$.

This is much too high when compared to experimental values (0.05 – 0.5 Fm^{-2}), and the problem lies with our choice of 80 for ε_r. This value presupposes that the water dipoles can re-orient freely under the influence of the electric field, but this is clearly not true for water molecules close to the electrode surface. In fact, in the inner Helmholtz layer, water molecules are subject to chemical and electrical forces, which determine preferential orientations depending on the metal charge. Furthermore, in the space between the inner and outer Helmholtz planes, water molecules have restrictions on their motion. Calculations suggest that for the inner layer of water molecules, an appropriate value of ε_r will be ca. 6 and for water molecules between inner and outer planes, a value of ca. 30 will be appropriate. In order to incorporate these values in the calculation, we must divide the double layer into two layers, which lie in series with each other. Under this geometrical arrangement, the total capacitance, C_H, will be given by

$$\frac{1}{C_H} = \frac{1}{C_{dipole}} + \frac{1}{C_{IHP-OHP}} = \frac{a_{H_2O}}{\varepsilon_0 \varepsilon_{dipole}} + \frac{a_{ion}}{2\varepsilon_0 \varepsilon_{IHP-OHP}} \tag{3.146}$$

where a_{H_2O} is the diameter of the water molecule, and a_{ion} the diameter of the largest solvated ion. Values of C_d calculated from this expression are very much closer to the experimental values. A second important consequence of (3.146) is that the value of C_d is, in fact, dominated by the first term, and the influence of the ionic diameter, a_{ion} is relatively small, as indeed is found experimentally. Equation (3.146) is still, however, not complete, since it is known that the value of C_d depends on the type of metal, a dependence that does not appear at all in the equation above. In fact, the incorporation of the metallic properties of the electrode into the value of C_H is an area of considerable current interest and research.

In dilute solutions, the capacity of the *diffuse* layer must be taken into account. The value of this capacitance is given by

$$C_{diffuse} = \kappa \varepsilon_r \varepsilon_0 \tag{3.147}$$

and as κ decreases, this term becomes increasingly significant in the expression for the total double layer capacitance

$$1/C_D = 1/C_H + 1/C_{diffuse} \tag{3.148}$$

As a result, in more dilute solutions, the value of C decreases, and thoroughgoing calculations reveal that the differential capacitance also decreases in dilute electrolytes, giving rise to a minimum in capacitance at E_{pzc}, as shown in Fig. 3.18.

Inside the double layer, very high electric fields may be sustained. For concentrated electrolytes, the thickness of the double layer will be ca. 0.1 nm; for a Galvani potential difference of 0.1 V, this corresponds to a field strength of 10^9 Vm^{-1}. This field strength is sufficiently large actually to weaken molecular bonds, and one example is the increase in dissociation constants of weak acids in the double layer, an effect termed *field dissociation*. Moreover, vibrational frequencies and intensities of adsorbed species can be affected by the electric field at the double layer. This effect of strong electric fields is known in Physics as *Stark effect* and can be observed in electrochemical cells using *in situ* infrared spectroscopy.

3.4.5
Electrocapillarity

Electrical charges of the same sign within a space-charge layer will repel one another, and if an electrical double layer forms at the boundary between an electronic and ionic conductor, inter-ionic interactions at the electrode surface will tend to extend the surface as far as possible. These interactions are visible in the case of a deformable electrode surface: thus, if a mercury droplet is covered with a layer of dilute sulphuric acid, then a double layer will form at the metal-electrolyte phase boundary, associated with which will be a Galvani potential. The resultant repulsion between the ions leads to a flattening of the drop as shown in Fig. 3.19.

The effect described in the previous paragraph corresponds to a lowering of the surface tension of mercury, and this lowering must be independent of the sign of the Galvani potential. If the potential difference between mercury and electrolyte is now changed by means of an external voltage, then the point of zero charge will correspond to a *maximum* in the surface tension, and this offers a method for finding the potential of zero charge straightforwardly. The experiment can be carried out as shown in Fig. 3.20: mercury is forced through a capillary at constant velocity, and the weight that each drop attains before it becomes detached from the capillary depends on the surface tension of the mercury itself. The drop time is easily measured, and this serves as a measure of the surface tension. The potential of the drop can be directly controlled with reference to the mercury pool in the flask, which acts as a calomel electrode.

The plot of surface tension, γ, against potential, as shown in Fig. 3.21 is called an *electrocapillary curve*. The shape of the electrocapillary curve is essentially the same for all anions such as NO_3^-, ClO_4^- and sulphate that do not adsorb on mercury itself, and in 1M KNO_3 solutions, as can be seen from Fig. 3.21, almost exactly parabolic curves are obtained; the maximum in this curve gives a value of $-$ 0.52 V vs. NCE for the potential of zero charge. If the anions do adsorb, however, the surface tension will be affected by the excess ionic charge at the electrode surface. Under these circumstances, the positive (right-hand) branch of the electrocapillary curve becomes lowered. In addition, given that the anions will only be desorbed at potentials below $-$ 0.52 V, the position of the potential of zero charge will therefore be displaced to a lower value; the electrocapillary curve becomes characteristically asymmetric.

The electrocapillary effect can be treated quantitatively: we consider a surface with a charge density q_M. Let the potential across this surface change by an infinitesimally small amount, $q_M \mathrm{d}E$. The work done per unit area is $q\mathrm{M}\mathrm{d}E$, and, at equilibrium, this must be offset exactly by the change in surface free energy per unit area. This latter is just $\mathrm{d}\gamma$, so

$$q_M \mathrm{d}E + \mathrm{d}\gamma = 0 \tag{3.149}$$

whence we obtain the Lippmann equation

$$\mathrm{d}\gamma/\mathrm{d}E = -q_M \tag{3.150}$$

given that $q_M = C(E - E_{zpc})$, where C is the integral capacitance per unit area, we can integrate (3.150) assuming C to be constant, and we obtain finally

$$\gamma_{pzc} - \gamma = -C(E - E_{pzc})^2/2 \tag{3.151}$$

where γ_{pzc} is the surface tension at the potential of zero charge. It can be seen that (3.151) is a parabola, and it is found that such a relationship between γ and E is accurately obeyed provided that no specific ionic adsorption occurs, and the ionic strength is reasonably high ($>$ ca. 0.1 M).

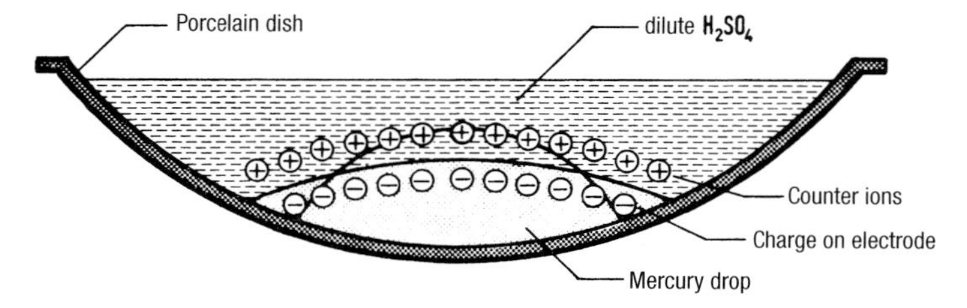

Fig. 3.19 Flattening of the surface of mercury on formation of an electrolyic double layer, represented for negative charging of the metal with respect to the solution.

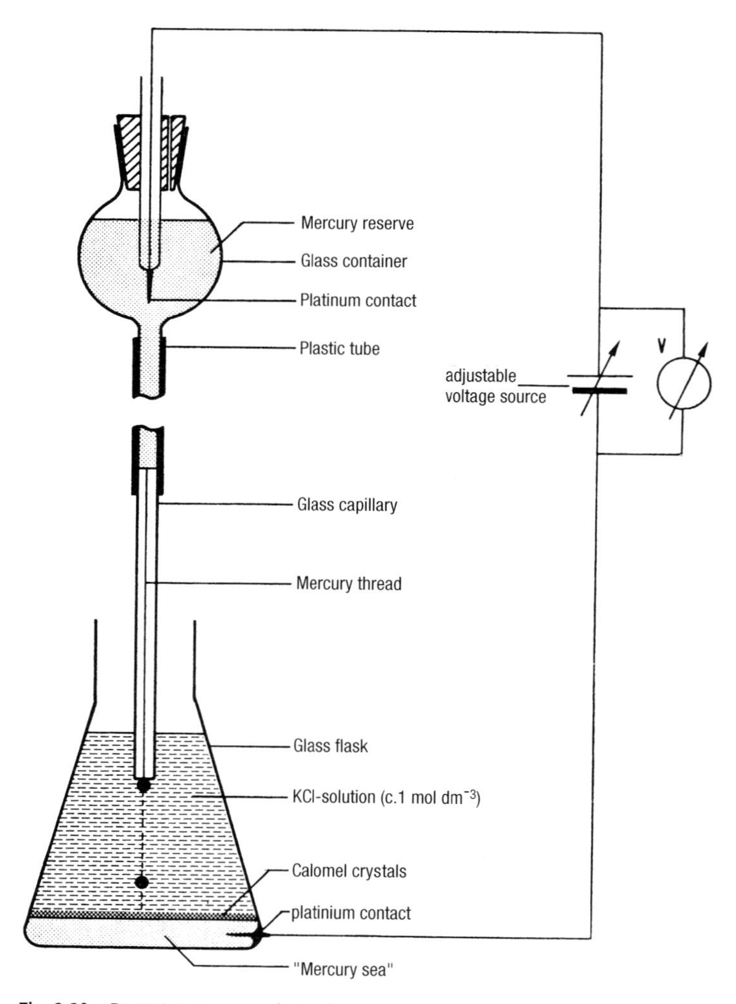

Fig. 3.20 Dropping mercury electrode.

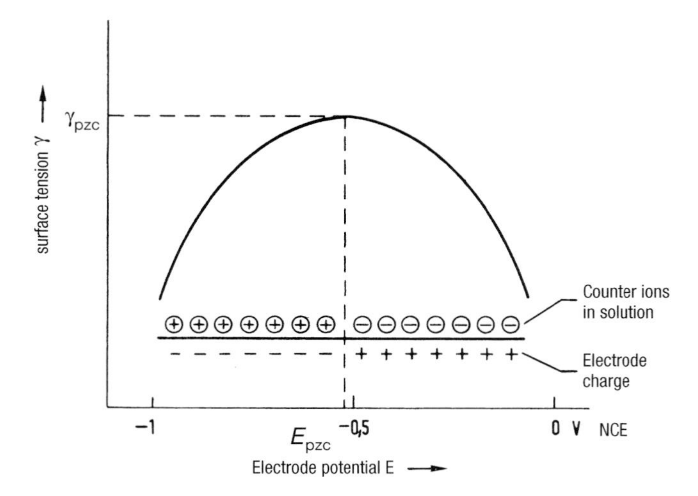

Fig. 3.21 Electrocapillary curve E_{pzc} is the potential of zero charge and γ_{pzc} the surface tension at the zero-charge potential.

Differentiation of the Lippmann equation with potential leads to

$$\frac{d^2\gamma}{dE^2} = -\frac{dq_M}{dE} = -C_d \tag{3.152}$$

where we are using (3.143), and C_d is now the differential capacitance *per unit area*; from this equation, the differential capacitance of the mercury droplet surface can be obtained as the curvature of the electrocapillary curve. This value is found closely to agree with that obtained by other measurements.

3.4.5.1 **Other Methods for the Determination of the Potential of Zero Charge (pzc)**

For solid electrodes, the method described above for the determination of the potential of zero charge clearly will not work. A direct method for pzc determination has been described by Sato and colleagues, using the fact that at any potential but that at the *pzc*, small *alterations* in potential lead to changes in surface tension and hence to small changes in surface stress.

Alternative direct methods include the measurement of the differential electrode capacitance with *ac* techniques as described in more detail in Section 5.2.2, with the minimum in this capacitance corresponding to the *pzc*.

Other methods that have been used for platinum electrodes include the measurement of charge in current transients produced when carbon monoxide becomes adsorbed by displacing adsorbed anions from the electrode surface. Since CO itself does not suffer oxidation or reduction at potentials below ca. 0.4 V NHE, the transients observed are associated to the displaced species. Anions undergoing specific adsorption may partially lose their charge during adsorption and recover it during desorption. Since, in general, anions become adsorbed at potentials positive to zero charge, the anion desorption-current passes through zero at potentials close to zero charge. In fact, as suggested by Weaver, this method gives the potential where the *total* electrode charge density equals zero and this is not just the potential at which the *free* charge of the metal is zero. While strictly speaking the latter is the *pzc*, the former, should be called *potential of total zero charge*. The method requires a small correction of potentials, since CO itself displaces the potential of zero charge of platinum due to electronic effects.

All experimental techniques for the determination of the pzc in solid electrodes confront different difficulties. Therefore, the dispersion of values reported for a given material is not surprising. For polycrystalline Pt, for example, values between 0.11 and 0.27 V vs. SHE can be found in the literature. It is noteworthy, that there is a correlation between the potential of zero charge of a metal and its electronic work function. Since the latter depends on the surface structure of the metal, the different planes of single crystal electrodes exhibit different values of the potential of zero charge.

3.4.6
Electrokinetic Effects – Electrophoresis, Electro-osmosis, Dorn-effect and Streaming Potential

Electrolytic double layers can also form at solid-liquid phase boundaries even if the solid concerned is not an electronic conductor. One example is the surface of a cation-exchange membrane: cations are free to migrate from such a membrane into solution, whereas the anionic groups being firmly bonded into the membrane remain behind. Similar considerations apply to inorganic materials such as glass, from which specific ions may again migrate into solution. However, the main origin of double layer formation on both inorganic and organic substrates is the specific adsorption of ions. In this case, no excess charge is held within the solid, but rather on the surface itself, and counter ions form both compact and diffuse layers in the solution itself, as shown in Fig. 3.22.

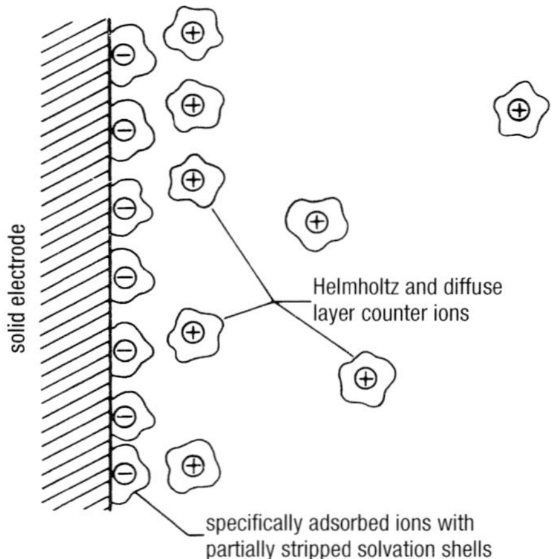

Fig. 3.22 Formation of an electrolytic double layer through specific adsorption of ions.

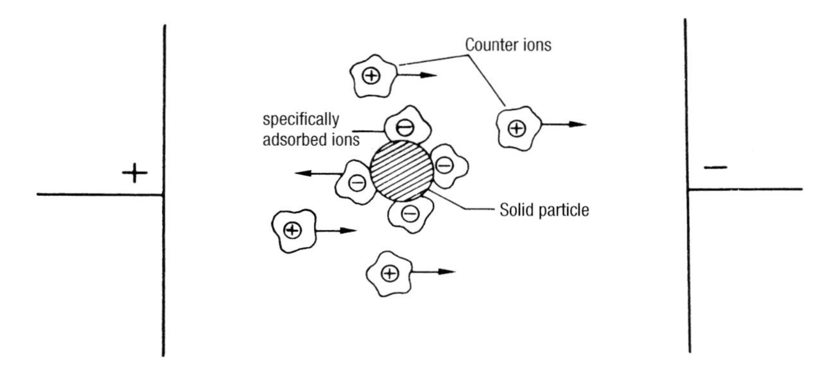

Fig. 3.23 Migration of charged particles in an electric field – electrophoresis.

Suspended solid particles or colloidally dispersed material may be treated in the same way. However, for colloidal particles, the ratio of adsorbed charge to particle mass is so large that the particles can actually be set in motion by an electric field, whilst the counter-ions migrate in the opposite direction, a process known as *electrophoresis* – see Fig. 3.23. The mobility of such colloidal particles is, in general, of the same order of magnitude as normal ionic mobilities, since the greater diameter of the particle is offset, at least to some extent, by its greater charge. In addition, since anions are predominantly the species adsorbed from aqueous solutions, the colloidal particles are frequently observed to migrate towards the positive electrode.

The actual extent to which charge is adsorbed by a colloidal particle will depend on the nature of the particle itself; different types of particle will, therefore, migrate at different speeds and this affords a method of separation which is much used in the separation of biological macromolecules. Technically, electrophoresis is extensively used in the deposition of particles of paint (electrophoretic painting) and in the separation of particles of kaolin from aqueous emulsions (electrophoretic drying).

If the colloidal particles undergo forced movement through an electrolyte solution, for example by the imposition of ultrasonic vibrations, then a potential difference is observed in the direction of motion as a consequence of the transport of the charge attached to the particle. This, which is the inverse of the electrophoretic effect, is termed the Dorn effect.

If a colloidal particle has equal amounts of both negative and positive charge adsorbed on its surface, no movement will be induced in an electric field. An important case of this is found for colloidal oxide particles that can adsorb both H_3O^+ and hydroxide ions. There is usually an accessible value of the pH at which the overall surface charge density is zero, and this value of pH is referred to as the 'isoelectric point', or, in the colloid literature, the 'point of zero zeta potential', the latter name arising from one of the techniques commonly used to determine the isoelectric point.

Associated with the migration of ions in an electric field is the transport of water that is bound in the hydration sheaths. If an electric field is applied along a fixed charged solid surface, the counter ions in the diffuse layer can move with the field, whereas the

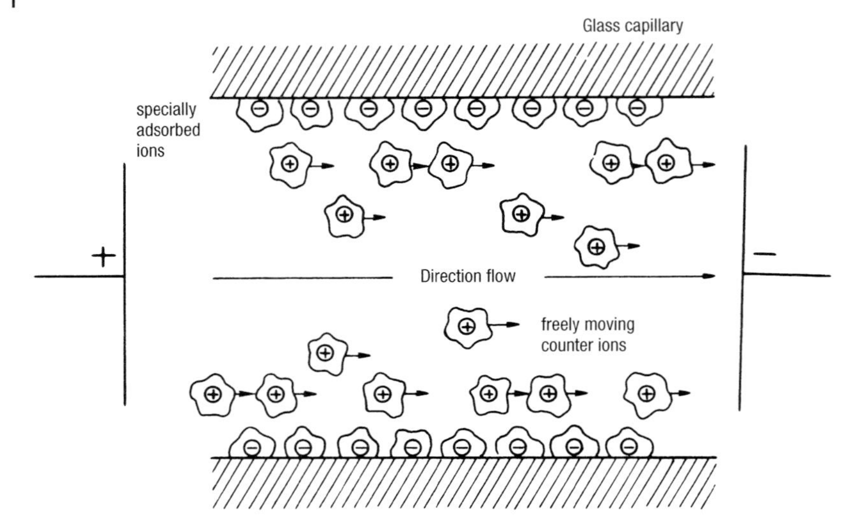

Fig. 3.24 Electro-osmotic current.

ions adsorbed on the surface cannot. The result is an ion current along the surface. This effect can be magnified by using a porous solid; a field applied across such a solid will also induce counter-ion movement in the pores, which manifests itself through the generation of an osmotic pressure; this is termed the electro-osmotic effect, and is illustrated in Fig. 3.24. The inverse of this effect is manifested by the forcing of a stream of ions through a capillary or a porous electrode, which gives rise to a potential difference termed the streaming potential.

All these effects can be treated from a unified standpoint with the help of the theory of irreversible thermodynamics. Such treatments are given in a number of advanced treatises given in the list of references.

3.4.7 Theoretical Studies of the Double Layer

We saw, in section 2.5.3, that considerable recent progress has been made in understanding the structure of water and other solvents with and without dissolved ions by using molecular modelling. Such models have, in recent years, been extended to the study of the electrolyte double layer, and results of considerable interest have emerged.

The earlier studies treated the metal as a smooth solid phase with no atomic structure, with the metal treated as a medium of infinite dielectric constant with a boundary located very close to, but not necessarily exactly coincident with the metal surface. The position of the dielectric boundary is of fundamental importance in these calculations, since a charged species with charge ze_0^+ will, in the absence of screening in the electrolyte phase, be attracted to the metal through the polarisation of the electron cloud. Elementary calculations in electrostatics show that this attraction can be modelled accurately by incorporation of a virtual "mirror charge" of opposite sign, located in the metal at the position of the mirror image of the real charge reflected in the plane of the dielectric discontinuity. To locate the position of this plane, calculations have been carried by several authors in which the metal is modelled as a collection of fixed positive charges surrounded by a free-electron gas – the "jellium" model. Such calculations suggest that the plane of dielectric discontinuity is shifted by about 0.5 – 1.0Å towards the liquid phase, at least at negative potentials, though at positive potentials it has been suggested that the plane may move in towards the metal.

More realistic "chemical" models have also been developed in which the solid electrode is modelled by a regular array of atoms which interact with solution-phase molecules through local potential functions. As for the jellium model, the long-range Coulombic interactions with the electron cloud can be treated through the image charge model. The local pairwise potentials between metal atoms and solvent molecules are usually obtained by model quantum chemical calculations of varying levels of sophistication, and the overall interaction potential with the surface will typically reflect the corrugated structure of that surface, the surface atoms of which are usually allowed to relax from their positions in the isolated metal.

A further complication in electrochemical systems is the presence of an electric field at the surface, which was initially modelled by incorporating a uniform electric field. Such an approach takes no account of screening in the electrolyte; whilst this can be modelled in principle using the Goüy-Chapman theory, this theory does not work at high electrolyte concentrations and cannot easily accommodate the fact that the solvent molecules are not only affected by the mean field of the ions but by localised solvation phenomena. Modern approaches usually place a homogeneous surface charge density on the electrode and modify the number of ions in solution to give overall charge neutrality.

The most carefully studied interface is that between pure water and various electrodes, and theoretical investigations have suggested the following conclusions:

1. At metals, even in the absence of any imposed surface electronic charge, the density profile of water in the first adlayer is strongly structured, showing a very marked maximum with a clearly visible weaker maximum in the second layer, a much weaker structure in the third layer and essentially bulk behaviour outside this. By contrast the *lateral* water structure becomes very weak even after the first layer. This structure is dominated by local potential interactions between the first layer and the metal surface, with hydrogen bonded structure giving rise to the rapidly decreasing structure further out. This appears to be consistent with X-ray reflectivity measurements on dilute NaF solutions (Nature **368** (1994) 444).
2. For *liquid* metals, and mercury in particular the water structure is weakened considerably relative to the solid-metal/water interface, primarily due to the reduction in the corrugated atomic structure of the metal surface itself.
3. On incorporation of a surface electronic charge, the water charge density perpendicular to the surface is not greatly affected for surface electronic charge densities less than about 10 μC cm^{-2}, but if the charge density is increased above this the water molecules begin to show much longer-range structure, eventually under-

going a ferroelectric transition in which all the dipoles become oriented in the same direction.

4. The orientation of the water molecules near the metal surface in the absence of a surface charge is usually such that the dipoles are orientated over a broad range of angles into the solution, with the H-H vector normally *parallel* to the surface, but with *very few molecules* oriented with dipole moments parallel to the surface *normal* in spite of a preference for isolated water molecules to adsorb with oxygen closest to the surface. This is because the adsorbed layer tends to adopt a structure that maximises the number of hydrogen bonds in the adlayer and packing forces also tend to maximise the number of molecules in each adlayer. This analysis has shown that models of water at the interface with two or more *discrete* orientations are rather unrealistic. On introduction of a surface charge, there is a marked change in orientation, as might be expected, especially once the dipole interaction energy becomes comparable to the hydrogen bonding energy. With a positive charge density on the electrode, the maximum in the orientational distribution of dipole moments moves steadily towards the surface normal and the distribution also narrows; for negative charges, the mean orientation initially becomes more parallel to the surface and then moves to a rather broad distribution with the hydrogen atoms closer to the metal. This preferred orientation persists to some extent in the second water layer but is then lost at larger distances from the electrode surface.
5. These changes in dipole orientation have a strong effect on the electric field in the water layer; even the direction of the field can change in the highly local region of the first two adlayers. Furthermore, the structure of the adlayers nearest the surface severely effects the mobility of the water molecules, which is strongly decreased, especially in the closest adlayer, where it becomes more oscillatory and where reorientation dynamics are also slowed down to give an almost solid-like structure.
6. Calculations using the Car-Parrinello method, in which the electronic structure is calculated quantum mechanically but the water molecules satisfy the classical equations of motion, show that within this framework also there is clear preferred orientation in the first adlayer, and some recent calculations have also suggested the possibility of the first adlayer being split into two distinguishable sublayers, a model originally suggested by some early UHV experiments (Surf. Sci. **123** (1982) 305). However most simulations do not show this effect.

Theoretical studies have also been carried out on ionic solutions at the boundary with electrodes, and it is fair to say that the relatively simple picture painted above, in which ions can be sharply divided into those that specifically adsorb on the surface and those that remain hydrated with their closest distance of approach in the OHP, is not reproduced, especially by the more thoroughgoing calculations. These calculations suggest that a major contribution to the energetics of adsorption is the energy required to remove a water molecule from the surface to create a 'hole' for the ion to adsorb. Molecular dynamic calculations for NaCl solutions carried out by Spohr (Electrochimica Acta **49** (2003) 23 and references therein) also show that the layered water structure close to the electrode surface also effects the ionic charge distribution, with the ionic densities not showing the expected monotonic decrease expected from Goüy-Chapman theory. As might be expected, if the electrode is charged, the counter-ions exhibit the stronger maxima, but there is a tendency, especially in more concentrated solutions, for the ions to 'over-compensate' the surface charge, leading to the appearance of ions of the same charge as the electrode in the next water layer. One unexpected result from these simulations is that as ions approach the electrode surface the mean hydration number remains roughly constant out to ~ 6 but then shows an initial *increase*, particularly for the larger ions, as the ion nears the surface and encounters more ordered adlayers. The coordination number only falls when the ion penetrates into the first adlayer. As with the water molecules themselves, once the ions penetrate the first two water layers, their *lateral* mobility is much reduced.

The central problem is turning these qualitative conclusions into quantitative results is the difficulty of defining a reliable interaction energy between the ions and the metal electrode. If purely electrostatic forces are used, together with an appropriate repulsion term, then cations tend to adsorb preferentially to anions because fewer water molecules are displaced by the smaller cations. This is not in accord with the experimental data described above, and points up the need to describe the electronic interaction much more accurately, perhaps through developments in the Car-Parrinello approach. At this point in time, the ion-electrode interactions can only be described with semi-empirical potentials, and qualitative behaviour of the ions studied through adjustment of the potential parameters. We can anticipate considerable advances in this area over the next few years.

3.5 Potential and Phase Boundary Behaviour at Semiconductor Electrodes

Semiconductors differ significantly in their electrochemical properties from metals, and an understanding of these differences requires us to develop a model of the solid-state bonding of semiconductors in more detail than we have for metals.

3.5.1 Metallic Conductors. Semiconductors and Insulators

Metals are usually characterised by a very high electrical conductivity, which is due to the existence of a large number of electrons freely moving in the metal lattice. These electrons originate from the component atoms of the lattice, and form a state of matter with marked non-classical properties. The origin of the mobility of the electrons lies in the very strong interactions between neighbouring atoms in the lattice. Atomic orbitals on these sites overlap to form a continuum of energy levels, termed a *band*, with associated electron kinetic energies ranging from zero to very high values. In a classical system, all the electrons would occupy the lowest few energy levels, giving rise to a Boltzmann distribution as shown in Fig. 3.25. Increasing the temperature would lead

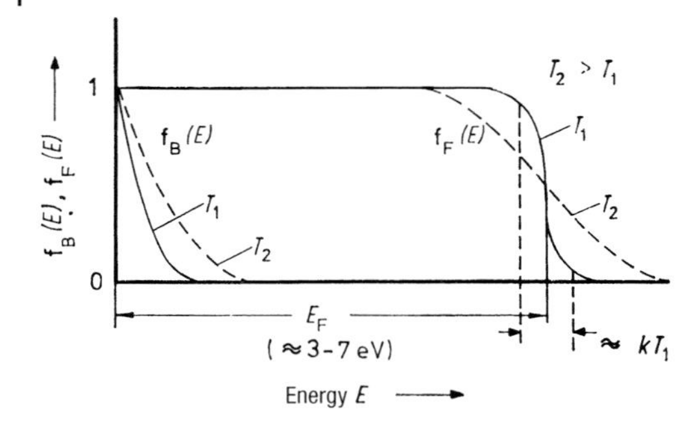

Fig. 3.25 Fermi and Boltzmann distributions at two different temperatures, T_1 and T_2; E_F is the Fermi energy.

to higher energy states becoming populated, but the overwhelming number of electrons would be in the lowest energy states.

In fact, the electron distribution does *not* follow the Boltzmann law, since the occupancy of each level is constrained by Pauli's principle, which in its simplest form restricts the occupancy of each energy orbital to *two* electrons with opposing spins. The effect of this is to force the occupancy of much higher energy levels, as shown also in Fig. 3.25, leading, at a temperature of 0 K, to a rectangular distribution of occupied energy levels. The topmost occupied energy level at 0 K is termed the *Fermi level*, and has an energy E_F relative to the lowest energy state in the band. At higher temperatures, *a small fraction* of the electrons, those nearest the Fermi level, become thermally excited, leading to a spread of occupancy in that energy region; however, the overwhelming majority of electrons are *not* affected by temperature increase unless very high temperatures are accessed.

In the presence of an electric field, *provided the band is only partially occupied*, some electrons can be excited into levels with a preferential motion in the direction of the field. The net effect is that current can flow, *but only if the band is partially occupied.* A typical energy band diagram for a metal is shown in Fig. 3.26; it can be seen that there

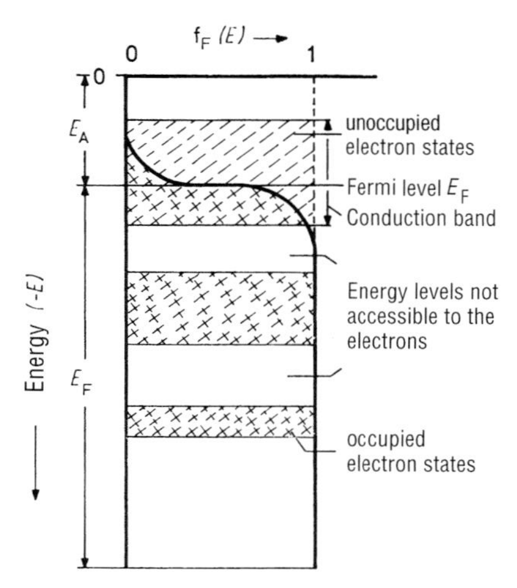

Fig. 3.26 Band scheme and occupation density for metal electrons.

are several *bands* of energy levels separated by unpopulated energy regions. The existence of several bands usually arises from the fact that discrete atomic energy levels are usually well separated in energy, and even after overlap of the constituent atomic orbitals with neighbouring atoms, the resultant band may not be sufficiently wide to cover the entire energy spectrum. A familiar example would be the band derived from the $2p$ atomic orbitals in metallic sodium, which form a band well separated in energy from the $3s$ band responsible for the conductivity.

It can be seen from Fig. 3.26 that the key feature determining the electrical properties of the material is the highest energy band. This is only partially occupied with an electron distribution at finite temperature shown by the cross hatching. *Provided overlap is sufficiently large*, this partial occupancy will then lead to the electrical conductivity described before. It is important to emphasise that not all partially occupied bands will necessarily lead to a high electrical conductivity. When overlap is poor, as between the partially occupied *d*-orbitals in transition-metal oxides, then there may be insufficient energy in the band to overcome electron repulsion effects as the electrons move, with the result that the electrons become effectively localised on specific sites. Well-known examples of this latter behaviour include NiO and CoO, which in fact behave as semiconductors.

The most important difference between semiconductors and metals lies in the band occupancy of the former. Unlike metals, the bands in semiconductors are, at least at 0 K, either completely occupied or completely empty, with the result that the conductivity of these materials near 0 K is extremely low. At finite temperatures, two mechanisms may exist to enhance the conductivity substantially. If the energy difference between the upper edge of the highest-energy fully occupied band, *the valence band*, and the lower edge of the lowest-energy unoccupied band, the *conduction band*, is small ($<$ ca. 1.0 eV), then *intrinsic* thermal excitation of electrons from the valence to the conduction bands can take place even at room temperature. The electrons in the conduction band will now be mobile; for the reasons given above, they will also be able to sustain a net current in the presence of an electric field. In addition, removal of electrons from the valence band *also leaves a partially filled band* and this can sustain electrical current since *the vacancies left behind in the valence band can be considered as positively charged 'holes' moving in a immobile sea of electrons*. The mobility of these holes is usually similar to the electrons in the topmost energy levels of the valence band, but their assumed charge means that they will migrate in a direction opposite to that of the electrons in the conduction band.

Obviously, for this first intrinsic conduction mechanism, there must be equal numbers of electrons and holes formed by thermal excitation, but a second extrinsic conduction mechanism exists that depends on the introduction of atoms of *different valence* into the lattice. If, for example, an atom of phosphorus is introduced into a silicon lattice, it will have one more valence electron than silicon normally has, and it can lose this electron very easily to the conduction band, leaving behind a positively charged P ion. The P atom is said to act as a *donor*, but since the ion left behind is *immobile*, there is now a marked preponderance of electrons, giving rise to *n*-type conductivity. In a similar way, atoms with a valence *lower* than silicon, such as boron, can be introduced. In this case, the B atom effectively traps an electron from the

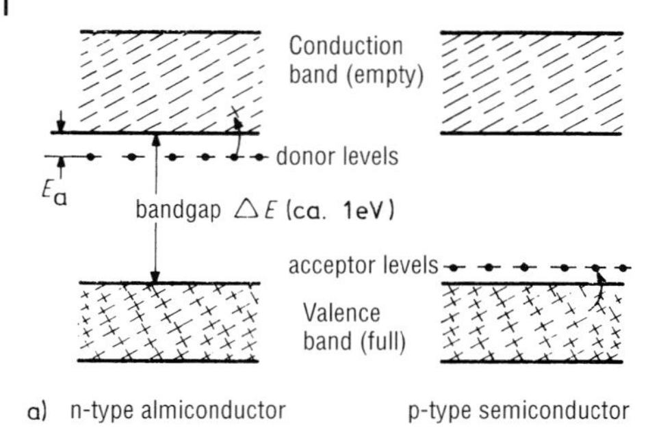

Fig. 3.27 Band scheme for a doped semiconductor with (a) excess electrons; (b) excess holes.

valence band, leaving behind a vacancy or *hole* in the band and leading to *p*-type conductivity.

This mechanism can more clearly be seen with the band schemes of Fig. 3.27; the donor levels are at an energy E_a below the conduction-band edge, and at 0 K are fully occupied. At higher temperatures, electron excitation takes place, and for P in silicon, this ionisation is complete well below 100 K. At room temperature, then, all these donor sites are empty, and the number of electrons in the conduction band is essentially equal to the number of donor sites. This number is quite variable: extremely pure silicon can now be made, and even at donor levels of 10^{13} cm^{-3}, corresponding to one P in ca. 10^9 silicon atoms, there are easily measurable effects on the conductivity. Normally, donor levels between 10^{15} and 10^{18} cm^{-3} would be encountered, giving conductivities ca. 10^7 to 10^4 lower than those encountered in metals. The Fermi level in semiconductors normally lies between conduction and valence bands. For *intrinsic* semiconductors (i.e. those relying on direct thermal excitation from valence to conduction band), the Fermi level will, in fact, lie close to the mid-point between the two bands, but for *extrinsic* conduction, the Fermi energy is found to lie close to one of the band edges. For *n*-type semiconductors, the Fermi level usually lies just below the conduction band edge, but may, for higher doping levels, actually fall inside the conduction band itself. In a similar way, for *p*-type extrinsic semiconductors, the Fermi level normally lies between the acceptor levels and the top of the valence band edge, as shown in Fig. 3.27.

3.5.2
Electrochemical Equilibria on Semiconductor Electrodes

A primary consequence of the limited number of conduction electrons in semiconductors is that the double-layer region has a completely different potential distribution, particularly if the potential on the electrode is such as to lead to a *decrease* in majority carriers near the surface. At a metal electrode, the electrical double layer is characterized by a potential drop mainly *on the solution side* associated with an inhomogeneous distribution of electrolyte ions of different charge, whilst on the metal side, the nett charge associated with this distribution is fully compensated by electronic charge density essentially localised at the metal surface. In semiconductors, the concentration of

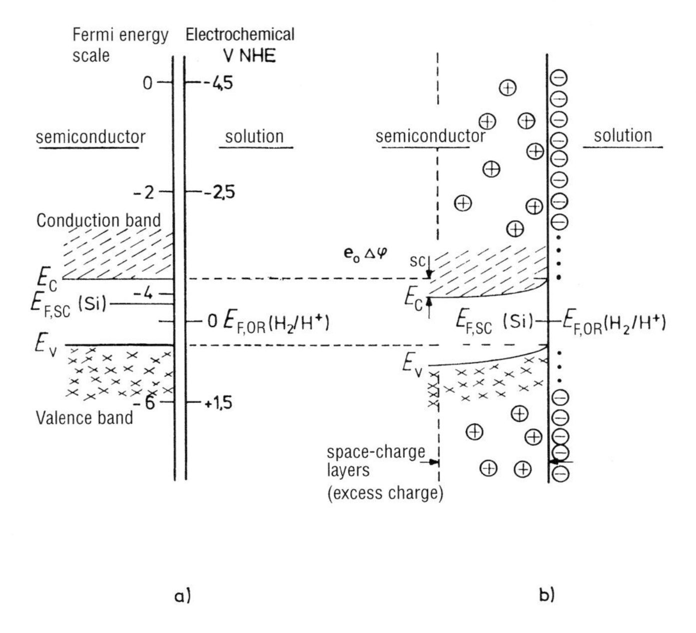

Fig. 3.28 Band scheme of a semiconducting electrode: (a) semiconductor and electrolyte phase independent of each other; (b) band bending through electrolyte contact. $E_{F,sc}$ is the Fermi level of the semiconductor; $E_{F,OR}$ is the Fermi level of the redox couple in solution; E_C and E_V are the energies of conduction and valence bands.

charge carriers is not only lower than in a metal, it is also lower than the concentration of ions in the electrolyte This low charge density in the semiconductor means that in the absence of localised surface charge arising from surface electronic states, the main part of the potential drop at the interface occurs within the electrode itself, as a consequence of the Poisson equation (eq. 3.130) and only a small part of the potential change across the interface occurs in the solution.

The corresponding charge distribution when a semiconductor and electrolyte are brought together is shown in Fig. 3.28, for an *n*-type Si semiconductor and an aqueous electrolyte with the main electrochemical couple being the H_2/H^+ couple as shown. Were the electrode a metal, and the kinetics of the hydrogen couple sufficiently rapid to dominate the equilibrium, the metal would adopt a potential close to 0 V. Given that the Fermi level corresponds in a metal approximately to the highest occupied levels in the band, the condition for equilibrium can be thought of as equalising the electrochemical potential of the electrons on both sides of the interface:

$$\tilde{\mu}_{e^-,M} = \tilde{\mu}_{e^-,S} \tag{3.153}$$

This, in turn, means that the energy of the electrons in the highest occupied levels of the band must have the same free energy as the redox couple. There is a conceptual problem here, which arises because the Fermi energy of a solid is usually defined with respect to the 'vacuum level', i.e. the energy of an electron removed a macroscopic distance from the solid and at rest zero potential energy, whereas the free energy of the redox couple is referenced to the standard hydrogen electrode. As explained in Section 3.1, it is not possible formally to *measure* the absolute energy of a redox couple in solution against the vacuum level, but it is possible to estimate this energy *theoretically* with moderate confidence, and the Fermi level of the SHE is believed to be close to $-$ 4.5 eV vs. vacuum level. Since the absolute positions of the electronic bands in silicon are well known from photoelectron studies, this allows us to put the energy scales in solid and liquid on the same basis, and this is shown in Fig. 3.28(a).

Provided the electron transfer from semiconductor to the H^+/H_2 redox couple remains the dominant kinetic process, once again we expect that the Fermi level of the semiconductor will equalise with the H^+/H_2 level when the silicon and electrolyte are placed in contact. This process is shown in Fig. 3.28(b), and it is evident that since $E_F(Si) > E_F(H^+/H_2)$, the semiconductor must acquire a positive charge relative to the solution to equalise the Fermi levels. As pointed out before, the potential distribution at the interface now extends over a substantial spatial region inside the semiconductor, *the space-charge layer*. Furthermore, to a good approximation in this idealised example, the energy of the conduction band at the *surface does not alter as the potential on the semiconductor electrode changes*. The potential accommodated inside the semiconductor, $\Delta\varphi_{sc}$, can immediately be seen in Fig. 3.28 from the fact that the Fermi level inside the semiconductor *must be at a constant energy*. If this were not the case, electrons would migrate until the Fermi level everywhere were equalised. Thus, the energy difference between $E_F(Si)$ and the conduction band at the surface will be larger (in our example) than that energy difference in the bulk by an amount equal to $e_0\Delta\varphi_{sc}$, the space-charge potential.

Finally, it is evident that there will be a potential, usually denoted by V_{fb}, at which there will be *no* potential drop inside the semiconductor. This potential is termed the *flat-band potential*, and its significance is that it can be measured experimentally using standard *ac* techniques as described elsewhere. All potentials can be referred to this flat-band potential, allowing us to determine the absolute potential drop in the semiconductor.

It should be noticed that the potential distribution shown in Fig. 3.27 does not hold if, for example, a *negative* potential is applied to an n-type semiconductor electrode or where the Nernst potential of the kinetically dominant redox couple lies within the conduction band energy levels. This is because the concentration of electrons at and near the surface can now build up to a level determined by the density of states in the conduction band, and this is comparable to that in a metal. This is termed *forward bias* in the physics literature, and in forward bias, the semiconductor electrode will have a potential distribution at the interface more reminiscent of a metal.

3.6
Simple Applications of Potential Difference Measurements

3.6.1
The Experimental Determination of Standard Potentials and Mean Activity Coefficients

The experimental determination of standard EMFs leads to a series of physico-chemical and thermodynamic quantities such as solubility products, acid and base dissociation constants and activity coefficients, and we will exemplify this type of measurement with reference to the silver-silver chloride electrode

$$AgCl + e^- \rightleftharpoons -Ag + Cl^- \tag{3.154}$$

which can be studied in cells of the form

$$Pt|H_2|HCl|AgCl|Ag \tag{3.155}$$

The potentials of the silver-silver-chloride and hydrogen electrodes are given by equations of the form

$$E_{Ag|AgCl|Cl^-} = E^0_{Ag|AgCl|Cl^-} - (RT/F)\ln a_{Cl^-} \tag{3.156}$$

and (for $p_{H_2} = 1$ atm.)

$$E_{H_2|H^+} = E^0_{H_2|H^+} + (RT/F)\ln a_{H_3O^+} \tag{3.157}$$

By placing both hydrogen and silver-silver chloride electrodes in the same solution, a diffusion potential will not form, and the measured potential of the cell, E is

$$\begin{aligned} E &= E_{Ag|AgCl|Cl^-} - E_{H_2|H^+} \\ &= E^0_{Ag|AgCl|Cl^-} - E^0_{H_2|H^+} - \frac{RT}{F}\ln\left[a_{Cl^-}\cdot a_{H_3O^+}\right] \end{aligned} \tag{3.158}$$

From the discussion of standard electrodes in Section 3.1, we recall that $E^0_{H_2|H^+} = 0$ by definition, so finally

$$E = E^0_{Ag|AgCl|Cl^-} - \frac{RT}{F}\ln\left[a_{Cl^-}\cdot a_{H_3O^+}\right] \tag{3.159}$$

or, from the definition of the mean ionic activity of the HCl solution, $a_\pm^2 = a_{H_3O}\cdot a_{Cl^-}$ and

$$E = E^0_{Ag|AgCl|Cl^-} - \frac{2RT}{F}\ln a_\pm^{HCl} \tag{3.160}$$

Using the equation $a_\pm^{HCl} = \gamma_\pm^{HCl} m_{HCl}/m^0$, we finally obtain, at 298 K,

$$E + 0.05136\ln(m_{HCl}/m^0) = E^0_{Ag|AgCl|Cl^-} - 0.05136\ln\gamma_\pm^{HCl} \tag{3.161}$$

From equation (2.76) for the Debye-Hückel limiting law, it will be seen that $\ln\gamma_\pm$ is proportional to $\sqrt{m/m^0}$. The measurement of E at different concentrations of HCl, provided that some of these are low enough for the Debye-Hückel law to hold, allow us to determine $E^0_{Ag|AgCl|Cl^-}$ as the intercept of the limiting slope of a plot of

$E + 0.05136 \ln (m_{HCl}/m^0)$ vs. $\sqrt{m_{HCl}/m^0}$ as shown in Fig. 3.29. The value obtained by this technique is 0.2224 V vs. SHE. If data from higher concentrations are also incorporated, it becomes necessary to use a better approximation than the limiting Debye-Hückel law, and one of the empirical equations (2.81) or (2.82) can be employed.

Mean activity coefficients can be determined from this same analysis. Transformation of equation (3.161) allows us to write

$$0.05136 \ln \gamma_{\pm}^{HCl} = E^0_{Ag|AgCl|Cl^-} - E - 0.05136 \ln(m_{HCl}/m^0) \tag{3.162}$$

Provided $E^0_{Ag|AgCl|Cl^-}$ has already been determined, then measurements of E allow us to calculate the mean activity coefficient, $\gamma_{\pm}^{HCl}$. The resultant values of $\gamma_{\pm}^{HCl}(c)$ have already been tabulated in Table 2.10, and the process described can be extended to the determination of the mean activity coefficient of *any* electrolyte that can be incorporated into this type of cell, with one cation-responsive and one anion-responsive electrode.

In many cases, cells cannot be constructed between the NHE and the electrode whose standard potential is of interest without incorporating a salt bridge. A simple example is the $Cu|Cu^{2+}$ electrode; a hydrogen electrode cannot be placed directly into the Cu^{2+} solution, since Cu^{2+} is reduced to copper metal at such an electrode. In such cases, given the errors involved in measurements with liquid-junction potential, another standard electrode must be used whose potential with respect to the NHE is accurately known. Thus, in the case of the $Cu|Cu^{2+}$ electrode, the following cell can be used:

$$Ag|AgCl(s)|CuCl_2\ (aq)|Cu \tag{3.163}$$

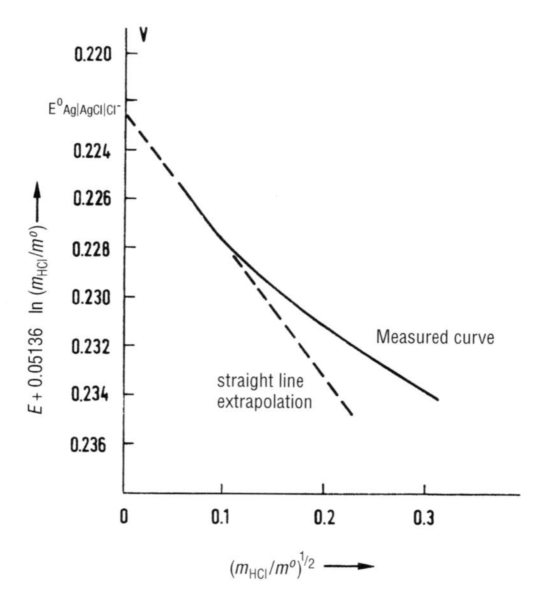

Fig. 3.29 The determination of the standard potential of the silver-silver chloride electrode from measurements with a $Ag|AgCl(s)|HCl|Pt|H_2$ cell at concentrations of HCl in the range 0.0025M to 0.01M.

whose EMF is

$$\begin{aligned} E &= E_{Cu|Cu^{2+}} - E_{Ag|AgCl|Cl^-} \\ &= E^0_{Cu|Cu^{2+}} - E^0_{Ag|AgCl|Cl^-} + \frac{RT}{2F}\ln a_{Cu^{2+}} + \frac{RT}{F}\ln a_{Cl^-} \\ &= E^0_{Cu|Cu^{2+}} - E^0_{Ag|AgCl|Cl^-} + \frac{3RT}{2F}\ln a_{\pm}^{CuCl_2} \end{aligned} \tag{3.164}$$

since, from Section 2.5.1, $\left(a_{\pm}^{CuCl_2}\right)^3 = a_{Cu^{2+}} \cdot \left(a_{Cl^-}\right)^2$.

From the expression $a_{\pm}^{CuCl_2} = \gamma_{\pm}^{CuCl_2}\left[\left(m_{Cu^{2+}}/m^0\right) \cdot \left(m_{Cl^-}/m^0\right)^2\right]^{\frac{1}{3}}$, we finally obtain

$$\begin{aligned} &E + E^0_{Ag|AgCl|Cl^-} - \frac{3RT}{2F}\ln\left\{\left[\left(m_{Cu^{2+}}/m^0\right) \cdot \left(m_{Cl^-}/m^0\right)^2\right]^{\frac{1}{3}}\right\} = \\ &E_{Cu|Cu^{2+}} + \frac{3RT}{2F}\ln \gamma_{\pm}^{CuCl_2} \end{aligned} \tag{3.165}$$

from which $E_{Cu|Cu^{2+}}$ can be obtained by the same sort of plot as before, provided a concentration region can be accessed for which (2. 81) or (2.82) is valid.

3.6.2
Solubility Products of Slightly Soluble Salts

Equation (3.156) permits us to determine the solubility product of AgCl from standard potential measurements, since

$$\frac{RT}{F}\ln K_s^{AgCl} = E^0_{Ag|AgCl|Cl^-} - E^0_{Ag|Ag^+} \tag{3.166}$$

In fact, the thermodynamic solubility product of any solid that can form an electrode of the second kind can be determined by measurement of the corresponding standard potential, provided the standard potential of the corresponding metal-ion electrode is known. With $E_{Ag|AgCl|Cl^-} = 0.2224$V vs. SHE and $E_{Ag|Ag^+} = 0.7996$ V vs. SHE (see Table 3.2), the solubility product of AgCl can be calculated to be 1.784×10^{-10} at 25°C.

3.6.3
The Determination of the Ionic Product of Water

The ionic product of water, K_w, is defined as

$$K_W = a_{H_3O^+} \cdot a_{OH^-} \tag{3.167}$$

and can be determined from a cell of the form

$$Pt|H_2(p = 1\text{ atm})|KOH(a_{OH^-}), KCl(a_{Cl^-})|AgCl|Ag \tag{3.168}$$

The EMF of this cell, E, is given by

$$\begin{aligned} E &= E_{Ag|AgCl|Cl^-} - E_{H^+|H_2} \\ &= E^0_{Ag|AgCl|Cl^-} - \frac{RT}{F}\ln a_{Cl^-} - \frac{RT}{F}\ln a_{H_3O^+} \end{aligned} \tag{3.169}$$

From the definition of K_w, we have

$$E = E^0_{Ag|AgCl|Cl^-} - \frac{RT}{F}\ln a_{Cl^-} + \frac{RT}{F}\ln a_{OH^-} - \frac{RT}{F}\ln K_W \tag{3.170}$$

whence

$$\frac{F}{RT}(E - E^0_{Ag|AgCl|Cl^-}) + \ln\left\{\frac{(m_{Cl^-}/m^0)}{(m_{OH^-}/m^0)}\right\} = -\ln K_W - \ln\frac{\gamma_{Cl^-}}{\gamma_{OH^-}} \tag{3.171}$$

and a plot of the left-hand side of (3.171) against ionic strength will yield $\ln K_w$ as intercept, since the logarithmic term on the right-hand side tends to zero rapidly as $I \to 0$. At 25°C, the value of $K_w = 1.008 \times 10^{-14}$ is obtained, very similar to that obtained from the residual conductivity of pure water discussed above.

3.6.4
Dissociation Constants of Weak Acids

The dissociation constants of a number of weak acids may be obtained from cells of the form

$$Pt|H_2(p = 1\text{ atm.})|HA, NaA, NaCl|AgCl|Ag \tag{3.172}$$

in which the hydrogen ions that determine the potential of the hydrogen electrode are supplied by dissociation of the acid, and the molality of the chloride ions determines the potential of the silver-silver chloride electrode. The measured EMF of this cell, E is given by

$$\begin{aligned} E &= E_{Ag|AgCl|Cl^-} - E_{H^+|H_2} \\ &= E^0_{Ag|AgCl|Cl^-} - \frac{RT}{F}\ln a_{Cl^-} - \frac{RT}{F}\ln a_{H_3O^+} \\ &= E^0_{Ag|AgCl|Cl^-} - \frac{RT}{F}\ln\left\{\gamma_{H_3O^+}\cdot\gamma_{Cl^-}\cdot(m_{H_3O^+}/m^0)(m_{Cl^-}/m^0)\right\} \end{aligned} \tag{3.173}$$

and from the definition of the acid dissociation constant, K_a^{HA}

$$K_a^{HA} = \frac{\gamma_{H_3O^+}\cdot\gamma_{A^-}\cdot(m_{H_3O^+}/m^0)(m_{A^-}/m^0)}{[\gamma_{HA}\cdot(m_{HA}/m^0)]} \tag{3.174}$$

we have, on rearranging (3.173) and incorporating (3.174)

$$E - E^0_{\text{Ag|AgCl|Cl}^-} + \frac{RT}{F}\ln\left\{\frac{(m_{\text{HA}}/m^0)(m_{\text{NaCl}}/m^0)}{(m_{\text{NaA}}/m^0)}\right\} = -\frac{RT}{F}\ln K_a^{\text{HA}} - \frac{RT}{F}\ln\{\gamma_{\text{Cl}^-}\cdot\gamma_{\text{HA}}/\gamma_{\text{A}^-}\} \quad 3.175$$

where we have assumed that the acid is weak so that $m_{A^-} = m_{\text{NaA}} + m_{\text{H}^+} \approx m_{\text{NaA}}$ and m_{HA} is the molality of the initially added weak acid. The right hand side of equation (3.175) will show only a slight dependence on ionic strength, arising from the different radii of Cl^- and A^-. At sufficiently dilute solution, the right-hand side will, in fact, become independent of ionic strength, and extrapolation to $I \rightarrow 0$ is straightforward. This cell, sometimes referred to as a 'Harned' cell in the literature, has been extensively used to find K_a^{HA} values for weak acids: as an example, for acetic acid the value obtained for K_a^{HA} simply by neglecting the activity coefficient term and working with molalities of ~ 0.05 mol kg^{-1}, a value of 1.729×10^{-5} is found. By extrapolation to zero ionic strength, this was refined to 1.754×10^{-5}, in very good agreement with that obtained by conductivity measurements. The cell can be used for stronger acids such as formic acid, for which our assumption that $m_{A^-} \approx m_{\text{NaA}}$ is no longer valid. In this case, a series of successive approximations is necessary.

This type of cell may also be used for determining the acid dissociation constant of the ammonium ion, NH_4^+:

$$\text{NH}_4^+ \rightleftharpoons \text{NH}_3 + \text{H}^+; \quad K_a^{am} = a_{\text{H}^+}\cdot\frac{a_{\text{NH}_3}}{a_{\text{NH}_4^+}}$$

The cell $\text{Pt}|\text{H}_2(p = 1\text{atm.})|\text{NH}_4\text{Cl}\ (m_1),\ \text{NH}_3\ (m_2),\ \text{KCl}\ (m_3)|\text{AgCl}|\text{Ag}$ has the *EMF*

$$\begin{aligned}
E &= E^0_{\text{Ag|AgCl|Cl}^-} - \frac{RT}{F}\ln\{a_{\text{H}^+}\cdot a_{\text{Cl}^-}\}\\
&= E^0_{\text{Ag|AgCl|Cl}^-} - \frac{RT}{F}\ln\{\gamma_{\text{H}^+}\cdot\gamma_{\text{Cl}^-}\cdot(m_{\text{H}^+}/m^0)(m_{\text{Cl}^-}/m^0)\}\\
&= E^0_{\text{Ag|AgCl|Cl}^-} - \frac{RT}{F}\ln K_a^{am} - \frac{RT}{F}\ln\left\{\frac{(m_{\text{NH}_4^+}/m^0)(m_{\text{Cl}^-}/m^0)}{(m_{\text{NH}_3}/m^0)}\right\} - \\
&\quad - \frac{RT}{F}\ln\left\{\frac{\gamma_{\text{NH}_4^+}\cdot\gamma_{\text{Cl}^-}}{\gamma_{\text{NH}_3}}\right\}\\
&= E^0_{\text{Ag|AgCl|Cl}^-} - \frac{RT}{F}\ln K_a^{am} - \frac{RT}{F}\ln\left\{\frac{m_1/m^0)([m_1 + m_3]/m^0)}{(m_2/m^0)}\right\} - \\
&\quad - \frac{RT}{F}\ln\left\{\frac{\gamma_{\text{NH}_4^+}\cdot\gamma_{\text{Cl}^-}}{\gamma_{\text{NH}_3}}\right\}
\end{aligned}$$

By working in the molality region 0.01 to 0.1 mol kg^{-1}, where Guggenheim's formula (2.82) is valid, an extrapolation to zero ionic strength will yield K_a^{am} to very high accuracy.

3.6.5
The Determination of the Thermodynamic State Functions ($\Delta_r G^0$, $\Delta_r H^0$ and $\Delta_r S^0$) and the Corresponding Equilibrium Constants for Chemical Reactions

In Section 3.1, we saw that the connection between the standard free energy of a reaction and the *EMF* of a galvanic cell with the same overall chemical reaction was

$$\Delta_r G^0 = -nFE^0 \tag{3.176}$$

where the reaction takes place with both reactants and products in their standard states (gas pressures of 1 atmosphere; unit mean ionic activities).

In a similar way, the entropy change for a chemical reaction may be obtained from the EMF value of the galvanic cell with the same overall cell reaction from the equation

$$\Delta_r S^0 = nF \cdot \left(\frac{\partial E^0}{\partial T}\right)_p \tag{3.177}$$

For the standard reaction enthalpy $\Delta_r H^0$, we have, from the relationship $\Delta_r G^0 = \Delta_r H^0 - T\Delta_r S^0$

$$\Delta_r H^0 = -nF \cdot \left\{ E^0 - T\left(\frac{\partial E^0}{\partial T}\right)_p \right\} \tag{3.178}$$

Finally, the thermodynamic equilibrium constant, K_{eq}, of a chemical reaction can also be determined from the EMF of the corresponding electrochemical cell, since

$$\Delta_r G^0 = -RT \ln K_{eq} \tag{3.179}$$

from which we find

$$K_{eq} = \exp\left\{\frac{nFE^0}{RT}\right\} \tag{3.180}$$

To exemplify this, we consider a specific reaction, that of the Daniell Cell

$$\mathrm{Zn}|\mathrm{Zn}^{2+}(\mathrm{aq.})||\mathrm{Cu}^{2+}(\mathrm{aq.})|\mathrm{Cu} \tag{3.181}$$

for which the corresponding electrode reactions are:

$$\begin{array}{r} \mathrm{Cu}^{2+} + 2e^- \rightarrow \mathrm{Cu}^0 \qquad\qquad \\ \mathrm{Zn}^0 \rightarrow \mathrm{Zn}^{2+} + 2e^- \\ \hline \mathrm{Cu}^{2+} + \mathrm{Zn}^0 \rightarrow \mathrm{Cu}^0 + \mathrm{Zn}^{2+} \end{array} \tag{3.182}$$

A value of 1.103 V can be determined for the standard EMF of such a cell using the methods described in Section 3.6.1, from which the free energy of the reaction $\mathrm{Cu}^{2+} + \mathrm{Zn}^0 \rightarrow \mathrm{Cu}^0 + \mathrm{Zn}^{2+}$ can be calculated (with $n = 2$) as:

$$\Delta_r G^0 = -2 \times 96485 \times 1.103 = -212.846\ \mathrm{kJmol}^{-1} \tag{3.183}$$

The temperature dependence of the standard EMF of the Daniell cell in the temperature region near 25°C can be measured to be $-0.83 \cdot 10^{-4}$ VK^{-1}. From (3.177) we find

$$\Delta_r S^0 = 2 \times 96485 \times (-0.83410 - 4) = -16.02 \text{ JK}^{-1}\text{mol}^{-1} \tag{3.184}$$

and from equation (3.178) for $\Delta_r H^0$

$$\Delta_r H^0 = \Delta_r G^0 + 298 \times \Delta_r S^0 = -212867 - 4774 = \\ -217641 \text{ Jmol}^{-1} \equiv -217.641 \text{ kJmol}^{-1} \tag{3.185}$$

and finally the equilibrium constant is given by

$$K_{eq} = \exp\{2 \times 96485 \times 1.103/(8.314 \times 298)\} = 2.041 \cdot 10^{37} \tag{3.186}$$

3.6.6
pH Measurement with the Hydrogen Electrode

The concept of pH was introduced briefly in chapter 2, but we now wish to extend the definition and consider in more detail the problems inherent in the pH scale. If the activity of the H_3O^+ ion appears in the overall chemical equilibrium governing a galvanic cell, the equilibrium potential of that cell can be used, at least in principle, as a measure of the pH-value. The simplest, and in many ways the most important of such equilibria is that of the hydrogen electrode $Pt|H_2|H_3O^+$ with an electrode reaction

$$H_3O^+ + e^- \rightleftharpoons {}^1/_2 H_2 + H_2O \tag{3.187}$$

For the hydrogen electrode equilibrium potential at $p = 1$ atm., $E^0_{H^+|H_2} = 0$.

$$E_{H^+|H_2} = \frac{RT}{F} \ln a_{H_3O^+} \text{ vs.SHE} \tag{3.188}$$

$$\equiv -0.0591 \text{ pH at } 298 \text{ K} \tag{3.189}$$

where we *define* $\text{pH} = -\log_{10} a_{H_3O^+}$.

In order to use this type of electrode for the determination of the pH of an unknown solution, pH_x, it must be combined with a suitable reference electrode to form a complete electrochemical cell. Thus, if we use a saturated calomel electrode $Hg|Hg_2Cl_2|Cl^-$ (saturated KCl) as the reference electrode, then the corresponding cell takes the form

$$Hg|Hg_2Cl_2|KCl(sat.)\,\|\,\text{solution of pH} = pH_x|H_2|Pt \tag{3.190}$$

It is evident that this cell will have a liquid-liquid junction potential, $\Delta\varphi_{diff}$ between the reference electrode solution and the solution whose pH is to be determined. This junction potential can be minimised as described in Section 3.2, as denoted by the symbol $\|$ in (3.190), and the cell is shown schematically in Fig. 3.30. The EMF of cell (3.190) is evidently

$$E = E_{\mathrm{H^+|H_2}} - E_{\mathrm{SCE}} + \Delta\varphi_{diff} = -0.0591\ \mathrm{pH}_x - E_{\mathrm{SCE}} + \Delta\varphi_{diff} \tag{3.191}$$

and

$$\mathrm{pH}_x = -\frac{E + E_{\mathrm{SCE}} - \Delta\varphi_{diff}}{0.0591} \tag{3.192}$$

As an example, if the measured EMF is $-$ 0.8627 V, E_{SCE} is taken as 0.2415 V (see Table 3.3), and $\Delta\varphi_{diff}$ has been reduced to below $\pm$ 2 mV, then

$$\mathrm{pH}_x = -(-0.8627 + 0.2415 \pm 0.002)/0.0591 = 10.51 \pm 0.03 \tag{3.193}$$

Although this is apparently satisfactory, there are some real problems with this approach that must be addressed. The definition:

$$\mathrm{pH} \equiv -\log_{10} a_{\mathrm{H_3O^+}} \tag{3.194}$$

is clearly purely notional since single-ion activities cannot be determined experimentally. Furthermore, in order to carry out a pH measurement in practice, avoiding the use of E_{SCE}, the equipment shown in Fig. 3.30, consisting of the cell:

$$\mathrm{Hg|Hg_2Cl_2|KCl}_{aq}\ \text{standard buffer solution}\mathrm{|H_2|Pt} \tag{3.195}$$

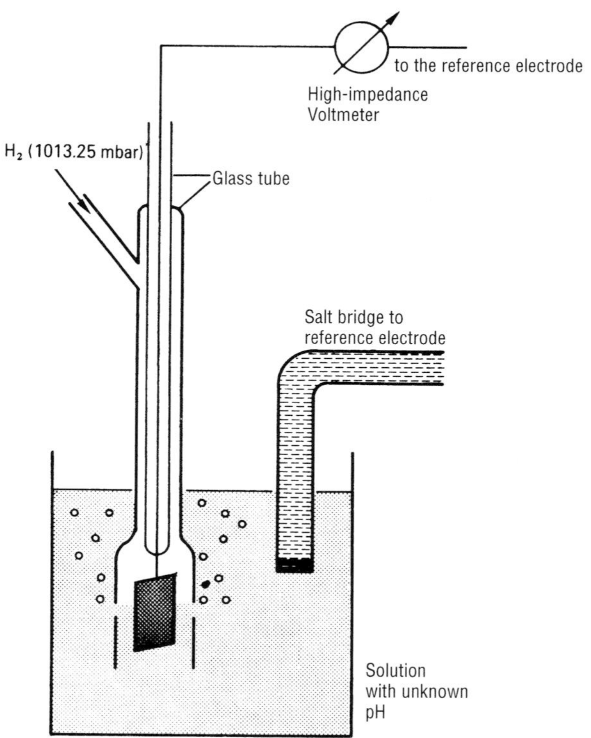

Fig. 3.30 Measurement of pH with a hydrogen electrode.

is first calibrated with one or more standard buffer solutions before a reading is taken with the unknown replacing the standard buffer. The pH of the unknown is then calculated with respect to the standard by assuming that a shift of – 59.1 mV (at 25°C) corresponds to a positive change in pH of unity. This process is usually automated nowadays by using a high-impedance voltmeter directly calibrated in pH. The calibration is carried out in the easily accessible pH range from 2–12 using commercially available standard buffer solutions, and the effects of working at different temperatures can easily be incorporated into the measurement with the aid of compensation from a calibrated dial on the equipment.

If the activity coefficients in the standard buffer solutions and the unknown are very close to each other (which will be the case if the standard and unknown(s) have similar ionic compositions and concentrations) and liquid-junction potentials cancel, then the measurement will correspond to the operational pH scale, so-called because this scale is defined in terms of the operation of the cell. However, this operational scale still leaves open the question of how we establish the connection between the composition of the standard buffers and the associated pH-values.

In order to do this, we need some convention that will allow us to circumvent the problem of single-ion activities. To do this, we return to the Harned cell described in Section 3.6.4. In effect, this cell contains a buffer solution together with a solution of known molality of Cl^- ions. If we write the cell in the form

$$\mathrm{Ag|AgCl|KCl,\,HA,\,KA|H_2|Pt} \tag{3.196}$$

then the EMF has the form

$$\begin{aligned} E &= -E^0_{\mathrm{Ag|AgCl|Cl^-}} + \frac{2.303RT}{F}\log_{10}(a_{\mathrm{H_3O^+}}\cdot a_{\mathrm{Cl^-}}) \\ &= -E^0_{\mathrm{Ag|AgCl|Cl^-}} - \frac{2.303RT}{F}\mathrm{pH} + \frac{2.303RT}{F}\log_{10}(m_{\mathrm{Cl^-}}/m^0)\cdot\gamma_{\mathrm{Cl^-}} \\ &= -E^0_{\mathrm{Ag|AgCl|Cl^-}} - 0.0591\mathrm{pH} + 0.0591\log_{10}(m_{\mathrm{Cl^-}}/m^0)\cdot\gamma_{\mathrm{Cl^-}} \end{aligned} \tag{3.197}$$

where, at 298 K, 2.303 RT/F =0.0591 V. In (3.197), the only quantity, aside from pH, that is not known is γ_{Cl^-}. It is now conventional, following Bates and Guggenheim, to use the extended Debye-Hückel expression for this activity coefficient (2.81) with an assumed value for the radius of the ion. Referring back to (2.81), it is evident that we can write

$$\ln\gamma_{\mathrm{Cl^-}} = -\frac{A\mathrm{I}^{\frac{1}{2}}}{1 + Ba_0\mathrm{I}^{\frac{1}{2}}} \tag{3.198}$$

where A and B are constants and a_0 is the radius of the ion. The Bates-Guggenheim convention is to take the product Ba_0 as equal to 1.5 for the chloride ion. Thus, in the region of ionic strength where (3.198) is valid (up to 0.1) the value of the pH can be determined for a range of buffer solutions, and standard solutions devised with accurately known pH values on the Bates-Guggenheim scale.

Thus, returning to the cell in (3.190), we can now see that the determination of the pH of a solution involves (i) the preparation of standard buffer solutions whose pH has been calculated from measurements in a Harned cell and use of the Bates-Guggenheim convention; (ii) measurement of the EMF, E_s, of the cell (3.190) in which the solution is a standard whose pH (= pH_s) is close to that of the unknown solution; (iii) measurement of the EMF, E_x, of the cell in which the standard solution is replaced by the unknown of $pH = pH_x$. Then

$$E_S = -0.0591 \cdot pH_S - E_{SCE} + \Delta\varphi_{diff} \quad (3.199)$$

$$E_x = -0.0591 \cdot pH_x - E_{SCE} + \Delta\varphi'_{diff} \quad (3.200)$$

Provided the values of $\Delta\varphi_{diff}$ and $\Delta\varphi'_{diff}$ are very similar for the two solutions, then

$$pH_x = pH_S + (E_S - E_x)/0.0591 \quad (3.201)$$

which encapsulates the *operational* definition of pH.

The hydrogen electrode of cell (3.190) can be used as a pH sensor over the complete range of pH values; it does not, however, work satisfactorily in all solutions. Thus, solutions containing CN^- ions, H_2S or As(III) disturb the electrode by poisoning it. Also, certain heavy-metal ions, nitrate ions and certain organic substances are all reduced by hydrogen in the presence of finely divided platinum, as found in platinum black, and this reduction process alters the pH in the neighbourhood of the electrode. However, by attention to the points above, and by frequent re-platinisation (re-deposition of a fresh finely-divided platinum surface) and accurate control of the measurement temperature, reliable and reproducible results can be obtained with the hydrogen electrode in many aqueous solutions.

In addition to the hydrogen electrode, there are other redox electrodes or electrodes of the second type that can be used as pH sensors. In alkaline media, the pH sensitivity of the mercury-mercury oxide electrode can be exploited:

$$HgO + H_2O + 2e^- \rightleftharpoons Hg + 2OH^- \quad (3.202)$$

with $E^0 = +0.097$ V vs. SHE (cf. Table 3.3).

The electrode predominantly applicable in acid solutions in the quinhydrone-electrode, $Pt|C_6H_4O_2, C_6H_4(OH)_2|H_3O^+$, with electrode reaction

$$O{=}C_6H_4{=}O + 2H_3O^+ + 2e^- \longrightarrow HO{-}C_6H_4{-}OH + 2H_2O$$

for which $E^0 = 0.899$ V vs. SHE.

However, none of these electrodes, or the hydrogen electrode itself, is any longer in daily use in research laboratories and in industry. All have been replaced almost universally by the glass pH-electrode, which is described in the next section.

3.6.7
pH Measurement with the Glass Electrode

If a glass surface (e.g. formed from a SiO_2-CaO-Na_2O glass) is immersed in water, some of the cations, which are bound in the silica network, are exchanged with

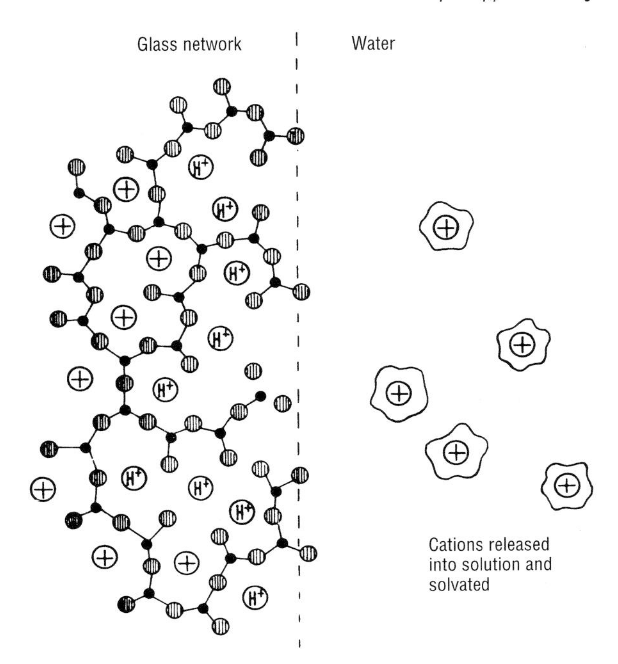

Fig. 3.31 Schematic representation of the exchange of cations for protons in the surface region of a glass membrane; shaded circles, oxygen; solid circles, silicon; circles with + charges, unsolvated cations in the glass or solvated cations in solution.

H_3O^+ ions in a thin layer at the surface. This so-called "swelling process", illustrated in Fig. 3.31, is complete after 24–48 hours and the resultant "swollen layer" is termed the Haugaard Layer (HL): it is 5 to 500 nm thick.

If the HL is now brought into contact with a solution containing H_3O^+ ions, the activity, $a_{H_3O^+}$, of the hydrogen ions and their chemical potentials, $\mu_{H_3O^+}$, in the two phases (HL and liquid) will be different. From section 3.3, exchange of H_3O^+ ions between solution and Haugaard layer will take place until equation (3.15) is satisfied, i.e. until the electrochemical potentials in both phases are equal.

$$\tilde{\mu}^{\mathrm{HL}}_{\mathrm{H_3O^+}} = \tilde{\mu}^{sol}_{\mathrm{H_3O^+}} = \mu^{0*,\mathrm{HL}}_{\mathrm{H_3O^+}} + RT\ln a^{\mathrm{HL}}_{\mathrm{H_3O^+}} + F\varphi^{\mathrm{HL}}$$
$$= \mu^{0*,sol}_{\mathrm{H_3O^+}} + RT\ln a^{sol}_{\mathrm{H_3O^+}} + F\varphi^{sol} \tag{3.203}$$

if, $\mu^{0*,\mathrm{HL}}_{\mathrm{H_3O^+}} = \mu^{0*,sol}_{\mathrm{H_3O^+}}$ and both are invariant with respect to change of solution, then the potential difference between HL and solution can be written

$$\Delta\varphi = \varphi^{sol} - \varphi^{\mathrm{HL}} = \frac{RT}{F}\ln\left\{\frac{a^{\mathrm{HL}}_{\mathrm{H_3O^+}}}{a^{sol}_{\mathrm{H_3O^+}}}\right\} \tag{3.204}$$

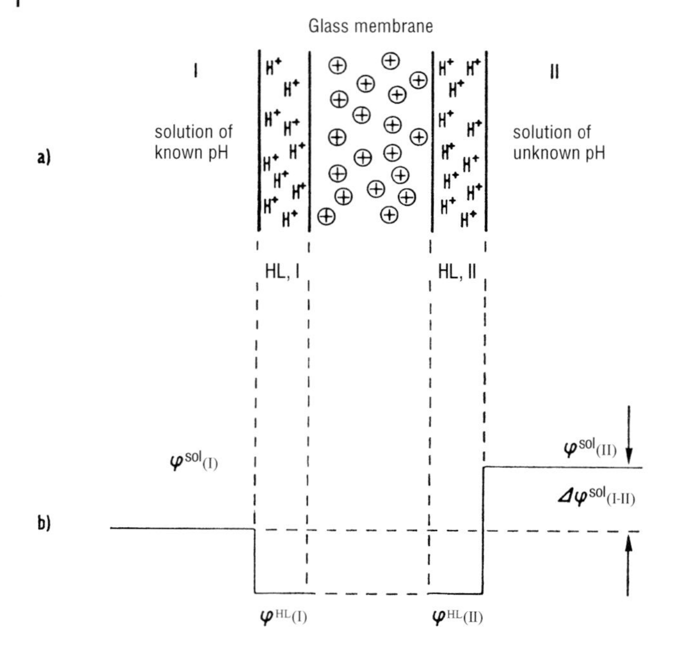

Fig. 3.32 Operation of the glass electrode: (a) schematic representation of a glass membrane between two solutions of different pH. HL, I and HL, II are the Haugaard layers at the two membrane surfaces; (b) potential variation through the membrane.

and $\Delta\varphi$ changes, from equation (3.204) at the rate of 59.1 mV for every decade change in H_3O^+ activity at 25°C.

This effect can be used for the determination of an unknown pH value: a thin glass membrane (of thickness ca. 0.5 mm), both of whose sides have formed swollen layers, is used to separate the solution of unknown pH, II, from a second solution, I, whose pH is assumed constant, as shown in Fig. 3.32. Electrical connection between the two faces of the membrane is achieved through the finite ionic conductivity of the glass, and equilibria across the two faces leads to the expression

$$\Delta\varphi_0(\mathrm{I}-\mathrm{II}) = \frac{RT}{F}\ln\left\{\frac{a_{\mathrm{H_3O^+}}(\mathrm{HL,I})}{a_{\mathrm{H_3O^+}}(\mathrm{I})}\right\} - \frac{RT}{F}\ln\left\{\frac{a_{\mathrm{H_3O^+}}(\mathrm{HL,II})}{a_{\mathrm{H_3O^+}}(\mathrm{II})}\right\} \tag{3.205}$$

Putting $\log_{10} a_{\mathrm{H_3O^+}}(\mathrm{I}) = -\mathrm{pH_I}$ and $\log_{10} a_{\mathrm{H_3O^+}}(\mathrm{II}) = -\mathrm{pH_{II}}$

$$\Delta\varphi_0(\mathrm{I}-\mathrm{II}) = \frac{RT}{F}\ln\left\{\frac{a_{\mathrm{H_3O^+}}(\mathrm{HL,I})}{a_{\mathrm{H_3O^+}}(\mathrm{HL,II})}\right\} + 0.0591(\mathrm{pH_I}-\mathrm{pH_{II}}) \tag{3.206}$$

The expression $\frac{RT}{F}\ln\left\{\frac{a_{\mathrm{H_3O^+}}(\mathrm{HL,I})}{a_{\mathrm{H_3O^+}}(\mathrm{HL,II})}\right\}$ that appears in (3.206) is termed the 'asymmetry potential', $\Delta\varphi_{as}$, and, if needed, may be determined by making a measurement in which the solution (II) is replaced by solution (I), so that $pH_I = pH_{II}$, and identical reference electrodes are used in the two solutions; under these circumstances, the potential $\Delta\varphi_0(I - II)$ reduces to $\Delta\varphi_{as}$. The origin of the asymmetry potential is the differing mechanical stresses and chemical composition of the silicate framework on both sides of the membrane caused during manufacture. Depending on the type of glass, the form of the membrane (planar or curved) and the previous history (preparation procedure, treatment temperature, previous solutions contacted etc), values of $\Delta\varphi_{as}$ of between 0 and 50 mV are commonly encountered.

Practical measurements are carried out using a spherical membrane of about 1 cm diameter which is sealed to the lower end of a thicker glass tube and filled with a buffer solution of known pH value, as shown in Fig. 3.33. The usual pH for the inner buffer solution is 7.0, and KCl is dissolved in the buffer in order to enable a definite potential difference to be developed at the inner electrode. This electrode is normally a silver-silver chloride, calomel or 'Thalamid®' electrode, the latter being an electrode of the second type constructed as $Tl(Hg)|TlCl|Cl^-$, which gives a potential difference of -0.555 V vs. SHE at a chloride concentration of 1.000M. Internal electrode, buffer solution and glass membrane are together known as a *glass electrode*. The pH of a solution can now be measured by immersing this glass electrode in turn in the unknown

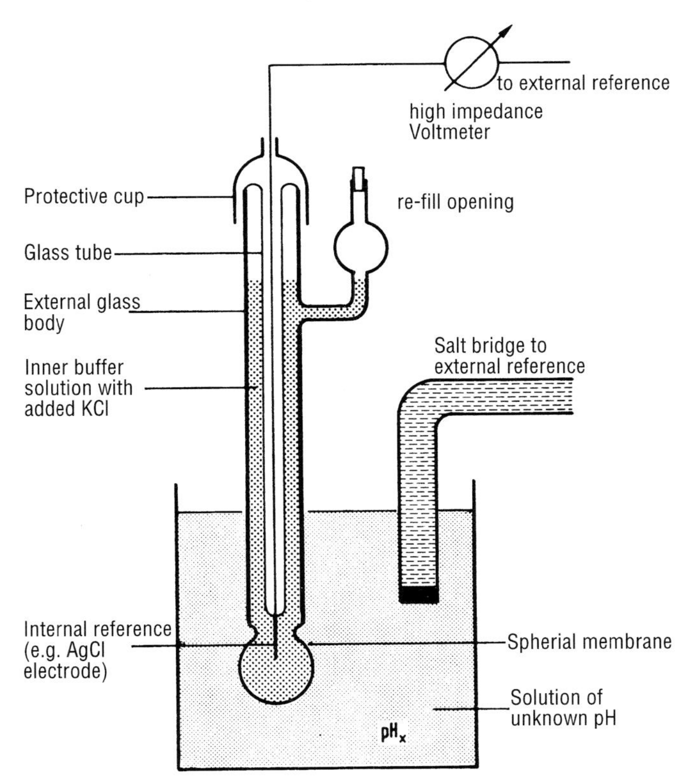

Fig. 3.33 pH-measurement with a glass electrode.

solution and in the standard pH buffer, measuring the potential difference in both cases between it and an external reference electrode immersed in the same solution.

Schematically, we can represent this arrangement by:

$$\mathrm{Hg|Hg_2Cl_2|KCl(satd.)||solution\ (II)|glass|solution(I),\ KCl|AgCl|Ag}$$

if the standard potential of the SCE is E_{SCE} and of the silver-silver chloride electrode is E^0_{SSE}, then the emf of the cell shown is

$$\begin{aligned} E &= E^0_{\mathrm{SSE}} - E_{\mathrm{SCE}} + \Delta\varphi_0(\mathrm{I} - \mathrm{II}) \\ &= E^0_{\mathrm{SSE}} - E_{\mathrm{SCE}} + \Delta\varphi_{as} + 0.0591(\mathrm{pH_I} - \mathrm{pH_{II}}) \end{aligned} \tag{3.207}$$

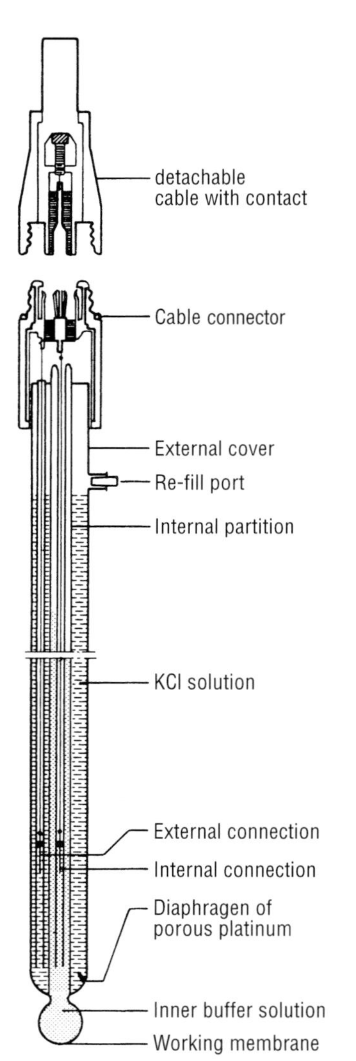

Fig. 3.34 Combined pH electrode and reference electrode, allowing direct pH measurement in any solution save those of very high or low ionic concentration.

If $E = E_s$ for solution (II) being a standard buffer and $E = E_x$ for solution (II) being the unknown, it follows once again that

$$pH_x = pH_S + (E_S - E_x)/0.0591 \quad (3.208)$$

Measurements with the glass electrode can be carried out over a wide pH range, although in alkaline solution, exchange of cations with the HL leads to a so-called 'alkaline error'. This can be minimised by the use of special glasses, allowing accurate measurement over the pH range 1–14. There is also a smaller 'acid error' (for $pH < 1$) that needs to be considered in precise work.

In precision measurements, the electrode should be re-calibrated every day against standard buffers if possible, since $\Delta\varphi_{as}$ is not stable over time. Strongly alkaline solutions and solutions of HF attack the glass of the membrane and destroy the electrode. In addition, the membrane will be severely affected if it is allowed to dry out, so the electrode should be stored in distilled water when not in use. The EMF must be measured using a voltmeter of very high input impedance ($10^{11} - 10^{12}\Omega$) since the glass membrane itself has an extremely high impedance ($10^8\Omega$) and even small ionic currents through the HL will disturb the equilibrium between this layer and the solution. Because of the use of high-impedance systems, the leads must be carefully shielded against electrical pick-up.

A further development, shown in Fig. 3.34, is the combination or dual 'electrode' in which glass and external reference electrodes are placed in concentric glass tubes. For special measurement problems, a series of special types of glass electrodes have been developed; these include shockproof conical membranes for soil investigations, flat membranes for the determination of solid samples such as cheese and miniaturised needle-like systems for measurements in vivo.

3.6.8
The Principle of Potentiometric Titrations

If, during a chemical reaction, there is a change in the activity of an ion i that can be sensed through the change in potential of a suitable electrode, then the progress of the reaction can be followed through this potential change. It follows that EMF measurements, like conductivity measurements, can serve to determine the equivalence point or end-point of a titration. Both conventional electrodes, and the types of ion-sensitive electrodes described in Section 10.3.3 can be used to follow the process.

To follow an acid-base titration, for example, a hydrogen electrode or a pH-sensitive glass electrode may be used as the indicator electrode. In both cases, as the titration is carried out, for example, by addition of alkali to an acid substrate, the potential difference measured will decrease at the rate of 59.1 mV per decade lowering in H_3O^+ activity. As long as the acid is in excess, the pH will vary only slightly with addition of base; however, near the equivalence point, $a_{H_3O^+}$ falls very rapidly, and the potential difference also falls rapidly before levelling out again, in excess base. Thus, the measured potential difference will show a step-like behaviour, as seen in Fig. 3.35, in which the change in potential of the pH-electrode is calculated for titra-

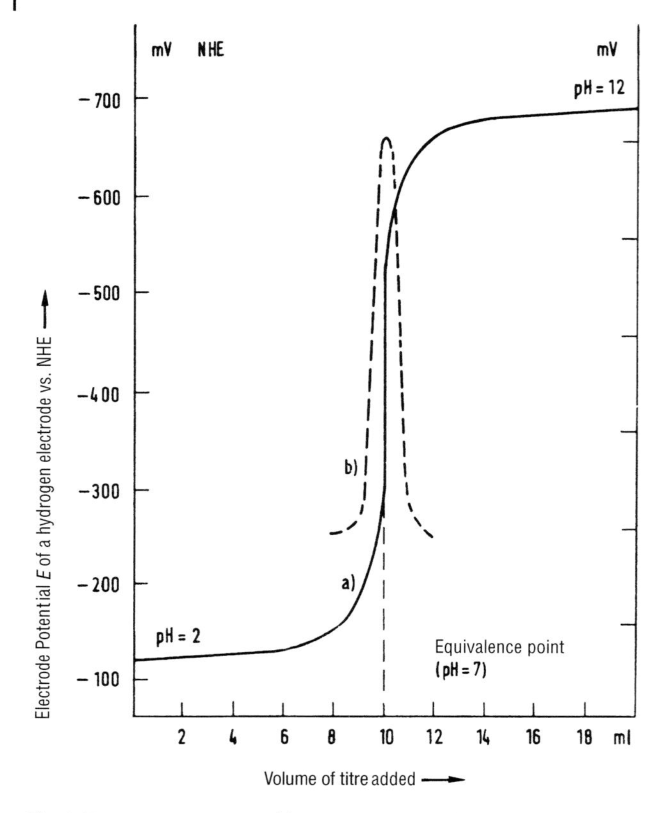

Fig. 3.35 Potentiometric acid-base titration: (a) schematic representation of the titration of 100 cm^3 of a strong acid of concentration 0.01 M with a strong base of concentration 0.1 M; (b) differential potential change on addition of aliquots of titrant, showing a marked peak at the end-point.

tion of 100 cm^3 of 0.01M strong acid with 0.1M strong base. The end-point of the titration corresponds to the point at which the potential changes most rapidly.

A more precise measure of the end-point than the position of steepest change in the potential difference can be obtained by plotting the differential of the potential difference with titre. This is shown as the dotted curve in Fig. 3.35, and it can be seen to have a sharp maximum at the end point.

The change of potential in the case of precipitation, complexation and redox titrations is very similar to that shown in Fig. 3.35. As an example of a precipitation titration, consider the determination of chloride ion by silver nitrate

$$Ag^+ + Cl^- \rightarrow AgCl(s) \tag{3.209}$$

Here, the equilibrium-potential of the Ag|Ag$^+$ electrode can be followed as a function of addition of the titrant. At the beginning of the titration, the concentration of silver in solution will, in effect, be determined by the solubility product of silver chloride, and, at the end-point, the Ag^+ concentration will rise very rapidly as the last of the chloride is precipitated. The height of the potential step can be further enhanced if the titration is carried out in a water-acetone mixture, in which the solubility product of silver chloride is lower.

Potentiometric titrations have the great advantage, in common with conductimetric titrations, of being possible in turbid, coloured and very dilute solutions. Further advantages of potentiometric titrations are the generally very sharp end-points and the ease of automation; a large number of commercial rigs are available. The range of applications is enormous, and accurate methods have been developed for many electronanalytical processes, including the determination of traces of water in organic solvents.

References for Chapter Three

Further background reading in electrochemistry can be found in:

"Encyclopedia of Electrochemistry", Eds. A.J. Bard and H. Stratmann, Wiley-VCH, Weinheim, 2003.

J.O'M. Bockris and A.K.N. Reddy: "Modern Electrochemistry", 2nd Ed. Vol **2A** Plenum Press, New York, 2000.

A.J. Bard and L.R. Faulkner: "Electrochemical Methods", John Wiley and Sons, 2001.

R.A. Robinson and R.H. Stokes: "Electrolyte Solutions", Butterworth, London, 1959.

Discussions of Reference Electrodes and their construction can be found in:

D.I.G. Ives and G.J. Janz, "Reference Electrodes", Academic Press, New York, 1961.

D.T. Sawyer and J.L. Roberts: "Experimental Electrochemistry for Chemists", John Wiley and Sons, 1974.

Tabulations of Electrode potentials can be found in:

A.J. Bard, R. Parsons and J. Jordan: "Standard Potentials in Aqueous Solution", Dekker, New York, 1985.

A.K. Covington and T. Dickinson: "Physical Chemistry of Organic Solvents", Plenum Press, New York, 1973.

More extended discussions of the Structure of the Electrode/electrolyte interface can be found in:

"Comprehensive Treatise of Electrochemistry", vol. **1**, eds. J. O'M. Bockris, B. Conway and E. Yeager, Plenum Press, New York, 1980.

R. Parsons, "The Metal-Liquid Electrolyte Interface", Solid State Ionics **94** (1997) 91.

"Interfacial Electrochemistry" ed. A. Wieckowski, Marcel Dekker, New York, 1999.

"Adsorption of Molecules at Metal electrodes", ed. J. Lipkowski and P.N. Ross, VCH, Weinheim, 1992.

E. Spohr in "Advances in Electrochemical Science and Engineering", vol. **6**, Wiley-VCH, Weinheim, 1999.

More detailed treatments of Semiconductor electrochemistry can be found in:

A. Hamnett: "Semiconductor Electrochemistry", in "Comprehensive Chemical Kinetics", ed. R.G. Compton, Elsevier, 1987.

S.R. Morrison: "Electrochemistry at Semiconductor and Oxidised Metal Electrodes", Plenum, New York, 1980.

R. Memming, "Semiconductor Electrochemistry", Wiley-VCH, Weinheim, 1999.

Further treatments of colloidal phenomena can be found in:

R.J. Hunter, "Foundations of Colloid Science", 2 vols., Oxford University Press, Oxford, 1987.

D.H. Everett, "Basic Principles of Colloid Science", Royal Society of Chemistry, London, 1988.

4
Electrical Potentials and Electrical Current

Earlier discussion has focussed, by and large, on the situation of electrochemical equilibrium within the electrolyte and at the phase boundary between electrolyte and electrode, and we must now turn to the situation in which we depart from equilibrium by permitting the flow of electrical current through the cell. Such current flow leads not only to a potential drop across the electrolyte, which affects the cell voltage by virtue of an Ohmic drop $I \cdot R_E$, where R_E is the internal resistance of the electrolyte between the electrodes, but each electrode exhibits a characteristic current-voltage behaviour; the overall cell voltage will, in general, reflect both these effects.

4.1
Cell Voltage and Electrode Potential during Current Flow: an Overview

Fig. 4.1 shows an electrochemical cell during passage of a current with the corresponding cell resistances. As we shall see, the electrolyte resistance, R_E, is an *ohmic resistance*, which does not depend on the magnitude or direction of the current and for which, therefore, the potential drop is proportional to the current passed. However, the resistances R_I and R_{II} are not ohmic; they depend on the magnitude and direction of the current.

The rest potential, E_r, of the electrodes and the equilibrium *EMF* of the cell, which is the difference between the rest potentials of the two electrodes, will, in the ideal case be obtainable from thermodynamic data. Tables of such data will allow us to calculate the *standard EMF*, E^0, for the cell from the value of the standard free energy, $\Delta_f G^0$, of the corresponding chemical reaction, and the value of the *EMF* of the cell can then be obtained from the Nernst equation for any concentrations of reactants and products.

On passage of a current, the potentials all shift to new values. In the example of Fig. 4.1, the cell contains as electrolyte aqueous hydrochloric acid and it is divided into two compartments by a semi-permeable membrane. In one compartment, hydrogen is dissolved, and in the other chlorine, and the electrodes themselves can be made of platinum. In this case, the potential determining electrode reactions under rest conditions are:

$$Cl_2 + 2e^- \rightleftharpoons 2Cl^- \quad (4.1)$$

Electrochemistry. Carl H. Hamann, Andrew Hamnett, Wolf Vielstich

ISBN: 978-3-527-31069-2

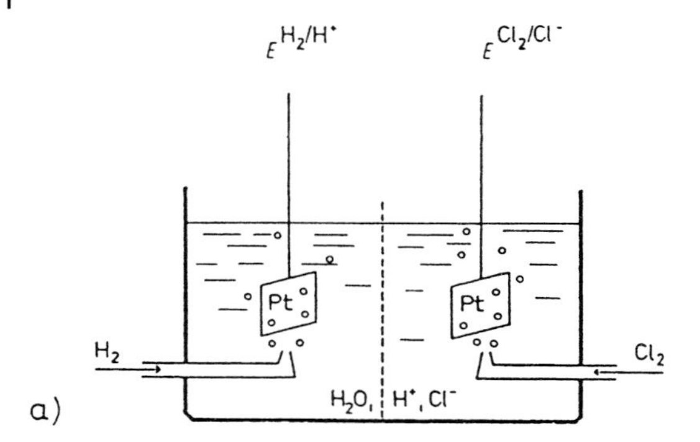

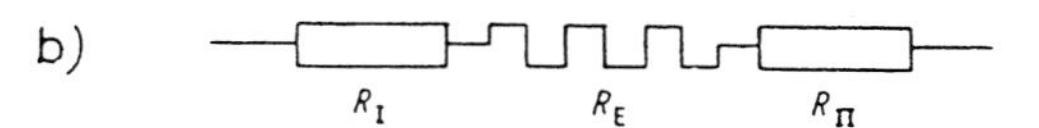

Fig. 4.1 (a) Example of an electrochemical cell; (b) simple equivalent circuit.

and

$$H_2 \rightleftharpoons 2H^+ + 2e^- \tag{4.2}$$

If the hydrochloric acid has unit mean activity, and the equilibrium pressure of both gases is one atmosphere, then standard thermodynamic tables show that the cell voltage will be $E^{0,Cl_2/Cl^-} - E^{0,H^+/H_2} = +1.37V$; the chloride electrode will appear as the positive pole.

If the cell is operated as a battery by joining the two electrodes through a resistor, electrons will flow from the hydrogen electrode to the chlorine electrode and the equilibrium established before the electrodes were connected will be destroyed. The potential of the hydrogen electrode will shift in a *positive* direction as the reaction

$$H_2 \rightarrow 2H^+ + 2e^-$$

proceeds. This reaction corresponds to electrons being transferred from the solution to the electrode, which, under these circumstances, is termed the *anode*. Correspondingly, the potential of the chlorine electrode will be *reduced* as electrochemical reduction of Cl_2 takes place at the *cathode*:

$$Cl_2 + 2e^- \rightarrow 2Cl^-$$

These two effects are shown in Fig. 4.2a, and it is evident from this figure that the overall cell voltage will decrease with increasing current. Note that Fig. 4.2 is schematic: the current-voltage behaviour is *not* normally linear as shown but exponential as explained below.

If instead of operation as a battery, the cell is operated in electrolysis mode, by imposing a voltage that is larger than the equilibrium emf, then the two cell reactions will be driven in the opposite direction. The anode is now the chlorine electrode, and the electrode reaction becomes

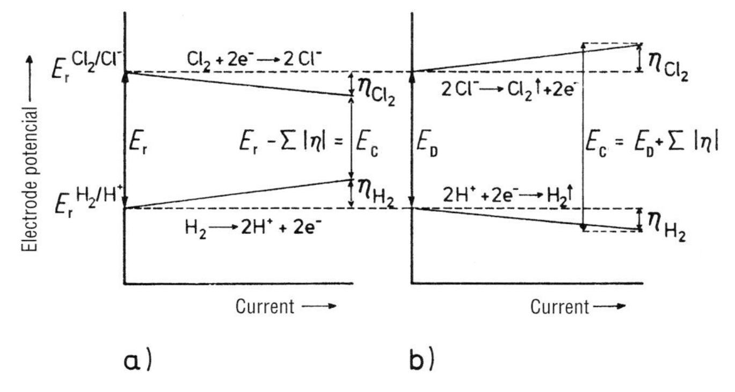

Fig. 4.2 Schematic representation of the variation of cell voltage with current for the case of (a) a galvanic (or "fuel") cell based on the Cl_2/H_2 reaction; (b) an electrolysis cell using hydrochloric acid as the electrolyte.

$$2Cl^- \rightarrow Cl_2 + 2e^- \tag{4.3}$$

and the *cathode* is now the hydrogen electrode, where the corresponding reaction is

$$2H^+ + 2e^- \rightarrow H_2 \tag{4.4}$$

In this case, as the current increases, the voltage across the cell will also increase, with the potential of the hydrogen electrode becoming more *negative* and the potential of the chlorine electrode more *positive*, as shown schematically in Fig. 4.2b.

4.1.1 The Concept of Overpotential

The magnitude of the deviation of the electrode potential at anode or cathode from the equilibrium value E_r is termed the *overpotential*, defined as

$$\eta = E - E_r \tag{4.5}$$

and its sign is obviously determined by whether E is greater than or less than E_r. The values of η are also shown schematically in Figs. 4.2a,b.

At low currents, the rate of change of electrode potential with current is associated with the limiting rate of electron transfer across the phase boundary between the electronically conducting electrode and the ionically conducting solution, and is termed the *electron transfer overpotential*. The electron transfer rate at a given overpotential has been found to depend on the nature of the species participating in the reaction, and the properties of the electrolyte and the electrode itself (such as, for example, the chemical nature of the metal). At higher current densities, the primary electron transfer rate is usually no longer limiting; instead, limitations arise through slow transport of reactants from solution to the electrode surface or conversely the slow transport of product away from the electrode (*diffusion overpotential*) or through the inability of chemical reactions coupled to the electron transfer step to keep pace (*reaction overpotential*).

Examples of the latter include adsorption or desorption of species participating in the reaction or the participation of chemical reactions before or after the electron transfer step itself. One such process occurs in the evolution of hydrogen from a solution of a weak acid, HA: in this case, the process (4.4) must be preceded by the acid dissociation reaction taking place in solution

$$HA \rightleftharpoons H^+ + A^- \tag{4.6}$$

4.1.2
The Measurement of Overpotential; the Current-Potential Curve for a Single Electrode

The equilibrium or rest potential, $E^{0,Cl_2|Cl^-}$, of the chlorine electrode under standard conditions can obviously be measured quite easily in the experimental arrangement described above, by measuring the *EMF* of the cell using a high-impedance voltmeter, since the hydrogen electrode is conventionally taken as our zero point of reference, as described in chapter 3. However, the measurement of the potential E of the chlorine electrode when a current is flowing is more tricky, since the measured cell voltage will include contributions from the overpotentials of both the chlorine and hydrogen electrodes.

The potential of an electrode that is passing a current (the *working* electrode) cannot, in general, be determined by measurement with respect to the other current-carrying electrode in the cell (the *counter* electrode). We need, as a potential reference, to introduce a third electrode into the cell as a *reference* electrode. Such electrodes have already been described in chapter 3, and in practical measurements, a reference electrode of the second type (such as a silver-silver chloride or calomel electrode) is com-

a)

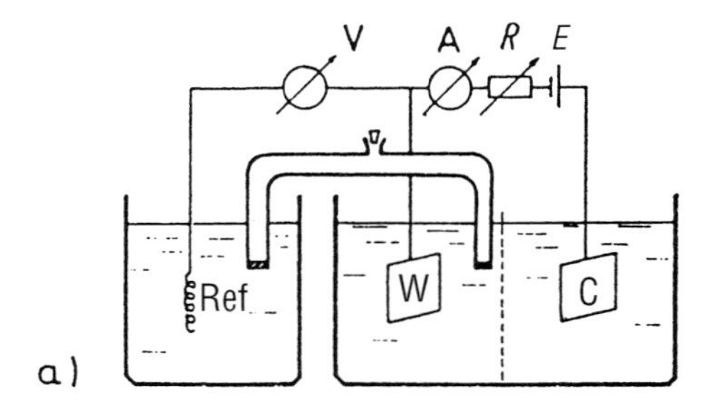

b)

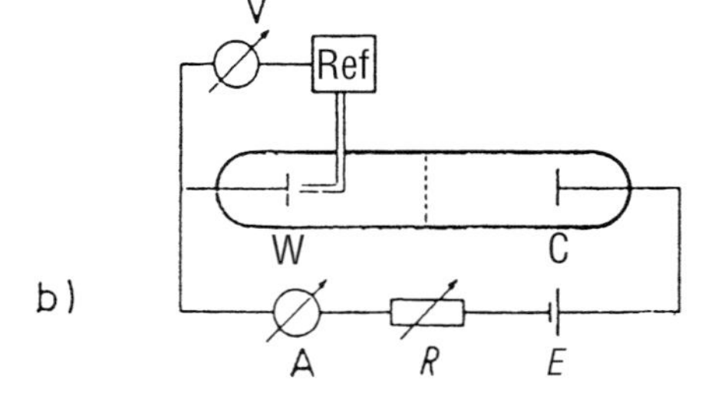

Fig. 4.3 (a) The experimental arrangement for the measurement of the potential of an electrode that is passing a finite current; (b) a simple circuit representation. W is the working electrode; C the counter electrode and Ref the reference electrode.

monly used, which might be connected to the main electrolyte of the cell by means of a salt bridge, as shown in Fig. 4.3a. The potential difference between the chlorine electrode and this reference electrode now allows us to determine the overpotential purely of the working electrode itself. Obviously, to avoid any overpotential contribution from the reference electrode, no significant current must pass through this electrode, which can be achieved with a high impedance voltmeter as shown in Fig. 4.3b. This three-electrode arrangement is essential for all measurements of electrode characteristics save where the cell over-voltage is entirely dominated by the working electrode, with negligible contributions from the solution resistance and counter electrode. Under these circumstances, a two-electrode measurement may be used, as is commonly the case in analytical work.

One important source of error in the measurements described so far is the fact that there will be a voltage drop in the electrolyte arising from Ohm's law. The reference electrode will sample the potential at the end of the salt bridge in the cell, and since current flows from working to counter electrodes in the cell, it is evident that this end should be placed immediately in front of the working electrode. It is also clear that the end should be as well defined as possible, and it is common to draw it into a capillary tip called a Luggin capillary. Corrections to the current- potential curve for the finite solution resistance will be discussed further below.

The variation of current with potential measured as above is termed the *current-potential-characteristic* of the electrode, and such a characteristic, as measured for the chlorine electrode above, is shown in Fig. 4.4. It is evident that the approximation of a *linear* relationship between I and E used in Fig. 4.2 is not valid for potentials more than a few mV from the rest potential.

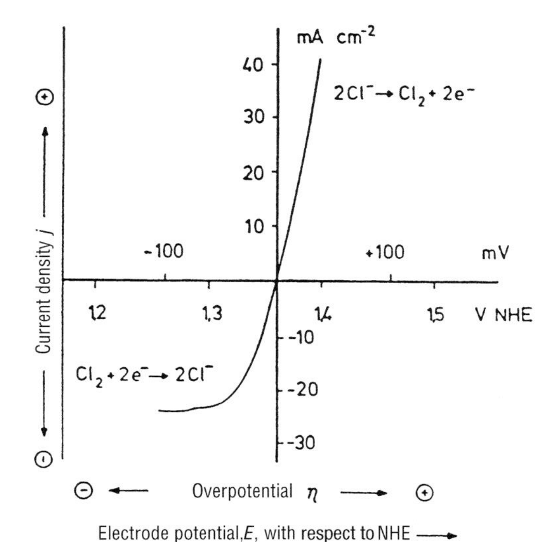

Fig. 4.4 The current-potential characteristic for a chlorine-sparged platinum electrode of 1 cm^2 area in aqueous HCl ($p_{Cl_2} = 1$ bar; $a_{HCl} = 1$) showing both anodic and cathodic current branches.

4.2
The Electron-transfer Region of the Current-Potential Curve

Of the various factors affecting the current-potential curve, including electron transfer, transport and chemical processes, limitations due to the rate of electron transfer are unavoidable, especially close to the rest potential, and we now turn to general considerations of the effect of the rate of electron transfer on the overall current-potential characteristic of the electrode.

For the general process

$$\mathrm{Ox} + \mathrm{ne}^- \rightleftharpoons \mathrm{Red} \tag{4.7}$$

at equilibrium, the forward (*cathodic*) and reverse (*anodic*) processes must be taking place at the same rate. In general, these rates are *not* zero, and the equilibrium is a *dynamic* one.

If an overpotential is now applied to the electrode, either the anodic reaction rate will be increased and the cathodic reaction rate slowed down or vice versa. As a result, a nett chemical conversion process takes place at the electrode, corresponding to which a nett current flows through the external circuit. The actual electron transfer process through the phase boundary between electronic and ionic conductors takes place, in general, through the *quantum mechanical tunnelling effect*. In a cathodic reaction, an electron transfers from the conduction band of the metal directly into an unoccupied molecular orbital of an ion or neutral molecule either adsorbed on the electrode or situated in the electrolytic double layer; an anodic process is the reverse of this, with an electron transferred from an occupied molecular orbital of an ion or neutral molecule within the electrolyte or adsorbed on the electrode surface to an empty level within the metal conduction band.

4.2.1
Understanding the Origin of the Current-Potential Curve in the Electron-transfer-limited Region with the Help of the Arrhenius Equation

Initially, we will restrict ourselves to metallic electrodes and to electrolyte solutions with high concentrations of salt, so that the change in potential between electrode and electrolyte is essentially restricted to within the Helmholtz layer. For semiconducting electrodes, the potential drop takes place predominantly *within the electrode,* as we saw in Section 3.6, and for more dilute electrolytes, the region of potential change can extend far further into the solution than the Helmholtz layer, as dicussed in chapter 2. In neither case can the equations developed below be used without modification.

If the potential on the electrode is altered from E_1 to E_2, then the primary change in the potential distribution, under the restrictions given above, will be accommodated within the double layer; we can assume that the potential in the inner part of the solution does not change, and the potential on the metal will change in value but remain constant throughout the bulk of the electrode. Note that this change in potential will effect all charged particles both in the metal and in the double layer, since for a charge ze_0, the electrical energy associated with a potential E is ze_0E.

Consider now a chemical process of the form $A + B \rightarrow$ products. In activated complex theory, the reaction between A & B is represented as passing through *an activated complex,* (AB), which can be thought of as in equilibrium with the reactants, but which can also decompose to the reaction products at a finite rate, determined essentially by the appropriate bond relaxation processes. If the free energy difference for the forward reaction between the initial reactants A + B and (AB) is ΔG_{f}, the rate constant for the forward reaction can be written $k_{\mathrm{f}} = k_f^0 \exp(-\Delta G_f^{/} RT)$, and the overall rate of the reaction is given by

$$v = k_{\mathrm{f}}^0 c_{\mathrm{A}} c_{\mathrm{B}} \exp(-\Delta G_{\mathrm{f}}^{/} RT) \tag{4.8}$$

where the concentrations c_{A}, c_{B} are in principle those of A and B at the electrode surface, though for very low current densities, such concentrations will be little perturbed from their bulk values. It can be seen, then, that any alteration in ΔG_{f} will affect the reaction rate; if ΔG_{f} is *lowered,* the reaction rate will *increase,* and if ΔG_{f} is *increased,* the reaction rate will *decrease.*

The way in which alteration of the potential affects ΔG_{f} is shown schematically in Fig. 4.5, which illustrates the variation in energy of our redox system as a function of the reaction coordinate for two potentials E_1 and E_2 on the electrode, where in the figure, E_2 is more negative than E_1. On the left-hand side of the diagram is the free energy for the left-hand side of equation (4.7), i.e. the free energy of

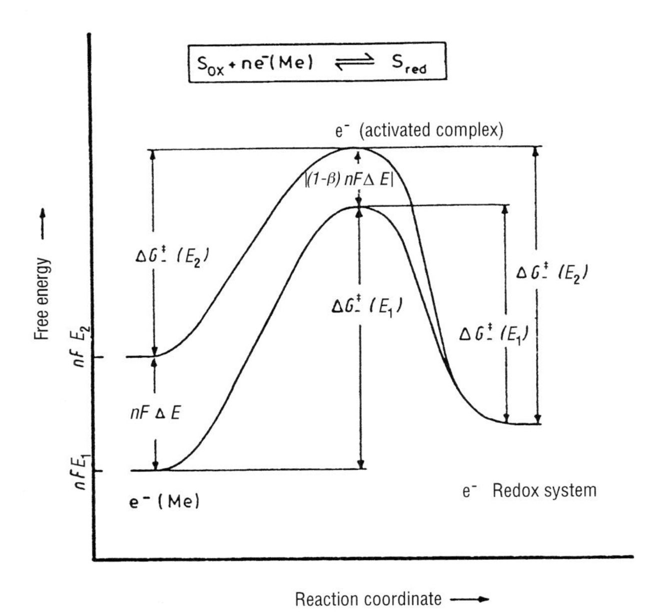

Fig. 4.5 Internal energy of an electron during its transition from the electrode to a component of a suitable redox couple in solution and during the reverse process. The change in potential from E_1 to E_2 corresponds in this figure to a potential shift to more negative values.

$Ox + ne^{-}_{M}$, and on the right hand side of the diagram is the free energy of Red. If we assume that the electron transfer process is an *outer-sphere one*, that is neither Ox nor Red becomes adsorbed on the electrode surface, then neither Ox nor Red will penetrate into the region in which the potential changes. If follows that the only change in free energy on altering the potential on the metal is the free energy of the electrons within the metal, which alters by an amount $(-e_0) \cdot (E_2 - E_1)$ per electron, or $-nF \cdot \Delta E$ for n moles of electrons.

Not all this change in ΔG appears for the activated complex, since, as shown in Fig. 4.5, there is *no* change in ΔG for the right-hand side of equation (4.7) on changing the potential. For simplicity, it is usually assumed that only a fraction of the total free energy change $(1 - \beta)|nF \cdot \Delta E|$ actually appears for the activated complex. Thus, for the *cathodic reaction* as shown, with the activation energy written as ΔG_{-}, and bearing in mind that $\Delta E < 0$ in this figure:

$$\Delta G^{(}_{-} E_2) = \Delta G^{(}_{-} E_1) + \beta nF \cdot \Delta E \tag{4.9}$$

and for the anodic reaction

$$\Delta G^{(}_{+} E_2) = \Delta G^{(}_{+} E_1) - (1 - \beta) nF \cdot \Delta E \tag{4.10}$$

The fraction β is termed the asymmetry parameter, since the more asymmetric the free energy potential curve with respect to reactants and products, the smaller or larger β will be. Only if the curve is quite symmetric will β be close to 0.5. It should be noted that the alternative convention of fraction (1-β) for the cathodic reaction and β for the anodic reaction is also in common usage. The convention in equations (4.9) and (4.10) is, however, now well-established in textbooks, but the reader should carefully check any source of asymmetry parameters in the literature for the convention actually used. In the case illustrated in Fig. 4.5, where $E_2 < E_1$, equation (4.10) predicts that the free energy of activation for the *anodic* reaction will *increase* (since $\Delta E < 0$) and equation (4.9) predicts that the free energy of the *cathodic* reaction will *decrease*.

We can proceed by calculating the rate of the cathodic and anodic reactions using equation (4.8); if the flux density for Ox at the surface is $J^{-}(E_1)$ at potential E_1, then

$$J^{-}(E_1) = -c_{Ox} k_0^{\prime -} \cdot \exp(-\Delta G^{(}_{-} E_1)/RT) \tag{4.11}$$

where the units of J^{-} are mol sec^{-1} per m^2 of the electrode surface and c_{Ox} is the concentration of oxidised species (mol m^{-3}), which we assume is maintained constant by rapid transfer of Ox from solution to the surface. The electron concentration in the metal is assumed to be constant since the electron mobility in the metal is very high, and it is taken into the electrochemical rate constant, $k_0^{\prime}$, which then has the units of m sec^{-1}. Particularly in the older literature, units are in *cm* rather than *m*, and readers should beware of the habit of using concentration units as mol dm^{-3}.

From the flux of material, the corresponding current density, *j*, can be immediately calculated as every molecule of Ox requires n electrons, so $j = $ nFJ; the units of j are A m^{-2}. The cathodic current density is then given by

$$j^{-}(E_1) = -nFc_{Ox} k_0^{\prime -} \cdot \exp(-\Delta G^{(}_{-} E_1)/RT) \tag{4.12}$$

where the negative sign indicates transfer of electrons *to* the solution species. On alteration of the potential from E_1 to E_2, we have

$$j^{-}(E_2) = -nFc_{\text{Ox}}k_0^{\prime -} \cdot \exp(-\{\Delta G^{(}_{-} E_1) + \beta nF \cdot \Delta E\}/RT) \tag{4.13}$$

If the potential E_1 is set equal to zero for the reference potential scale used in the experiment, then since the reference electrode does not depend on factors such as c_{Ox}, the value of $\exp(-\Delta G^{(}_{-} E_1)/RT)$ can be treated as a constant and incorporated into our rate constant without any loss of generality. Replacing E_2 and ΔE by an arbitrary electrode potential, E, we finally have

$$j^{-}(E) = -nFc_{\text{Ox}}k_0^{-} \cdot \exp(-\beta nFE/RT) \tag{4.14}$$

where the electrochemical rate constant clearly now has a value that depends on our choice of the potential zero and hence the reference scale used.

By an analogous process, we obtain for the anodic current

$$j^{+}(E_2) = nFc_{\text{Red}}k_0^{\prime +} \cdot \exp(-\{\Delta G^{(}_{+} E_1) - (1-\beta)nF \cdot \Delta E\}/RT) \tag{4.15}$$

where the current now has a *positive* sign, and again choosing the zero of potential as the zero of the reference electrode we find

$$j^{+}(E) = nFc_{\text{Red}}k_0^{+} \cdot \exp(+(1-\beta)nFE\}/RT) \tag{4.16}$$

Equations (4.14) and (4.16) describe the two partial current densities of opposite algebraic sign at each potential. It follows that at the rest potential, E_r, when the nett current is zero, the two partial currents must be numerically equal, and both have the magnitude of the so-called *exchange current density*, j_0. It is important to recognise, as already emphasised, that the rest potential corresponds to a *dynamic* equilibrium and not to zero activity. Experimentally this can be easily demonstrated by, for example, immersing a silver electrode into a solution containing radioactive silver ions; even though no nett current is passed through the silver, radioactive silver atoms are found incorporated into the metal electrode after removal from the solution. At the rest potential, then, we have

$$j^{-}(E_r) = -j_0 = -nFc_{\text{Ox}}k_0^{-} \cdot \exp(-\beta nFE_r/RT) \tag{4.17}$$

and

$$j^{+}(E_r) = +j_0 = +nFc_{\text{Red}}k_0^{+} \cdot \exp(+(1-\beta)nFE_r\}/RT) \tag{4.18}$$

and equating these two, we find, after some rearrangement, that

$$E_r = \frac{RT}{nF}\ln\frac{k_0^{+}}{k_0^{-}} + \frac{RT}{nF}\ln\frac{c_{\text{Ox}}}{c_{\text{Red}}} \tag{4.19}$$

which is evidently the Nernst equation of chapter 3, provided that we put $E^0 = \frac{RT}{nF}\ln\frac{k_0^{+}}{k_0^{-}}$ and we also remember that our whole derivation has been carried out for a large excess of electrolyte salt, under which circumstances the activity coefficients of Ox and Red will be concentration independent.

If we now write the actual electrode potential E as $E_r + \eta$, where η is the overpotential and E_r the rest potential as defined above, then

$$j^-(E) = -nFc_{\text{Ox}}k_0^- \cdot \exp\{-\beta nFE_r/RT - \beta nF\eta/RT\} \quad (4.20)$$

and

$$j^+(E) = nFc_{\text{Red}}k_0^+ \cdot \exp\{+(1-\beta)nFE_r/RT + (1-\beta)nF\eta/RT\} \quad (4.21)$$

from (4.14) and (4.16). From the definition of the exchange current density, as in equations (4.17) and (4.18)

$$j^-(E) = -j_0 \cdot \exp\{-\beta nF\eta/RT\} \quad (4.22)$$

$$j^+(E) = j_0 \cdot \exp\{+(1-\beta)nF\eta/RT\} \quad (4.22a)$$

The nett current is the algebraic sum of the anodic and cathodic currents:

$$j = j^- + j^+ = j_0[\exp\{(1-\beta)nF\eta/RT\} - \exp\{-\beta nF\eta/RT\}] \quad (4.23)$$

which is the celebrated Butler-Volmer equation.

4.2.2
The Meaning of the Exchange Current Density j_0 and the Asymmetry Parameter β

Equations (4.22) and (4.23) show that the partial current densities will increase or decrease *exponentially* with overpotential, the steepness of the rise or fall depending on the asymmetry parameter, β. For $\beta = 0.5$, the anodic and cathodic branches of the total current will be symmetrical, but for β near unity, the cathodic branch would rise much more steeply than the anodic branch as shown in Fig. 4.6 (dotted line), whereas for β near zero, the inverse is found (dashed line in 4-6). Experimentally, β is usually found between 0.4 and 0.6.

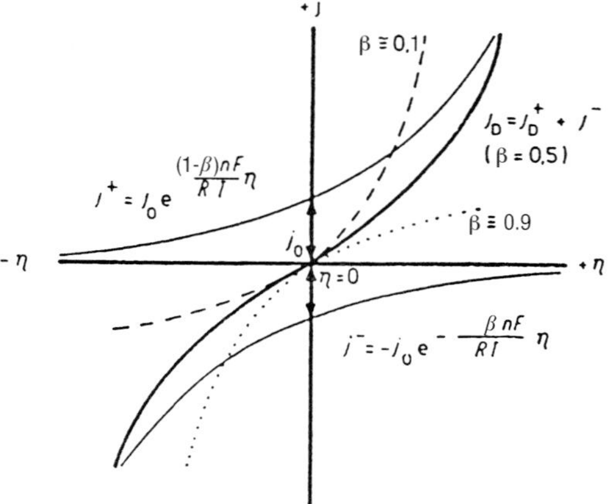

Fig. 4.6 Graphical representation of the equations 4.22 and 4.22a and the Butler-Volmer equation, 4.23.

By contrast, the exchange current density is a multiplicative factor, affecting both branches of the curve equally. Its value, in terms of the fundamental free energies of activation for the forward and back reactions, can be written from (4.12) - (4.14) as:

$$j_0 = nFc_{Ox}k_0^{\prime -} \exp\left\{-\frac{\Delta G_-^{(}E_r)}{RT}\right\} = nFc_{Ox}k_0^{\prime +} \exp\left\{-\frac{\Delta G_+^{(}E_r)}{RT}\right\} \tag{4.24}$$

and clearly depends on the concentration of the species participating in the electrochemical reaction and the free energy of activation at the rest potential. Its magnitude is a central quantity in electrochemistry, and in particular, it is the value of j_0 that is increased when the electrochemical reaction is catalysed.

For $|\eta| \gg RT/nF$ ($= 25.7/n$ mV at 25°C), then the smaller of the anodic and cathodic current values can be neglected. Thus, for a cathodic reaction, $\eta < 0$, and if $\eta \ll -RT/nF$, the total current can be written to good approximation as:

$$j = -j_0 \exp\{-\beta nF\eta/RT\} \tag{4.25}$$

and taking logarithms to the base 10, we obtain

$$\eta = \left\{\frac{2.303RT}{\beta nF}\right\} \cdot \log_{10} j_0 - \left\{\frac{2.303RT}{\beta nF}\right\} \cdot \log_{10} |j| \tag{4.26}$$

which has the form

$$\eta = A + B \cdot \log_{10} |j| \tag{4.27}$$

This semilogarithmic equation is known as the *Tafel* equation, and clearly the magnitude of B, known as the Tafel slope, has the value $-2.303RT/\beta nF$. For $\beta = 0.5$ and $n = 1$, this corresponds to -118 mV at 25°C. The expression (4.26) can be transformed into the form

$$\log_{10} |j| = \log_{10} j_0 + \{\beta nF/2.303RT\} \cdot |\eta| \tag{4.28}$$

which allows us, from a plot of $\log_{10} |j|$ vs. $|\eta|$, to obtain the value of j_0 directly from the intercept on the potential axis and β from the slope of the graph.

For high anodic overpotentials, $\eta \gg RT/nF$, provided we remain within the region where the current is entirely determined by the electron transfer rate, we have

$$j = j_0 \exp\{(1-\beta)nF\eta/RT\} \tag{4.29}$$

from which we obtain by manipulation

$$\eta = -\{2.303RT/(1-\beta)nF\} \cdot \log_{10} j_0 + \{2.303RT/(1-\beta)nF\} \cdot \log_{10} j \tag{4.30}$$

or

$$\log_{10} j = \log_{10} j_0 + \{(1-\beta)nF/2.303RT\} \cdot \eta \tag{4.31}$$

Both the anodic and cathodic plots are shown in Fig. 4.7.

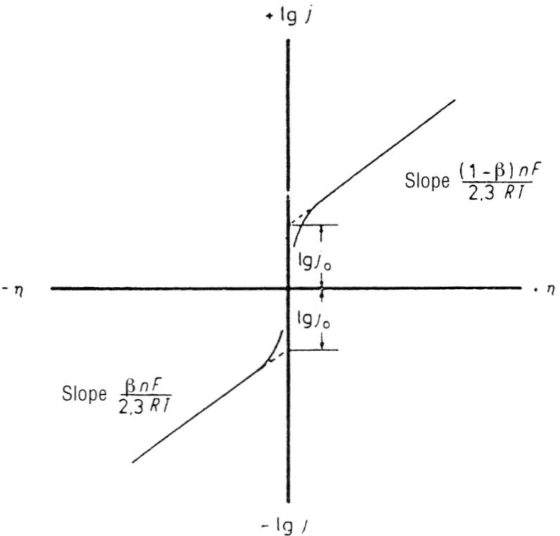

Fig. 4.7 Graphical representation of the Tafel equation for $\beta = 0.5$, with the current plotted on a logarithmic scale.

At very small overpotentials ($\eta <$ ca. 10 mV), the exponents in the Butler-Volmer equation are small enough to allow us to expand both exponential terms; remembering that $e^x \approx 1 + x$ for small x, we find

$$j = j_0\{1 + ((1-\beta)nF\eta/RT) - [1 + \beta nF\eta/RT]\} \tag{4.32}$$

$$= j_0 \cdot nF\eta/RT \tag{4.33}$$

In the region, therefore, of $|\eta| \leq 10$mV, the current-potential plot will be linear, with a slope that depends only on the value of j_0 and not on β. Thus j_0 can be determined solely from a knowledge of the current at low overpotentials. Furthermore, since j has the units of current density on (4.33) and η the units of volts, it follows that (RT/j_0nF) must have the units of Ωm^2 or Ωcm^2; it is usually referred to as the electron transfer resistance.

The exchange current density can vary substantially in magnitude: for example, j_0 has the value 10^{-9} A cm^{-2} for the oxygen reduction reaction, $^1/_2\, O_2 + 2H^+ + 2e^- \rightarrow H_2O$, on platinum and ca. 1 A cm^{-2} for the silver dissolution reaction, $Ag^+ + e^- \rightarrow Ag^0$. In fact, given that the linear region of the current-potential plot only exists for $|\eta| < 10$ mV, practical considerations of current measurement lead to the conclusion that j_0 cannot be measured by this method if it is less than ca. 10^{-5} A cm^{-2}.

A comparison of Figs. 4.4 and 4.6 shows that the *anodic* part of the current potential curve for the chlorine electrode does indeed show the behaviour expected from the Butler-Volmer equation above, but the *cathodic* part deviates to smaller values from the predictions of equation (4.25) at more negative overpotentials. In general, this type of deviation is found for all current-potential curves for large enough value of the overpotential, since at sufficiently high current densities, other limitations on the overall reaction rate become significant, as already discussed. In these high current-density regions, the logarithmic plots of (4.28) and (4.31) will, therefore, also fail to allow us to deduce either j_0 or β. In order to derive these parameters, other effects such as mass transport must either be corrected for, or the experiment must be conducted in such a way as to eliminate them. Such procedures are described in more detail below.

4.2.3
The Concentration Dependence of the Exchange-current Density

The exchange-current density j_0 is a measure of the rate of electron transfer at equilibrium, and is defined in equations (4.17) and (4.18). By substituting the value of the rest potential, E_r, from (4.19) into these two equations, remembering the definition of E^0 above, and setting $n = 1$ for simplicity, we have

$$j_0 = Fc_{\text{Red}}k_0^+ \cdot \exp\left\{\frac{(1-\beta)FE^0}{RT} + (1-\beta)\ln\frac{c_{\text{Ox}}}{c_{\text{Red}}}\right\} \tag{4.34}$$

$$= Fk_0^+\left(c_{\text{Ox}}^{1-\beta}c_{\text{Red}}^{\beta}\right) \cdot \exp\left\{\frac{(1-\beta)FE^0}{RT}\right\} \tag{4.35}$$

and for the cathodic branch

$$-j_0 = -FK_0^-\left(c_{\text{Ox}}^{1-\beta}c_{\text{Red}}^{\beta}\right) \cdot \exp\left\{-\frac{\beta FE^0}{RT}\right\} \tag{4.36}$$

By equating (4.35) and (4.36) we have the important relationship

$$Fk_0^+ \cdot \exp\left\{\frac{(1-\beta)FE^0}{RT}\right\} = Fk_0^- \cdot \exp\left\{-\frac{\beta FE^0}{RT}\right\} \equiv Fk_0 \tag{4.37}$$

where k_0 is defined by (4.37). Then we can write

$$j_0 = Fk_0\left(c_{\text{Ox}}^{1-\beta} \cdot c_{\text{Red}}^{\beta}\right) \tag{4.38}$$

If n electrons are transferred in a single step, then $j_0 = nFK_0\left(c_{\text{Ox}}^{1-\beta}c_{\text{Red}}^{\beta}\right)$. The *standard* exchange current density, j_0^0, is defined as the current density when the concentrations of reagents Ox and Red are in their standard states. If both Ox and Red are in solution, this implies that c_{Ox} and c_{Red} are of unit molarity, whereas if either one is a gas, the concentration will correspond to the standard condition of one atmosphere pressure at the relevant temperature, and for a solid, the concentration, like the activity, is taken as unity. Some examples are given in Table 4.1.

Table 4.1

System	Electrolyte	Temperature	Electrode	j_0/ A cm^{-2}	j_0^0/ A cm^{-2}	β
Fe^{3+}/Fe^{2+} (0.005M)	1M H_2SO_4	25°C	Pt	$2 \cdot 10^{-3}$	$4 \cdot 10^{-1}$	0.42
$K_3Fe(CN)_6/K_4Fe(CN)_6$ (0.02M)	0.5M K_2SO4	25°C	Pt	$5 \cdot 10^{-2}$	5	0.51
$Ag/10^{-3}$M Ag^+	1M $HClO_4$	25°C	Ag	$1.5 \cdot 10^{-1}$	13.4	0.35
$Cd/10^{-2}$M Cd^{2+}	0.4M K_2SO_4	25°C	Cd	$1.5 \cdot 10^{-3}$	$1.9 \cdot 10^{-2}$	0.45
$Cd(Hg)/1.4.10^{-3}$M Cd^{2+}	0.5M Na_2SO_4	25°C	Cd(Hg)	$2.5 \cdot 10^{-2}$	4.8	0.2
$Zn(Hg)/2.10^{-2}$M Zn^{2+}	1M $HClO_4$	0°C	Zn(Hg)	$5.5 \cdot 10^{-3}$	0.10	0.25
Ti^{4+}/Ti^{3+} (10^{-3}M)	1M acetic acid	25°C	Pt	$9 \cdot 10^{-4}$	0.9	0.45
H_2/OH^-	1M KOH	25°C	Pt	10^{-3}	10^{-3}	0.5
H_2/H^+	1M H_2SO_4	25°C	Hg	10^{-12}	10^{-12}	0.5
H_2/H^+	1M H_2SO_4	25°C	Pt	10^{-3}	10^{-3}	0.5
O_2/OH^-	1M KOH	25°C	Pt	10^{-6}	10^{-6}	0.7
O_2/H^+	1M H_2SO_4	25°C	Pt	10^{-6}	10^{-6}	0.75

4.2.4 Electrode Reactions with Consecutive Transfer of Several Electrons

For the overall reaction:

$$S_{Ox} + ne^- = S_{red}$$

the simultaneous transfer of n electrons is only likely if S_{ox} is a metal ion whose lower oxidation states are unstable. However, where such states are stable, electron transfer normally takes place in several consecutive steps, one example being the reduction of Cu(II) ions, which can take place as:

$$Cu^{2+} + e^- \rightarrow Cu^+$$

$$Cu^+ + e^- \rightarrow Cu^0$$

Quite generally, the realisation of multi-electron transfer processes will be through *stepwise* transfer of single electrons, and two cases can be distinguished: the first corresponds to all the elementary electron transfer steps having the same values of j_0 and β, and the second case would correspond to a succession of steps with differing j_0 and β values.

Case A

The overall n-electron reaction proceeds through p identical steps, each transferring $m = n/p$ electrons, where most commonly $m = 1$. An example is the electrochemical reduction of water to hydrogen in acid solution, where one well-established mechanistic possibility is:

$$\begin{aligned}&H^+ + e^- \rightarrow H_{ads}\\&H^+ + e^- \rightarrow H_{ads}\\&2H_{ads} \rightarrow H_2\end{aligned} \tag{4.39}$$

Clearly, in this case, $p = 2$, $n = 2$ and $m = 1$.

In general, for this case, the charge-transfer relationship has the form

$$j = j_0\left\{\exp\frac{(1-\beta)(n/p)F\eta}{RT} - \exp\left[-\frac{\beta(n/p)F\eta}{RT}\right]\right\} \tag{4.40}$$

and for small values of η,

$$j = \frac{nF}{pRT} j_0 \eta \tag{4.41}$$

An informative example of case A is the oxidation of hydrazine, N_2H_4 in alkaline electrolyte on platinum electrodes. The nett reaction is:

$$N_2H_4 + 4OH^- \rightarrow N_2 + 4H_2O + 4e^- \tag{4.42}$$

where n is clearly 4. Measurement of the Tafel slope under non-stationary conditions (to avoid interference from adsorbed reaction products) gives 118 mV at room temperature, implying a value of unity for (n/p) in equation (4.40) if β is close to 0.5. This, in turn, can be understood in terms of four successive identical one-electron transfer reactions, in which the four hydrogen atoms of hydrazine are successively oxidised as:

$$\begin{aligned}&(N_2H_4)_{ads} \rightarrow (N_2H_3)_{ads} + H_{ads}\\&H_{ads} + OH^- \rightarrow H_2O + e^- \quad \text{etc.}\end{aligned}$$

However, if the Tafel slope is measured under steady-state conditions of current flow, a value of 28 mV is obtained, corresponding to $(n/p) = 4$. However, it is most unlikely that this latter interpretation is correct, since it would imply that four electrons and four OH^- ions were transferred simultaneously, a most improbable conclusion. It is, in fact, far more probable that none of the electron-transfer steps is rate-limiting, and the rate-limiting process is a chemical step following the final electron transfer. In this case, the rate-limiting process is, in fact, desorption of chemisorbed N_2.

Case B

The overall reaction consists of a sequence of different electron transfer steps:

$$\begin{aligned}&S_{Red} \leftrightarrows S_{int,1} + e^-\\&S_{int,1} \leftrightarrows S_{int,2} + e^-\\&S_{int,2} \leftrightarrows S_{int,3} + e^- \quad \text{etc.}\end{aligned} \tag{4.43}$$

with symmetry factors β_1, β_2, β_3... exchange currents $j_{0,1}$, $j_{0,2}$, $j_{0,3}$... and equilibrium potentials E_1^0, E_2^0, E_3^0.... In general, this sequence can only be analysed mathematically in any simple way if the intermediate products $S_{int,i}$ are either bonded to the surface or consumed by a rapid subsequent reaction and do not diffuse into the bulk of the solution.

The simplest case is of two sequential one electron processes:

$$S_{Red} = S_{int} + e^-$$
$$S_{int} = S_{Ox} + e^- \tag{4.44}$$

(an example being $Tl^+ = Tl^{2+} + e^-$; $Tl^{2+} = Tl^{3+} + e^-$). If an overall steady-state current density j is flowing, then the current corresponding to each individual reaction $j_1(E) = j_2(E) = 0.5j$. We then have, for each reaction, using equations (4.14) and (4.16):

$$0.5j = Fc_{Red}k^+_{0,1}\exp\left\{\frac{(1-\beta_1)FE}{RT}\right\} - Fc_{int}(E)k^-_{0,1}\exp\left\{-\frac{\beta_1 FE}{RT}\right\} \tag{4.45}$$

$$0.5j = Fc_{int}(E)k^+_{0,2}\exp\left\{\frac{(1-\beta_2)FE}{RT}\right\} - Fc_{Ox}k^-_{0,2}\exp\left\{-\frac{\beta_2 FE}{RT}\right\} \tag{4.46}$$

We assume that c_{Ox} and c_{Red} are constant, or equivalently that electron transfer is rate limiting, but that $c_{int}(E)$ is explicitly a function of potential. At the rest potential, c_{int} takes the value c^0_{int}, and by developing the mathematics in exact analogy to equations (4.20) and (4.21) we find that

$$0.5j = j_{0,1}\left\{\exp\left[\frac{(1-\beta_1)F\eta}{RT}\right] - \frac{c_{int}(\eta)}{c^0_{int}}\exp\left[-\frac{\beta_1 F\eta}{RT}\right]\right\} \tag{4.47}$$

$$0.5j = j_{0,2}\left\{\frac{c_{int}(\eta)}{c^0_{int}}\exp\left[\frac{(1-\beta_2)F\eta}{RT}\right] - \exp\left[-\frac{\beta_2 F\eta}{RT}\right]\right\} \tag{4.48}$$

and eliminating $\left(c_{int}(\eta)/c^0_{int}\right)$ we find:

$$j(\eta) = \frac{2j_{0,1}j_{0,2}\left\{\exp\left[\frac{(2-\beta_1-\beta_2)F\eta}{RT}\right] - \exp\left[-\frac{(\beta_1+\beta_2)F\eta}{RT}\right]\right\}}{j_{0,2}\exp\left[\frac{(1-\beta_2)F\eta}{RT}\right] - j_{0,1}\exp\left[-\frac{\beta_1 F\eta}{RT}\right]} \tag{4.49}$$

This expression can be simplified at high anodic or cathodic potentials: at high anodic potentials, the positive exponentials will dominate and the current density reduces to:

$$j(\eta) = 2j_{0,1}\exp\left[\frac{(1-\beta_1)F\eta}{RT}\right] \quad \textit{for } \eta \gg \frac{RT}{F}$$

or

$$\log_{10} j = \log_{10} 2j_{0,1} + \frac{(1-\beta_1)F\eta}{2.303RT} \tag{4.50}$$

and at cathodic potentials, for $\eta \ll -RT/F$

$$j(\eta) = 2j_{0,2}\exp\left[-\frac{\beta_2 F\eta}{RT}\right]$$

or

$$\log_{10}|j| = \log_{10} 2j_{0,2} + \frac{\beta_2 F\eta}{2.303RT} \tag{4.51}$$

It is evident that in this case, the values obtained for the exchange current densities by extrapolation of the anodic and cathodic Tafel equations will, in general, differ as will the asymmetry factors β_1 and β_2 obtained from the slopes, since they belong to two different reactions.

4.2.5
Electron Transfer with Coupled Chemical Equilibria; the Electrochemical Reaction Order

In many cases, the overall electrochemical reaction involves additional species, S_1, S_2... in a reaction of the form

$$S_{Red} + \nu_1 S_1 + \nu_2 S_2 \rightleftharpoons \nu_m S_m + \nu_j S_j + S_{Ox} + ne^- \tag{4.52}$$

where the ν_i are termed stoichiometric coefficients. A simple example is the anodic dissolution of silver in cyanide solution:

$$Ag + 3CN^- \rightleftharpoons [Ag(CN)_3]^{2-} + e^-$$

In this case, the actual electron transfer step is both preceded and followed by chemical equilibria. Thus, on *reduction*, the mechanism at higher CN^- concentrations has been established (see below) as:

$$\begin{aligned} &[Ag(CN)_3]^2 \rightleftharpoons [Ag(CN)_2]^- + CN^- \\ &[Ag(CN)_2]^- + e^- \rightleftharpoons [Ag(CN)_2]^{2-} \\ &[Ag(CN)_2]^{2-} \rightleftharpoons Ag^0 + 2CN^- \end{aligned} \tag{4.53}$$

We recognise that the stoichiometric coefficients of participating species for the overall electrode reaction may differ from those for the several equilibria making up the nett reaction, as indeed is obvious from (4.53). The stoichiometric factors involved in the formation of S_{Ox} and S_{Red} are termed the electrochemical reaction orders, since it is these factors that appear in the rate equations. To see this, we consider the equilibria

$$S_{Ox} \rightleftharpoons \sum_{j=1}^{m} z_j^{Ox} S_j \tag{4.54}$$

and

$$S_{red} \rightleftharpoons \sum_{j=1}^{m} z_j^{Red} S_j \tag{4.55}$$

The concentration of S_{Ox} is clearly given by the equilibrium constant for (4.54). Neglecting activity coefficients, we find

$$c_{Ox} = K_{Ox} \cdot \prod c_j^{z_j^{Ox}} \tag{4.56}$$

and

$$c_{Red} = K_{Red} \cdot \prod c_j^{z_j^{Red}} \tag{4.57}$$

and the rate of the reaction $S_{Ox} + me^- \rightleftharpoons S_{Red}$ can then be obtained by substituting the first of these expressions into (4.14):

$$j^-(E) = -mFK_{Ox}\left(\prod c_j^{z_j^{Ox}}\right)k_0^- \cdot \exp\left[-\frac{\beta mFE}{RT}\right] \tag{4.58}$$

whence we find

$$\frac{\partial \ln |j^-|}{\partial \ln c_k} = z_k^{Ox} \tag{4.59}$$

and by analogy, we find

$$\frac{\partial \ln j^+}{\partial \ln c_k} = z_k^{Red} \tag{4.60}$$

where the partial derivatives imply that the potential E and all other concentrations are kept constant. Thus, the electrochemical reaction orders can be determined by measuring the change in current in the electron-transfer-limited region of the current-potential curve with concentration of reagent.

Because the electrochemical reaction orders play so important a role in the determination of reaction mechanisms in complex electrochemical processes, their accurate measurement is essential. A second approach is the extrapolation of the Tafel slope to E_r to determine the exchange-current density. From (4.58), we find

$$\ln j_0 = \ln mFK_{Ox}k_0^- + \ln \prod c_j^{z_j^{Ox}} - \frac{\beta mFE_r}{RT} \tag{4.61}$$

This expression can be differentiated with respect to any c_j, but great care must be taken here since E_r itself may be a function of c_j. For the overall reaction (4.52), the Nernst equation will take the form

$$E_r = E^0 + \frac{RT}{nF} \cdot \sum_{j=1}^{m} \nu_j \ln c_j \tag{4.62}$$

and incorporating this into (4.61) and differentiating with respect to c_j, we find

$$\frac{\partial \ln j_0}{\partial \ln c_k} = z_k^{Ox} - \beta \cdot (m/n).\nu_k \tag{4.63a}$$

By analogy, we find that

$$\frac{\partial \ln j_0}{\partial \ln c_k} = z_k^{Red} + (1-\beta) \cdot (m/n) \cdot \nu_k \tag{4.63b}$$

Since the values of β, m, n and ν_k can be determined independently, the concentration dependence of the exchange-current density then determines the reaction order for any component. Following this procedure, the reduction mechanism (4.53) was established by varying the CN^- concentration. In this case, assuming $m = n = 1$, equations (4.63a) and (4.63b) can be written:

$$\frac{\partial \ln j_0}{\partial \ln c_{CN}^-} = z_{CN^-}^{Ox} - \beta \nu_{CN^-} = z_{CN^-}^{Red} + (1-\beta)\nu_{CN^-}$$

and the value of ν_{CN^-} from the nett electrochemical reaction $[Ag(CN)_3]^{2-} + e^- \rightarrow Ag + 3CN^-$ is given by

$$\nu_{CN^-} = \frac{F}{RT} \cdot \frac{\partial E_0}{\partial \ln c_{CN^-}} = -3$$

from the Nernst equation. At high concentrations, $\frac{\partial \ln j_0}{\partial \ln c_{CN^-}} = +0.44$, consistent with the above mechanism, where clearly $z^{red}_{CN^-} = 2$, $z^{ox}_{CN^-} = -1$ and $\beta \approx 0.48$. At lower concentrations of cyanide than 0.2M a clear change in mechanism takes place, with $\frac{\partial \ln j_0}{\partial \ln c_{CN^-}} = -0.25$. This is consistent with $z^{red}_{CN^-} = 1$, $z^{ox}_{CN^-} = -2$ and $\beta \approx 0.58$, suggesting that the mechanism is now:

$$[Ag(CN)_3]^{2-} = [AgCN] + 2CN^-$$
$$[AgCN] + e^- = [AgCN]^-$$
$$[AgCN]^- = Ag + CN^-$$

*4.2.5.1 Systematic Treatment of Coupled Reactions

The results described in sections 4.2.4 and 4.2.5 show that the measurement of Tafel slopes and reaction orders can often lead to important mechanistic insights, and for this reason, the theoretical treatment of complex coupled electrochemical and chemical reactions has been very fully developed, and some complex general results obtained. However, in practice, it is usually better to work out the values of reaction order and Tafel slope for specific mechanisms, since it is the values of z^{ox}_k, z^{red}_k and the Tafel slope that are diagnostic, and we need a simple way of determining these. We will follow a treatment described by Oldham and Ryland ("Fundamentals of Electrochemical Science", Academic Press, 1994), which allows us to write down a kinetic analysis for any arbitrary electrochemical reaction. We write down the electrochemical mechanism in the rather general form of the previous section and then follow a set of simple rules:

1. If an electrochemical step is rate limiting, we write it in the form

   $$R + \beta e^- \rightarrow S - (1 - \beta)e^-$$

   else if the rate-limiting step is purely chemical, we write it as

   $$R \rightarrow S$$

2. All the steps prior to the rds are written with all species on the left-hand side, adjusting the signs of the stoicheiometric coefficients accordingly, and in a similar way, all the steps after the rds are written with all species on the right-hand side, again with adjustment in sign.
3. We now multiply or divide each of the equations by (invariably small) integers, such that when the entire set of equations is added then the stoichio-kinetic equation (that is the overall equation describing the reaction) is recovered *and all intermediates formed prior or post the reaction cancel.*

4. The rate equation for the overall reaction can now be written down with the Tafel slopes derived from the coefficient of the electron "concentration" on each side and the concentrations of species raised to the powers corresponding to their stoicheiometric coefficients in the stiochio-kinetic equation.

This will become easier with some examples. For the case of Tl^{3+} reduction above, using the same notation, if the first step is rate limiting we have:

$$\begin{aligned} \mathrm{Tl}^{3+} + \beta_2 \mathrm{e}^- &\rightarrow \mathrm{Tl}^{2+} - (1-\beta_2)\mathrm{e}^- \\ &= \mathrm{Tl}^{+} - \mathrm{e}^- - \mathrm{Tl}^{2+} \end{aligned}$$

Adding the two equations to eliminate Tl^{2+}, we obtain

$$\mathrm{Tl}^{3+} + \beta_2 \mathrm{e}^- = \mathrm{Tl}^{+} - (2-\beta_2)\mathrm{e}^-$$

and our current is

$$i_{\mathrm{T}} = Fk\left[c_{\mathrm{Tl}^+(aq)} e^{(2-\beta_2)F\eta/RT} - c_{\mathrm{Tl}^{3+}(aq)} e^{-\beta_2 F\eta/RT}\right]$$

whereas if the second step is rate limiting (with β_1) as the asymmetry parameter

$$i_{\mathrm{T}} = Fk[c_{\mathrm{Tl}^+(aq)} e^{(1-\beta_1)F\eta/RT} - c_{\mathrm{Tl}^{3+}(aq)} e^{-(1+\beta_1)F\eta/RT}]$$

It is evident that at high anodic potentials, the second step will be rate-limiting (since the exponent is smaller) and similarly the first step must be rate limiting at high cathodic potentials, in agreement with the analysis above.

An interesting case is the reduction of I_3^- in aqueous solution:

$$\begin{aligned} &\mathrm{I}_3^-(\mathrm{aq}) = \mathrm{I}_2(\mathrm{ads}) + \mathrm{I}^-(\mathrm{aq}) \\ &\mathrm{I}_2(\mathrm{ads}) = 2\mathrm{I}(\mathrm{ads}) \\ &\mathrm{I}(\mathrm{ads}) + \mathrm{e}^- \rightarrow \mathrm{I}^-(\mathrm{ads}) \end{aligned}$$

Let us follow the rules above: we have

$$\begin{aligned} &\mathrm{I}_3^-(\mathrm{aq}) - \mathrm{I}_2(\mathrm{ads}) - \mathrm{I}^-(\mathrm{aq}) = \\ &\mathrm{I}_2(\mathrm{ads}) - 2\mathrm{I}(\mathrm{ads}) = \\ &\mathrm{I}(\mathrm{ads}) + \beta\mathrm{e}^- \rightarrow \mathrm{I}^-(\mathrm{aq}) - (1-\beta)\mathrm{e}^- \end{aligned}$$

We now need to eliminate the intermediate species I_2(ads) and I(ads), and it is evident that we can do this by multiplying each of the first two equations by $^1/_2$; on addition of the first two equations, each divided by 2 to the last we have:

$$^1/_2\ \mathrm{I}_3^-(\mathrm{aq}) - {}^1/_2\ \mathrm{I}^-(\mathrm{aq}) + \beta\mathrm{e}^- \rightarrow \mathrm{I}^-(\mathrm{aq}) - (1-\beta)\mathrm{e}^-$$

Note that we do *not* cancel the I^-(aq) on the two sides of this stoichio-kinetic equation. The final expression for the current can now be written:

$$i_T = Fk\left[c_{1^-(\mathrm{aq})} e^{(1-\beta)F\eta/RT} - c^0\sqrt{\frac{c_{\mathrm{I}_3^-(\mathrm{aq})}}{c_{\mathrm{I}^-(\mathrm{aq})}}} e^{-\beta F\eta/RT}\right] = Fkc_{\mathrm{I}_3^-(\mathrm{aq})}^{\frac{(1-\beta)}{2}}\left[e^{(1-\beta)F\eta' RT} - e^{-\beta F\eta'/RT}\right]$$

where as above, c^0 is added to ensure the correct units and η again refers to the overpotential with respect to Nernstian potential E^0, or more accurately in this case the formal potential, E^0, measured with respect to unit molarities rather than activities, and η' refers to the overpotential with respect to the rest potential E_r. This is, in fact, close to the behaviour observed in practice, though perhaps a little surprisingly the value of β is only 0.21 rather than the expected ~ 0.5.

In a similar way, we can analyse the deposition of $[Ag(CN)_3]^{2-}$; for the high cyanide concentration mechanism:

$$[Ag(CN)_3]^{2-} - [Ag(CN)_2]^- - CN^- =$$
$$[Ag(CN)_2]^- + \beta e^- \rightarrow [Ag(CN)_2]^{2-} - (1-\beta)e^-$$
$$= Ag + 2CN^- - [Ag(CN)_2]^{2-}$$

and adding these we remove the two intermediate cyano-complexes to obtain

$$[Ag(CN)_3]^{2-} - CN^- + \beta e^- \rightarrow Ag + 2CN^- - (1-\beta)e^-$$

giving us immediately

$$i_T = Fk\left[\left(\frac{c_{CN^-}}{c^0}\right)^2 e^{(1-\beta)F\eta/RT} - \left(\frac{c_{CN^-}}{c^0}\right)^{-1} e^{-\beta F\eta/RT}\right]$$

in agreement with the analysis above, and where again η is defined with respect to the formal Nernstian potential, $E^{0'}$.

As another example, we briefly explore the hydrogen evolution reaction on Palladium metal, for which the reductive Tafel coefficient takes the unusual value of 2. Consider the mechanism:

$$H^+(aq) + e^- = H(ads)$$
$$2H(ads) \rightarrow H_2(aq)$$

Writing down the sequence, we have

$$H^+(aq) + e^- - H(ads) =$$
$$2H(ads) \rightarrow H_2(aq)$$

and multiplying the first equation by 2 and adding to the second to eliminate the intermediate H(ads) we have

$$2H^+(aq) + 2e^- \rightarrow H_2(aq)$$

and the relevant rate equation is now:

$$i_T = Fk\left[c_{H_2(aq)} - \frac{c^2_{H^+(aq)}}{c^0} e^{-2F\eta/RT}\right]$$

which again reproduces the behaviour on Pd, again provided η is measured with respect to the Nernstian potential at unit activities of the reactant and product.

Implicit in the above analysis is the assumption that the activity of any adsorbed species is solely a function of its coverage, and that the reactivity of any adsorbed species does not depend on the coverage save through a concentration term of the type θ or $(1-\theta)$. Thus for the hydrogen reaction above, the rates of the forward and reverse steps for the formation of H_{ads} from H^+ can be written:

$$v_1 = k_1 c_{H^+}(1-\theta_H)e^{-\beta_1 F\eta/RT}$$
$$v_{-1} = k_{-1}\theta_H e^{(1-\beta_1)F\eta/RT}$$

However, for many adsorbates, interaction between the adsorbed species leads to a dependence of the enthalpy of adsorption on coverage, which often takes a simple linear expression of the form:

$$\Delta H_\theta = \Delta H_0 + \gamma\theta$$

which leads to the Temkin adsorption isotherm discussed in more detail in section 4.5.1. This isotherm normally obtains only for coverages in the range 0.2 – 0.8, but in that range has a substantial effect on the form of the kinetics. This is because the activation energy for the formation of the adsorbed layer itself becomes coverage dependent. We might expect that the activation energy for the formation of an additional increment in coverage when the coverage is θ would have the form:

$$U_\theta = U_0 + \xi\gamma\theta$$

where ξ plays a role analogous to the asymmetry parameter β. To see how this effects the kinetic scheme consider again a simple discharge process:

$$X^+ + e^- = X_{ads}$$
$$2X_{ads} \rightarrow X_2$$

The rates of the forward and back reactions are now:

$$v_{+1} = k_1(1-\theta_X)c_{X^+}e^{-\beta F\eta/RT}e^{-\xi\gamma\theta/RT}$$
$$v_{-1} = k_{-1}\theta_X e^{(1-\beta)F\eta/RT}e^{(1-\xi)\gamma\theta/RT}$$

and if we assume that the second reaction above is rate-limiting, so that the forward and reverse steps in the first reaction are in quasi-equilibrium, then we find that

$$\gamma\theta_X/RT = -F\eta/RT + \ln(k_1/k_{-1}) + \ln[(1-\theta_X)/\theta_X] + \ln c_{X^+}$$

Now, provided that γ is not too small, the change in $\gamma\theta/RT$ will be much larger than the change in the term $\ln[(1-\theta_X)/\theta_X]$, at least over the coverage range 0.2 – 0.8, and we can write

$$\gamma\theta_X/RT \approx -F\eta/RT + \ln(k_1/k_{-1}) + \ln c_{X^+}$$

The rds in the scheme above will have the rate $v_2 = k_2\theta_X^2 e^{2\xi\gamma\theta/RT}$, and assuming that the coverage is given by the approximate result above, we have, for the cathodic current:

$$i_{cath} = 2Fk_2\theta_X^2 e^{2\xi\gamma\theta/RT} \approx 2Fk_2' \cdot e^{-2\xi\eta F/RT}c_{X^+}^{2\xi}$$

Comparing this to the expression when the normal Langmuir isotherm applies:

$$i_{\text{cath}} = 2Fk_2\theta_X^2 = Fk_2' \cdot e^{-2F\eta/RT} c_{X^+}^2$$

it can be seen that if $\xi \approx {}^1/_2$ the Tafel slope will double in the Temkin case and the power dependence on the concentration of X^+ will halve. Essentially this result derives from the fact that the presence of the Temkin term severely damps the changes in θ that would otherwise take place, reducing the change in current with potential and reducing also the effect of the variation in concentration.

The importance of this analysis is that even in rather simple cases, it can lead to fractional powers of concentration. If we take the simple mechanism

$$X^+ + e^- = X_{ads}$$

$$X_{ads} + e^- \rightarrow Y$$

and assume once again that the adsorption of X_{ads} follows a Temkin isotherm, then following the analysis above, we have, for the current:

$$i_{\text{cath}} = 2Fk_2\theta_X e^{\xi\gamma\theta/RT} \approx 2Fk_2' \cdot e^{-\xi\eta F/RT} c_{X^+}^{\xi}$$

Where, for $\xi \sim 0.5$, we see that we obtain a Tafel slope of 120 mV, apparently consistent with the *first* step being rate-determining, but with a current dependent on the *square-root* of the concentration of X^+. This type of behaviour has been reported for the oxidation of methanol on platinum.

4.2.6
Further Theoretical Considerations of Electron Transfer

The rate of simple *chemical* reactions can now be calculated with some confidence either within the framework of activated-complex theory or directly from quantum-mechanical first principles, and theories that might lead to analogous predictions for simple electron-transfer reactions at the electrode-electrolyte interface have been the subject of much recent investigation. Such theories have hitherto been concerned primarily with greatly simplified models for the interaction of an ion in solution with an inert electrode surface. Specific adsorption of electroactive species has been excluded, and electron transfer is envisaged only as taking place between the strongly solvated ion in the outer Helmholtz layer and the metal electrode. The electron transfer process itself can only be understood through the formalism of quantum mechanics, since the transfer itself is a tunnelling phenomenon that has no simple analogue in classical mechanics.

Our initial treatment above was concerned merely with the potential dependence of the overall free energy of activation, ΔG. By considering the physical situation of the electrode double layer, this free energy of activation can be identified with the *reorganisation energy* of the solvation sheath around the ion. We will carry through this idea in detail for the simple case of the strongly solvated Fe^{3+}/Fe^{2+} couple, following the change in ligand-ion distance as the critical reaction variable during the transfer process.

4.2.6.1 Calculation of the Critical Ligand-Ion Separation x_S

In aqueous solution, the reduction of Fe^{3+} can be conceived as a reaction of two aquo-complexes of the form:

$$[Fe(H_2O)_6]^{2+} = [Fe(H_2O)_6]^{3+} + e_M^- \tag{4.64}$$

The H_2O-molecules of these aquo-complexes constitute the inner solvation shell of the ions, which are, in turn, surrounded by an external solvation shell of more or less uncoordinated water molecules forming part of the water continuum and described in chapter 2. Owing to the difference in the solvation energies, the radius of the Fe^{3+} aquo-complex is smaller than that of Fe^{2+}, which implies that the mean distance of the vibrating water molecules at their normal equilibrium point must change during the electron transfer. Similarly, changes must take place in the outer solvation shell during electron transfer, all of which implies that the solvation shells themselves inhibit electron transfer. This inhibition by the surrounding solvent molecules in the inner and outer solvation shells can be characterised by an activation free energy ΔG.

Given that the tunnelling process itself requires *no activation energy*, and that tunnelling will take place at some particular configuration of solvent molecules around the ion, the entire activation energy referred to above must be associated with ligand/solvent movement.

Furthermore, from the Franck-Condon principle, the electron tunnelling process will take place on a time-scale rapid compared to nuclear motion, so that the ligand and solvent molecules will be essentially stationary during the actual process of electron transfer.

We now consider the aquo-complexes above, and describe by x the distance of the centre of mass of one of the water molecules constituting the inner solvation shell from the central ion. The binding interaction of these molecules leads to vibrations of frequency $f = \omega/2\pi$ taking place about an equilibrium point x_0, and if the harmonic approximation is valid, the potential energy change U_{pot} associated with the ligand vibration can be written in parabolic form (see Fig. 4.8) as

$$U_{pot} = {}^1/_2\, M\omega^2 (x - x_0)^2 + B + U_{el} \tag{4.65}$$

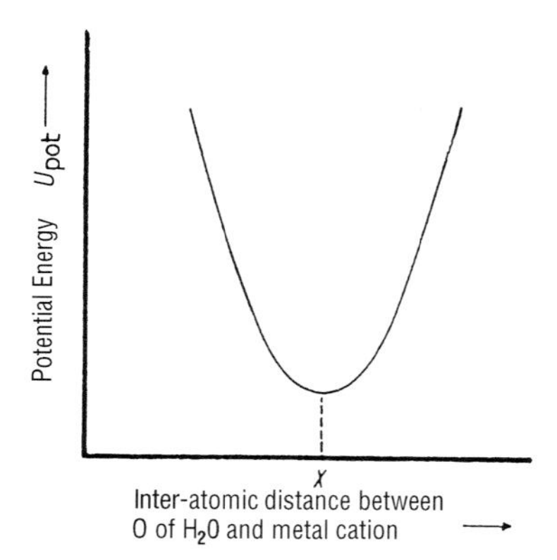

Fig. 4.8 Parabolic potential curve (in the harmonic oscillator approximation) for the for the binding between water molecules and the central metal ion in an aquo-complex.

where M is the mass of the ligands, B is the binding energy of the ligands and U_{el} is the electrical energy of the ion-electrode system. The total energy of the system will also contain the kinetic energy of the ligands, written in the form $p^2/2M$, where p is the momentum of the molecules during vibrations:

$$U_{tot} = p^2/2M + {}^1\!/_2\, M\omega^2(x - x_0)^2 + B + U_{el} \tag{4.66}$$

It is possible to write two such equations for the initial state, i, (corresponding to the reduced aquo-complex $[Fe(H_2O)_6]^{2+}$) and the final state, f, corresponding to the oxidised aquo-complex and the electron now present in the metal electrode. Clearly

$$U^i_{tot} = \frac{p^2}{2M} + {}^1\!/_2\, M\omega^2(x - x_0)^2 + B^i + U^i_{el} \tag{4.67}$$

with a corresponding equation for state f, and with the assumption that the frequency of vibration does not alter between initial and final states of the aquo-complex. During electron transfer, the system moves, as shown in Fig. 4.9 from an equilibrium situation centred at x^i_0, along the parabolic curve labelled U^i_{pot} to the point x_s where electron transfer takes place; following this, the system will move along the curve labelled U^f_{pot} to the new equilibrium situation centred on x^f_0.

The point at which electron transfer takes place clearly corresponds to the condition $U^i_{pot} = U^f_{pot}$; equating equations (4.67) for the states i and f we find that

$$x_S = \frac{B^f + U^f_{el} - B^i - U^i_{el} + (M\omega^2/2)([x^f_0]^2 - [x^i_0]^2)}{M\omega^2(x^f_0 - x^i_0)} \tag{4.68}$$

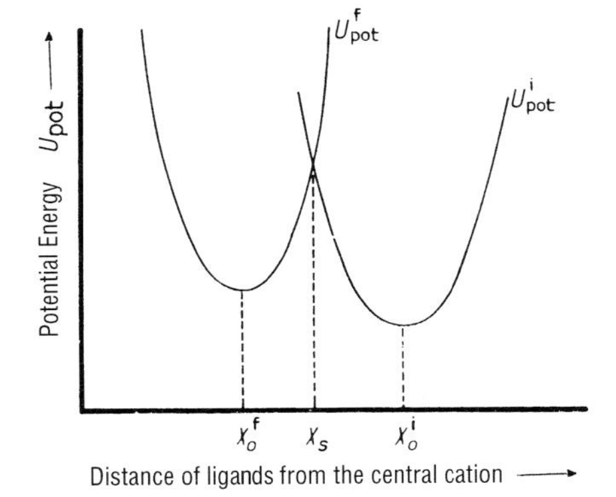

Fig. 4.9 Potential energy of a redox system as a function of ligand-metal separation.

4.2.6.2 Reorganisation Energy and the Electrochemical Rate Constant

The activation energy, E_{act}, is defined as the minimum additional energy above the zero-point energy that is needed for a system to pass from initial to final state in a chemical reaction. In terms of equation (4.67), the energy of the initial reactants at $x = x_s$ is given by

$$U_i = \frac{p^2}{2M} + \frac{M\omega^2}{2}(x_S - x_0^i)^2 + U_{el}^i + B^i \qquad (4.69)$$

where $U_{el}^i + B^i$ is the zero-point energy of the initial state. The *minimum* energy required to reach the point x_s is clearly that corresponding to $p = 0$. By substituting for x_s from equation (4.68), we find

$$U_{act} = \frac{M\omega^2}{2}(x_S - x_0^i)^2 = \frac{(U_S + U_{el}^f - U_{el}^i + B^f - B^i)^2}{4U_S} \qquad (4.70)$$

where U_s has the value $\frac{M\omega^2}{2}(x_0^f - x_0^i)^2$, and is termed the *reorganisation energy*, since it is the additional energy required to deform the complex from initial to final value of x. It is common to find the symbol λ for U_s, and model calculations suggest that U_s normally has values in the neighbourhood of 1 eV (10^5 Jmol^{-1}) for the simplest redox processes.

In our simple model, the expression in equation (4.70) corresponds to the activation energy for a redox process in which only the interaction between the central ion and the ligands in the primary solvation shell is considered, and this only in the form of the totally-symmetrical vibration. In reality, the rate of the electron-transfer reaction is also influenced by the motion of molecules in the outer solvation shell, as well as by other vibrational modes of the inner shell. These can be incorporated into the model provided that each type of motion can be treated within the simple harmonic approximation. The total energy of the system will then consist of the *kinetic* energy of all the atoms in motion together with the *potential* energy arising from each vibrational degree of freedom. It is no longer possible to picture the motion as in Fig. 4.9, as a one-dimensional translation over an energy barrier, since the total energy is a function of a large number of normal coordinates describing the motion of the entire system. Instead, we have two potential energy *surfaces* for the initial and final states of the redox system, whose intersection describes the reaction *hypersurface*. The reaction pathway will proceed now via the saddle point, which is the minimum of the total potential energy subject to the condition $U_{pot}^i = U_{pot}^f$ as above.

This is a standard problem, and essentially the same result is found as in equation (4.70), save that the B^i and B^f now become the sum over all binding energies of the central ion in the initial and final states and U_s is now given by

$$U_S = \sum_j \frac{M_j\omega_j^2}{2}(x_{j,0}^f - x_{j,0}^i)^2 \qquad (4.71)$$

where M_j is the effective mass of the j-th mode and ω_j the corresponding frequency, and we still retain the approximation that these frequencies are all the same for the initial and final states.

With the help of U_s, an expression for the rate constant for the reaction

$$[Fe(H_2O)_6]^{2+} \rightarrow [Fe(H_2O)_6]^{3+} + e_M^- \quad (4.72)$$

can be written

$$k_f = k_f^0 \exp\left(-\frac{U_{act}}{k_B T}\right) = A \cdot \exp\left[-\frac{(U_S + U_{el}^f - U_{el}^i + B^f - B^i)^2}{4U_S k_B T}\right] \quad (4.73)$$

where the magnitude of the so-called frequency factor A will be explored below, and e_M^- refers to an electron in the metal electrode. The rate constant for the back reaction is obtained by interchanging the indices i and f in (4.73); it will be observed that under these circumstances, U_s remains the same, and we obtain

$$k_b = A \cdot \exp\left[-\frac{(U_S + U_{el}^i - U_{el}^f + B^i - B^f)^2}{4U_S k_B T}\right] \quad (4.74)$$

4.2.6.3 Exchange-current Density and Current-Voltage Curves

We now want to derive an expression for the exchange-current density from (4.73) and, for simplicity, we assume that the reaction of interest is (4.72), with equal concentrations, c, of Fe^{2+} and Fe^{3+}.

We must incorporate into (4.73) and (4.74) the effect of the potential difference between the electrode and the outer Helmholtz-layer, ϕ, which we can do by incorporating into the electronic energy of the $Fe^{3+} + e_M^-$ system a potential-dependent term of the form:

$$U_{el}^f = U_{el,0}^f - e_0 \Delta\phi \quad (4.75)$$

where the minus sign in (4.74) arises through the negative charge on the electron. Inserting this into (4.73) and (4.74) gives:

$$j^+ = FAc \cdot \exp\left[-\frac{(U_S + U_{el,0}^f - U_{el}^i + B^f - B^i - e_0\Delta\phi)^2}{4U_S k_B T}\right] \quad (4.76)$$

$$j^- = -FAc \cdot \exp\left[-\frac{(U_S + U_{el}^i - U_{el,0}^f + B^i - B^f + e_0\Delta\phi)^2}{4U_S k_B T}\right] \quad (4.77)$$

At the equilibrium potential, $\Delta\phi_0$, the rates of these two reactions are equal, which implies that the terms in brackets in the two equations must also be equal when $\Delta\phi = \Delta\phi_0$. From this, we see that

$$e_0 \Delta\phi_0 = U_{el,0}^f - U_{el}^i + B^f - B^i \quad (4.78)$$

and if we introduce the overpotential, $\eta = \Delta\phi - \Delta\phi_0$, we see

$$j^+ = FAc \cdot \exp\left[-\frac{(U_S - e_0\eta)^2}{4k_B T U_S}\right] \quad (4.79)$$

$$j^- = -FAc \cdot \exp\left[-\frac{(U_S - e_0\eta)^2}{4k_B T U_S}\right] \tag{4.80}$$

from which we obtain the exchange-current density as

$$j_0 = FAc \cdot \exp\left[-\frac{U_S}{4k_B T}\right] \tag{4.81}$$

It can be seen that the activation energy of the exchange-current density is $U_s/4$. If the overpotential is small, such that $e_0\eta \ll U_s$ (and recalling that U_s lies in the region of $\sim$ 1eV), the quadratic form of (4.79) and (4.80) can be expanded with neglect of terms in η^2; recalling also that $e_0/k_B = F/R$, then we obtain

$$j = j^+ + j^- = FAc \cdot \exp\left(-\frac{U_S}{4k_B T}\right)\left\{\exp\left(\frac{F\eta}{2RT}\right) - \exp\left(-\frac{F\eta}{2RT}\right)\right\} \tag{4.82}$$

which is the familiar Butler-Volmer equation with a symmetry factor $\beta = 1/2$. This result arises from the strongly simplified molecular model that we have used above, and in particular the assumption that the values of ω_j are the same for all normal modes in both the oxidised and reduced states of the system. Relaxation of this assumption leads to values of β differing from 1/2.

For higher overpotentials, the approximations leading to (4.82) are no longer valid, and the full form of (4.79) or (4.80) must be used, for which $\log_{10} j$ is no longer proportional to η. In fact, the full form of these equations have proved very difficult to verify, since at higher current densities, as will be seen below, transport of electroactive material to the electrode normally becomes rate-limiting. Only recently have experimental data in the high overpotential region been obtained that convincingly demonstrate that equations (4.79) and (4.80) have the correct form.

4.2.7
Determination of Activation Parameters and the Temperature Dependence of Electrochemical Reactions

The expression for the exchange-current density in (4.24) takes the form

$$j_0 = nFc_{Ox}k_0^{'-} \exp\left\{-\frac{\Delta H_-^{(}E_r)}{RT}\right\} \cdot \exp\left\{\frac{\Delta S_-^{(}E_r)}{R}\right\}$$

from which, neglecting the temperature dependence of the pre-exponential terms and of ΔH_- and ΔS_-, we see that

$$\frac{\partial \ln j_0}{\partial(1/T)} = -\Delta H_-^{\bar{R}}$$

allowing us to determine the enthalpy of activation by plotting $\ln j_0$ vs. 1/T. From (4.24), it is also evident that within the same assumptions, $\frac{\partial \ln j_0}{\partial(1/T)} = -\Delta H_+^{\bar{R}}$, from which clearly $\Delta H_-^{(}E_r) = \Delta H_+^{(}E_r)$.

The experimental values of activation enthalpy vary considerably, even for the same electrochemical reaction on different electrodes; as an example, for the hydrogen electrode, values of ca. 70 $kJmol^{-1}$ are found on mercury and ca. 25 $kJmol^{-1}$ on nickel. An important point is that the activation energy determined from the measured current density at some fixed overpotential will *not* be the same as that determined from the exchange current density itself, as can be seen from (4.29), which clearly shows that the apparent activation enthalpy measured from $\left(\frac{\partial \ln j}{\partial(1/T)}\right)_\eta$ will be less than the true ΔH by an amount equal to $\beta nF\eta$. For $n = 1$ and $\beta = 0.5$, this corresponds to c. 50 $kJmol^{-1}V^{-1}$, an appreciable correction at high overpotential.

Finally, it should be remembered that for many electrochemical reactions, particularly at higher overpotentials, the rate-limiting step may well be diffusion of material to or from the electrode rather than the electron transfer process. Diffusion coefficients in aqueous solution show a much lower temperature dependence than normal chemical reactions, increasing only at the rate of 2 – 3 % per degree K, with the result that electrochemical reactions in the diffusion control region show little temperature sensitivity.

4.3 The Concentration Overpotential – The Effect of Transport of Material on the Current-Voltage Curve

In section 4.1, we established that the course of any electrochemical reaction depended on the interaction of charge transfer at the electrode, transport of electroactive material to and from the electrode and the possible occurrence of any coupled chemical reactions. Having considered the effects of charge transfer, we now turn to the interaction of this with transport in simple electrode reactions.

Once an electrochemical reaction is initiated, the concentrations of those species taking part in the reaction will normally be higher or lower at the electrode/electrolyte interface than in the bulk of the solution. Fig. 4.10 shows the formation of concentration profiles arising through the effects of simple diffusion being, in this case, insufficient to replenish the copper ions removed during the electroplating reaction. In more complex cases, the effects of forced or natural convection must be taken into account, the former arising from stirring and the latter from density differences in the liquid. If the electroactive species also make a significant contribution (more than a few percent) to the ionic conduction of the liquid, the effects of migration in the electric field at or near the electrode must also be taken into account.

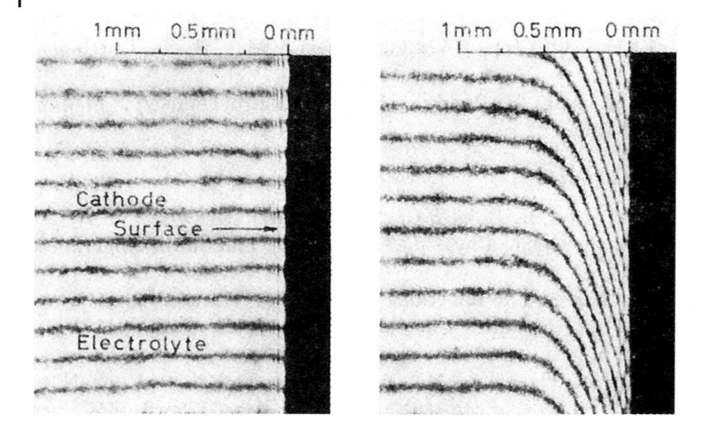

Fig. 4.10 Interferogram of the cathodic diffusion layer for deposition of copper from 0.05M $CuSO_4$: (a) at the beginning of the electrolysis process; (b) during current flow.

4.3.1
The Relationship between the Concentration Overpotential and the Butler-Volmer Equation

The key point at issue here is the fact that, save at very low current densities, transport to or from the electrode surface of those species which participate in the overall electrochemical process will not be sufficiently rapid to maintain the same concentration at the electrode surface as in the bulk. Consider a *reactant* with bulk concentration c^0 and electrode surface concentration c^s. If transport processes are rapid compared to charge transfer, then c^0 will be very little different from c^s, but under circumstances, where there is appreciable current flow, we normally find that $c^s < c^0$. This in turn leads to the necessity of increasing the potential in order to maintain the desired current density, and this increased potential is termed the *concentration overpotential.* The total overpotential can be written:

$$\eta_{tot} = \eta_{el} + \eta_{co} \tag{4.83}$$

where η_{el} refers to the charge-transfer overpotential discussed above and η_{co} to the concentration overpotential.

If, for simplicity, we assume that we have a high positive overpotential, so that $j \approx j^+$, then from (4.21), for $n = 1$, we have

$$j(\eta) = Fc^s_{Red} \cdot k_0^+ \cdot \exp\left[\frac{(1-\beta)FE_r}{RT} + \frac{(1-\beta)F\eta}{RT}\right] \tag{4.84}$$

and introducing the exchange-current density, when $c^s_{red} = c^0_{red}$, we have, from (4.18),

$$j(\eta) = j_0 \cdot \frac{c^s_{Red}}{c^0_{Red}} \cdot \exp\left[\frac{(1-\beta)F\eta}{RT}\right] \tag{4.85}$$

This can be re-written as:

$$\eta = \frac{RT}{(1-\beta)F} \ln\left\{\frac{j}{j_0} + \ln\frac{c^0_{\text{Red}}}{c^{\text{s}}_{\text{Red}}}\right\} \tag{4.86}$$

and a comparison of this with (4.83) shows that the concentration overpotential can be written as

$$\eta_{co} = \frac{RT}{(1-\beta)F} \cdot \ln\frac{c^0_{\text{Red}}}{c^{\text{s}}_{\text{Red}}} \tag{4.87}$$

In a similar way, for a cathodic process, we find

$$\eta = -\frac{RT}{\beta F} \ln\left\{\frac{|j|}{j_0} + \ln\frac{c^0_{\text{Ox}}}{c^{\text{s}}_{\text{Ox}}}\right\} \tag{4.88}$$

and

$$\eta_{co} = -\frac{RT}{\beta F} \cdot \ln\frac{c^0_{\text{Ox}}}{c^{\text{s}}_{\text{Ox}}} \tag{4.89}$$

4.3.2
Diffusion Overpotential and the Diffusion Layer

To proceed quantitatively, the various possible transport processes need to be considered separately, and we now turn to *diffusion*. Fig. 4.11 shows the concentration profile of a metal ion with distance from the electrode surface following the initiation of electro-deposition. If no current passes, the concentration profile is *flat*, with a constant value c^0. After turning on the current, the surface concentration decreases to the value c^{s}, as shown in curve 2 of (4.11). It can be seen that the concentration profile thus formed extends into the solution near the electrode surface by an amount δ_{N}, termed the Nernst diffusion layer thickness. The magnitude of δ_{N} is defined by the inter-

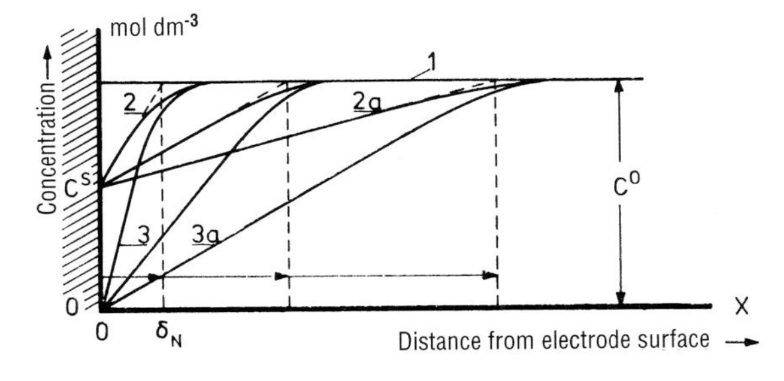

Fig. 4.11 Concentration distribution of an electrochemically active material in unstirred solution during deposition; the initial concentration was c^0, and evolution in time for two distinct values of c^{s} are shown, the second of the two distinct values being very close to zero (curves 3).

section of the tangent to the concentration profile near the electrode surface with the horizontal line c^0, and it is *time dependent*, increasing with the total charge passed until it reaches an approximately constant value termed the stationary diffusion layer thickness as shown in curve 2a. In unstirred solutions with no defined hydrodynamics, however, the value of δ_N is rather indeterminate, with the small density differences arising as electrolysis proceeds giving rise, in turn, to microscopic convection effects. This limits the stationary value of δ_N to ca. 0.5 mm, and attainment of the steady state is likely to take up to a minute. Where the solution is stirred, δ_N takes on a well-defined value, depending on the nature of the electrode, and a steady state is usually attained within one second, with δ_N values in the order of 10^{-4} cm.

If a sufficiently high overpotential is chosen, the value of c^s decreases to an extremely small value, and the current density effectively ceases to be determined by the electron-transfer rate and becomes entirely determined by the rate of diffusion. From Fick's first law, remembering that $j = nFJ$ (where J is the flux of electroactive material at the surface),

$$j = nFD\left(\frac{\partial c}{\partial x}\right)_{x=0} = nFD\frac{c^0 - c^s}{\delta_N} \tag{4.90}$$

which evidently tends for δ_N constant to a limiting time-independent value as $c^s \to 0$. This limiting value is termed the diffusion-limited current density, and has the value:

$$j_{\text{lim}} = nFD \cdot \frac{c^0}{\delta_N} \tag{4.91}$$

The corresponding overpotential region is termed the *limiting-current* region.

Equation (4.90) allows us to define the *diffusion overpotential*, η_d. Consider a simple reaction, such as metal deposition. The concentration of the metal ion M^{z+} at the surface will, if thermodynamic equilibrium is attained, be related to the standard electrode potential through the expression

$$E = E^0 + (RT/zF) \cdot \ln(c^0) \tag{4.92}$$

and if the electrochemical rate is extremely fast, so that Nernstian equilibrium is always maintained, then evidently

$$\eta_d = (RT/zF) \cdot \ln(c^s/c^0) \tag{4.93}$$

From (4.90) and (4.91), we see that

$$\frac{j}{j_{\text{lim}}} = 1 - \frac{c^s}{c^0} = 1 - \exp\left[\frac{zF\eta_d}{RT}\right] \tag{4.94}$$

whence

$$\eta_d = \frac{RT}{zF} \cdot \ln\left(1 - \frac{j}{j_{\text{lim}}}\right) \tag{4.95}$$

Equation (4.95) would describe the relationship between overpotential and current density *under the condition that the electron transfer rate constant was always extremely*

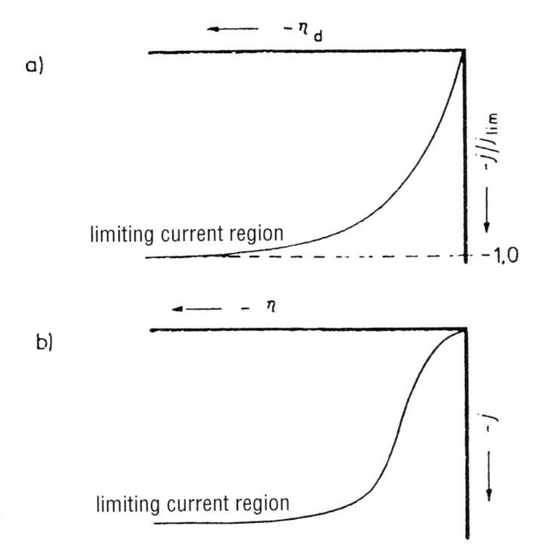

Fig. 4.12 (a) Dependence of the normalised cathodic current density, j/j_{lim} on the diffusion overpotential η_d; (b) experimental current-voltage curve derived from combining electron-transfer and diffusion limitations.

fast. The relationship is shown in Fig. 4.12(a), and whilst it may hold close to the limiting current values, it is evident from the typical experimental curve of Fig. 4.12(b) that it does not hold, save in exceptional circumstances, at low overpotentials, where, as expected, the rate of the electron-transfer process limits the reaction rate. In the intermediate region, the full equations given in (4.86) to (4.89) must be used.

4.3.3 Current-Time Behaviour at Constant Potential and Constant Surface Concentration c^S

If the situation envisaged at the end of the previous section were to obtain, such that the electron-transfer rate was always very fast, and the value of concentration c^s was time-independent, and determined by the Nernst equation, then prior to establishment of the micro-convection driven steady state, we would expect δ_N to increase steadily with time and the current density to fall. Quantitatively, to describe this we must use Fick's second law of diffusion, which can be expressed in the form

$$\frac{\partial c}{\partial t} = D\frac{\partial^2 c}{\partial x^2} \tag{4.96}$$

where, as before, x is the distance from the electrode, and the boundary conditions are:

$$t = 0,\ x \geq 0,\ c = c^0;$$
$$t > 0,\ x \to \infty,\ c = c^0; \quad t > 0,\ x = 0,\ c = c^S.$$

Solving (4.96) by standard methods described elsewhere (see references at the end of this chapter), we find

$$c(x,t) = c^0 - \left(c^0 - c^S\right)\left(1 - \mathrm{erf}\left\{x/\sqrt{4Dt}\right\}\right) \tag{4.97}$$

where erf{y} is a standard mathematical function called the error function tabulated in e.g. M. Abramovitch/Stegun: "Handbook of Mathematical Functions", Dover, New York, 1965). For small values of $x/(4Dt)^{1/2}$, we have

$$c(x,t) = c^0 - (c^0 - c^s)\left(1 - \frac{x}{\sqrt{\pi D t}}\right) \tag{4.98}$$

from which we can see that the slope of the concentration profile at the point $x = 0$ is

$$\left(\frac{\partial c}{\partial x}\right)_{x=0} = \frac{c^0 - c^s}{(\pi D t)^{\frac{1}{2}}} \tag{4.99}$$

and we immediately obtain from (4.90)

$$\delta_N = (\pi D t)^{\frac{1}{2}} \tag{4.100}$$

which will hold so long as the diffusion layer is not disturbed by the onset of natural convection (usually 30 – 60 sec.). From (4.90), it is evident that

$$j = nF\left(\frac{D}{\pi}\right)^{\frac{1}{2}} \cdot \frac{c^0 - c^s}{t^{1/2}} \tag{4.101}$$

which is shown graphically in Fig. 4.13. Under the circumstances in which the solution (4.96) is valid, it follows that a plot of j vs. $t^{-1/2}$ will be linear; if we take measurements in the potential region where $c^s \rightarrow 0$, the slope of this line will allow us to calculate the diffusion coefficient D. It should be pointed out that (4.101) is obviously *not* valid at extremely short times; the reason for this is that at very short times the concentration profile will be very steep, and physically it is not possible for the current to be limited initially by diffusion. In these very early times, the electron-transfer pro-

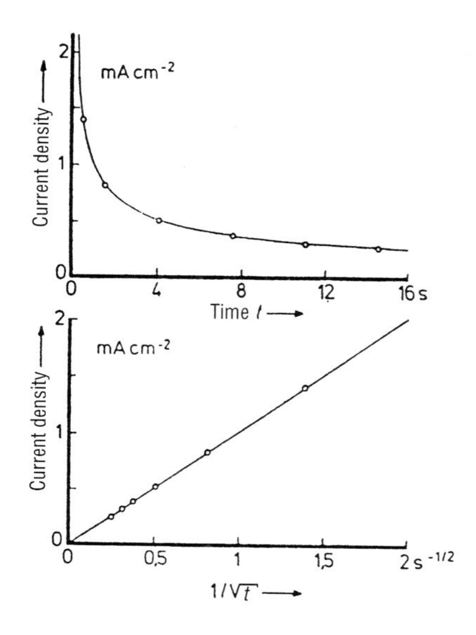

Fig. 4.13 The time-dependence of the current density at constant electrode potential (a) plotted vs. time; (b) plotted vs. $1/\sqrt{t}$.

cess will always be limiting. Establishment of a régime in which diffusion is limiting will normally take a fraction of a second.

4.3.4
Potential-Time Behaviour at Constant Current: Galvanostatic Electrolysis

Under circumstances in which diffusion controls the current, it follows that at constant current, the slope $(\partial c/\partial x)_{x=0}$ must also be constant. The boundary conditions for the solution of (4.96) are now

$$t = 0,\ x \geq 0,\ c = c^0$$

$$t > 0,\ x \rightarrow \infty,\ c = c^0; \quad t > 0,\ \left(\frac{\partial c}{\partial x}\right)_{x=0} = \frac{j}{nFD}$$

where j is the (constant) current flowing. Solving this equation for our electro-deposition reaction now leads to the behaviour shown in Fig. 4.14, in which the growing concentration profiles at different times are parallel. As the electrolysis proceeds, however, the concentration of the species being deposited decreases effectively to zero at the electrode surface, the total time taken being termed the transition time τ. Beyond this time, the current being forced through the system can no longer be sustained purely by the electro-deposition reaction, and the potential must now swing rapidly to allow another reaction (typically decomposition of the solvent) to take place. The experimental time-potential plot for the electro-deposition of Cd^{2+} from solution is shown in Fig. 4.15, where the second reaction is hydrogen evolution.

The solution of (4.96) under the conditions of constant current is:

$$c^s(t) = c^0 - \frac{2j}{nF} \cdot \left(\frac{t}{\pi D}\right)^{\frac{1}{2}} \tag{4.102}$$

and the transition-time, which corresponds to $c^s = 0$, is given by

$$j \cdot \tau^{\frac{1}{2}} = \frac{nF(\pi D)^{\frac{1}{2}} c^0}{2} \tag{4.103}$$

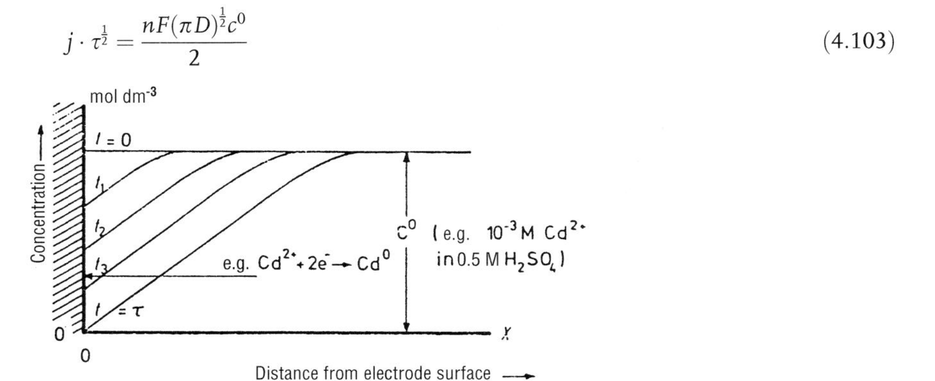

Fig. 4.14 Evolution of the concentration of an electrochemically active material during its deposition from an unstirred solution under galvanostatic conditions. The *transition time* is denoted by τ.

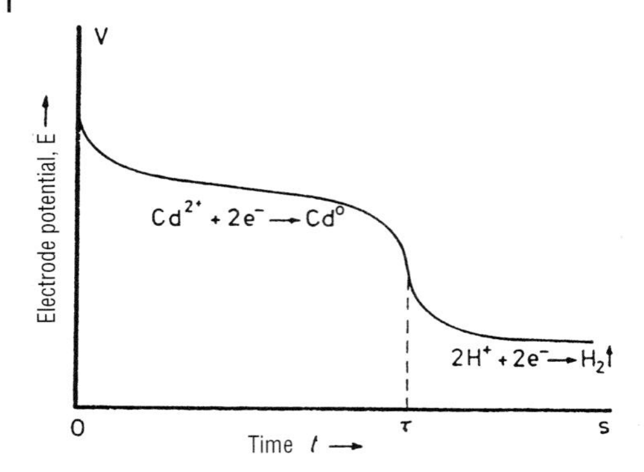

Fig. 4.15 Potential-time behaviour during galvanostatic electrolysis corresponding to Fig. 4.14.

The value of τ is clearly dependent on the current density, the metal-ion concentration and the diffusion coefficient. For the electrodeposition of Cd^{2+} from 10^{-3} M solution with a current density of 10 mA cm^{-2}, for example, $\tau \approx 3.1$ ms. By varying j, the product $j \cdot \tau^{1/2}$ will be found to be a constant; given D, this can be used analytically to find c^0.

4.3.5
Transport by Convection, Rotating Electrodes

If convective flow is imposed on a liquid near an electrode, a steady state can be rapidly established giving rise to a well-defined value of δ_N. The stronger the convection, the smaller the value of δ_N, and as long as we are well removed from the current-limiting region, the effects of transport can be neglected in comparison with the rate of electron transfer. Such vigorous convection can be effected by, for example, the rapid evolution of gas bubbles from the surface or, in a more controlled manner, by forming the electrode as a metal rod and rotating it rapidly.

As the current increases, however, the effects of convective transport will become more marked, and some more exact means of stripping out the transport contribution from the overall current-voltage behaviour is needed. The exact calculation of δ_N will depend on both the *type* and *magnitude* of the convection, and has been carried out for

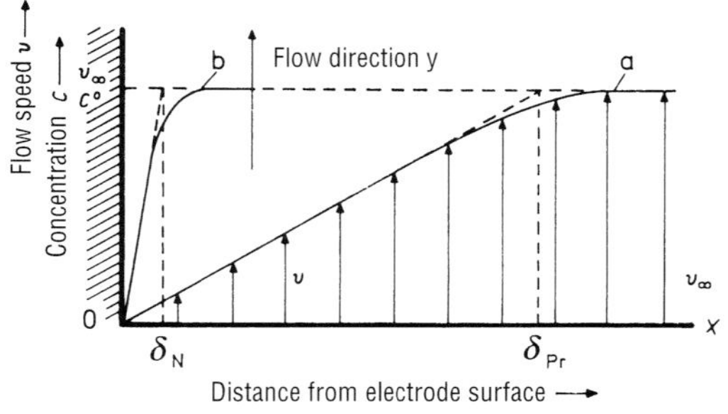

Fig. 4.16 (a) Flow speed and (b) concentration profiles at a flat electrode surface with laminar electrolyte flow.

rotating disk and ring electrodes as described below. Before turning to these solutions, however, it is instructive to consider the simple case of an electrode of arbitrary shape across whose surface a flow of liquid is maintained *parallel* to the surface and *perpendicular* to the diffusion direction. If the flow is *laminar* (ie there is no turbulence) and of velocity υ_{∞} well out from the electrode surface, then, owing to friction, at the electrode surface itself the velocity will be *zero*. A velocity profile will then be established, as shown in Fig. 4.16, with a characteristic length, δ_{Pr}, termed the Prandtl layer thickness, within which changes in fluid velocity are accommodated. This thickness is not the same as the diffusion length, δ_{N}, and it is possible to show that at steady-state in this configuration, $\delta_{\mathrm{N}} \approx \delta_{\mathrm{Pr}} \cdot (\nu/D)^{-1/3}$, where ν is the *kinematic viscosity* (defined as the ratio of the normal viscosity η to the density ρ). For water, with $\eta \approx 10^{-3}$ kgm^{-1}s^{-1} and $\rho = 10^{3}$ kgm^{-3}, and normal diffusion coefficients in the order 10^{-9} m^{2}s^{-1}, we find $\delta_{\mathrm{N}} \approx 10^{-1} \cdot \delta_{\mathrm{Pr}}$.

4.3.5.1 Rotating Ring and Ring-Disc Electrodes

A system of greater electrochemical importance is the rotating-disk electrode shown in Fig. 4.17, which rotates with angular velocity ω in solution, drawing up liquid along the rotation axis and flinging it out radially. Quantitative calculation of the convection pattern set up is possible using the general equation

$$\frac{\partial c}{\partial t} = D \cdot \mathrm{div}\,\mathbf{grad}(c) - \upsilon \cdot \mathbf{grad}(c) \qquad (4.104)$$

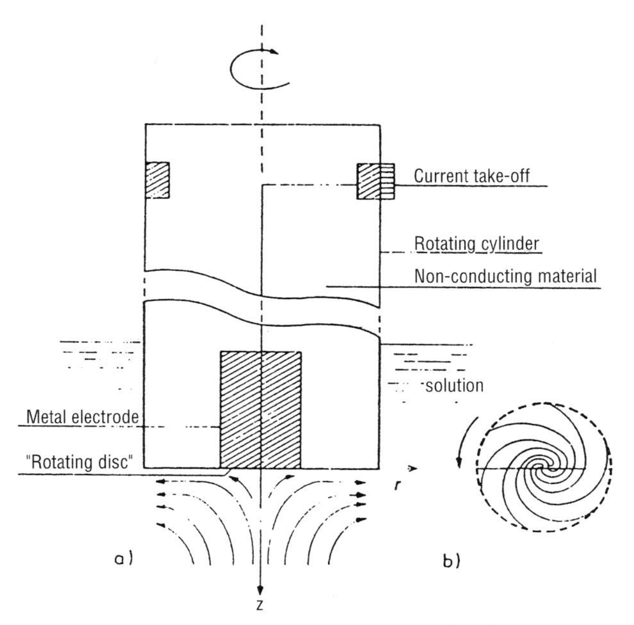

Fig. 4.17 (a) Schematic construction of a rotating-disc electrode and the radial flow of liquid during rotation; (b) schematic representation of the azimuthal flow at the disc during rotation.

where v is the vector velocity of the liquid, $\mathbf{grad}(c) \equiv \mathbf{i}\partial c/\partial x + \mathbf{j}\partial c/\partial y + \mathbf{k}\partial c/\partial z$, with $\mathbf{i}$, $\mathbf{j}$, $\mathbf{k}$ unit vectors in the x, y, z directions, and $\text{div}\mathbf{grad}(c) \equiv \partial^2 c/\partial x^2 + \partial^2 c/\partial y^2 + \partial^2 c/\partial z^2$. Transforming this into cylindrical coordinates and neglecting the radial diffusion term, we find, at the steady state, when $\partial c/\partial t = 0$

$$D\frac{\partial^2 c}{\partial z^2} = \upsilon_r \frac{\partial c}{\partial r} + \upsilon_z \frac{\partial c}{\partial z} \tag{4.105}$$

where υ_r and υ_z are the radial and axial components of the liquid velocity. To solve this equation we need expressions for υ_r and υ_z and appropriate boundary conditions. The latter are straightforward: if the electrode reaction consumes the material rapidly, then at $z = 0$, $c = 0$ and for $z \to \infty$, $c \to c^0$. The convection pattern was solved by Cochran (see refs. in the article by Riddiford), who showed that close to the disk surface (i.e., for small z)

$$\upsilon_z \approx -az^2 \tag{4.106}$$

$$\upsilon_r \approx arz \tag{4.107}$$

where $a = 0.510\omega^{3/2}\nu^{-1/2}D^{-1/3}$. Putting these formulae into (4.105) and solving for the concentration c on the assumption that $c(z = 0) = 0$ (i.e. we are in the limiting current region), we obtain, for the flux of material at the surface of the electrode:

$$\left(\frac{\partial c}{\partial z}\right)_{z=0} = 0.62D^{-1/3}\nu^{-1/6}\omega^{1/2}c^0 \tag{4.108}$$

where c^0 is the concentration of the electroactive species in the bulk of the solution, and where, as before, ν is the kinematic viscosity. Note that $(\partial c/\partial z)_{z=0}$ is independent of r, the radial coordinate, showing that the surface is uniformly accessible in this convective configuration. From this expression, comparing with (4.90) and remembering that $c^s = 0$ in our analysis, we see that δ_N for the disk is also independent of the radius, having the value:

$$\delta_N = 1.61D^{1/3}\nu^{1/6}\omega^{-1/2} \tag{4.109}$$

Note that under these circumstances, convective diffusion takes place equally over the entire disk surface. Taking typical values of $\nu(10^{-6}\ \text{m}^2\text{s}^{-1})$, $D\ (10^{-9}\ \text{m}^2\text{s}^{-1})$ and $\omega(628\ \text{s}^{-1}$, corresponding to 100 Hz rotation rate), we find, from (4.109), $\delta_N \approx 6.10^{-6}$ m or 6.10^{-4} cm. Given this, it is clear that the rotating disk itself must be well polished and free of scratches to ensure that the surface is truly flat on the $6 - \mu$m scale. The limiting current density on the disk is, from (4.91):

$$j_{\text{lim}} = \frac{nFDc^0}{\delta_N} = 0.62nFD^{2/3}\nu^{-1/6}\omega^{1/2}c^0 \tag{4.110}$$

and from this, the value of either D or c^0 may be obtained by plotting j_{lim} vs. $\omega^{1/2}$; linear plots are obtained as shown in Fig. 4.18, which also shows the overall variation of current density with electrode potential for a series of rotation speeds. In this latter case, it can be seen that at low overpotentials, the current follows the Butler-Volmer equation, following which, as the overpotential increases, there is a transition region in

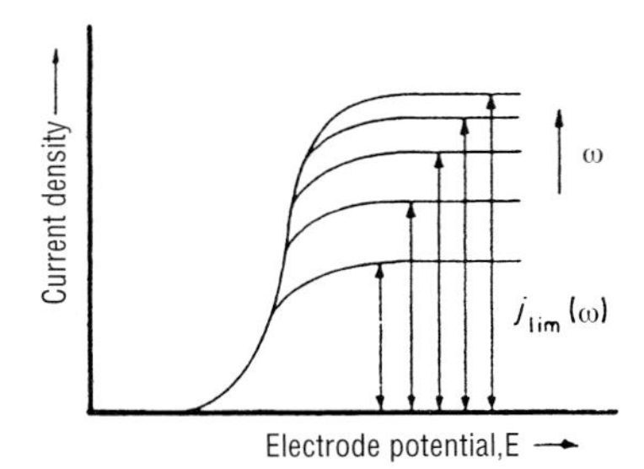

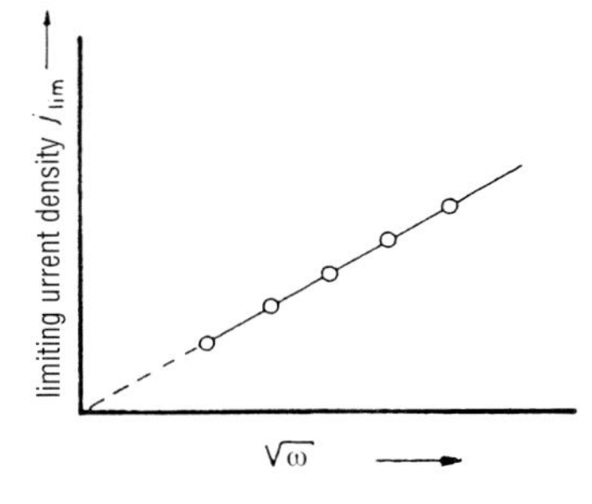

Fig. 4.18 Limiting electrochemical currents at a rotating disc electrode with angular rotation speed ω; (a) variation of the limiting current with potential at different values of ω; (b) plot of the limiting current against $\sqrt{\omega}$.

which both electron transfer and mass-transport rates play a role, before the limiting-current region, where only mass transport is rate determining.

Analysis of the transition region can be carried out straightforwardly. If we have a reversible electrochemical process, then the experimental current density will have the form

$$j(\eta) = nF(k^+ c^{\text{s}}_{\text{Red}} - k^- c^{\text{s}}_{\text{Ox}}) \tag{4.111}$$

where $k^+ \equiv k_0^+ \cdot \exp\left(\dfrac{(1-\beta)nF\eta}{RT}\right)$ and $k^- \equiv k_0^- \cdot \exp\left[-\dfrac{\beta nF\eta}{RT}\right]$. If mass transport were not rate limiting, the concentrations c^{s} would be identical to the bulk values c^0. The current that would be obtained under these circumstances, corresponding to infinite rotation speed and hence to pure electron-transfer limitation, can be represented by:

$$j_\infty(\eta) = nF(k^+ c^0_{\text{Red}} - k^- c^0_{\text{Ox}}) \tag{4.112}$$

It is also the case that the current must be related to the flux of material to the surface; in fact:

$$j = -\frac{nFD_{\text{Ox}}(c^0_{\text{Ox}} - c^{\text{s}}_{\text{Ox}})}{\delta_{\text{N,Ox}}} = \frac{nFD_{\text{Red}}(c^0_{\text{Red}} - c^{\text{s}}_{\text{Red}})}{\delta_{\text{N,Red}}} \tag{4.113}$$

where, as above, $\delta_{\text{N,ox}} = 1.61 D_{\text{ox}}^{1/3} \nu^{1/6} \omega^{-1/2}$ with a similar expression for $\delta_{\text{N,red}}$. Writing $A_{\text{ox}} = 1.61 \nu^{1/6} D_{\text{ox}}^{1/3}$, and similarly for A_{red}, we clearly see that

$$\frac{1}{j(\eta)} = \frac{1}{j_\infty(\eta)} \cdot \left\{ 1 + \frac{\left(\frac{k^+ A_{\mathrm{Red}}}{D_{\mathrm{Red}}} + \frac{k^- A_{\mathrm{Ox}}}{D_{\mathrm{Ox}}} \right)}{\omega^{\frac{1}{2}}} \right\}$$

or equivalently:

$$\frac{1}{j(\eta)} = \frac{1}{j_\infty(\eta)} + \left(\frac{1}{j_\infty(\eta)} \right) \cdot \frac{\text{const.}}{\omega^{1/2}} \tag{4.114}$$

It follows that in the transition region, a plot of $1/j$ vs. $\omega^{-1/2}$ will be linear, allowing us to determine j_∞ from the intercept, as shown in Fig. 4.19 at a series of potentials. From these values, β and j_0 may be estimated using the Tafel equation (4.27). Equation (4.114) is usually referred to in the literature as the Koutecky-Levich equation, though there seems little doubt that the equation was first used by Frumkin and Tedoradse. At high overpotentials, the limiting current will become potential independent since it will be solely determined by the transport effects, and the expression (4.114) simplifies to

$$1/j = 1/j_{\mathrm{lim}} = \delta_{\mathrm{N}}/nFDc = \frac{\text{const.}}{\omega^{\frac{1}{2}}} \tag{4.114a}$$

In this region the linear Frumkin-Tedoradse plots go essentially through the origin unless exceptionally rapid rotation rates are used.

For the more complex geometry of the *ring-disc electrode*, where a central disc (of radius r_1) is surrounded by a concentric ring of inner radius r_2 ($> r_1$) and an outer radius r_3, separated from the inner disc by an insulating ring ($r_1 < r < r_2$), then the value of δ_{N} at the ring is now a function of r, taking the form

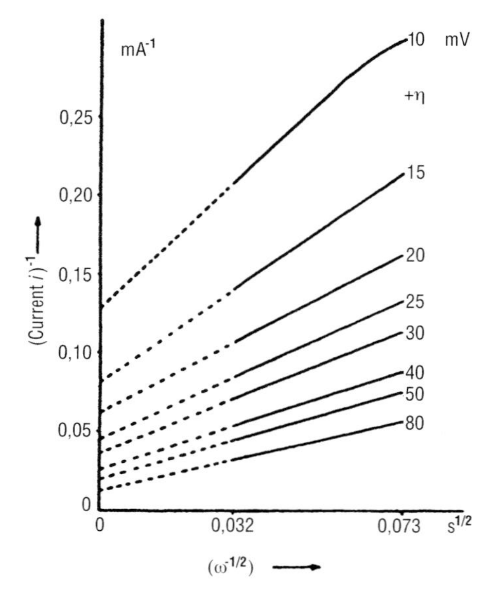

Fig. 4.19 Plot of inverse current in the rising part of the currentpotential curve for a 3.5 mm radius rotating disc vs. $1/\sqrt{\omega}$ at different electrode potentials. The electrolyte is 0.01M $[Fe(CN)_6]^{3-/4-}$ in 0.5M K_2SO_4.

$$\delta_N = \frac{1.61 D^{1/3} \nu^{1/6} \omega^{-1/2} (r_3^3 - r_2^3)^{1/3}}{r}; \quad r_2 \le r \le r_3 \tag{4.109a}$$

The limiting current (not current density) at the ring must be found by integrating the current density j_{lim} over the area of the ring. The final result has the form:

$$i_{lim,R} = 0.62\pi n F D^{2/3} \nu^{-1/6} \omega^{1/2} c^0 (r_3^3 - r_2^3)^{2/3} \tag{4.115}$$

Given that $i_{lim,D} = (2\pi r_1^2) \cdot j_{lim,D}$, we find

$$\frac{i_{lim,R}}{i_{lim,D}} = \frac{(r_3^3 - r_2^3)^{2/3}}{r_1^2} \tag{4.116}$$

which depends solely on the geometric properties of the system. We will consider the uses of the ring-disc electrode further below.

4.3.5.2 Rotating Films with Flat Surfaces for the Study of Porous Metal Structures

The equations developed above depend on three underlying assumptions: (i) the small deviations in the material transport at the edge of the disc can be ignored; (ii) the cell walls are sufficiently distant that they do not influence the motion of the rotating fluid and disc; (iii) the disc surface is absolutely flat. We now explore how the rotating disc can be used to study porous films where the last of these assumption might be expected to break down.

Measurements of porous catalyst layers (nano-particle structures) with the help of a rotating disc are possible if, for example, over such a layer a thin film of Nafion is introduced through which the electroactive substrate can diffuse *without* convection. The surface of the Nafion film is flat, and this allows the use of the equations above to account for transport to the surface of the film.

We must now modify equation (4.114) to take account of this further transport term of diffusion through the film:

$$1/j(\eta) = 1/j_\infty(\eta) + (1/j_\infty(\eta))\left(\frac{\text{const.}}{\omega^{\frac{1}{2}}}\right) + 1/j_{diff,f} \tag{4.114b}$$

where the first term on the right hand side of this equation corresponds to the electron-transfer kinetics, the second to the material transport in the electrolyte and the third to the transport of material by diffusion in the Nafion film:

$$j_{diff,f} = nFD_f c_f / L \tag{4.114c}$$

where D_f is the diffusion coefficient of substrate within the film, c_f the concentration of the substrate at the film-electrolyte boundary and L the film thickness. For $\omega \to \infty$, transport of substrate in the Nafion film will be rate-limiting, and a plot of $1/j$ vs. $\omega^{-1/2}$ will intercept the y-axis at a value of $1/j = 1/j_{diff,f} = L/nFD_f c_f$.

Fig. 4.19a shows an example for the case of the H_2-oxidation reaction on an electrode consisting of Pt dispersed on active carbon and held together with Nafion as a binder and covered with a thin Nafion layer of variable thickness L. The experimentally determined intercepts on the $1/j$ axis of the Frumkin-Tedoradse lines are, as expected, proportional to L, but only down to a film thickness of about 1 μm. In the lower part of the figure, the limiting current density is plotted against the reciprocal of the film thickness, and it is apparent that a limiting relationship is reached at about 0.5 μm. Below this thickness, diffusion is sufficiently rapid that it is no longer limiting and the electrode kinetics become the main limiting factor in the current. In the example shown, the limiting current density at 200 mV RHE is found to be 40 mA cm^{-2}, quite close to the measured limiting current density of 60 mA cm^{-2} on a flat Pt surface at the same overpotential. This experiment, with a Pt loading of 7 μg Pt cm^{-2}, allows us to calculate the maximum obtainable mass-specific current density for a H_2-based fuel cell; the result [T.J. Schmidt, H.A. Gasteiger, G.D. Stab, P.M. Urban, D.M. Kolb and R.J. Behm, J. Electrochem. Soc. **145** (1998) 2354] suggest that this maximum is of order 1 A mg^{-1} Pt (compare chapter 9).

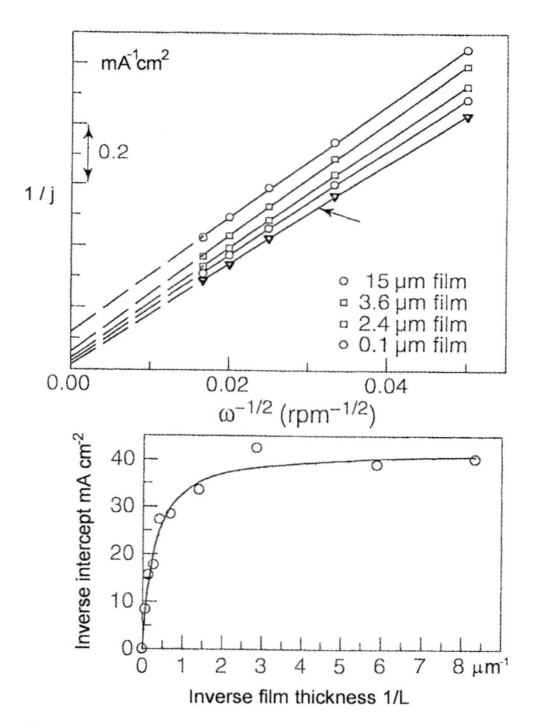

Fig. 4.19a H_2-oxidation at 200 mV vs. RHE on a Pt/active-carbon-powder electrode with Nafion as binder cast onto a rotating disc and covered with different thicknesses of Nafion film. The platinum loading was 7 μg Pt cm^{-2}, the electrolyte 0.5M H_2SO_4 and temperature 298 K. Upper figure: plot of $1/j$ vs. $1/\sqrt{\omega}$ (Frumkin-Tedoradse type plot); lower figure: dependence of the current density at the intercept on the inverse film thickness, $1/L$.

4.3.6
Mass Transport Through Migration – The Nernst-Planck Equation

We have not considered the transport of ions by *migration* up to now since, in order to make a significant contribution to the overall transport process, either the concentration of supporting electrolyte must be zero or, at the maximum, comparable in concentration to the electrochemically active species being transported. If we consider the first of these cases for simplicity, and consider ions of type i with an electrical charge $z_i e_0$ in an electrical field $\boldsymbol{E} = -i\partial V/\partial x$, travelling with a speed v_i, the migration current density is given by:

$$j_{i,\mathrm{M}} = z_i F c_i v_i = -z_i F c_i u_i \cdot \frac{\partial V}{\partial x} \qquad (4.117)$$

where we have used the mobility of the ion as defined in chapter 2. By comparison, the diffusion current density has the value:

$$j_{i,\mathrm{D}} = -z_i F D_i \cdot \frac{\partial c_i}{\partial x} \qquad (4.118)$$

The total current density now has the value

$$j_i = j_{i,\mathrm{M}} + j_{i,\mathrm{D}} = -z_i F\left(c_i u_i \cdot \frac{\partial V}{\partial x} + D_i \cdot \frac{\partial c_i}{\partial x}\right) \qquad (4.119)$$

This is one form of the Nernst-Planck equation, and it clearly shows the separability of migration and diffusion, and it also indicates clearly that the first of these can only make a substantial contribution if the potential gradient is sufficiently large. In most cases of practical interest, this condition is not satisfied, since the solutions used in electrochemical systems are too conducting, and diffusion dominates, even to the extent of ensuring that negatively charged ions may diffuse to the *cathode*. A good example of this latter behaviour is the electrodeposition of silver from solutions containing the $[Ag(CN)_2]^-$-ion.

Quantitatively, if we consider a solution containing, say, 0.1 *M* electroactive substance with a supporting electrolyte added to give an overall conductivity of 0.2 $^{-1}$ cm^{-1} (which would correspond to ca. 5 *M* NaCl), the limiting diffusional current density is $FDc^0/\delta_\mathrm{N} \approx 10$ mA cm^{-2} assuming $\delta_\mathrm{N} \approx 10^{-2}$ cm. At this current density, in a solution of conductivity 0.2 $^{-1}$ cm^{-1}, the potential gradient is $0.01/0.2 = 0.05$ Vcm^{-1}, and from (4.117), taking the ionic mobility as $5 \cdot 10^{-4}$ cm^{2}V^{-1}s^{-1}, we have $j_\mathrm{M} \approx 0.25$ mA cm^{-2}, which is only 2.5 % of the diffusion-limited value. It follows that there must be an approximately hundred-fold excess of supporting electrolyte if migration effects are to be ignored.

4.3.7
Spherical Diffusion

The diffusion layer thickness, δ_N, for a curved surface whose radius of curvature, r_0, is much larger than δ_N can be calculated without serious error using expressions derived for a planar electrode (see 4.100), but this approximation ceases to be valid as r_0 decreases. To see this, it is necessary to solve the diffusion equation for spherical geometry; for the boundary condition $c^s = 0$, the concentration, $c(r, t)$ at radius r and time t is given by:

$$c(r, t) = c^0 \left\{ 1 - \frac{r_0}{r} \operatorname{erfc}\left[\frac{r - r_0}{(4Dt)^{1/2}} \right] \right\} \tag{4.120}$$

where erfc$\{z\}$ is the complement of the error function defined above (4.97), and is simply 1 – erf$\{z\}$. A comparison of the concentrations of electroactive species for the cases of planar and spherical diffusion ($r_0 = 10^{-4}$ cm) following a step in potential into the diffusion controlled region, as shown in Fig. 4.20, demonstrates the fact that a *stationary* current density is established quite quickly in the spherical diffusion case. From (4.120) in fact, we obtain, on differentiating at $r = r_0$

$$j_{\text{lim,sph}} = nFDc^0 \left[\frac{1}{(\pi Dt)^{1/2}} + \frac{1}{r_0} \right] \tag{4.121}$$

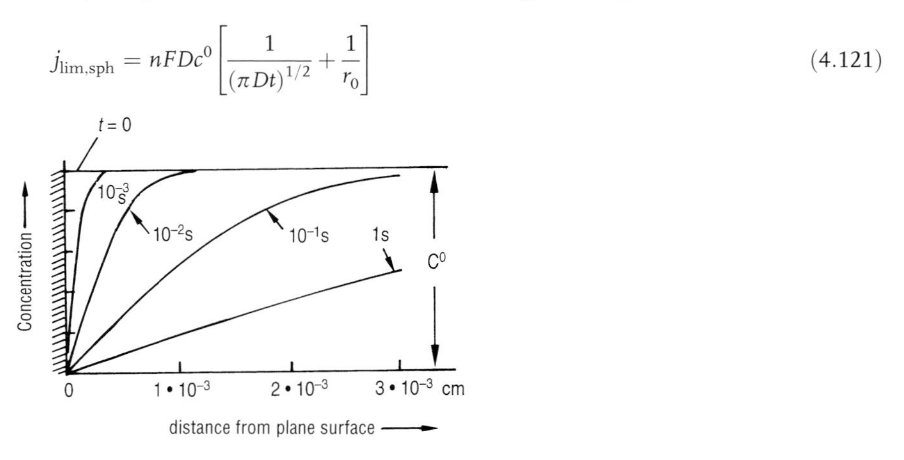

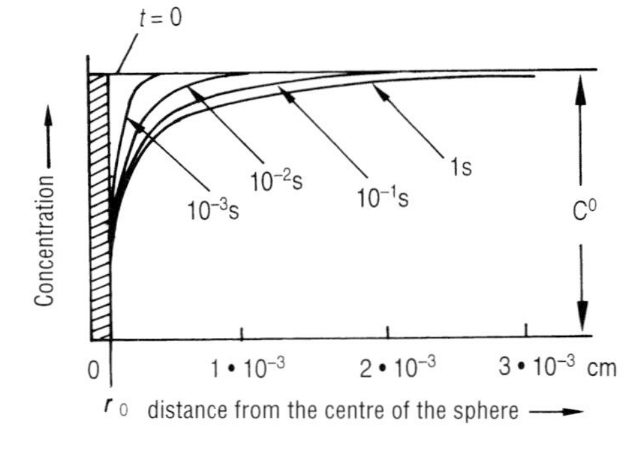

Fig. 4.20 The evolution of the concentration of an electroactive material at (a) a flat electrode; (b) a spherical electrode ($r_0 = 10^{-4}$ cm) with time. At time $t = 0$, the potential is stepped into the limiting current region. The initial concentration is c^0 and the diffusion coefficient is 10^{-5} cm^{2}s^{-1}.

At short time, this reduces to (4.101) with $c^s = 0$, but at longer times

$$j_{\text{lim,sph}} \approx \frac{nFDc^0}{r_0} \tag{4.122}$$

giving

$$\frac{j_{\text{lim,sph}}}{j_{\text{lim,plan}}} \approx 1 + \frac{(\pi Dt)^{1/2}}{r_0} \tag{4.123}$$

For large values of r_0, the ratio of limiting currents will be close to unity, but for $r_0 = 10^{-4}$ cm and $t = 1$s the ratio is ca. 50 (for $D = 10^{-5}$ cm^{2}s^{-1}). Of course, the actual measured current is very small, since the overall surface area is small; for a sphere of radius 10^{-4} cm, for example, the limiting *current* is only ca. 1 nA. The advantage of such small currents is, however, that two-electrode operation becomes possible, since many reference electrodes show negligible polarisation. This has led to a new generation of sensors based on two-electrode operation employing micro and *ultra-micro* ($r_0 \leq 20\ \mu$m) electrodes.

4.3.8 Micro-electrodes

Micro-electrodes are defined as having at least one electrode dimension in the micron range, the commonest type being fabricated by embedding a wire of a few micron

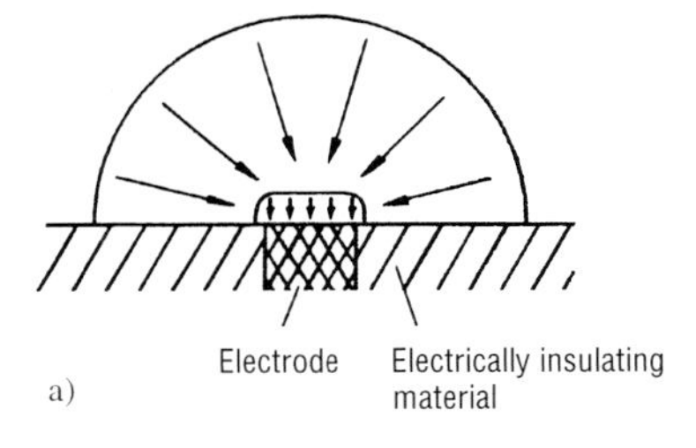

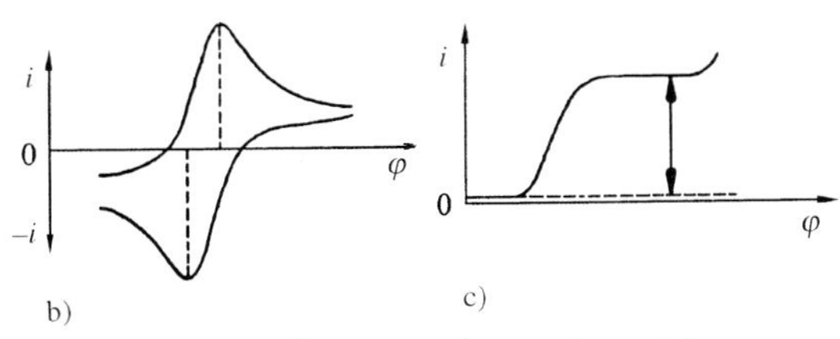

Fig. 4.21 Transport of reactant and current/potential curves at micro-electrodes: (a) transition from linear to spherical diffusion at more extended potential excursions; (b) cyclic voltammogram for rapid potential sweep rates (0.5 – 100 Vs^{-1}); (c) cyclic voltammogram for a slow potential sweep (5 – 100 mVs^{-1}).

diameter in an insulating material, such as a thermo-setting resin, which is then polished down to expose the end of the wire, as illustrated in Fig. 4.21(a). The importance of these electrodes is (i) very little current passes so they can be used in highly resistive electrolytes; (ii) their small size makes them suitable for analytical measurements on very small volumes of analyte, and (iii) there is a transition from linear to *spherical* diffusion at increasing time of measurement, as shown in Figs. 4.21(b) and 4.21(c). As indicated above, the latter diffusion profile is characterised not only by much enhanced transport rates to the electrode but by the onset of *stationary* diffusion at long times. Thus, in cyclic voltammetric measurements, high scan rates give rise to CV's of normal appearance, but at lower scan rates, the current rises to a limiting value which is proportional to the concentration of electroactive reagent.

4.3.8.1 Time Dependence of the Current at a Microelectrode

From (4.121), assuming fast electrochemical kinetics and that the flat surface can be approximated by a hemisphere, the total current, i is obtained by multiplying (4.121) by the area of the hemisphere, $2\pi r^2$:

$$i = -nFc^0 D\left\{2r^2\left[\frac{\pi}{Dt}\right]^{\frac{1}{2}} + 2\pi r\right\} \tag{4.124}$$

At long times, the first term in the brackets vanishes, giving

$$i = -2\pi r n F c^0 D \tag{4.124a}$$

or, in terms of current density, $j = i/2\pi r^2$

$$j = -nFc^0 D/r \tag{4.124b}$$

From (4.124), the time at which the first term becomes small compared to the second is when $2r^2[\pi/Dt]^{1/2} < 2\pi r$, or when

$$t > r^2/\pi D \tag{4.124c}$$

For a radius r of 10 μm, this will be true for values of t in excess of ca. 0.1 s.

4.3.8.2 Further Advantages of Microelectrodes

By mounting the microelectrode on a micromanipulator, and maintaining the separation between the microelectrode and a large (i.e. macroscopic) planar electrode at a few microns, the current distribution at the macro-electrode can be measured. This device is termed a *scanning electrochemical microscope*, and was first described by Bard: it has proved very powerful in characterising heterogeneous electrode reactions such as corrosion.

The very small surface area of the microelectrode also means that the double-layer capacitance is extremely small. In turn, this leads to a very small RC time constant, allowing a dramatic improvement in the ratio of faradaic to capacitive currents in analytical measurements.

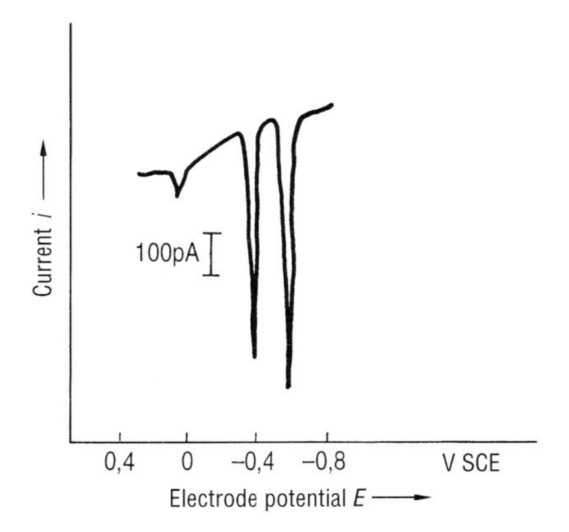

Fig. 4.22 Inverse polarographic determination of 8.7×10^{-8} M Pb^{2+} and 1.6×10^{-7} M Cd^{2+} at a mercury drop of 7 μm radius following deposition at −0.8 V vs. SCE; sweep rate 50 mV s^{-1} [from S. Pons and M. Fleischmann, Anal. Chem. **59** (1987) 1391A].

The improved mass-transport properties facilitate the measurement of higher exchange-current densities and electron-transfer rate constants, as well as allowing the study of fast coupled chemical reactions. This improved mass transport can also be exploited in the improvement of detection limits in electro-analytical measurements, as shown in Fig. 4.22.

The very small currents also allow meaningful measurements in poorly conducting solutions. This has allowed the development of electrochemistry in such solvents as CH_2Cl_2 and even superfluid CO_2. It also allows the use of much more dilute electrolytes, as shown in Fig. 4.23, where the oxidation of ferrocene in acetonitrile on a gold microelectrode and on a platinum *macro*electrode disc are compared.

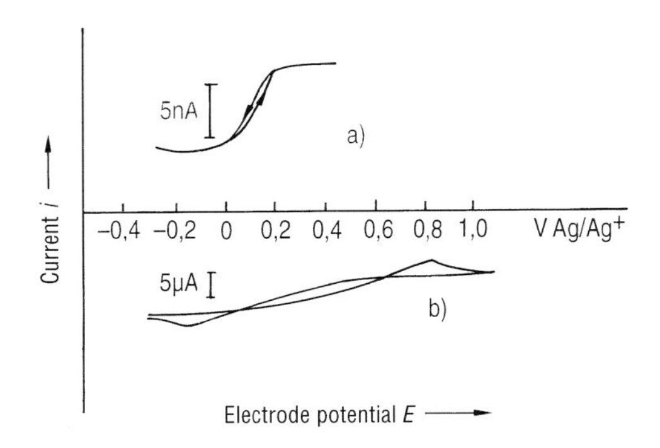

Fig. 4.23 Cyclic voltammogram for the oxidation of ferrocene in acetonitrile (1.1 mM with 0.01 M tetra-butyl ammonium perchlorate as the supporting electrolyte) on: (a) a gold microelectrode of 6.5 μm radius; (b) a platinum disc of radius 0.4 mm [from R.M. Wightman and D.O. Wipf: "Voltammetry at Ultramicroelectrodes", Electroanal. Chem. **15**, ed. A.J. Bard, Marcel Dekker, New York].

4.4 The Effect of Simultaneous Chemical Processes on the Current Voltage Curve

In addition to charge transfer and mass transport limitations on the rate of an electrochemical reaction, the presence of pre- or post-electron transfer chemical equilibria may also restrict the rate, particularly in those regions where electron transfer and mass transport are expected to be intrinsically rapid. Such chemical reactions may be either *homogeneous* or *heterogeneous*, the former taking place primarily in a thin liquid layer near the electrode, and the latter restricted to the adsorption layer on the electrode surface.

One example of a pre-electron transfer chemical equilibrium is the dissociation of a weak acid, HA, which may precede the evolution of hydrogen by electron transfer to the H^+ generated. The overall process may be described in terms of a series of elementary steps:

1. Transport of HA present in the main part of the solution through convection and diffusion to the reaction layer of thickness δ_R.
2. Dissociation of HA in the reaction layer to form protons and anions in order to compensate for the removal of protons by electrochemical reduction in stage 3 below

   $$\mathrm{HA} \rightarrow \mathrm{H}^+ + \mathrm{A}^-; \text{ rate constant } k_d$$

   It should be emphasised that owing to the finite rate at which dissociation takes place, the concentration of H^+ ions is always smaller than would be maintained in the absence of any electrochemical process. In addition, although water ionisation might, in principle, also take place to maintain the H^+ concentration, this process is sufficiently slow to be neglected.
3. Electron transfer to protons situated in the double layer, which are reduced to atomic hydrogen adsorbed on the electrode surface

   $$\mathrm{H}^+ + \mathrm{e}^- \rightarrow \mathrm{H}_{ads}$$

4. Further reaction of H_{ads} takes place to yield hydrogen, either through

   $$\mathrm{H}^+ + \mathrm{e}^- + \mathrm{H}_{ads} \rightarrow \mathrm{H}_2$$

   or

   $$2\mathrm{H}_{ads} \rightarrow \mathrm{H}_2$$

By increasing the mass transport rate, and by working at high overpotentials, the rate of step 2 becomes steadily more rate-limiting.

4.4.1
Reaction Overpotential, Reaction-limited Current and Reaction Layer Thickness

If, through the presence of an associated chemical reaction, the concentration of some species S that takes part in the electron transfer step is reduced, then there will appear a component of the overpotential associated with the reduction of the concentration from c^0 to c^s. This is termed the reaction overpotential, η_r. We can derive an expression for η_r by considering the Nernst equation: if the stoicheiometric number for the species S is ν, then arguments similar to those that led to (4.93) show that

$$\eta_r = \frac{\nu RT}{nF} \ln \frac{c^s}{c^0} \tag{4.125}$$

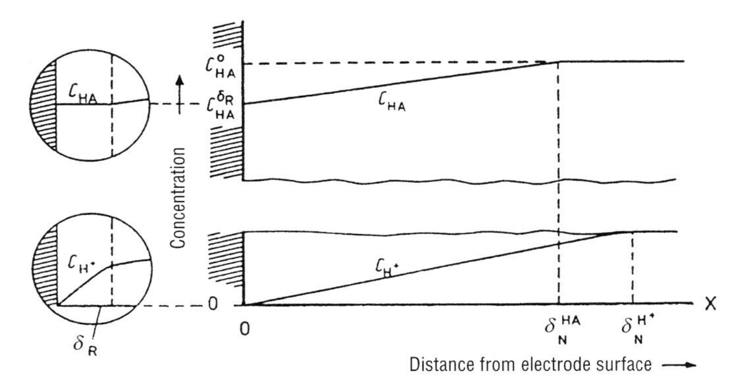

Fig. 4.24 Dependence of proton and undissociated acid concentrations during the anodic evolution of hydrogen from a solution containing a weak acid.

We can clarify the concentration profile of S near the electrode surface by returning to our example of the dissociation of a weak acid. The reaction layer can be defined through the condition that any proton formed by ionisation of HA moves towards the electrode and is discharged more rapidly than it can recombine with the anion A^-. We shall show below that this layer is *very* thin compared to normal diffusion layer thicknesses. In the region in front of the electrode, the concentration of the weak acid is reduced through dissociation and loss of H^+, and a diffusion layer of thickness δ_N^{HA} is set up, which is similar in thickness to the diffusion layer for H^+, $\delta_N^{H^+}$. In Fig. 4.24, assuming that the acid is *weak*, and that the reaction is layer very thin, we can assume that the concentration of HA in this layer is effectively constant as c^s_{HA}. Within this reaction layer as well, c_{H^+} decreases from its value of $c_{H^+}^{\delta_R}$ at δ_R to zero at the electrode surface.

The limiting current is clearly given by:

$$j_{\text{lim,r}} = nFD_{HA}\frac{c^0_{HA} - c^{\delta_R}_{HA}}{\delta^{HA}_N} \tag{4.126}$$

since this is the limiting flux of HA into the electrode region, and we presume that HA is responsible for all the H^+ discharged. Evidently, the limiting current is also given by

$$j_{\text{lim,r}} = \frac{nFD_{H^+}c^{\delta_R}_{H^+}}{\delta_R} \tag{4.126a}$$

The calculation of the diffusion layer thickness is carried through by solving the appropriate diffusion equation modified by chemical kinetic terms. At steady state:

$$\frac{\partial c_{H^+}}{\partial t} = D_{H^+}\cdot\frac{\partial^2 c_{H^+}}{\partial t^2} - k_f c_{HA} + k_b c_{A^-} c_{H^+} = 0 \tag{4.127}$$

where k_f is the rate constant for acid dissociation and k_b that for recombination. A similar expression can be obtained for H^+:

$$\frac{\partial c_{H^+}}{\partial t} = D_{H^+} \cdot \frac{\partial^2 c_{H^+}}{\partial t^2} + k_f c_{HA} - k_b c_{A^-} c_{H^+} = 0 \tag{4.128}$$

with the boundary conditions:

$$x = 0,\ c_{H^+} = 0,\ \frac{\partial c_{HA}}{\partial x} = 0;\ x = \delta_R;\ c_{HA}^{\delta_R} = c_{HA}^S;$$

$$x > \delta_R,\ c_{H^+} = \frac{k_f c_{HA}}{k_b c_{A^-}} \tag{4.129}$$

The last of these conditions corresponds to the condition that save in the reaction layer itself, the equilibrium concentration of H^+ is maintained by the dissociation and recombination of the weak acid.

Making the substitution

$$c' = \frac{k_f c_{HA}}{k_b c_{A^-}} - c_{H^+} \tag{4.130}$$

it is clear that c' will correspond to the difference between the equilibrium proton concentration and that actually found in the reaction layer. Evidently, at $x = 0$, $c' = \frac{k_f c_{HA}}{k_b c_{A^-}}$ and for $x > \delta_R$, $c' = 0$.

By substituting into (4.127) and (4.128), we obtain the differential equation for c' in the region $x < \delta_R$ as:

$$\frac{\partial^2 c'}{\partial x^2} = \left\{ \frac{k_f}{c_{A^-} D_{HA} k_b} + \frac{1}{D_{H^+}} \right\} \cdot k_b c_{A^-} c' \tag{4.131}$$

which can be further simplified since for a weak acid, $k_f / c_{A^-} k_b \ll 1$. As a result

$$\frac{\partial^2 c'}{\partial x^2} \approx \frac{k_b c_{A^-} c'}{D_{H^+}} \tag{4.132}$$

Given that c_{A^-} is essentially a constant for $x < \delta_R$, this is easily solved, giving

$$c' = \left(\frac{k_f c_{HA}^S}{k_b c_{A^-}} \right) \cdot \exp\left[-\left(\frac{k_b c_{A^-}}{D_{H^+}} \right)^{\frac{1}{2}} x \right] \tag{4.133}$$

which clearly tends to zero exponentially with increasing x. We can associate δ_R with the coefficient of x in the exponent of (4.133), since for values of $x > \left(D_{H^+}/c_{A^-} k_b\right)^{1/2}$, c' rapidly approaches zero: hence

$$\delta_R \approx \left(D_{H^+}/c_{A^-} k_b\right)^{1/2} \tag{4.134}$$

The value of δ_R can be established for acetic acid from the known data: $D_{H^+} \approx 10^{-8}\ m^2 s^{-1}$, $c_{A^-} \approx 10^{-1}$ M and $k_b \approx 10^{10}$ mol.s^{-1}, from which we find $\delta_R \approx 3 \cdot 10^{-9}$ m, some three orders of magnitude smaller than δ_N values associated with strong convection.

It can easily be shown that (4.134) is the distance from the electrode within which protons can diffuse to the surface more rapidly than they recombine: the half-life of a proton in an electrolyte of A^- concentration c_{A^-} is $\tau_{H^+} = (\ln 2)/(k_b c_{A^-})$ and the distance diffused in this time will be $(2D_{H^+}\tau_{H^+})^{\frac{1}{2}} = 1.2(D_{H^+}/(k_b c_{A^-}))^{\frac{1}{2}} = 1.2\delta_R$.

By substituting back into the the flux equations, it is possible to show that the limiting current at the rotating disc is given by

$$\frac{i_{\text{lim}}}{\sqrt{\omega}} = \frac{i_{\text{lim,diff}}}{\sqrt{\omega}} - \frac{D^{1/6}(c_{A^-})^{1/2} i_{\text{lim}}}{1.62\nu^{1/6} K_{\text{eq}} \cdot \sqrt{k_b}}$$

where K_{eq} is the equilibrium constant for the acid dissociation process (which is, of course, equal to k_f/k_b), and $i_{\text{lim,diff}}$ the diffusion-limited current in the absence of any kinetic limit imposed by the chemical reaction ($k_f \to \infty$). It follows that a plot of $i_{\text{lim}}/\sqrt{\omega}$ vs. i_{lim} should allow k_b and hence k_f to be determined from K_{eq}. This has been verified for acetic acid, with a value of k_f of $5.4 \times 10^5\ \text{s}^{-1}$ being obtained.

In conclusion, homogeneous chemical reactions can be fairly easily treated in terms of their effects on the current, provided the potential and convection regions are chosen such that the primary limitation is indeed the coupled chemical process. Heterogeneous chemical reactions are much more difficult to characterise in this way, partly owing to the sensitivity of any surface adsorption process to the condition and history of the surface and the existence of adventitious poisons in the electrolyte. The reproducibility of current measurements when coupled heterogeneous reactions are present is, therefore, much poorer unless great care is taken, and we turn our attention to such processes below.

4.5 Adsorption Processes

In addition to the simple redox processes that we have considered up to now, in which electron transfer takes place between the electrode surface and species retained in the outer Helmholtz plane, there is a very wide range of electrochemical processes that involve physisorption or chemisorption of reactants and/or products on the electrode surface itself. This is true for all *electrocatalytic* processes, for which adsorption is a prerequisite. A typical example is the oxidation of hydrogen: in this case, each approaching hydrogen molecule is dissociatively chemisorbed on the surface, the necessary free energy being found from the enthalpy of adsorption. From the microscopic standpoint, surfaces are rarely homogeneous, and adsorption often involves interaction with *active sites* on the surface, such as isolated atomic clusters on otherwise flat planes or defects in the surface lattice. Such surface sites play an important role in the kinetics of adsorption.

In general, there is a relationship between the amount of material adsorbed on a surface and the concentration of that material in solution, c^0. This relationship varies with temperature, but if measured at a single temperature is termed the *adsorption isotherm* for that temperature. It is found that the form of the adsorption isotherm is dependent both on the interaction of neighbouring adsorbed species with each other and with the underlying substrate, and these forms therefore differ according to whether the adsorbed species is molecular, atomic or ionic. In practice, however, the differences predicted theoretically are often rather subtle and difficult to measure, and rather crude approximations due to Langmuir and Frumkin are frequently found to be sufficiently accurate.

4.5.1 Forms of Adsorption Isotherms

To simplify the mathematical treatment, it is usually assumed that adsorption is limited to a single monolayer. If we further assume that every adsorption site on the surface is equivalent, and that the enthalpy of adsorption at any one site is independent of the occupancy of any neighbouring site, we have an extremely simple model due to Langmuir. Thermodynamically, equilibrium will be established if the electrochemical potential of the adsorbate is equal to the electrochemical potential of the solute:

$$\tilde{\mu}_{\mathrm{ads}} = \tilde{\mu}_{\mathrm{sol}} \tag{4.135}$$

Neglecting any effects due to difference in electrical potential between inner Helmholtz layer and solution allows us to replace electrochemical by chemical potential, though this approximation will clearly be a poor one for the adsorption of ions or strongly dipolar molecules, and more complex isotherms then result. The chemical potential of the solution is given by:

$$\mu_{\mathrm{sol}} = \mu^0_{\mathrm{sol}} + RT\ln\frac{c^0}{c^*} \tag{4.136}$$

(with c^* a standard concentration, usually taken as 1 M) and the chemical potential of the adsorbate, with the assumptions made above, is easily shown to be

$$\mu_{\mathrm{ads}} = \mu^0_{\mathrm{ads}} + RT\ln\frac{\theta_0}{1-\theta_0} \tag{4.137}$$

where θ_0 is the *equilibrium* coverage of the adsorbate, defined as the ratio of the number of sites actually occupied by the adsorbate to the total number of possible adsorption sites.

Equating (4.136) and (4.137) we find:

$$\frac{\theta_0}{1-\theta_0} = \frac{c^0}{c^*}\exp\left[-\frac{\mu^0_{\mathrm{ads}} - \mu^0_{\mathrm{sol}}}{RT}\right] \tag{4.138}$$

or, writing $\Delta G^0_{\mathrm{ads}} = \mu^0_{\mathrm{ads}} - \mu^0_{\mathrm{sol}}$,

$$\frac{\theta_0}{1-\theta_0} = \frac{c^0}{c^*}\exp\left[-\frac{\Delta G^0_{\mathrm{ads}}}{RT}\right] \tag{4.139}$$

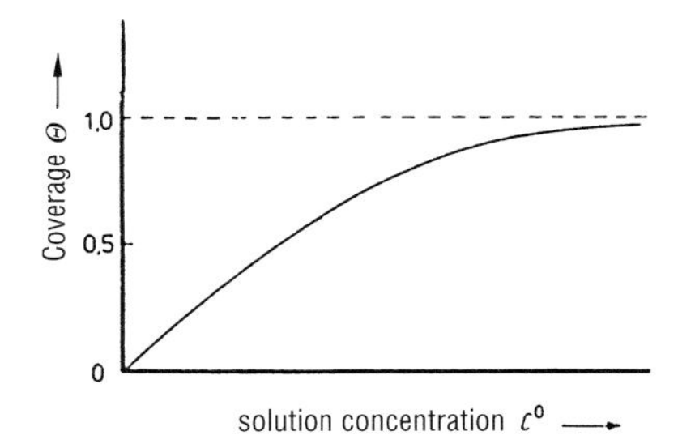

Fig. 4.25 Langmuir adsorption isotherm: coverage θ as a function of the solution concentration c^0 of adsorbate.

which is termed Langmuir's Isotherm. Re-writing this to express θ_0 as a function of (c^0/c^*), we find

$$\theta_0 = \frac{(c^0/c^*)\exp(-\Delta G^0_{ads}/RT)}{1+(c^0/c^*)\exp(-\Delta G^0_{ads}/RT)} \tag{4.140}$$

It can be seen that for (c^0/c^*) small, θ_0 will rise linearly with c^0, but at higher concentrations, the coverage will asymptotically tend towards unity as shown in Fig. 4.25.

The weakest assumption in Langmuir's treatment is the third, that the enthalpy of adsorption is independent of the occupancy of neighbouring sites. Whilst this assumption is reasonable for very low or very high coverage, for intermediate coverages, between 0.2 and 0.8, it frequently fails. A first approximation improvement to the model is then to allow ΔG^0_{ads} to depend linearly on coverage:

$$\Delta G_{ads} = \Delta G^0_{ads} + \gamma\theta_0 \tag{4.141}$$

where $\gamma > 0$ implies increasing repulsion between adsorbate species. Inserting this into (4.139) gives the Frumkin Isotherm:

$$\frac{\theta_0}{1-\theta_0} = \frac{c^0}{c^*}\exp\left[-\frac{\Delta G^0_{ads}}{RT} - \frac{\gamma\theta_0}{RT}\right] \tag{4.142}$$

For θ_0 values in the region of 0.5, this can often be further simplified since $\theta_0/(1-\theta_0)$ lies close to unity. Taking logarithms of (4.142) we see that

$$RT\ln(c^0/c^*) \approx \gamma\theta_0 + \Delta G^0_{ads} \tag{4.143}$$

which is termed the Temkin Isotherm, already introduced in section 4.2.5.

The *rate* of adsorption, υ_{ads}, can also be calculated within the Langmuir-Temkin framework. Within the Langmuir model,

$$\upsilon_{ads} = k_{ads}c^0(1-\theta) = k^0_{ads}c^0(1-\theta)\exp(-\Delta G^{/}_{ads}RT) \tag{4.144}$$

with a similar expression for υ_{des}. Equating υ_{ads} and υ_{des} gives the equilibrium coverage, θ_0, in a form identical to that of equation (4.140), provided we write $\Delta G_{ads} = \Delta G^{-}_{des}\Delta G_{ads}$. Given that

$$d\theta/dt = \upsilon_{ads} - \upsilon_{des} = k_{ads}c^0(1-\theta/\theta_0) \tag{4.145}$$

integration gives

$$\theta = \theta_0(1 - \exp[-k_{ads}c^0 t/\theta_0]) \tag{4.146}$$

Rather similar expressions can be written for the rates of adsorption and desorption in the Frumkin case. For example

$$\upsilon_{ads} = k^0_{ads}c^0(1 - \theta)\exp[-(\Delta G^{+}_{ads}\beta'\gamma\theta)/RT] \tag{4.147}$$

where β' is an asymmetry parameter, lying between 0 and 1, and γ is the interaction parameter defined above. Similarly

$$\upsilon_{des} = k^0_{des}\theta\exp[-(\Delta G^{-}_{des}(1 - \beta')\gamma\theta)/RT] \tag{4.148}$$

from which (4.142) can be recovered by equating υ_{ads} to υ_{des}. Integration of the rate equation is now less straightforward: if we assume that adsorption is slow and desorption very slow, so that $\theta_0 \to 1$, then we can write, for intermediate values of θ, approximately

$$\mathrm{d}\theta/\mathrm{d}t \approx k_{ads}c^0\exp(-\beta'\gamma\theta/RT) \tag{4.149}$$

from which we obtain, at long times

$$\theta \approx K + (1/\beta'\gamma)\cdot\ln t \tag{4.150}$$

where K is a constant.

For the adsorption of ions and molecules with large dipole moments, it is no longer possible to neglect the dependence of the free energy of adsorption on the electrode potential, nor can the simple assumption of linear change in ΔG_{ads} with coverage be sustained. The simplest assumption that can be made is that the coverage of *ions* increases with opposite charge density on the electrode, q_M, but it is also necessary to take account of interionic repulsions, the free energy of which has been calculated to depend on $\theta^{1/2}$ if image effects are ignored, and $\theta^{3/2}$ if they are included. Inclusion of these two terms leads to isotherms of the general form:

$$\ln\left[\frac{\theta}{1-\theta}\right] \approx K + \ln c^0 + Aq_M - B\theta^y \tag{4.151}$$

where A is a constant that depends on the gain in electrostatic energy on specific adsorption of an ionic species and B depends on the adsorbate/substrate dipole interactions. For neutral dipolar molecules, B depends on the dipole moment, and y has values of 3/2 – 5/2. The value of A may be unfavourable regardless of the sign of q_M since the displacement of water molecules, with large dipole moments, is rarely favourable. This leads to the frequent observation that the coverage of neutral molecules on the surface is usually a *maximum* when $q_M = 0$, ie at the point of zero charge.

4.5.2
Adsorption Enthalpies and Pauling's Equation

In a first approximation, we can treat chemisorption as chemical bond formation, and this allows us to estimate the enthalpy of adsorption using simple bond-energy formulae such as that of Pauling. A concrete example is that of hydrogen adsorption, where the process

$$H_2 \rightleftharpoons H_{ads} \tag{4.152}$$

has an enthalpy, ΔH_{ads} given by

$$\Delta H_{ads} = D_{HH} - 2D_{MH} \tag{4.153}$$

where D_{HH} is the dissociation energy of H_2 (and is a positive quantity) and D_{MH} the dissociation energy of the metal-hydrogen bond at the surface. The value of the latter can be approximated by Pauling's formula:

$$D_{MH} \approx \frac{D_{MM} + D_{HH}}{2} + 97{,}000(\chi_M - \chi_H)^2 \tag{4.154}$$

where χ_M, χ_H are the Pauling electronegativities of metal and hydrogen, and D_{MM} is a single metal-metal bond dissociation energy, which, depending on the metal lattice, will be between 1/4 and 1/6 of the metal sublimation energy and all energies are expressed in J mol^{-1}.

Inserting (4.154) into (4.153) we obtain

$$\Delta H_{ads} \approx -D_{MM} - 97{,}000(\chi_M - \chi_H)^2 \tag{4.155}$$

4.5.3
Current-Potential Behaviour and Adsorption-limited Current

Consider the following electrochemical process, in which electron transfer is preceded by chemisorption:

$$S_{ox} \underset{k_{des}^{ox}}{\overset{k_{ads}^{ox}}{\rightleftarrows}} S_{ox}^{ads}$$

$$S_{ox}^{ads} + e^- = S_{Red} \tag{4.156}$$

If mass transport from the solution to the near electrode region is rapid, and we assume that (4.156) is sufficiently rapid to maintain the equilibrium coverage θ_0 of S_{ox}^{ads}, then the current density is given by:

$$j = j^+ + j^- = F(1-\theta_0)c_{Red}^0 k_0^+ \exp\left[\frac{(1-\beta)F\eta}{RT}\right] - F\theta_0 k_0^- \exp\left[-\frac{\beta F\eta}{RT}\right] \tag{4.157}$$

where the factor $(1-\theta_0)$ in the first term arises because a fraction θ_0 of the sites are already occupied and cannot, therefore, accept another adsorbed S_{ox}. Provided θ_0 is given by the appropriate equilibrium isotherm, (4.157) provides a complete descrip-

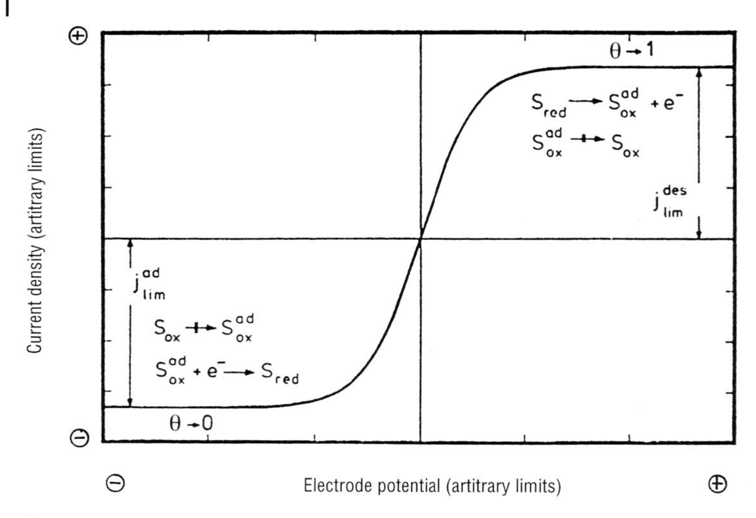

Fig. 4.26 Current density-overpotential curve combining charge transfer and adsorption processes, and leading to convection independent adsorption and desorption limiting currents.

tion, but as the overpotential for reduction of S_{ox} increases, the rate of adsorption of S_{ox} may become insufficient to maintain θ_0 at its equilibrium value; under these circumstances an *adsorption overpotential* will appear, limiting the current. Evidently, the coverage will become very small under these circumstances, and if the rate of adsorption is now given by $k_{ads}c_{ox}^0$ (*i.e.* we can assume that $(1 - \theta_0) \approx 1$), then the limiting cathodic current will be $-Fk_{ads}c_{ox}^0$. The anodic current will also be limited, since under these circumstances, the rate of *desorption* of S_{ox} will become limiting. In fact, it is evident that the limiting anodic current will be Fk_{des}. This behaviour is shown in Fig. 4.26.

4.5.4 Dependence of Exchange Current Density on Adsorption Enthalpy, the Volcano Curve

It is well known from gas-phase catalytic studies that if the adsorption enthalpy of an intermediate is too high, the overall reaction-rate will not be optimal, and that the most rapid reactions involve intermediate values of adsorption enthalpy. The reason for this is that if the adsorption enthalpy is too small, adsorption will be rate limiting, but if the adsorption enthalpy is too high, the intermediate is strongly stabilised, and its subsequent reactions become slow. In electrochemical reactions, this effect leads to a small *observed* exchange current density, j_0, since in general the adsorption-desorption steps and the electron transfer step cannot be uncoupled. If, therefore, j_0 is plotted against $|\Delta H_{ads}|$ values for a series of electrodes or electrocatalysts, such plots usually show a maximum at some intermediate ΔH_{ads} value, and the curve is often referred to, for this reason, as a 'volcano' curve. Fig. 4.27 shows an example: we show here a plot of the exchange current density, j_0, for hydrogen evolution on a large number of metals

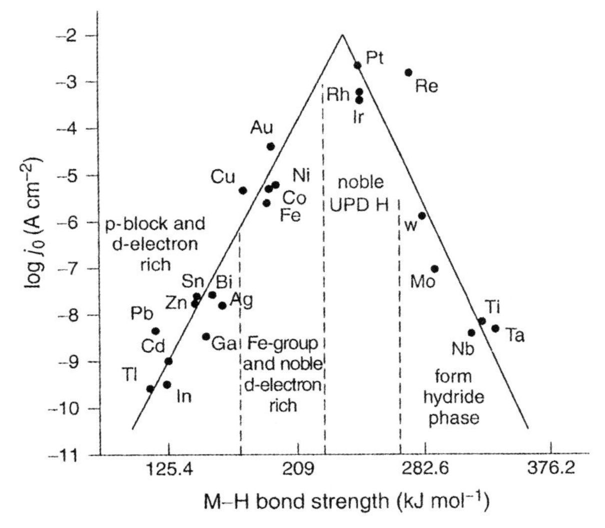

Fig. 4.27 Volcano curve for the exchange-current density of the hydrogen evolution reaction on different metals as a function of the binding energy, D_{MH} [B.E. Conway, G. Jerkiewicz, Elechtrochim. Acta 45 (2000) 4075].

against the binding energy of the metal hydrogen bond, D_{MH}, which is in turn connected to the adsorption enthalpy. The course of the curve corresponds to the model above.

4.6
Electrocrystallisation – Metal Deposition and Dissolution

An electrochemical process that leads to the build up or dissolution of a metallic overlayer on a surface is termed electrocrystallisation. Since both charge and material exchange take place across the double layer, such processes can give rise to complex behaviour.

In general, metallic electrodes are *polycrystalline,* with individual crystallites exposing specific surfaces to the solution. There is no *long-range* order, and the intercrystalline areas may show a variety of defects such as grain boundaries, severe lattice imperfections, inclusions, adsorbed molecules or even oxide layers. Such defects play a major role in corrosion, and will be considered further in section 4.7; in this section, however, we shall consider some simple but important elementary processes in metal deposition.

4.6.1 Simple Model of Metal Deposition

The equilibrium we wish to consider has the overall form

$$\text{Me(lattice)} + m\text{X} = [\text{Me}^{z+}\text{X}_m](\text{solution}) + ze^- \tag{4.158}$$

where X is a neutral dipolar molecule or an anion that can form a complex or solvate with Me^{z+}. The overall process of deposition, the reverse of (4.158), will take place on a single-crystal surface through one of two mechanisms, shown in Figs. 4.28(a) and 4.28(b). The first of these, mechanism **A**, has discharge of the ion from the outer Helmholtz plane taking place to give an *ad-atom*, an atom which still retains some partial charge and solvation but which is confined to the surface. Such an ion may be far more mobile than a completely discharged species, and can diffuse rapidly to a step-site as shown. The surface then forms by extension of these steps. A second possibility, mechanism **B**, is that preferred discharge of the ion actually takes place at the step sites themselves. Under these circumstances, diffusion of Me^{z+} takes place in solution laterally, and the step extends itself by direct incorporation of metal ions.

The *origin* of these steps can again be traced to two processes: in the first, two-dimensional nucleation takes place on an otherwise flat surface to give a step that can expand by either of the mechanisms given above. Alternatively, as we shall see in section 4.6.2, growth of steps arising from a *screw dislocation* may take place giving rise to a spiral defect that can wind upwards without disappearing.

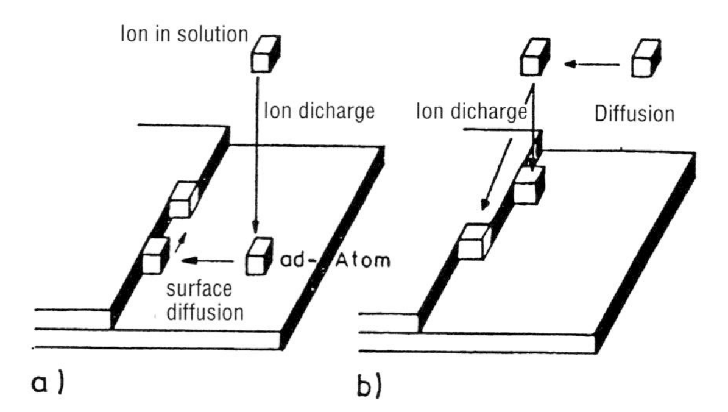

Fig. 4.28 Schematic representation of the elementary steps for metal deposition: (a) discharge of an ad-atom and surface diffusion to a vacant lattice site; (b) direct discharge at a vacant lattice site.

4.6.1.1 Electrodeposition with Surface Diffusion

During passage of current, the concentration of surface ad-atoms will be a function of the distance x from the nearby surface steps. A simple model that can be used to predict the basic types of behaviour starts from the idea that such steps form a series of monolayer edges on the surface with a line density of L_s. The mean distance apart of the steps will be $\sim 1/L_s$, and the furthest point that an adatom will be from any step is then $1/2\ L_s \equiv x_0$. It follows that for electrodeposition, the maximum coverage of ad-atoms will be at $x = x_0$, provided deposition takes place through mechanism **A**. Similarly, if dissolution takes place through the reverse of mechanism **A**, that is, metal atoms leave the edges of monolayers and translate as ad-atoms on the flat underlying surface before dissolution, then the concentration of such ad-atoms will be *lower* far away from the edges. In general, in the absence of current, there will be some equilibrium coverage of ad-atoms, θ_0^{M}, which, if surface diffusion is rate limiting, will also be the coverage of ad-atoms close to the growth or lattice dissolution edges. At values of x close to x_0, the coverage for deposition or dissolution, $\theta_{x=x_0}^{\mathrm{M}}$, will differ from this, and the associated electrocrystallisation overpotential will have the form:

$$\eta_{\mathrm{C}} = \frac{RT}{zF} \cdot \ln\left(\frac{\theta_0^{\mathrm{M}}}{\theta_{x=x_0}^{M}}\right) \tag{4.159}$$

where, for deposition, when $\theta_{x=x_0}^{\mathrm{M}} > \theta_0^{\mathrm{M}}$, η_{C} will be negative.

Quantitative models can be developed to calculate θ_{M} at arbitrary x, based on the diffusion-kinetic equation

$$\frac{\partial\theta^{\mathrm{M}}}{\partial t} = D_{\mathrm{ad}}\frac{\partial^2\theta^{\mathrm{M}}}{\partial x^2} + \frac{j^-(\eta)}{zF} - \frac{j^+(\eta)\theta^{\mathrm{M}}}{zF\theta^{\mathrm{M}}} \tag{4.160}$$

where the second term on the right hand side is the rate at which ad-atoms are deposited and the last term is the rate of dissolution, and we assume that $\theta^{\mathrm{M}} \ll 1$. Note that only the dissolution term depends explicitly on θ^{M}, and $j^-(\eta), j^+(\eta)$ will have the forms

$$j^-(\eta) = j_{0,\mathrm{C}} \cdot \exp\left[-\frac{\beta zF\eta}{RT}\right] \tag{4.161}$$

$$j^+(\eta) = j_{0,\mathrm{C}} \cdot \exp\left[\frac{(1-\beta)zF\eta}{RT}\right] \tag{4.162}$$

at steady state, (4.160) can be explicitly solved for the line model described above to yield

$$j = j_{0,\mathrm{C}}\left\{\exp\left[\frac{(1-\beta)zF\eta}{RT}\right] - \exp\left[-\frac{\beta zF\eta}{RT}\right]\right\} \cdot \frac{\lambda_0}{x_0}\tanh\left(\frac{x_0}{\lambda_0}\right) \tag{4.163}$$

where

$$\lambda_0 = \left(\frac{zFD_{\mathrm{ad}}\theta_0^{\mathrm{M}}}{j_{0,\mathrm{C}}}\right)^{\frac{1}{2}} \cdot \exp\left[-\frac{(1-\beta)zF\eta}{2RT}\right] \tag{4.164}$$

and has the units of *length*. Evidently, if $\lambda_0 \gg x_0$, so that $\tanh(x_0/\lambda_0) \to x_0/\lambda_0$, the parameters x_0 and λ_0 cancel from (4.163) and the rate-limiting step becomes simple deposition on the surface sites. If, by contrast, $\lambda_0 \ll x_0$, a marked plateau is found for θ^{M} near x_0 and surface diffusion becomes rate-limiting. At sufficiently high values of (x_0/λ_0), $\tanh(x_0/\lambda_0) \to 1$, and

$$j \to j_{0,\mathrm{C}} \cdot 2\mathrm{L_S}\lambda_0\{\exp[(1-\beta)zF\eta/RT] - \exp[-\beta zF\eta/RT]\} \tag{4.165}$$

from the definition of x_0. We can see here that the current density becomes dependent on the line density of step defects on the surface.

4.6.1.2 Direct Discharge onto Line Defects

Here, the model of mechanism **B** predicts that discharge onto line defects takes place at a preferential rate compared to discharge onto flat surfaces as ad-atoms. The corresponding current density takes the form:

$$j \rightarrow j_{0,\mathrm{L}} \cdot \mathrm{L_S}\{\exp[(1-\beta)zF\eta_{\mathrm{D}}/RT] - \exp[-\beta zF\eta_{\mathrm{D}}/RT]\} \tag{4.166}$$

which is similar to (4.165) in form, but where we must distinguish the current density associated with line defects from that associated with deposition onto planar areas, and remember that $j_{0,\mathrm{L}}$ will have units of A cm^{-1} or A m^{-1}. Note that in both cases, the current density scales linearly with $\mathrm{L_s}$.

4.6.1.3 Two-dimensional Nucleation

If the formation of deposited monolayers is hindered by, for example, the very low density of nucleation sites, then the rate-limiting step may become the formation of fresh surface nucleation sites. An overpotential is normally required to initiate growth, and for overpotentials smaller in magnitude, a thermodynamic excess ionic concentration can form in the immediate neighbourhood of the electrode surface. In the case of silver, this overpotential amounts to a few mV, and if the overpotential is further increased, current flows relatively easily.

If electro-deposition is effected galvanostatically, then as soon as nucleation is initiated, the overpotential falls whilst the edges of the nuclei grow. As these edges fuse to form a new layer, the overpotential must rise again to initiate new nucleation on the next layer, and the overpotential in this type of experiment shows characteristic oscillations.

We can quantify this reasonably easily for two-dimensional film growth. The basic idea is that nucleation is difficult on a completely flat surface, since although the metal depositing will be stable in the bulk at any potential below the thermodynamic deposition potential, nuclei with only a small number of atoms will have a large *surface* free energy compared to the bulk free energy. The effect of this is that at potentials below the deposition potential, assuming a hemispherical cluster, the free energy has the form $\Delta G = Nze_0\eta + \gamma aN^{2/3}$ where N is the number of atoms in the nucleus, η the overpotential (negative), γ the surface energy and a is a geometric factor. The value of ΔG will reach a maximum value of $\dfrac{4(\gamma a)^3}{27(ze_0\eta)^2}$ corresponding to the free energy of activation for the formation of nuclei. Evidently, if η is very small, then the free energy of activation for nucleus formation will be large, and few nuclei will form. A similar analysis for the two-dimensional case treated below shows that the free energy of activation in this case $\Delta G_c = \dfrac{(\gamma a)^2}{4ze_0|\eta|}$. It is extremely difficult to fabricate surfaces so flat that this model is applicable, but Budewski and co-workers [Electrochim. Acta 11 (1966) 1697] have been able to obtain very flat Ag(100) surfaces on which deposition of silver atoms takes place through a mechanism involving a single nucleation event for every monolayer: once a nucleus has formed, it spreads across the surface, since the probability of a second nucleation event is small.

In fact, as we have seen, surfaces are generally far from flat, and there will be defects of various kinds on the surface. We anticipate that at some of the point defects on the surface, termed *active sites*, nucleation will be favoured as compared to the completely flat surface and if nucleation at such sites dominates over nucleus formation by spontaneous aggregation, then the number of growing nuclei can be related to the total number of active sites on the surface. If, for simplicity, we further assume that all the active sites are essentially identical, and that there are M_0 such sites per unit area, and k_N is the rate constant for the formation of a nucleus at one of the active sites, then the number, $M(t)$, of growing nuclei at time t will be given by a simple first-order equation of the form

$$M(t) = M_0(1 - e^{-k_N t})$$

There are two extreme cases: if $k_N t \gg 1$ then the effect will be of instantaneous nucleation at all sites. If, on the other hand, $k_N t \ll 1$, then progressive nucleation will take place, and at early times $M(t) \approx k_N M_0 t$.

To progress, we can imagine that all our growing nuclei are circular monolayers. Let the radius of one such monolayer at time t be $r(t)$ and that the nucleus grows only by adding atoms at the periphery of the circle with a rate constant k. If the number of atoms in the nucleus is $N(t)$, then evidently

$$\frac{dN(t)}{dt} = k \cdot 2\pi r(t)$$

If ρ denotes the number of atoms per unit area in the nucleus, and S is the area of the nucleus, then provided the circle can grow without restriction, we have, assuming $N(t) = \rho S = \pi r^2 \rho$, $r(t) = kt/\rho$.

Of course, in a real system, the nuclei must eventually start to coalesce, and we need some relationship between the actual area covered and the area that would be covered were the nuclei able to grow without restriction. This relationship was first worked out by Avrami (J. Chem. Phys. **9** (1941) 177 and refs. therein), who showed that *actual* fractional area covered, S, is related to the area that would be covered, S_{ext}, by the Avrami equation:

$$S = 1 - \exp(-S_{ext})$$

Where complete coverage corresponds to $S = 1$.

For instantaneous nucleation, we obviously have, at time t

$$S_{ext}(t) = M_0 \pi r^2(t) = \frac{\pi M_0 k^2}{\rho^2} t^2$$

And from Avrami's theorem

$$S(t) = 1 - \exp\left(-\frac{\pi M_0 k^2}{\rho^2} t^2\right)$$

And the current density, $j(t) = ze_0 \frac{dN}{dt} = ze_0 \rho \frac{dS}{dt} = \frac{2\pi z e_0 M_0 k^2}{\rho} t \exp\left(-\frac{\pi M_0 k^2}{\rho^2} t^2\right)$.

For progressive nucleation, we suppose that a nucleus is born at time t_b. Then the radius at time $t > t_b$ is given by $r(t) = k(t - t_b)/\rho$ and the area of this nucleus $A(t) = \pi \frac{k^2}{\rho^2}(t - t_b)^2$. In order to find the total extended area S_{ext}, we must integrate over all values of t_b from 0 to t, and we get

$$S_{ext}(t) = k_N M_0 \frac{\pi k^2}{3\rho^2} t^3$$

And finally the current density now has the form:

$$j = ze_0 k_N M_0 \pi \frac{k^2}{\rho} t^2 \exp\left(-k_N M_0 \frac{\pi k^2}{3\rho^2} t^3\right)$$

The two forms of current density for instantaneous and progressive nucleation are easily distinguished experimentally, and the theory above has been applied to a wide variety of two-dimensional films on surfaces.

4.6.2
Crystal Growth in the Presence of Screw Dislocations

It is possible for a surface to grow continuously without nucleation in the presence of screw dislocations, as shown in Fig. 4.29. Growth in the presence of such a defect gives rise to pyramidal structures on the surface, which spiral up out of the surface. In practice, most surfaces possess significant numbers of such dislocations, and deposition can thus lead to appreciable surface roughening since deposition will be favoured at points of high field strength. This effect will be particularly pronounced at low ionic concentrations or in solvents of low intrinsic polarity, leading to the formation of nee-

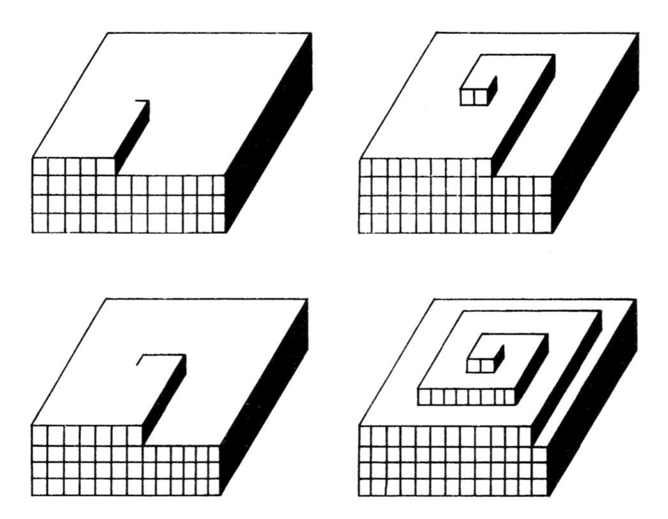

Fig. 4.29 Growth of a crystal surface through the presence of a screw dislocation (spiral growth in pyramid form).

dle-like growths termed *dendrites*. By contrast, in concentrated highly conducting solutions, electro-deposition can lead to very smooth surfaces, particularly if a strongly adsorbed inhibitor is added to prevent dislocation propagation.

4.6.3
Under-potential Deposition

The deposition of up to a monolayer of a metal M onto the surface of a different metal M′ is frequently found to take place at a potential positive of the thermodynamic potential for the reaction

$$M^{z+} + ze^- \rightarrow M \tag{4.168}$$

an effect termed **u**nder**p**otential **d**eposition (upd). Thermodynamically, the origin of this behaviour lies in the fact that the chemical potential of the metal in the monolayer, μ_{ML}, is lower than the chemical potential μ^0_M of the bulk metal:

$$\mu^0_M > \mu_{ML} = \mu^0_{ML} + RT \ln a_\theta \tag{4.169}$$

where a_θ is the activity of layer of M at coverage θ.

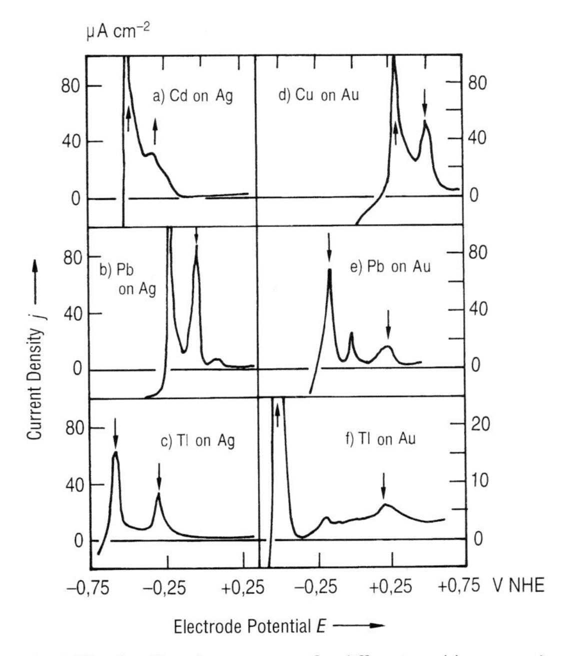

Fig. 4.30 Anodic voltammogram for different upd layers on Ag and Au in 1M Na_2SO_4 (pH 3) (a and c) and in 1M $NaClO_4$ (pH 3) (b,d,e and f); the concentration of the divalent metal ions deposited in each case is 0.0002M, and the scan rate is 20 mV s^{-1} [from D.M. Kolb, M. Prasnysk and H. Gerischer: J. Electroanal. Chem. **54** (1977) 25].

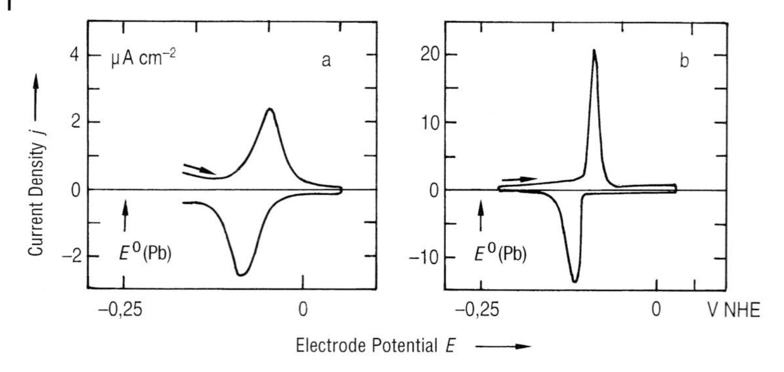

Fig. 4.31 Cyclic voltammogram of lead underdeposition on (a) polycrystalline silver; (b) on Ag(111); 2×10^{-4}M Pb^{2+} in 0.5M $HClO_4$ (pH 2), scan rate 1 mV s^{-1} [from the same source as 4.30].

At equilibrium for the electrode reaction, $M^{z+} + ze^- \rightarrow ML(\theta)$, if E^0 (M) is the corresponding potential for (4.168), we have

$$E_{ML} - E^0(M) = \frac{\mu_M^0 - \mu_{ML}^0}{zF} - \frac{RT}{zF}\ln[a_\theta] \qquad (4.170)$$

which is positive from (4.169) and the fact that $a_\theta < 1$.

Experimentally, upd can be seen on a substrate by slowly scanning the potential from positive values towards $E^0(M)$. In the upd potential region, the deposition of a sub-monolayer of M is signalled by the appearance of one or more current maxima before the rise in current associated with the onset of bulk deposition of M. On reversing the potential scan direction, the reverse process of bulk followed by surface dissolution takes place, as shown in Fig. 4.30, which shows upd peaks for Cd, Pb and Tl metals on Ag and Au substrates. Experimentally, it is also found that the sharpness and narrowness of the upd peaks is strongly amplified if the polycrystalline substrates of Fig. 4.30 are replaced by single-crystal surfaces, as shown in Fig. 4.31.

Extensive studies have shown that the value of $\Delta E \equiv E_{ML}^0 - E^0(M)$ can be up to one volt, depending on the identity of M and M′, corresponding to very large differences in energy (of up to 100 kJ mol^{-1}). A systematic investigation of ΔE for different pairs of M and M′ has shown a strong correlation with the difference in electronic work functions, $\Delta\phi$, for the two metals; in fact

$$\Delta E \approx 0.5\Delta\phi \qquad (4.171)$$

4.6.4 The Kinetics of Metal Dissolution and Metal Passivation

Corresponding to the mechanisms **A** and **B** referred to above, there are two dissolution mechanisms. The first, a reversal of **A**, is the formation of ad-atoms by diffusion of an atom from a step into the planar region, followed by electron transfer to form a solvated M^{z+} ion. As indicated above, this can give rise to dissolution overpotential of the form

$$\eta_D = \frac{RT}{zF} \cdot \ln \frac{\theta_0^M}{\theta^M} \tag{4.172}$$

where now $\theta_0^M > \theta^M$, and $\eta_D > 0$. The reversal of mechanism **B** corresponds to *direct dissolution* from a step site, and in this case, the kinetics of electron transfer and metal dissolution are closely connected. In both cases, the concentration of M^{z+} near the metal surface will exceed the bulk value owing to the finite rate of diffusion, and this must be taken into account in deriving any expression for metal dissolution rates. In general, we have then, for the dissolution current:

$$j_{Me} = j_{0,Me} \left\{ \frac{\theta^M}{\theta_0^M} \exp\left[\frac{(1-\beta)zF\eta}{RT}\right] - \frac{c^s_{M^{z+}}}{c^0_{M^{z+}}} \exp\left[-\frac{\beta zF\eta}{RT}\right] \right\} \tag{4.173}$$

where $c^0_{M^{z+}}$ is the concentration of metal ions in the bulk of the solution and $c^s_{M^{z+}}$ that at the electrode surface.

If the rate of dissolution exceeds some critical value, the local concentration of metal ions near the surface may actually exceed the solubility limit of the metal hydroxide or oxide, and a layer of monolayer or greater thickness may then precipitate onto the electrode surface, abruptly lowering the anodic current density, a process termed *passivation*. The corresponding potential at which this takes place is termed the Flade potential, and the current-voltage curve under these circumstances is shown in Fig. 4.32. Above the Flade potential, only a very small residual current flows which, in the case of Fe, corresponds to transport of Fe^{3+} ions through the passivating layer of (mainly) hydrated γ-Fe_2O_3 from the iron surface through to the oxide-electrolyte interface, where it may contribute to the slow growth of passive film or simply dissolve into the solution.

The increase in current density at very positive potentials seen in Fig. 4.32 is not due primarily to the onset of further metal dissolution but to the appearance of a new electrode reaction, normally oxygen evolution through oxidation of H_2O which takes place on the oxide film. This process requires that the passive film formed has at least some electronic conductivity, which is the case on iron, nickel, cobalt or zinc; where the film conductivity is very low, as for aluminium, titanium or tantalum, no *trans-*

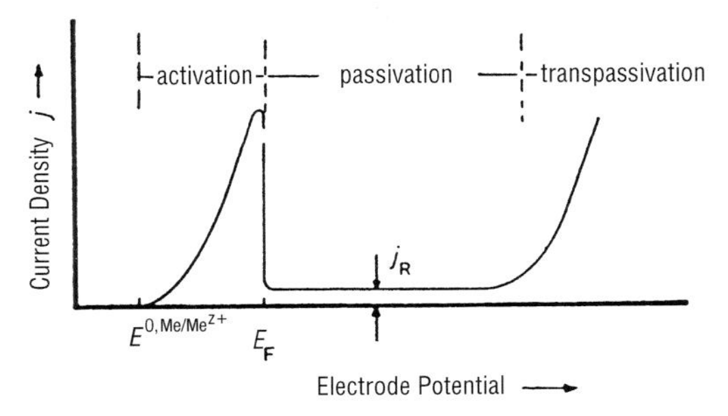

Fig. 4.32 Current-potential curve showing metal dissolution followed by the formation of a passivating layer (E_F is the Flade potential and j_R the rest current density in the passive region).

passive region is seen and very large voltages are necessary for any significant trans-passive current to flow.

The formation of a passive film depends both on the type of metal and strongly on the nature and concentration of the electrolyte. For example, aluminium forms a non-conducting Al_2O_3 passive layer in neutral solution, but dissolves freely in alkaline electrolytes. For iron in slightly acid solutions, the Flade potential, E_F, itself is pH dependent, corresponding to the pH dependence of the equilibrium potential for the reaction $2Fe + 3H_2O \rightarrow Fe_2O_3 + 6H^+ + 6e^-$. Above E_F, the current density in this case falls from about 100 $mAcm^{-2}$ to a few $mAcm^{-2}$. The transpassive region on iron is accessed for potentials above 1700 mV vs. NHE.

Removal of the passive film is possible by removing the passivating potential and observing the slow dissolution of the film, a process termed *reactivation*. This process can also be carried out galvanostatically, where it is observed that the potential decreases steadily to the Flade potential and then falls rapidly to the electrode potential for metal dissolution. This reactivation process can also be carried out by introduction of active complexants such as chloride ions, which play a major role in corrosion.

4.6.5 Electrochemical Materials Science and Electrochemical Surface Technology

4.6.5.1 Preparation of Metal Powders

We considered above the different mechanisms for metal electro-deposition, and, in particular, the dependence of the rate of nucleation on overpotential for single-crystal surfaces. At low overpotentials, this rate is low, leading to larger crystallites with extended crystal faces, giving a coarsely crystalline deposit under practical conditions. However, if the overpotential is increased, the rate of nucleation also increases, giving rise to a more finely crystalline deposit, and a further increase in overpotential can lead, in certain cases, to the electrodeposition of a finely dispersed powdery film that can be easily wiped off the substrate.

This condition obtains for copper, for example, when the overpotential rises to the value at which the current just attains its diffusion-limited value. Since this value depends strongly upon the hydrodynamic conditions at the electrode, it has proved possible to deposit copper powder at low current density in unstirred solutions, within which the natural convection is suppressed (eg 0.5 $mAcm^{-2}$ in 0.3M Cu^{2+} solution), and onto rotating disc electrodes at high current densities (eg 0.34 Acm^{-2} at 2000 rpm).

Which metals actually deposit as powders is determined by such factors as crystallisation energies, ad-atom surface-diffusion coefficients, and electron-transfer rates. However even where a powder does not form, it is often possible to obtain a sponge-like deposit by lowering the potential to the point at which hydrogen is simultaneously evolved. Furthermore, powder formation can be induced by adding inhibitors to the solution that prevent the further growth of the crystal nuclei, usually by adsorption onto high-energy sites on the crystal surface, such as step sites. The disadvantage of this approach is that current densities remain very low, even at high overpotentials, but the approach has been used to prepare powdered Cd and Zn, where the inhibitors are a wide variety of colloidal materials.

The most commercially important process is the preparation of copper powder, but electrolytic powders of iron, zinc, tin, cadmium, antimony, silver, nickel, manganese, tungsten, titanium and tantalum have all been made. Such powders possess a dendritic irregular appearance, fine grain size and high purity, properties highly desirable in metallurgical powders for the preparation of mouldings and sinters, such as gear wheels and self-lubricating bushes. Such powders have also been explored as catalysts, since they have a multiplicity of lattice defects and high-energy adsorption sites.

4.6.5.2 **Electrochemical Machining and Polishing**

If two metal electrodes are placed a very small distance apart (0.1 – 0.5 mm) and a concentrated electrolyte solution (such as KNO_3 or KCl) pumped through the gap, then very high current densities indeed can be obtained for metal dissolution (500 Acm^{-2} for a potential difference of 5–30 V). The rate of flow of the electrolyte must be sufficient both to remove the dissolution products from the anode and hydrogen from the cathode (and thereby prevent any back-reaction) and to remove excess Joule heat: flow rates of 5–50 ms^{-1} are usually employed. For those metals that passivate, it may be possible to operate in the transpassive region, provided oxygen evolution is not the dominant process, or it may be possible to depassivate the metal in a suitable electrolyte.

The rate of removal of metal, υ, from the surface is related to the current density, j, by the equation:

$$\upsilon = \frac{Mj}{\rho zF} \tag{4.174}$$

where ρ is the density of the metal; for iron at $j = 90$ Acm^{-2}, the rate of dissolution is 2 mm/min and for $j = 540$ Acm^{-2} it is 12 mm/min. However, such rates can only be maintained if the position of the cathode is continually re-adjusted to keep the electrode separation very small, and this is the basis of electrochemical machining (ECM), the principle of which is shown in Fig. 4.33. Dissolution is most rapid at the points at which the anode-cathode distance is the smallest, and this leads to the anode adopting the negative form of the cathode. The advantage of this process is that the most complicated forms can be prepared with very low material stresses: examples include drop-forgings for automobile crankshafts (from hardened tool-steel), fittings for nuclear-reactor fuel rods (from zirconium alloys), and turbine rotors.

Electro-machining works because any protuberances on the anode surface are dissolved first, since these are nearest to the cathode and hence are at points of lowest ohmic resistance between anode and cathode. In fact, the process can be used to polish surfaces: the best effects are obtained by using much lower current densities (0.01 – 0.5 A cm^{-2}) and electrode separations of at least 1 cm. The electrolytes used for electropolishing are usually concentrated acids.

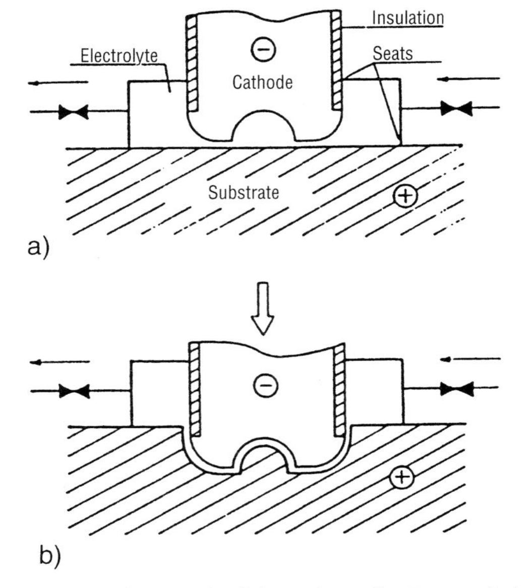

Fig. 4.33 The principle of electrochemical etching: (a) before the beginning of the process; (b) during the etching process.

4.6.5.3 Galvanoplastics

Galvanoplastics is a technique whereby a metallic construct can be prepared by metal electro-deposition on a suitably prepared negative form. The substrate can be made of any easily molded material, with non-metallic substances being made conducting through incorporation of graphite powder. The technique is particularly useful for the preparation of complex hollow structures through deposition of copper on a wax/carbon negative, followed by removal of the wax by melting. Historically, this was the method whereby car radiators were prepared.

Clearly, for this technique, a thick, pore-free, smooth (ie finely crystalline) deposit is required, and the conditions for this have already been described: low solution concentrations, high current density and addition of inhibitors such as gelatine. The formation of powdery deposits must be avoided at all costs, as must the formation of spongy deposits through co-evolution of hydrogen. The attainment of such conditions is the subject of an extensive patent literature, in which different inhibitors, or "brighteners" are described. Low concentrations of free metal ions can be achieved by the use of metal complexes in solution instead of the free metal cations, and suitable choice of such complexes can lead to very finely crystalline deposits.

4.6.5.4 Electrodeposition from Lyotropic Phases

A quite different approach to the electrodeposition of structured metallic layers has recently been pioneered by Bartlett and co-workers (Science **278** (1997) 838). These authors have prepared lyotropic electroplating baths by dissolving a mixture of the appropriate metal salt and a high concentration of non-ionic surfactant. At high concentrations, the molecules of the surfactant self-assemble into aggregates or *micelles* in which the hydrophilic parts of the molecule are on the outside and the hydrophobic tails are on the inside. The dimensions of such aggregates are normally between 2 and 10 nm. In the lyotropic phases further aggregation of these micelles takes place to form organised and regularly spaced structures in which they are separated by an aqueous phase again by distances of 2 – 10 nm. When the metal is electrodeposited from such phases, it grows on the surface in the aqueous regions between the micelles, giving it a porous periodic nanoscale structure that reflects the architecture of the lyotropic phase itself. After deposition, the surfactant can be washed away leaving the regularly structured mesoporous film.

Clearly with this method, considerable control can be exercised over the thickness of the walls between the pores by using different surfactants or by adding a swelling agent such as heptane. Also, the structure of the lyotropic phase itself can be altered by using different surfactants, plating mixture compositions or temperatures, giving a highly flexible method that can be tailored to template metals, metal alloys, metallic oxides and semiconductors. Applications up to now have included electrodes for supercapacitors and batteries, filters and catalytic sensors.

4.7 Mixed Electrodes and Corrosion

Characteristic of corrosion processes is the presence on the metal surface of two different simultaneous ongoing electrochemical processes, one giving rise to an anodic and the other to a cathodic partial current. The total current-potential curve is then constructed from these two partial currents, each with its own asymmetry parameter and exchange- current density. There will evidently be one potential for which the two partial currents are equal and opposite; although no nett current flows through the surface under these circumstances, this is not an equilibrium potential but a rest potential, and even at the rest potential, conversion of reactants to products is taking place at the surface. An electrode of this type is termed a mixed electrode, and the corresponding rest potential is termed the mixed potential.

The simplest example of a mixed electrode is a metal, such as zinc, in contact with an electrolyte containing a more noble metal, such as copper. The partial current curves then correspond to dissolution of the zinc and deposition of the copper. In *corrosion* processes, the anodic partial reaction is always metal oxidation, and the cathodic partial current may be due to reduction of H^+ to H_2 (acid corrosion) or of O_2 to H_2O (rust formation).

4.7.1
Mechanism of Acid Corrosion

A familiar example of this is the corrosion of zinc amalgam in aqueous acid, for which the nett reaction is

$$Zn(Hg) + 2H^+ \rightarrow Zn^{2+} + H_2(gas) \tag{4.175}$$

which consists of two partial electrochemical processes

$$Zn(Hg) \rightarrow Zn^{2+} + 2e^- \quad \text{and} \quad 2H^+ + 2e^- \rightarrow H_2(gas) \tag{4.176}$$

Both these processes take place simultaneously and independently of one another in different parts of the amalgam surface, and the partial currents for each process (shown as linear functions of potential for simplicity) and the total current are shown in Fig. 4.34. If no nett current flows through the surface, the corroding amalgam electrode must spontaneously adjust its potential to the mixed potential E_M at which $j_{H_2} = j_{Zn}$. Clearly this potential must lie between the corresponding standard electrode potentials for the Zn/Zn^{2+} and H_2/H^+ couples, and it is often referred to as the corrosion potential. The value of j_{Zn} at E_M is a measure of the rate of corrosion of the metal and is referred to as the corrosion current density j_{corr}.

The proof of the mechanism postulated in the previous paragraph requires the separate measurement of the current-potential curves for hydrogen evolution and metal dissolution. This has been carried out for the zinc amalgam case above and good agreement found. The importance of the separate determination of the two currents can be seen from the fact that zinc metal should not only corrode in acid solution but also in neutral solution on the basis of the relevant E^0 values for Zn/Zn^{2+} and H_2/H^+. In fact, zinc metal is stable in neutral solution, since the kinetics of the hydrogen evolution reaction on pure zinc are severely inhibited. It is possible, in this case, for impurity atoms such as copper to be deposited at specific sites on the zinc surface. At these sites, hydrogen evolution can take place rapidly, and this raises the overall rate of metal dissolution substantially.

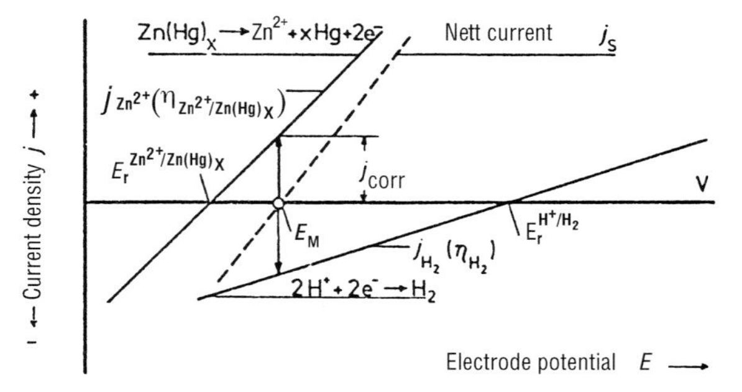

Fig. 4.34 Corrosion of zinc amalgam in aqueous acid with the formation of a mixed potential, E_M; j_{corr} is the corrosion current density.

4.7.2
Oxygen Corrosion

Instead of hydrogen evolution, oxygen reduction can take place at the metal surface through

$$ {}^{1}\!/_{2}O_2 + H_2O + 2e^- \rightarrow 2OH^- \tag{4.177} $$

and in the technically extremely important case of rust formation on iron, the hydroxide ions formed then react with Fe^{2+} to form a solid hydroxide which is the precursor to rust. The water necessary for rust formation comes normally from precipitation from atmospheric humidity; in absolutely dry conditions, iron does not rust at all.

The thermodynamic potential for oxygen reduction is given by $E^{0,O_2|H_2O} = 1.23 - 0.059\text{pH}$ vs. NHE, and some partial curves for iron and copper corrosion in oxygen/air are shown in Fig. 4.35. It is clear that iron will corrode far more rapidly in air than copper, and it is also evident that in neutral solution, acid corrosion of iron will be far less important than oxygen corrosion. It is also clear that the mixed potential for iron corrosion is in the region where oxygen reduction is *mass transport controlled.* Each oxygen molecule arriving at the iron surface will be immediately reduced, and this rate determines the corrosion current j_{corr}.

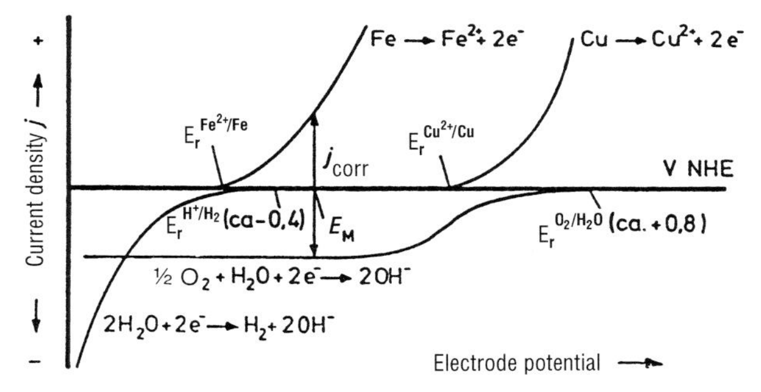

Fig. 4.35 Oxygen corrosion in neutral solution. The cathodic curve corresponds to oxygen reduction with hydrogen evolution being superimposed at negative potentials. The anodic partial current densities arise from metal dissolution: copper dissolution takes place in the potential region corresponding to the beginning of the oxygen reduction curve and iron dissolution takes place in the region where oxygen reduction is transport limited.

4.7.3
Potential-pH Diagrams or Pourbaix Diagrams

An accurate representation of the *thermodynamic stability conditions* of metals to acid or oxygen corrosion can be provided in the pH-diagrams first introduced by Pourbaix. An example is shown in Fig. 4.36 for zinc, in which consideration is given to all possible reactions of the zinc/water system, reflecting the fact that the predominant

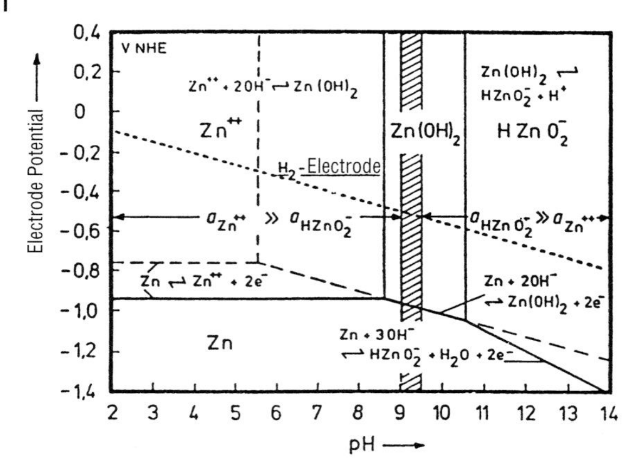

Fig. 4.36 Pourbaix or potential-pH diagram for the system zinc/water at 25°C.

solution species for zinc in acid is the Zn^{2+} ion and in alkali $HZnO_2^-$. Only in the neutral pH range is the solid hydroxide stable.

In Fig. 4.36, the solid horizontal line refers to an activity of 10^{-6} M for Zn^{2+} in solution, and the dashed line to a Zn^{2+} concentration of 1 M. The vertical solid lines correspond to the equilibrium points for the reactions

$$Zn^{2+} + 2OH^- \rightarrow Zn(OH)_2 \text{ and } Zn(OH)_2 \rightarrow HZnO_2^- + H^+$$

at activities of 10^{-6} M for each solution species. The vertical dashed line refers to the first of these equilibria for Zn^{2+} activities of 1 M. Dissolution of Zn in acid solution will clearly be pH independent, but dissolution as $Zn(OH)_2$

$$Zn + 2OH^- \rightarrow Zn(OH)_2 + 2e^-$$

will shift 59 mV for each pH unit, the same rate as the H_2/H^+ couple, which is shown as the dotted line in Fig. 4.36. Since this line lies above the Zn dissolution line at all pH values, it is evident that zinc is not thermodynamically stable at any pH value to dissolution. Pourbaix diagrams usually also show the O_2/H_2O line, which lies 1.23 V above the H_2/H^+ line; it is evident immediately that the driving force for oxygen corrosion is substantially larger than for acid corrosion.

4.7.4
Corrosion Protection

Metals susceptible to corrosion can often be protected by deposition of a thin film of a less susceptible metal, and examples include the gold, silver, copper or nickel plating of iron. In all these cases, the coating metal lies to the right of the substrate metal in the electrochemical series, the metal being coated with a "more noble" protection. However it is also possible to protect metals by coating with "less noble" species, the latter operating by either suppressing the cathodic corrosion partial reaction or inhibiting the anodic reaction: an example in the use of zinc-plating on iron ("galvanising"). For both types of film, electrodeposition conditions must lead to a pore-free finely- crystalline deposit, and such conditions are described above.

To understand how such protective films operate, consider now an iron substrate covered with a metal film; we assume that this film is, in turn, covered with a thin film of moisture or is in contact with a neutral electrolyte. This "electrode" takes up a mixed potential somewhere between the dissolution potential of the metal film and the reduction potential for oxygen. However, *should the protective layer become damaged, different behaviour is found for films that are more or less "noble" than the iron substrate.*

For more noble metal films, such as copper, once an area is damaged, the iron, which will adopt the potential of the overlaying copper film, is exposed to solution. Its potential will greatly exceed the dissolution potential for the reaction: $Fe \rightarrow Fe^{2+} + 2e^-$, and this reaction will then take place rapidly under the damaged site, with the corresponding cathodic reaction being the reduction of oxygen on the (much larger) copper surface. Very damagingly, corrosion of the underlying iron can continue under the initially damaged area of film, spreading until large areas of the surface can exfoliate. This effect is typically found for copper or nickel plated iron surfaces.

Corrosion-protection by a less "noble" coating is quite different, because the mixed potential adopted by the film actually lies negative of the dissolution potential of iron, as is the case for zinc-plating of the iron substrate. Under these circumstances, if the coating becomes damaged, the cathodic reaction will be oxygen reduction on the small exposed surface area of iron, balanced by metal dissolution from the much higher surface area of zinc.

4.7.4.1 **Cathodic Corrosion Protection**

The situation in which iron is held at a potential such that only cathodic reactions such as oxygen reduction or hydrogen evolution are possible (but no metal dissolution) is termed *cathodic protection*. Such cathodic protection can also be achieved by suspending a zinc or magnesium block in the same electrolyte and connecting it to the iron with a conducting wire. The reactions are now oxygen reduction at the (large) iron surface and metal dissolution from the Zn/Mg block, which serves as a sacrificial anode. This method of protection is used for the hulls of ships and for North-Sea oil rigs.

Similar protection may be achieved by making the iron structure the cathode by connection to a *dc* electrical source, and this process is used to protect metal pipework in marine or wetland environments: as the anodic reaction, oxygen evolution or dissolution of scrap iron can be used. Cathodic protection can also be used in reinforced concrete. Corrosion here can take place through attack at places where the steel is exposed by damage; the electrolyte can be quite aggressive, such as acid rain or salt water from gritted roads. Again, a sacrificial metal anode is normally used for the counter reaction.

Finally, for those metals that can form *passive layers*, anodic protection can be used, the potential being fixed with the aid of a potentiostat in the metal passivation region. This is used, for example, in the protection of the inner surfaces of steel holding tanks.

4.7.4.2 Inhibition of Corrosion Through Film Formation

The simplest case of inhibition is the coating of the susceptible metal substrate with a film of organic or inorganic material that prevents corrosion by preventing water access to the metal. Examples include coats of paint, lacquer, varnish or oil films, or the formation of layers by immersion of the metal in a suitable solution. A simple example of the latter is phosphatisation of iron: if the iron electrode is immersed in a solution containing zinc acid phosphate, $Zn(H_2PO_4)_2$, acid corrosion at the iron surface leads to H_2 evolution and formation of Fe^{2+}, with a concomitant pH shift to higher values near the surface. This shift leads to the formation of the highly insoluble $Zn_3(PO_4)_2$, which forms as a film at the iron surface, strongly inhibiting further corrosion. Iron is also incorporated into this film, through the reaction:

$$3Zn(H_2PO_4)_2 + 2Fe \rightarrow Zn_3(PO_4)_2 + 2Fe(H_2PO_4)_2 + 2H_2 \tag{4.178}$$

A similar process can be used to give lasting protection to aluminium: Al is protected by a thin layer of oxide, usually ca. 0.1 μm that forms in moist air, but the thickness of this can be increased to ca. 1 μm by anodisation at a constant current of ca. 10 $mAcm^{-2}$ up to a potential of ca. 50 V.

Inhibition in a narrower sense can be defined as corrosion protection by addition of materials that physisorb or chemisorb on the metal surface. There is a multiplicity of inorganic ions (e.g. nitrite or chromate), inorganic molecules (e.g. As_2O_3, CS_2, H_2S or CO), or organic compounds (e.g. phenols, alcohols or amines) that can be used. If such species act by blocking active sites for metal dissolution, they are termed *anodic inhibitors*. Similarly, *cathodic inhibitors* operate by increasing the overpotential for hydrogen evolution or oxygen reduction. Such inhibitors, even at concentrations of ca. 10^{-5} M, can reduce the corrosion current density by orders of magnitude.

4.7.4.3 Electrophoretic Coating of Metals

In this process, the metal, acting as either anode or cathode, is coated by a polymer film from solution. Polymer units can be solubilised in water if they can be made ionic by such processes as:

$$R^BCOOH + R_3N \rightarrow R^BCOO^-R_3NH^+ \tag{4.179}$$

or

$$R^BNR_2 + RCOOH \rightarrow R^BNR_2H^+RCOO^- \tag{4.180}$$

where R^B is a polymeric binding agent residue and R a general alkyl residue, the species R^BCOOH and R^BNR_2 are water *insoluble*, and the aim is to plate these as films on the metal. For (4.179), the reaction can be reversed close to an oxygen-evolving electrode, where the pH is lower, favouring the undissociated acid. Similarly, for (4.180), the reaction can be reversed near a *hydrogen*-evolving electrode, where the pH is higher, favouring formation of the free amine. Migration of the anion or cation takes place through electrophoresis as described above, and high field strengths are needed. The attraction of the process is that a very smooth final coating is formed: initially coating takes place at corners and edges, but since the film is insulating once formed, the field lines shift to give preferential coating at less energetically favoured sites. In addition, the coating process is rapid despite the low current densities: a coating of 10 μm forms within a few minutes.

Originally, anodic coating was used, but cathodic coating does possess important advantages: the pH in the neighbourhood of the metal is *alkaline* rather than *acidic*, and thus less aggressive. As a result, metal dissolution cannot take place, and metal ions are not, therefore, incorporated in the protective film. The use of cathodic electrophoretic coating is now very much commoner, and it is used, for example, in corrosion protection in the automobile industry.

4.8
Current Flows on Semiconductor Electrodes

In Section 3.5, we established that the potential distribution at the semiconductor electrolyte interface differed markedly from that at a conventional metal electrode; in particular, potential changes at the semiconductor electrode may be accommodated *not* within the Helmholtz layer, but within the bulk of the semiconductor near the surface, provided the interface is *reverse biased* or the semiconductor *intrinsic* with the Fermi level close to the mid-gap position. It is necessary to consider *four* partial current densities at the surface of a semiconductor, at least in principle: j_C^+ and j_C^- for current flow involving the electrons in the conduction band and j_V^+ and j_V^- for current flow involving holes in the valence band.

As an example, for the system H^+/H_2 in Fig. 4.37, we have the current flows:

$$H_{ads} \rightarrow H^+ + e_C^-; \quad j_C^+$$
$$H^+ + e_C^- \rightarrow H_{ads}; \quad j_C^-$$
$$H_{ads} + p_V^+ \rightarrow H^+; \quad j_V^+$$
$$H^+ \rightarrow H_{ads} + p_V^+; \quad j_V^-$$

where e_C^- is an electron in the conduction band and p_V^+ a hole in the valence band. If now the potential on the semiconductor electrode is changed from the equilibrium value E^0 to a new value E, where $\eta = E - E^0$, this potential change will again be accommodated *entirely within the space-charge layer* provided the conditions given above obtain. Under these circumstances, the energy-level of the electronic bands at the surface of the semiconductor change very little, and the primary effect of the potential change is to alter the position of the Fermi level within the bandgap. This is shown in Fig. 4.38, in which a positive potential is applied to an *n*-type semiconductor in contact with a solution containing a dissolved redox couple. Provided the potential change is not so great that the Fermi level would intersect conduction or valence band, this

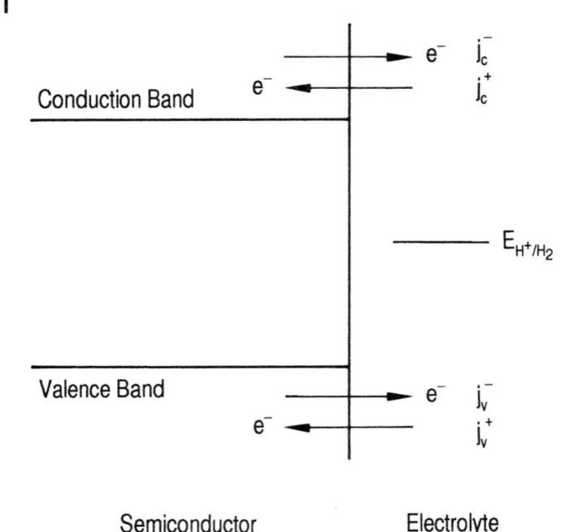

Fig. 4.37 Partial current flows for the semiconductor/electrolyte interface.

simple picture holds, and the primary effect of shifting the Fermi level with respect to the bands at the surface is to alter the surface density of electrons in the conduction band and holes in the valence band.

Specifically, if the Fermi level moves down in the diagram (corresponding to $\eta > 0$), the density of electrons in the CB *decreases* exponentially through the Boltzmann distribution referred to in Section 3.5, and the density of holes *increases*. Given that we expect the current to depend directly on the number of carriers where such carriers are transported to the redox couple, but the carrier density is sufficiently low that electrons can freely be injected into the CB and holes into the VB without any restrictions due to occupancy, we find:

$$j_C = j_C^+ + j_C^- = j_C^0 - j_C^0 \exp\left\{-\frac{e_0\eta}{k_B T}\right\} = j_C^0\left(1 - \exp\left\{-\frac{e_0\eta}{k_B T}\right\}\right) \tag{4.181}$$

$$j_V = j_V^+ + j_V^- = j_V^0 \exp\left\{+\frac{e_0\eta}{k_B T}\right\} - j_V^0 = j_V^0\left(\exp\left\{+\frac{e_0\eta}{k_B T}\right\} - 1\right) \tag{4.182}$$

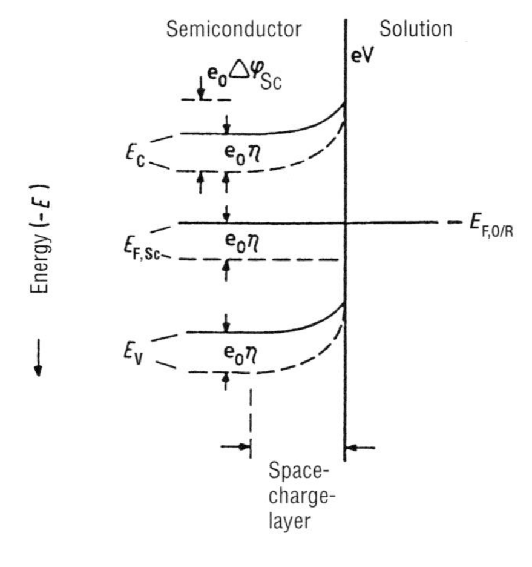

Fig. 4.38 Shift of the Fermi level, $E_{F,sc}$, of a semiconductor and the associated bending of the valence and conduction band energies in the interfacial region of the semiconductor in contact with a solution redox system, O/R, (a) at equilibrium (*continuous lines*); (b) after imposition of an overpotential η (*dashed lines*).

and the total current has the form:

$$j = j_C + j_V = j_C^0\left(1 - \exp\left\{-\frac{e_0\eta}{k_B T}\right\}\right) + j_V^0\left(\exp\left\{+\frac{e_0\eta}{k_B T}\right\} - 1\right) \qquad (4.183)$$

In general, particularly when the energy of the redox couple lies close to one or other of the band edges, there are marked differences between the exchange-current densities, and the measured current becomes dominated by one or other of the bands. This is particularly true of doped semiconductors, where the density of minority carriers (e.g. holes in an n- type material) becomes very small.

4.8.1 Photoeffects in Semiconductors

There are three types of photo-process observed on semiconductors:

1. Excitation of an electron-donor in the electrolyte, in which an electron is excited to a higher quantum state of the donor following which, if its energy now lies above E_C, electron transfer to the conduction band may take place.
2. Excitation of an electron-acceptor in the electrolyte, in which an electron is photoexcited to a higher level leaving behind a vacancy or hole. If the energy of this hole lies below E_V, electron transfer from the valence band to the hole may take place.
3. Excitation of an electron from the valence to the conduction band of the semiconductor, leading to the formation simultaneously of an electron in the conduction band and a hole in the valence band. This process obviously requires light whose energy exceeds the bandgap of the semiconductor, and if the hole and electron are generated in the interior of the semiconductor, they may either recombine or one or both may drift to the surface of the material where they may participate in electron transfer processes, as shown in Fig. 4.39.

In the presence of a depletion layer at the surface, the internal electric field acts to separate the photogenerated electron-hole pair, driving the electron one way and the hole the other. This is also seen in Fig. 4.40, in which an *n*-type semiconductor

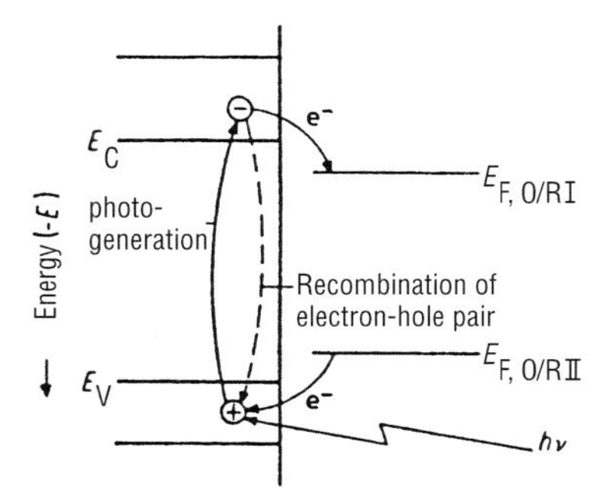

Fig. 4.39 Generation of electron-hole pairs on illumination of a semiconductor with light of energy greater than the bandgap. Also shown is the resultant charge transfer at the interface, with reduction taking place through electron capture by redox couple O/R_I and oxidation taking place through hole capture by redox couple O/R_{II} (equivalently electron transfer from R_{II} to the hole in the semiconductor valence band).

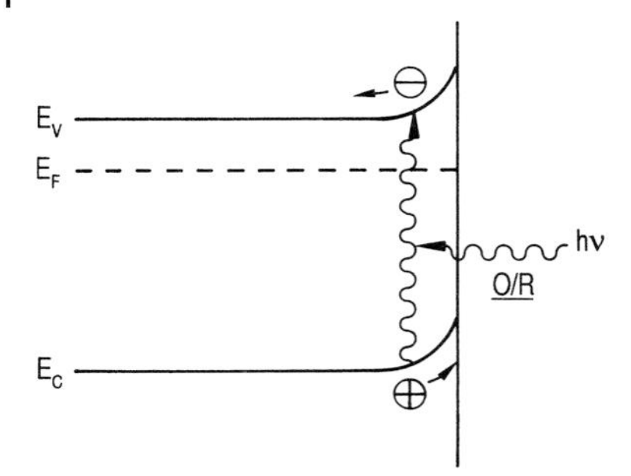

Fig. 4.40 Illumination of an *n*-type semiconductor electrode under reverse bias, showing that electrons and holes are driven in opposite directions by the electric field in the depletion layer, with the holes (minority carriers here) being driven to the surface where they can be captured by a suitable reduced form, R, of a solution redox couple O/R.

is illuminated with supra-bandgap light; the holes are driven to the surface and the electrons are driven into the interior. If the electrical circuit is completed by connection of the n-type semiconductor to a counter electrode, then it is evident that a photocurrent can be sustained *provided that there is a suitable redox couple in solution*, the reduced form of which can be oxidised by the holes arriving at the surface of the semiconductor and the oxidised form reduced by electrons at the counter electrode. If two different redox couples are present, there is the possibility of photochemical conversion taking place as:

$$\mathrm{R_I + p_V^+\ (semiconductor) \rightarrow O_I}$$
$$\mathrm{O_{II} + e^-(counter - electrode) \rightarrow R_{II}}$$

The nett process is

$$\mathrm{R_I + O_{II} \rightarrow R_{II} + O_I}$$

and a simple example is the oxidation of ethanoic acid:

$$\mathrm{CH_3COOH + 2H_2O + 8p_V^+ \rightarrow 2CO_2 + 8H^+}$$
$$\mathrm{2O_2 + 8e^- + 8H^+ \rightarrow 4H_2O}$$

in which the ethanoic acid is photodegraded to CO_2 and H_2O. This process finds application in the removal of organic impurities from effluents as described in more detail below.

4.8.2
Photoelectrochemistry

We have already seen that when a semiconductor-electrolyte interface is illuminated, a photocurrent may flow between the semiconducting electrode and a counter-electrode: such a process is termed *photoelectrochemical*. The magnitude of the photocurrent will, in general, depend on the electrode material, incident light wavelength, electrode potential and electrolyte composition, and three cases have been considered, of which the most important is the generation of electron-hole pairs in the bulk of the semiconductor. Provided that the energy of the incident photons is greater than the bandgap of the semiconductor, such electron-hole pairs will be formed, and, in the presence of a depletion layer within the semiconductor, the *majority* carriers will be transported into the interior of the electrode and the *minority* carriers will be transported to the surface. As explained above, what takes place then depends on the composition of the electrolyte: if facile transfer to a redox couple whose energy lies within the bandgap can take place, then a photocurrent will be seen. If carrier transfer is very slow, there will be a build-up of charge at the surface, changing the distribution of potential at the interface so as to *increase* that accommodated in the Helmholtz layer and diminishing that in the depletion layer. The effect of this "band-flattening" is to inhibit the separation of electron-hole pairs in the semiconductor and to facilitate bulk recombination.

An alternative scenario can exist in which the semiconductor is held at open circuit. Provided there is a fast redox couple in the solution whose energy lies within the bandgap, then in the absence of light, the Fermi level in the semiconductor will be of the same energy as the redox couple, and this effectively defines the electrode potential. On illumination, there is a temporary flow of minority carriers to the surface until the rate of photogeneration of carriers becomes equal to their rate of recombination. This process again leads to "band flattening"; the Fermi level in the semiconductor rises, and this is manifested as a change in the potential of the semiconductor electrode (the *photopotential*).

The generation of photocurrents and photopotentials has important commercial and technological consequences. As explained below, there are possible applications of such technology in solar-energy conversion, in solar driven electrosynthetic processes and in solar and UV-driven detoxification and disinfection of water supplies.

4.8.3 Photogalvanic Cells

In the simplest case, such a cell can be realised as follows: if an *n*-type semiconductor is immersed in an electrolyte containing a redox couple with redox energy $e_0 E^0$ (measured with reference to the same energy scale as the energies of the band edges of the semiconductor) lying between conduction and valence band energies, then on illumination holes will be transported to the surface where they can oxidise the reduced form of the couple. If the semiconductor is also connected by an external electrical wire to a second (counter) electrode immersed in the same electrolyte, electrons photogenerated in the semiconductor can be transferred to this counter electrode where they can *reduce* the oxidised form of the redox couple.

A concrete realisation of this device in shown in Fig. 4.41: the semiconductor is *n*-TiO_2 and the redox couple is Fe^{3+}/Fe^{2+}. In acid solution, the redox potential of this couple is 0.77 V, and the energy of the conduction band (referred to SHE) is close to 0.0 V. In the dark, the Fermi level of the TiO_2 will lie close to 0.77 V, as will the potential of the counter electrode, and there will be essentially zero photovoltage. On illumination (assuming no transport limitations at either electrode and very fast electron-transfer kinetics), if there is a simple wire connecting the electrodes, a photocur-

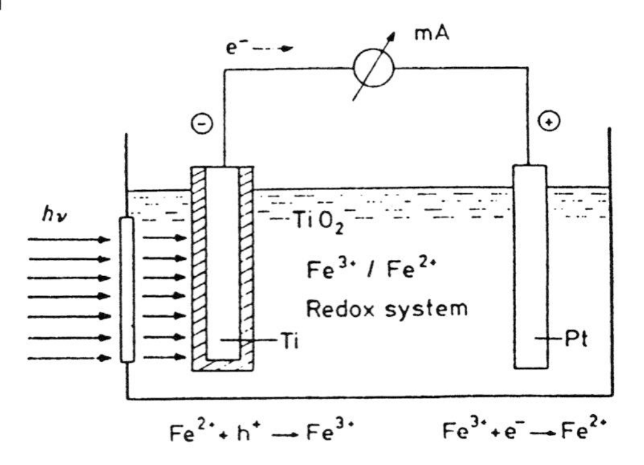

Fig. 4.41 Example of a photogalvanic cell with regeneration of the redox system at the counter electrode.

rent will flow, with the semiconductor serving as the anode and the counter electrode as the cathode. If a resistor, R_{pc}, is incorporated into the external cell, the potential of the anode will fall to a value $0.77 - i_{photo} R_{pc}$. The photovoltage across the cell will attain the value $i_{photo} R_{pc}$ and the power generated will be $(i_{photo})^2 R_{pc}$. If, for simplicity, we assume that the Fermi level lies very close to the conduction band edge in the bulk of the semiconductor (an assumption that will certainly be true at higher dopant concentrations), then the band bending is also $0.77 - i_{photo} R_{pc}$, and the *maximum* value of $i_{photo} R_{pc}$ will be 0.77 V; at this value there will be no band bending within the semiconductor, and the photocurrent will fall to zero. The functional dependence of i_{photo} on the band bending thus ensures that there will be a maximum in the power output of the photogalvanic cell, a maximum dependent on the parameters above.

Photogalvanic cells have been under development for some 30 years. The overall efficiency of such a device can be defined as the ratio of the output power to the input solar power, and clearly this will also show a maximum at a particular bandgap. At high bandgaps, such as that found in TiO_2 (ca. 3.0 eV), the fraction of the solar spectrum that can be absorbed is very low. However, if the bandgap is too low, a little reflection will show that the voltage that can be derived from the cell will be very small and the power output similarly small. The optimal bandgaps appear to be in the region of 1.4 to 1.6 eV, and materials such as MoS_2 and $Cd(Se_{1\text{-}x}S_x)$ have been used in photogalvanic cells showing overall energy conversion efficiencies of 2.5 – 12%.

4.8.4
Solar Energy Harvesting

An extremely important possible application of solar cells is in the photoelectrolysis of water. In such a cell, water, made ionically conducting with a suitable indifferent electrolyte such as KOH, provides the only redox couple. Referring again to Fig. 4.41, if there is no $Fe^{3+/2+}$ redox couple, then a possible hole reaction at the surface is $4p_v^+ + 4OH^- \rightarrow O_2 + 2H_2O$. Electrons generated within the semiconductor then pass to the counter-electrode as shown, where hydrogen is generated by the process $4e_c^- + 4H_2O \rightarrow 2H_2 + 4OH^-$. The nett process, $2H_2O \rightarrow 2H_2 + O_2$, corresponds to electrolysis. Clearly this process is highly attractive: if the energy in visible solar light can be converted cleanly into hydrogen and oxygen which can then be recombined in an appropriate fuel cell, then we have the basis for a sustainable hydrogen economy. Unfortunately, the cell shown in Fig. 4.41 will not work in this way; although oxygen can be evolved by hole reaction with water, the energy of the conduction band is insufficiently negative to drive the hydrogen evolution reaction, and a bias must be introduced externally to the cell. Clearly this bias could be eliminated by using oxide semiconductors with more negative conduction band energies, but this will increase the bandgap, further reducing the fraction of solar energy actually harvested. It has proved enormously difficult to identify a material that is both stable to oxygen evolution and can drive this electrochemical process whilst having a low enough electron affinity to drive the hydrogen reaction and a bandgap in the visible region of the spectrum.

Given the difficulty in directly converting light into chemical energy through photoelectrolysis, it is obviously possible to consider improving the simple photogalvanic cells described in section 4.8.3 to optimise solar-energy conversion efficiencies, using highly reversible electrochemical couples to harvest the solar energy and then using such cells to drive water electrolysis in suitable electrolysers. This type of device has proved far easier to fabricate, and efficiencies in excess of 10 % have been reported, efficiencies that compare not unfavourably with solid-state p-n junctions. However, long-term stability has proved a big headache.

This problem has been addressed in recent years by several methods, the most important of which is the combining of wide-bandgap oxide semiconductors with materials termed sensitisers. These latter are dyes, many of which were developed initially for the photographic industry. The mode of action of a semiconductor sensitiser is illustrated in Fig. 4.42: on illumination, the dye absorbs a photon, with concomitant

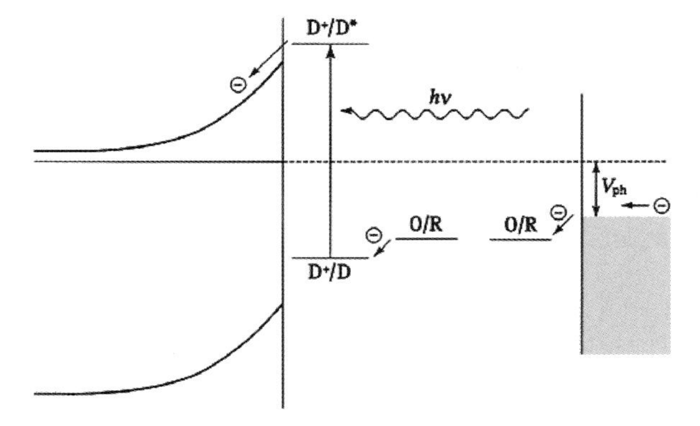

Fig. 4.42 The mode of action of a dye-sensitiser attached to the surface of a wide-bandgap semiconductor such as TiO_2. D, D* and D^+ refer to the dye in its neutral, optically excited and oxidised forms respectively, and the solution redox couple is denoted by O/R; V_{ph} refers to the photopotential obtained under operating conditions.

excitation of an electron from the highest occupied molecular orbital (HOMO) to the lowest unoccupied MO (LUMO). Provided the LUMO lies in energy just above the conduction band edge, electron injection into the conduction band of the semiconductor can take place, leaving an oxidised form of the dye on the surface. If a kinetically facile redox couple is present in solution, this oxidised form of the dye can be re-reduced, and the redox couple recycled at the cathode, allowing the cycle to begin again. The importance of this type of system is that very stable dyes, based on ruthenium are known, and these dyes can be adsorbed into colloidal TiO_2 films to form exceptionally efficient solar-conversion devices. Such devices were first fabricated by Grätzel and co-workers in 1991, and a typical dye is *cis*-X_2bis(2,2′-bipyridyl-4,4′-dicarboxylato) Ru(II), where $X = Cl^-$, Br^-, I^-, CN^-, or SCN^-. In the case of $X = SCN^-$, the solar conversion efficiency exceeds 10 %, and the quantum current efficiency is close to unity. Recently, further improvements for this type of dye in conjunction with a TiO_2 film electrode have led to the very efficient panchromatic sensitization of nanocrystalline TiO_2 solar cells over the whole visible range, extending into the near-IR region to $\sim$ 920 nm (see Fig. 4.43). The efficiency of photocurrent generation was practically 100 % over a wavelength range from 400–700 nm.

The comparative success of these “photovoltaic” cells has not prevented electrochemists dreaming of creating a single cell in which photoelectrolysis could be achieved. As our understanding of the process of photosynthesis in Nature has improved, it has become apparent that our objective of achieving both solar conversion and electrolysis using a single photon energy has been abandoned by Nature as simply too difficult. Instead, Nature uses a two-photon process, one to drive hydrogen generation (or, more accurately, carbon-dioxide reduction to carbohydrate) and one to drive oxygen evolution. Several possible devices have been suggested, particularly those in which two different semiconductor or dye-sensitised electrodes are separately illuminated, but efficiencies remained disappointingly low. However, more recently Khaselev and Turner [Science **280** (1998) 425] have reported a simple monolithic device comprising a fairly high bandgap (1.83 eV) semiconductor, p-$GaInP_2$ attached to a *p-n* junction based on GaAs (bandgap 1.42 eV). Hydrogen is evolved at the p-$GaInP_2$/electrolyte

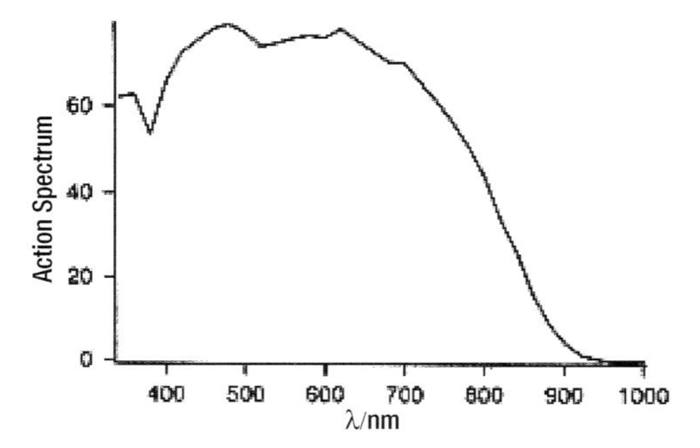

Fig. 4.43 Dye sensitisation in action: the photocurrent action spectrum of TiO_2 coated with a modern form of the dye referred to in the text [M.K. Nazeeruddin, P. Pechy, M. Grätzel, Chem. Commun. (1997) 1705–1706].

interface, provided the reverse bias is not too high (else complex surface chemistry leading to irreversible loss of activity is encountered), whilst light of lower energy than the bandgap of this material then passes through to the GaAs p-n junction allowing the formation of holes sufficiently energetic to oxidise water at a metal/solution contact, as shown in Fig. 4.44. This device is the first photoelectrolysis unit directly to exceed 12 % efficiency, and whilst it remains too costly to develop commercially at the moment, it is still a very attractive unit.

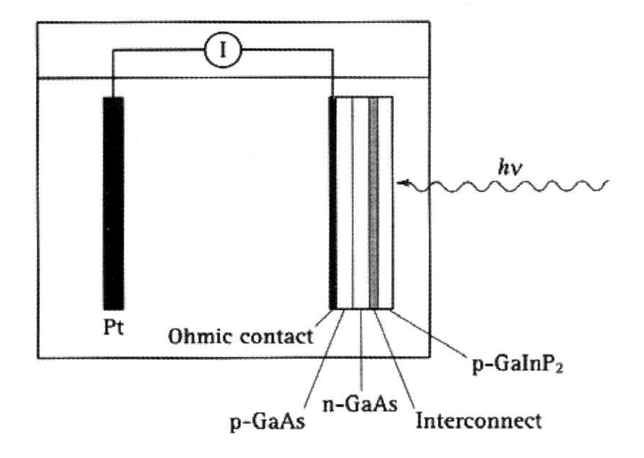

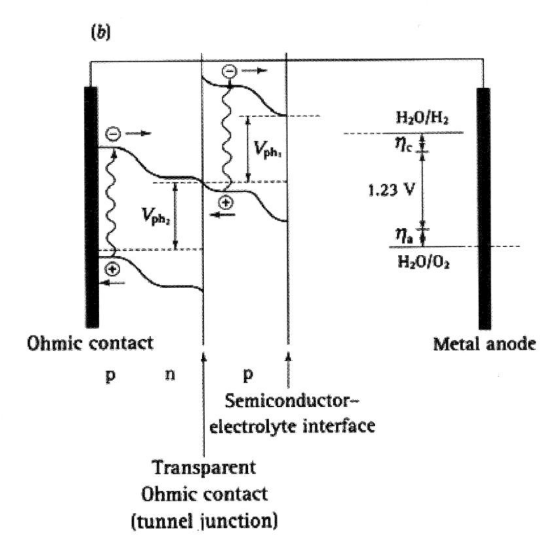

Fig. 4.44 (a) Schematic diagram of the construction of a two-photon photoelectrochemical cell for the electrolysis of water; (b) The principles of operation of such a device [O. Khaselev and J.A. Turner; Science **280** (1998) 425].

4.8.5
Detoxification using Photoelectrochemical Technology

The mechanism for the evolution of oxygen on an illuminated semiconductor surface, such as TiO_2, remains ill-understood, but is it likely that the first step involves hole capture by a surface hydroxyl species as:

$$p_v^+(s) + -OH^- \rightarrow -OH^{\cdot} \tag{4.184}$$

with subsequent formation of Ti-peroxy bonds and final evolution of O_2.

Detoxification of water is often achieved by using reagents such as ClO^-, generated *in situ* electrochemically (see section 8.4.1). The disadvantage of this process is that many organic substrates react with ClO^- with the formation of toxic chlorocarbons, and ideally we need a wide-spectrum oxidant that reacts cleanly with organic species to give complete mineralisation, i.e. complete conversion to CO_2, H_2O and inorganic species such as HCl. One such wide spectrum oxidant is the $OH^{\cdot}$ radical, which can be generated chemically from H_2O_2 and Fe^{3+} ions (the Fenton Reaction) or photochemically from O_3 and/or H_2O_2 and UV radiation, but which can also be generated photochemically using illuminated TiO_2 surfaces as described above. The cleanness of the chemistry of the $OH^{\cdot}$ radical has led to numerous detoxification processes being developed in recent years, which are known collectively as Advanced Oxidation Processes, and given the advantages of a heterogeneous catalytic process over homogeneous processes, attention has focussed on TiO_2.

The organic substrate is placed in the cell and the TiO_2 surface irradiated with supra-bandgap light. The $OH^{\cdot}$ radicals forming at the TiO_2 surface then attack the organic substrate, initially giving polyhydroxy species that further disintegrate as hydroxylation continues until complete conversion to CO_2 takes place. It is evident that given the short lifetime of the $OH^{\cdot}$ radical, the organic material must be close to the TiO_2 surface, ideally adsorbed, and this is borne out in kinetic studies, which show that the optimal efficiencies are often reached at low light intensities. The cell may be used in two forms: in the simplest, the electrode is kept at *open circuit* and oxygen sparged into the cell; under these circumstances, electrons will be transported to the surface as well, and reduction of oxygen takes place simultaneously. This has the advantage that partially reduced species such as the superoxide radical anion, $O_2^{\dot{-}}$, are also oxidants. The disadvantage is that both electrons and holes must be transported to the surface together, a process of low quantum efficiency. In spite of this, the open-circuit process forms the basis of a rapidly developing commercial treatment using TiO_2 *particulate slurries* suspended in water and illuminated.

Higher quantum efficiencies may be obtained if flat-plate TiO_2 electrodes are held at positive bias as in the photoelectrolysis cell; however the solution must have some ionic conductivity for this to be effective, and electrolyte must, therefore, be added. Furthermore, mass transport problems in dilute solutions are much more serious for this type of reactor than the slurry reactor, and no photo*electro*chemical reactors have yet been developed commercially. However, it is not necessary to have a substantial amount of electrolyte present; indeed even the adsorbed water present under ambient conditions may suffice for some surface chemistry, and the list of possible applications is very large indeed, including photocatalytic indoor air cleaners, photocatalytic automobile air cleaners, photocatalytically self-cleaning ceramic tiles for bathrooms and kitchens, photocatalytically self-cleaning lamp-shades for indoor lighting, photocatalytically self-cleaning Venetian window blinds, and photocatalytically self-cleaning glass covers for highway tunnel lamps.

It is not merely organic species that can be remediated in this way: the ability of TiO_2 films to kill bacteria and other microorganisms is also a very active area for research and development in many parts of the world, and examples include the bactericidal effect on *Escherichia coli* and the detoxification of the *E. coli* endotoxin. This topic is of particular interest because there have been several fatal outbreaks of food poisoning due to the O157 endotoxin. TiO_2-coated glass plates have been found to exhibit significant bactericidal and detoxification activity.

Related to this, and under active development in Japan are self-sterilizing ceramic tiles for hospitals, particularly operating rooms. These have been developed by the TOTO Company, the country's principal manufacturer of ceramic materials for restrooms. A critical problem in operating rooms is the presence of methicillin-resistant *Staphylococcus aureus* (MRSA). These are bacteria that have built up a resistance to methicillin and related antibiotics. Large quantities of disinfectants must be used to kill these bacteria, and the bactericidal effect is not continuous. In contrast, TiO_2-coated tiles are active continuously, as long as there is illumination. Newer coatings have also been developed that are even active in the dark – very small particles of silver are photocatalytically deposited on the surface.

4.9 Bioelectrochemistry

In recent years there has been a marked increase in interest in the driving of biochemical reactions using electrochemistry. Many biochemical processes involve the transfer of electrons or hydrogen atoms, and given the specificity of such reactions, a new generation of highly selective sensors has been developed, with particular emphasis on the detection of biologically active species such as glucose. The specificity of biological reactions is due to highly specific catalysts, termed enzymes. Enzymes are based on polypeptide sequences that adopt highly specific three-dimensional structures. Such structures contain an active site, but access to this site is highly restricted, ideally, to a single molecular substrate, which attaches itself to the site, undergoes conversion to product, and is then released. In this way, a number of redox carriers can co-exist in the cell, but each will have only one substrate that it can convert. However, from the electrochemical point of view, the inaccessibility of the site creates a serious problem, which we have not encountered up to now.

We saw, from equation 4.12, that the rate constant for electron transfer could be written $k_0'.\exp(-\Delta G^{(}E_1)/RT)$, and from equation 4.79 that ΔG can be written in the form $(U_s \pm e_0\eta)^2/4U_skT$ where U_s is the relaxation energy, and, as we saw, is a measure of the extent to which the reactant must be initally deformed geometrically for the potential energy curves of reactant and product to intersect before electron transfer can take place. For small molecules, the value of k_0', which measures the rate of electron tunnelling once the necessary molecular deformation has taken place, is rarely a limiting factor; almost all slow electron transfer processes for small molecules can be traced to large values of U_s. This is because small molecules can usually approach the electrode surface to at least the outer Helmholtz plane. However, tunnelling is exceedingly sensitive to tunnelling distance, and the probability of tunnelling decreases exponentially with this parameter. If d_H is the distance of the outer Helmholtz plane from the electrode, then the relative probability of electron transfer at a larger distance d is given as $e^{-\alpha(d-d_H)}$; α varies considerably, but in aqueous solution normally has values $\sim Å^{-1}$, suggesting that even for $d \sim 2d_H$ the probability of electron transfer will have fallen to a much lower level. This extreme sensitivity to distance is the basis for the technique of scanning tunnelling microscopy described in section 5.4.6, where it is exploited to study electrode surface morphology at atomic resolution.

However, for biological molecules this sensitivity to distance can give rise to serious problems, since the complex tertiary structure of enzymes is designed to keep the active centre inaccessible to all save highly selected substrates, and the result is that for very many biological catalysts, direct electron transfer from the electrode surface is either not possible or requires very specific surface preparation. For smaller redox-active molecules, such as cytochrome c (Cyt c), a redox protein, the active site is often more accessible, but only from one particular orientation; driving reversible electron transfer to this and related species then involves finding ways of orienting the molecule at the electrode surface to ensure that the active site is exposed very closely to the electrode without leading to unfolding of the molecule (with concomitant loss of biological activity). In the case of Cyt c, this orientation can be achieved by modifying the surface with pyridyl species; provided the N of the pyridine nucleus can be oriented into solution, then reversible bonding to the cytochrome molecule ensures that the latter is correctly oriented to facilitate electron transfer to and from the electrode surface.

4.9.1
The Biochemistry of Glucose Oxidase as a Typical Redox Enzyme

Enzymes are very large organic units with molecular weights frequently in excess of 100,000. The enzyme glucose oxidase, e.g., which acts as a catalyst for glucose oxidation (see equation (4.185)), is dimeric, with a total molecular weight of about 160,000. Each monomer incorporates one tightly bound *flavin adenine dinucleotide (FAD)* particle. The capacity of this FAD system to undergo two-electron two-proton reduction underpins its biochemical importance. Fig. 4.45 shows the reduction of a FAD unit to $FADH_2$ and the reoxidation via reaction with O_2 and formation of H_2O_2.

$$\beta\text{-D-glucose} + O_2 \xrightarrow{\text{glucose oxidase}} \text{D-glucono-1,5-lactone} + H_2O_2 \qquad (4.185)$$

The flavin unit in each of the monomers of glucose oxidase is located near the bottom of a deep cavity, access to which is provided by a large deep funnel-shaped pocket, one side of which is formed by the second molecule of the dimer. From the overall structure of the molecule as shown in Fig. 4.46, it is evident that electrochemical access to the FAD unit is strongly sterically impeded.

The mode of action of the enzyme is thought to be that glucose is first strongly bound to the oxidised form of the enzyme to form a complex that results in two hydrogen atoms transferred from the glucose (eqn. 4.185) to the FAD unit. The glucono lactone formed from the glucose then dissociates from the complex and subsequently hydrolyses to gluconic acid. The $FADH_2$ unit then chemisorbs molecular O_2, possibly

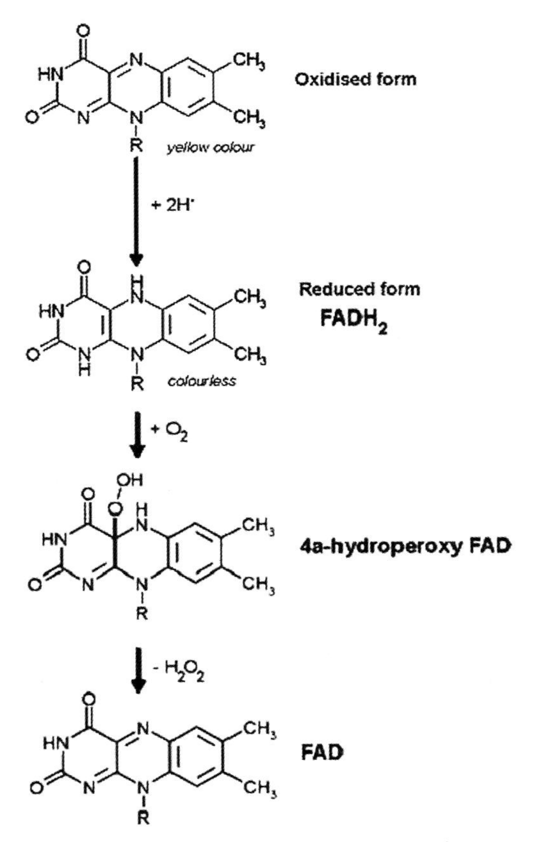

Fig. 4.45 Possible actions of a flavin (FAD) unit forming a complex with glucose.

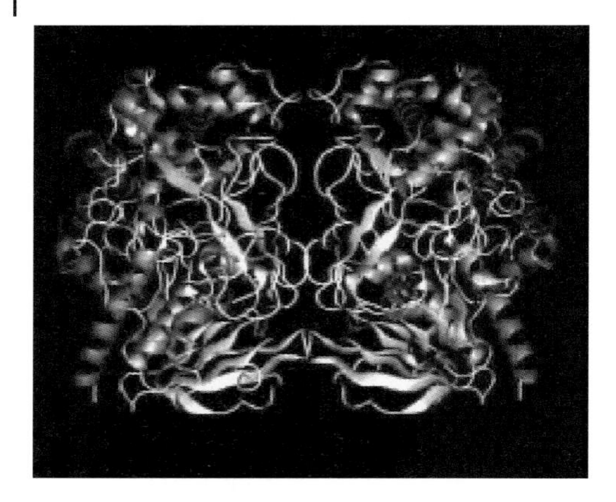

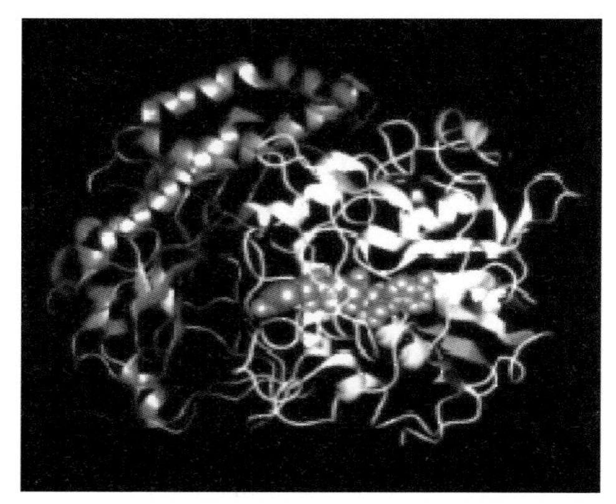

Fig. 4.46 (a) Overall topology of glucose oxidase, (b) subunit of glucose oxidase showing a FAD unit as ball and stick model.

as a bridged species, and this then results in two H-atom transfer to the O_2 and formation of H_2O_2 (Fig. 4.45). Whilst this is the re-oxidation route of the Glucose Oxidase *in vivo*, experiments in the laboratory have shown that $FADH_2$ can, in fact, be re-oxidised by a variety of one and two-electron systems such as p-phenylenediamine and a wide range of ferrocene derivatives. It is these derivatives that can be used to link the chemistry and electrochemistry of glucose oxidase.

4.9.2
The Electrochemistry of Selected Biochemical Species

Glucose oxidase does not show reversible electron transfer directly at an electrode surface. However, some promoters, such as relatively low molecular weight redox proteins, do allow reversible electron transfer to take place, provided that they can be appropriately oriented at the electrode, and this, in turn, can allow us to drive highly

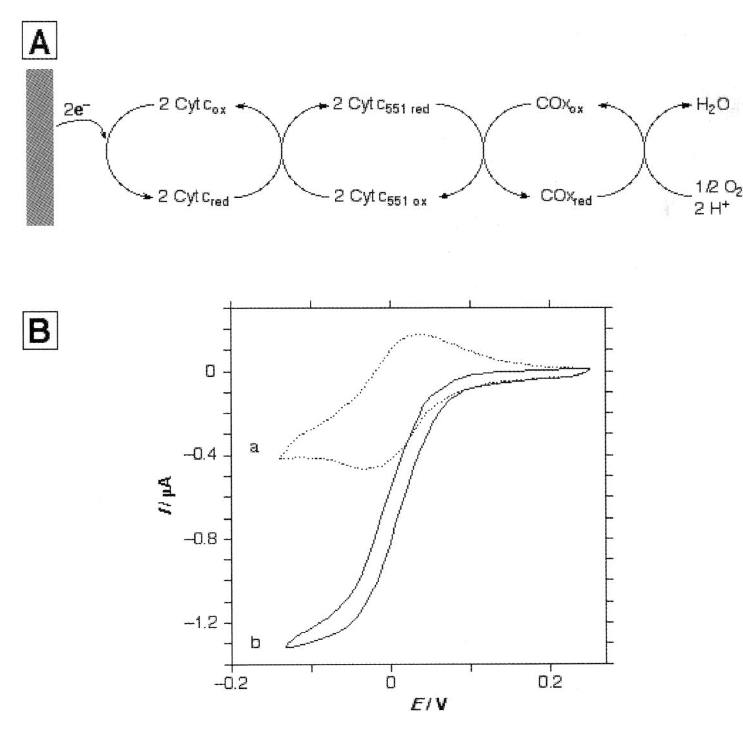

Fig. 4.47 (A) The bioelectrocatalyzed reduction of dioxygen by COx, mediated by Cyt c and Cyt c_{551} in a multistep electron-transfer process. (B) Cyclic voltammograms of a promoter-modified Au electrode (a) in the presence of Cyt c (5.3 mg mL^{-1}), and (b) after the addition of Cyt c_{551} (0.74 mg mL^{-1}) and COx (770 nM). Recorded in 0.02 M phosphate buffer, pH 7.0, scan rate 1 mV s^{-1}, in the presence of O_2 ((*from* http://chem.ch.huji.ac.il/~eugeniik/electrochemical_bio_tutorials .htm).

selective enzymatic reactions in solution. One example is the reduction of Cyt c, which in turn reduces the enzyme lactase that can, in turn, reduce O_2 cleanly to water. More complex sequences are possible: cytochrome oxidase (COx) cannot be directly re-reduced by Cyt c but if a second relay is added: Cyt c_{551}, then reduction of Cyt c at an electrode can again be directly coupled to reduction of oxygen as shown in the current-voltage curve (see section 5.2) of Fig. 4.47. It is clear that significant cathodic current only flows if all components of the relay are present. In a similar way, oxidative processes can also be coupled to enzymatic processes; thus lactate dehydrogenase can drive the oxidation of L-lactate and in turn be re-oxidised by the oxidised form of Cyt c. The oxidised form of Cyt c can also be used to regenerate $NADP^+$ which can then be connected to any $NADP^+$ dependent enzyme. This is important because the range of such enzymes is very large, but $NADP^+$ itself has extremely irreversible electrochemistry owing to its high reorganisation energy. Finally, by using negatively charged promoter layers, such as ω-thiocarboxylic acids, the positively charged Cyt c

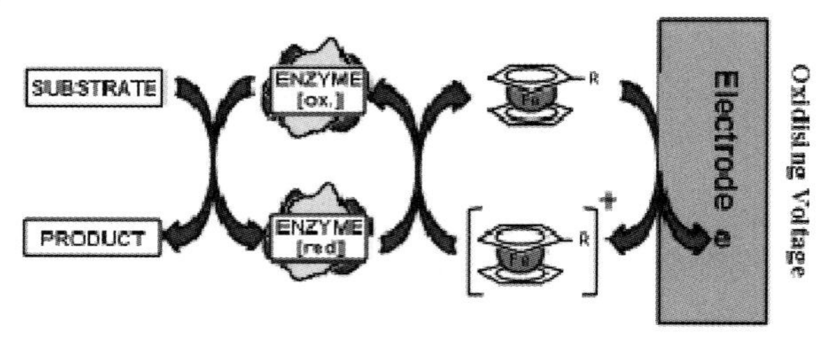

Fig. 4.48 Application of ferrocene derivatives as diffusional electron-transfer mediators for GOx and other oxidative enzymes. (*loc. cit.*)

molecules can be bound electrostatically at the electrode surface, allowing their continued function as catalysts but ensuring that all redox processes take place at the electrode itself.

To enhance electrochemical activity, low M.W. bio-mimics of Cyt c can be designed, one being microperoxidase-11, with a redox potential of −0.40 V vs. SCE. This species can be assembled as a monolayer on gold, but is sufficiently similar in structure to Cyt c to provide affinity interactions with haemproteins and cytochrome-dependent enzymes. One example is the formation of a complex of the μ-peroxidase-11 assembled monolayer with the native cytochrome-dependent nitrate reductase, allowing the construction of an amperometric nitrate sensor (see chapter 10).

However, as indicated above, direct electron transfer to most enzymes is either not possible or only possible under very specific and often difficult to reproduce electrode surface modifications. For Glucose oxidase (GOx) and similar flavin-based enzymes,

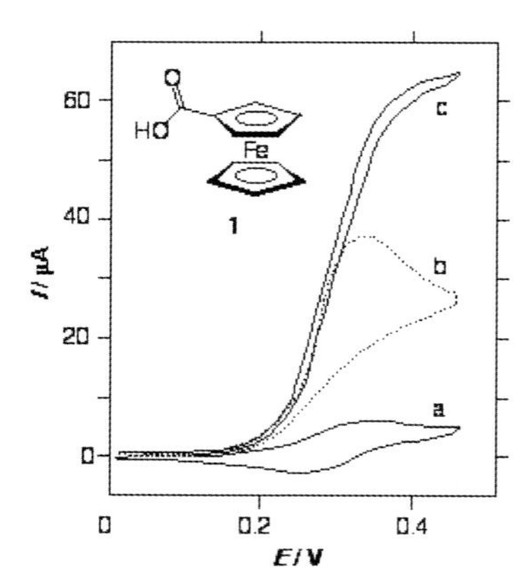

Fig. 4.49 Cyclic voltammograms for the bioelectrocatalyzed oxidation of glucose by GOx (38 μM), mediated by ferrocene carboxylic acid (1) (0.5 mM) in the presence of glucose at (a) 0 mM, (b) 0.5 mM and (c) 2.5 mM. Recorded in 0.085 M phosphate buffer, pH 7.0, under Ar, scan rate 5 mV s^{-1}. (*loc. cit.*)

with E^0 values ranging from − 0.49 to 0.19 V *vs.* SCE, mediated electron transfer has been extensively investigated using a wide range of small organic or inorganic molecules with fast electrode kinetics. Surprisingly, both one and two-electron oxidants are known, suggesting a facile route to re-oxidation through a semiquinoid intermediate, and the most satisfactory mediators have been found to be derivatives of ferrocene. This latter has the immense advantage of being rather easily fine tuned by substitution on either or both of the cyclopentadienyl rings, and all ferrocenes with E^0 values in excess of ca. 0.2 V *vs* SCE are capable of re-oxidising the reduced form of GOx, allowing the creation of a relay of the type shown in Fig. 4.48. The effectiveness of this relay can be judged from Fig. 4.49, which shows the biocatalysed oxidation of glucose by GOx in

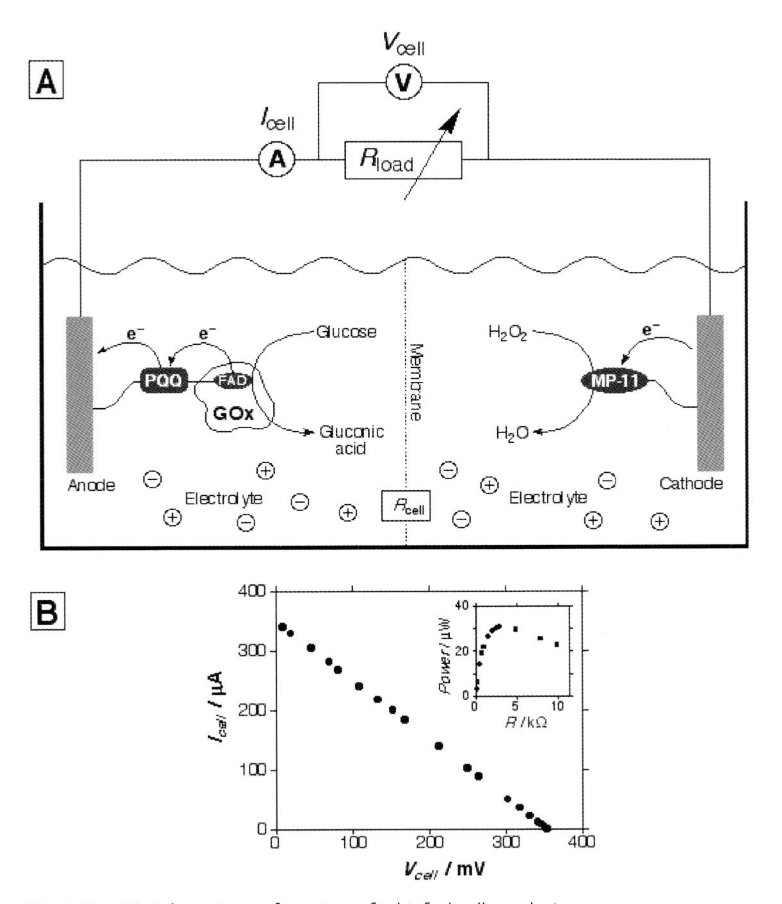

Fig. 4.50 (A) Schematic configuration of a biofuel cell employing glucose and H_2O_2 as fuel and oxidizer, and using PQQ-FAD/reconstituted GOx and MP-11-functionalized electrodes as biocatalytic anode and cathode, respectively. (B) Current-voltage behavior of the biofuel cell at different external loads. Inset: Electrical power extracted from the biofuel cell at different external loads. (*loc. cit.*)

the presence of a ferrocene relay. Curve a is typical of the cyclic voltammogram (section 5.2) of a reversible redox couple with E^0 value near 0.29 V vs. SCE. The other curves show clearly that the ferrocene is being recycled for further oxidation after rapid re-reduction by GOx. It is evident that this system can be elaborated into a glucose sensor.

The main applications of bioelectrochemical devices have been in biosensors, where their selectivity, especially in medical samples, has enormous benefits. However, there is increasing interest in bioreactors and biofuel cells; as a final example, the construction of a typical biofuel cell is shown in Fig. 4.50. Fuel Cells are described in detail in chapter 9, and have already been described in outline in chapter 1. The essential feature is the conversion of the chemical free energy of a reaction into electrical energy. The reaction of the fuel cell in Fig. 4.50 is the oxidation of glucose to gluconic acid by H_2O_2, and although currently attainable efficiencies are low (see chapter 9) the basic idea is being actively developed at the moment.

References for Chapter Four

For a further and more detailed account of the background quantum mechanics:

L. Landau and E. Lifschitz: "Quantum Mechanics", Pergamon Press, Oxford, 1966.

A.S. Davydov, "Quantum Mechanics", Pergamon Press, Oxford, 1965.

P.W. Atkins and R.S. Friedman: "Molecular Quantum Mechanics", Oxford University Press, Oxford, 2004.

R.D. Levine, "Molecular Reaction Dynamics", Cambridge University Press, Cambridge, 2005.

J. Baggott: "The Meaning of Quantum Theory", Oxford University Press, 1992.

More detailed accounts of the application of quantum mechanics to electrode processes can be found in:

J. Goodisman: "Electrochemistry: Theoretical Foundations", John Wiley, New York, 1987.

R.R. Doganadze: "Theory of Molecular Electrode Kinetics", in "Reactions of Molecules at Electrodes", Wiley Interscience, London, 1971, ed. N.S. Hush.

R.A. Marcus: "Electron Transfer Reactions in Chemistry; Theory and Experiment", *in* Rev. Mod. Phys. **65** (1993) 599 and references therein.

J. Ulstrup: "Charge Transfer Processes in Condensed Media", Springer-Verlag, Berlin, 1979.

V.G. Levich: "Kinetics of Reactions with Charge Transfer", in "Physical Chemistry", vol. IXb, Academic Press, New York, 1970, ed. Eyring, Henderson and Jost.

W. Schmickler: "Interfacial Electrochemistry", Oxford University Press, 1996.

"Interfacial Electrochemistry", ed. A. Wieckowski, Marcel Dekker, New York, 1999.

W. Schmickler and W. Vielstich: Electrochim Acta **18** (1973) 883; ibid. **21** (1976) 161.

H. Gerischer: Z, Phys. Chem. (Frankfurt) **26** (1960) 223, ibid. **26** (1960) 326, ibid. **27** (1961) 48.

P.P. Schmidt and H. Mark: J. Chem. Phys. **58** (1973) 4290.

"Microscopic Models of Electrode-electrolyte Interfaces", eds. J.W. Halley and L. Blum; Proceedings volume 93-5, The Electrochemical Society, Pennington, NJ, 1993.

More detailed accounts of the solution of the diffusion equation in electrochemistry can be found in:

D. Britz, "Digital Simulation in Electrochemistry", 3rd Ed., Springer-Verlag, Berlin, 2005.

J.O'M. Bockris and A.K.N. Reddy: "Modern Electrochemistry, 2nd ed.", Kluwer/Plenum Press, New York, 2000.

E. Kreysig, "Advanced Engineering Mathematics", John Wiley, New York, 1993.

H.S. Carslaw and J.C. Jaeger, "Conduction of Heat in Solids", Clarendon Press, Oxford, 1959.

Discussions of more complex electrochemical reactions can be found in:

J.O'M. Bockris and A.K.N. Reddy: "Modern Electrochemistry, 2nd ed.", Kluwer/Plenum Press, New York, 2000.

A.J. Bard and L.R. Faulkner, "Electrochemical Methods", John Wiley, New York, 1980.

P.A. Christensen and A. Hamnett, "Techniques and Mechanisms in Electrochemistry", Blackie and Son, Edinburgh, 1995.

Tabulated Data from:

B.E. Conway, "Electrochemical Data", Greenwood Press, Westport (Connecticut) 1969.

Discussions of the Rotating Disc Electrode can be found in:

V.G. Levich: "Physicochemical Hydrodynamics", Prentice Hall, Englewood Cliffs, NJ, 1962.

A.C. Riddiford, in "Advances in Electrochemistry and Electrochemical Engineering", volume 4, ed. P. Delahay and C.W. Tobias, J. Wiley, New York, 1966.

W.J. Albery and M.L. Hitchman, "Ring-Disc Electrodes", Clarendon Press, Oxford, 1971.

V. Pleskov and V. Yu. Filinovsky: "The Rotating Disc Electrode" Consultants Bureau, New York, 1976.

Spherical Microelectrodes can be found discussed in:

R.M. Wightman and D.O. Wipf: "Voltammetry at Ultramicroelectrodes", in Electroanalytical Chemistry, vol. 15,, ed. A.J. Bard, Marcel Dekker, New York, 1989.

Coupled electrochemical and acid dissociation processes are discussed further in:

W.J. Albery: "Electrode Kinetics", Clarendon Press, Oxford, 1975.

Adsorption Processes are further discussed in:

B.B. Damaskin, O.A. Petrii and V.V. Baratrov, in "Adsorption of Organic Compounds in Electrodes" (ed. K. Schwabe), Akadamie Verlag, Berlin, 1975.

B.N.W. Trapnell: "Chemisorption", Academic Press, New York, 1955.

J.O'M. Bockris and A.K.N. Reddy, loc. cit.

I.M. Campbell: "Catalysis at Surfaces", Chapman and Hall, London, 1988.

H.R. Thirsk and J.A. Harrison: "A Guide to the Study of Electrode Kinetics", Academic Press, New York, 1972.

B.B. Damaskin and V.E. Kazarinov: "The Adsorption of Organic Molecules" in "Comprehensive Treatise of Electrochemistry", vol. 1, "The Double Layer", eds. J.O'M. Bockris, B.E. Conway and E. Yeager, Plenum Press, New York, 1980.

"Adsorption of Molecules at Metal Electrodes", eds. J. Lipkowski and P.N. Ross, VCH, Weinheim, 1992.

Electrocrystallisation effects are further discussed in:

Southampton Electrochemistry Group: "Instrumental Methods in Electrochemistry", Ellis Horwood, Ltd., Chichester, 1985.

M. Fleischmann and H.R. Thirsk: Electrochim. Acta **21** (1960) 22.

E. Bosco and S.K. Rangarajan: J. Electroanal. Chem. **134** (1981) 213.

E. Budievskii in "Comprehensive Treatise of Electrochemistry", vol. 7, "The Double Layer", eds. J.O'M. Bockris, B.E. Conway and E. Yeager and R.E. White, Plenum Press, New York, 1983.

E. Budievskii, G. Staikov and W.J. Lorenz, "Electrochemical Phase Formation and Growth", VCH Publications, Weinheim, 1996.

"Kinetics of Ordering and Growth at Surfaces", ed. M.G. Lagally, Plenum Press, New York, 1990.

R. de. Levie in "Advances in Electrochemistry and Electrochemical Engineering" eds. H. Gerischer and C.W. Tobias, Interscience, New York, 1985.

Electrochemical Machining is further discussed in:

A.E. de Barr and D.A. Oliver, "Electrochemical Machining", MacDonald and Co. Publishers, London 1968.

Introductory accounts of corrosion can be found in:

U.R. Evans: "The Corrosion and Oxidation of Metals", Arnold, London, 1960.

J.M. West: "Basic Corrosion and Oxidation", 2nd. Ed., Ellis Horwood, Chichester, 1986.

M. Pourbaix: "Atlas of Electrochemical Equilibria in Aqueous Solutions", Pergamon Press, Oxford, 1966.

H. Gerischer and C.W. Tobias (eds.), "Advances in Electrochemical Science and Engineering", vol. 3, VCH, Weinheim, 1994.

An introduction to semiconductor electrochemistry can be found in:

A. Hamnett: "Semiconductor Electrochemistry" in "Comprehensive Chemical Kinetics", vol. 27, ed. R.G. Compton, Elsevier, Oxford, 1987.

S.R. Morrison: "Electrochemistry at Semiconductor and Oxidised Metal Electrodes", Plenum Press, New York, 1980.

R. Memming: "Semiconductor Electrochemistry", Wiley-VCH, Weinheim, 2001.

V.A. Myamlin and Yu.V. Pleskov: "Electrochemistry of Semiconductors", Plenum Press, New York, 1967.

A. Many, Y. Goldstein and N.B. Grover: "Semiconductor Surfaces", North Holland Press, Amsterdam, 1965.

Applications to solar energy conversion can be found in

A. Fujishima and K. Honda, Nature **238** (1972) 37.

J.G. Mavroides, J.A. Kafalas and D.F. Kolesar, Applied Phys. Lett. **28** (1976) 241.

A. Hamnett and P.A. Christensen in: "The New Chemistry", ed. Nina Hall, Cambridge University Press, 2000.

D.A. Tryk, A. Fujishima and K. Honda, Electrochim. Acta **45** (2000) 2363.

M.R. Hoffmann, S.T. Martin, W. Choi and D.W. Bahnemann, Chem. Rev. **95** (1995) 69.

M.K. Nazeeruddin, P. Pechy, M. Grätzel, Chem. Commun. (1997) 1705–1706.

A. Fujishima, K. Hashimoto and T. Watanabe: "TiO_2 Photocatalysis: Fundamentals and Applications, BKC, Tokyo, 1999.

V.M. Aroutiounian, V.M. Arakelyan and G.E. Shahnazaryan, Solar Energy **78** (2005) 581.

Bioelectrochemistry

This is a very fast moving area, and one where there is little in the form of conventional second-tier literature. A very valuable source of information is the Internet, and some useful links are:

http://chem.ch.huji.ac.il/~eugeniik/electrochemical_bio_tuto rials .htm *and* http://www-biol.paisley.ac.uk/marco/enzyme_electrode/Chapter 1/STA RT.HTM.

Biochemical fuel cells, E. Katz, A.N. Shipway and I. Willner, in Handbook of Fuel Cells, vol. 1, Eds. W. Vielstich, A. Lamm and H.A. Gasteiger, Wiley UK, Chichester 2003.

E.S. Forzani, G.A. Rivas and V.M. Solis: Amperometric determination of dopamine on vegetal-tissue enzymatic electrodes, J.Electroanal.Chem. 435(1997)77-84.

More conventional literature includes:

"Encyclopedia of Electrochemistry, vol. 9: Bioelectrochemistry", eds. A.J. Bard, M. Stratmann and G.S. Wilson, publ. Wiley-VCH, Weinheim, 2002.

J.M. Savéant: "Elements of Molecular and Biomolecular Electrochemistry", John Wiey, Hoboken, New Jersey, 2006.

A.M. Kuznetsov: "Charge Transfer in Physics, Chemistry and Biology: Physical Mechanisms of Elementary Processes and an Introduction to the Theory", Taylor and Francis, London, 1995.

C. Fry and S.E.M. Langley: "Ion selective Electrodes and Biological Systems", Harwood Academic, Chichester, England, 2001.

"Electroanalytical Methods for Biological Systems", eds. A. Brajter-Toth and J.Q. Chambers, publ. Marcel Dekker, New York, 2002.

J.O'M. Bockris and A.K.N. Reddy: "Modern Electrochemistry: Electrodics in Chemistry, Engineering, Biology and Environmental Science, vol. 2B", Kluwer Academic/Plenum Press, New York, 2001.

R.A. Bullen, T.C. Arnot, J.B. Lakeman and F.C. Walsh: "Biofuel Cells and their Development" *in* Biosensors and Bioelectronics **21** (2006) 2015.

There are many conferences currently in this area, including a recent Faraday Discussion (Vol. 116, 2000). These should be consulted for more details of recent discoveries.

5
Methods for the Study of the Electrode/Electrolyte Interface

There is a large number of methods available for the study of electrode processes, the most important of which, we will describe in this chapter. The data resulting from these methods will provide information about the reaction rate, the mechanism of the reaction, the types of possible intermediates and any participating adsorption processes.

A rapid oversight of any electrode reaction can be obtained from current-potential curves obtained either from stationary or quasi-stationary measurements. However, accurate measurement of the electrode kinetics requires the mass transport to be carefully controlled, and also requires corrections to be made for the ohmic potential drop between the working electrode and the tip of the reference electrode which is associated with the passage of electrical current.

It should be emphasised that the study of adsorption processes by purely electrochemical methods leads to very limited information about the identity and form of any electrode film or adsorbed intermediate. Such studies have been greatly facilitated in recent years by the development of a whole variety of novel optical techniques, which are described below. However, even these techniques, taken singly, rarely yield unambiguous information, particularly about complex reactions, and a combination of electrochemical and spectroscopic techniques must be used.

5.1
The Measurement of Stationary Current-Potential Curves

In principle, the apparatus described in Figure 4.3 can be used, with the variable resistance, *R*, being used to control the current. By altering *R*, pairs of current-potential values can be measured for the working electrode and the current-potential curve constructed. However, this method is both laborious and of low accuracy, particularly at lower current densities, and modern studies invariably use a *potentiostat*.

Electrochemistry. Carl H. Hamann, Andrew Hamnett, Wolf Vielstich

ISBN: 978-3-527-31069-2

5.1.1
The Potentiostat

A schematic diagram of a potentiostat is shown in Fig. 5.1. The central electronic component is the operational amplifier. Modern operational amplifiers have three important properties: the first is that its two voltage inputs, here E_d and E_m, have very high impedance, so they draw very little current, and the second is that its output voltage $E_{out} = -A\delta V$, where δV is the difference in the two input voltages and A the amplifier gain. In practice, A is very large indeed, which means that the output voltage will also become very large unless the operational amplifier maintains δV at an extremely small level. In addition, the output impedance is very low, allowing current to flow freely. These properties allow us to control the three electrodes in our cell: under normal operation, the working electrode is maintained at ground, and a potential, E_d, is applied between this ground and one of the inputs of the amplifier. The second imput of the amplifier is connected to the reference electrode, and the properties of the amplifier then ensure that the applied potential actually equals the potential difference between working and reference electrodes *in the cell,* whilst the high input impedance prevents any significant current being drawn through the reference electrode. The *output* of the amplifier is connected to the counter electrode, and the low output impedance then permits current to flow between working and counter electrodes. We are not normally interested in measuring the potential of the counter electrode, provided it remains small enough for δV above to be negligible, but we are

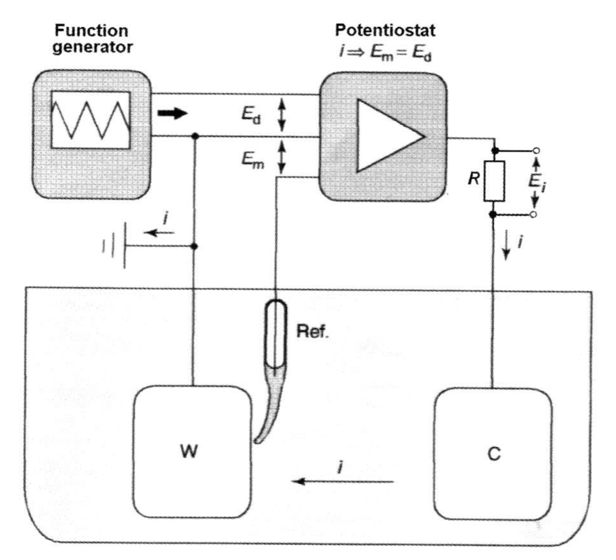

Fig. 5.1 Basic potentiostatic circuit. W, C and Ref: working, counter and reference electrodes of the electrochemical cell; E_d, applied driving voltage; E_m, measured voltage. The potentiostat delivers a current equalising E_d and E_m. The resistance R converts the current into the voltage E_i to allow its measurement.

interested in monitoring the current, and this is done by passing the current through a resistor, R, and measuring the voltage drop, E_i, across that resistor using a second type of operational amplifier termed a *current follower*. The time taken by operational amplifiers to equalise their input voltages is less than microseconds in typical modern potentiostats, thus enabling the accurate determination of current for oscillating input voltage frequencies up to 10 kHz. The two voltages, E_d and E_i are usually fed to the x-y inputs of either a chart recorder or, for fast transient measurements, to an oscilloscope; alternatively they may be stored in digital form in a computer.

5.1.2 Determination of Kinetic Data by Potential Step Methods

The principle behind methods in which the current or potential is suddenly stepped (over a timescale of microseconds or less) from a stationary to a non-stationary state is that over short periods following the step, the kinetically limiting processes will be electron transfer rather than diffusional in nature. In the case of a potential step, the measurement of the resultant current transient yields information both about the rate of electron transfer and the diffusion rate of material to the surface.

The current and concentration transients are shown schematically in Fig. 5.1a, and assuming, for simplicity, that only electron transfer and diffusion of material to the electrode are important (i.e. neglecting adsorption and desorption steps and coupled

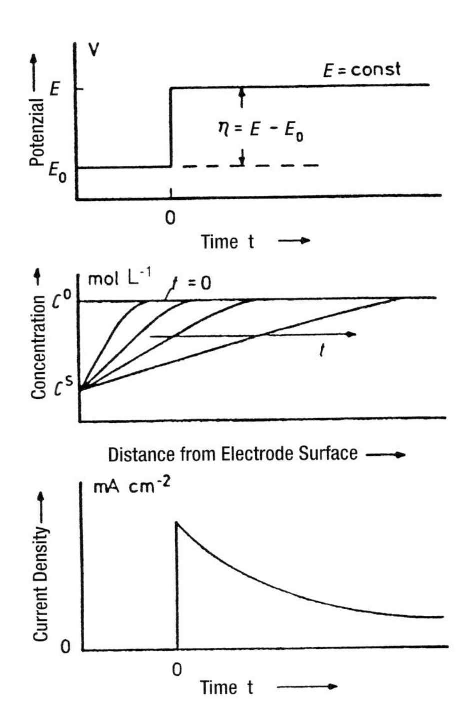

Fig. 5.1a Increase in diffusion layer thickness and evolution of the current with time following a jump in potential (schematic).

chemical reactions), we can write down the diffusion equations for the participating species, Ox and Red, assuming an electrochemical reaction

$$Ox + e^- \rightarrow Red$$

in the form

$$\frac{\partial c_{ox}}{\partial t} = D_{ox}\frac{\partial^2 c_{ox}}{\partial x^2}; \quad \frac{\partial c_{red}}{\partial t} = D_{red}\frac{\partial^2 c_{red}}{\partial x^2} \tag{5.1}$$

and with the electron transfer rate determined by the concentration depending Butler-Volmer equation

$$j(\eta) = j_0\left\{\left(\frac{c^s_{red}}{c^0_{red}}\right)\exp\left[\frac{(1-\beta)F\eta}{RT}\right] - \left(\frac{c^s_{ox}}{c^0_{ox}}\right)\exp\left[-\frac{\beta F\eta}{RT}\right]\right\} \tag{5.2}$$

The boundary conditions for a step from stationary state ($\eta = 0$) to an overpotential η are:

$$\begin{aligned}
&t = 0 \quad x \geq 0 \quad c_{red} = c^0_{red};\ c_{ox} = c^0_{ox}\\
&t \geq 0 \quad x \rightarrow \infty \quad c_{red} = c^0_{red};\ c_{ox} = c^0_{ox}\\
&t > 0 \quad x = 0 \quad c_{red} = c^s_{red};\ c_{ox} = c^s_{ox}\\
&j = -FD_{ox}\left(\frac{\partial c_{ox}}{\partial x}\right)_{x=0} = FD_{red}\left(\frac{\partial c_{red}}{\partial x}\right)_{x=0}
\end{aligned} \tag{5.3}$$

The solution can be expressed as:

$$j(\eta, t) = j_0\left\{\exp\left[\frac{(1-\beta)F\eta}{RT}\right] - \exp\left[-\frac{\beta F\eta}{RT}\right]\right\}e^{\lambda^2 t}\operatorname{erfc}(\lambda\sqrt{t}) \tag{5.4}$$

where

$$\operatorname{erfc}(\lambda\sqrt{t}) = 1 - \frac{2}{\sqrt{\pi}}\int_0^{\lambda\sqrt{t}} e^{-x^2}\,dx, \quad \lambda = \frac{j_0}{F}\left\{\frac{\exp[(1-\beta)F\eta/RT]}{c^0_{red}\sqrt{D_{red}}} + \frac{\exp[-\beta F\eta/RT]}{c^0_{ox}\sqrt{D_{ox}}}\right\}. \tag{5.5}$$

In other words, the current j can be expressed as the product of two functions $j_e(\eta) \times g(t, \eta)$, where the time independent term corresponds to the electron-transfer-limited current and $g(t, \eta) \rightarrow 1$ as $t \rightarrow 0$. In other words, immediately after the potential jump, the current corresponds purely to electron-transfer-limited behaviour. At long times, $g(t, \eta) \rightarrow \frac{1}{\lambda\sqrt{\pi t}}$, and we recover a purely diffusion-limited current behaviour. For very short times, $\lambda\sqrt{t} << 1$, we find $e^{\lambda^2 t}\operatorname{erfc}(\lambda\sqrt{t}) \rightarrow 1 - 2\lambda\sqrt{t/\pi}$, and we find

$$j(\eta, t) = j_0\left\{\exp\left[\frac{(1-\beta)F\eta}{RT}\right] - \exp\left[-\frac{\beta F\eta}{RT}\right]\right\}(1 - 2\lambda\sqrt{t/\pi}) = j_e - 2j_e\frac{\lambda}{\sqrt{\pi}}\sqrt{t} \tag{5.6}$$

from which we see that plots of $j(\eta, t)$ vs. $\sqrt{t}$ allow us to obtain both β and j_0 (see Fig. 6.17). Unfortunately, there is a limiting time of the order 1 ms below which interference from the capacitive charging current makes accurate determination of the faradaic current $j(\eta, t)$ impossible. Given that $\lambda \sim j_0$, this means that potential step methods cannot be used for large exchange-current densities, and the limit is 20–50 mAcm^{-2}.

In addition to the potential-step method, it is also possible to step the current in a galvanostatic experiment, and then to measure the evolution of electrode potential with time. With initial boundary conditions and assumptions similar to those above, and with $c^0_{ox} = c^0_{red} = c^0$ we find an approximate solution:

$$E(j,t) = j\frac{RT}{F}\left\{\frac{1}{j_0} + \frac{2}{\sqrt{\pi}Fc^0}\left(\frac{1}{D_{ox}^{\frac{1}{2}}} + \frac{1}{D_{red}^{\frac{1}{2}}}\right)t^{\frac{1}{2}}\right\} \tag{5.7}$$

where E is the cell potential measured with respect to the rest potential, and j is the galvanostatic current step. The solution is valid for times below 0.1τ, where τ is the transition time defined in Section 4.3.4. As for the potential-step method, extrapolation to very short times is prevented by the double-layer charging.

5.1.3 Measurements with Controlled Mass Transport

We have already encountered the fact that at higher currents, the overall current density will become limited by the rate at which material can be transported to the electrode surface. For *natural convection* which can arise in any unstirred solution through concentration differences, temperature gradients etc, the current will be restricted to ca. 10 μA cm^{-2} for an electroactive solution species present at a concentration of 10^{-3} M. To obtain higher currents, *forced convection* must be employed, which can be achieved, for example, through vigorous gas bubbling or, more reproducibly, by using a cyclindrical rotating electrode, as shown in Fig. 4.17.

We have already seen in Section 4.3.5 how the rotating disc electrode can be used to measure electrochemical rate constants, and in Section 4.4 how the rate constant of a chemical step prior to electron transfer can also be determined. However, in order to probe chemical processes *following* electron transfer and to identify electrochemically formed intermediates, we must use double or multiple electrode arrangements, of which the most frequently employed are the ring-disc and ring-ring configurations (see below Fig. 5.2a). In such configurations, material formed at the inner electrode is transported to the outer electrode through the radial component of the convection current, and any changes that take place during transport can be monitored at the second electrode. It is, of course, necessary to eliminate possible interference from the reactant, which can normally be achieved by holding the outer electrode at a potential at which the reactant is inactive but the product(s) reduced or oxidised rapidly. By dividing the outer ring into segments, several reaction products can simultaneously be detected.

Quantitative calculation of the current at the outer ring has been carried out, and such calculations have shown that even in the absence of any further chemical reac-

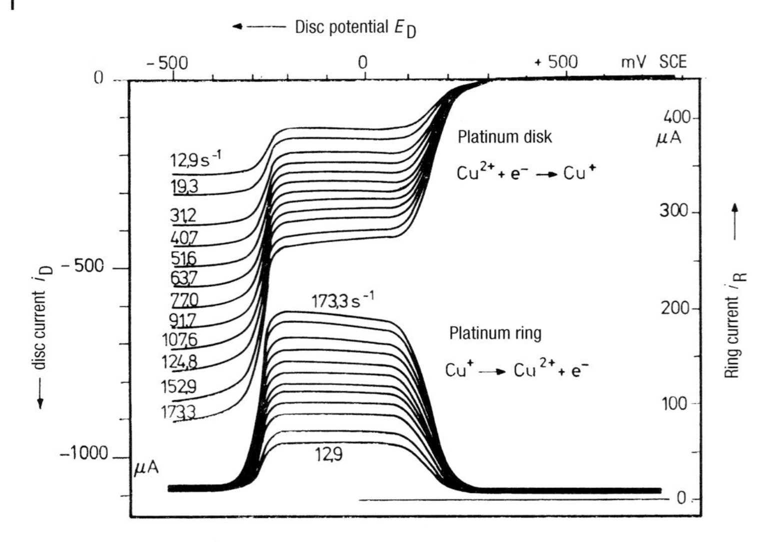

Fig. 5.2 Current-potential curves at a rotating ring-disc electrode in 10^{-3}M $CuCl_2$/0.5 M KCl with rotation frequency varying as indicated. The potential scan rate at the disc is 10 mV s^{-1} and the ring is held constant at + 0.4 V vs. SCE.

tions involving the reaction product, not all the product will be detected at the ring, since some can diffuse into the bulk of the solution. Theory shows, however, that regardless of the rotation rate, *a constant fraction of the product is detected at the outer ring electrode*. This fraction is termed the *collection efficiency*, *N*, and *N* is solely a function of the geometry of the system; specifically, for a rotating ring-disc system, *N* is a function solely of the radius of the disc and the inner and outer radii of the ring. In practice, N is measured rather than calculated for a specific system using a well-defined electrochemical process such as the oxidation of $[Fe(CN)_6]^{4-}$ to $[Fe(CN)_6]^{3-}$ at the disc, the oxidised species being reduced again at the ring.

An illustration of the use of the ring-disc electrode is given in Fig. 5.2, which shows the results of measurements on the system $Cu^{2+}/Cu^{+}/Cu$. At ca. +250 mV *vs.* SCE,

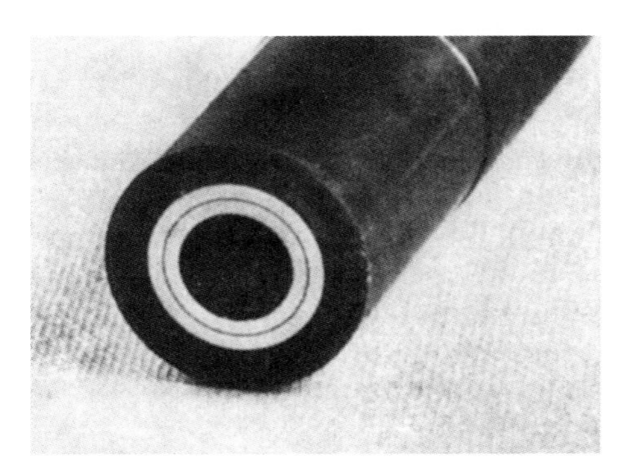

Fig. 5.2a Photograph of a ring-ring-double electrode. The metal rings are separated by a ring of isolating material.

the reduction of Cu^{2+} to Cu^+ starts at a Pt disc in 10^{-3}M $CuCl_2$/0.5 M KCl, and this reaches the diffusion limit below -150 mV. The further reduction of Cu^+ ions to metallic copper clearly begins below -250 mV vs. SCE, reaching a second diffusion limit at -350 mV. Between $+250$ and -250 mV, the Cu^+ ions formed at the disc are swept into solution and can be detected at the platinum ring, where they are *re-oxidised* to Cu^{2+} at a potential of $+400$ mV.

If the concentration of product formed at the disc is reduced through a subsequent chemical reaction before the product reaches the ring, the observed ring-current will be smaller than expected. If N_r is the apparent collection efficiency, and N the geometric collection efficiency, not only will N_r be smaller than N, it will be a function of rotation speed, since the longer the time taken for transport from disc to ring, the less product will be detected. Assuming that the chemical reaction does not lead to an electro-active species, analysis shows that

$$\frac{1}{N_r} = \frac{i_D}{i_R} = \frac{1}{N} + 1.28\left(\frac{\nu}{D}\right)^{\frac{1}{3}} \frac{k/\omega}{N} \tag{5.8}$$

where k is the rate constant for the removal of the disc product (assumed first order in the product), and as in Section 4.3.5, ν is the kinematic viscosity of the solvent and D the diffusion coefficient of the product, i_D the disc current and i_R the ring current.

An example of this type of analysis is the electrogeneration of Fe^{2+} ions in a solution containing H_2O_2. At the disc, we have $Fe^{3+} + e^- \rightarrow Fe^{2+}$, and at the ring $Fe^{2+} \rightarrow Fe^{3+} + e^-$, whilst in the solution the oxidation reaction $2Fe^{2+} + H_2O_2 + 2H^+ \rightarrow 2Fe^{3+} + 2H_2O$ takes place with a rate constant of 75 $dm^3 mol^{-1} s^{-1}$ in acid, as derived from (5.8).

For rapid reactions, the reaction path must be as short as possible, and double ring systems are used (Fig. 5.2a). By suitable choice of geometry, product lifetimes of as low as 10^{-5} s can be measured.

5.1.4
Stationary Measurement of Very Rapid Reactions with Turbulent Flow

The analysis of convection above pre-supposed that the liquid flow could be treated as *laminar*, that is as having a well-defined and calculable flow pattern. However, if, in a given cell, the flow rate of the liquid is increased, then at a particular flow velocity, the liquid ceases to show laminar flow, and there is a transition to *turbulent* flow. For a liquid of kinematic viscosity ν flowing across a surface which has some characteristic length l with a speed υ, then this transition takes place at a specific value of the so-called Reynolds number $Re \equiv \upsilon \cdot l/\nu$. Turbulent flow is characterised not only by macroscopic effects such as vortices, but more significantly by fluctuations of the velocity of liquid elements on a microscopic scale. If, for example, the flow is primarily in the y-direction, with a mean velocity $\bar{\upsilon}_y$, then superimposed on this will be fluctuations in the instantaneous velocities in the x- , y- and z-directions. These fluctuations contribute substantially to the mass transport both in the bulk of the solution and near the electrode surface, though they must fall to zero at the interface with any solid owing to

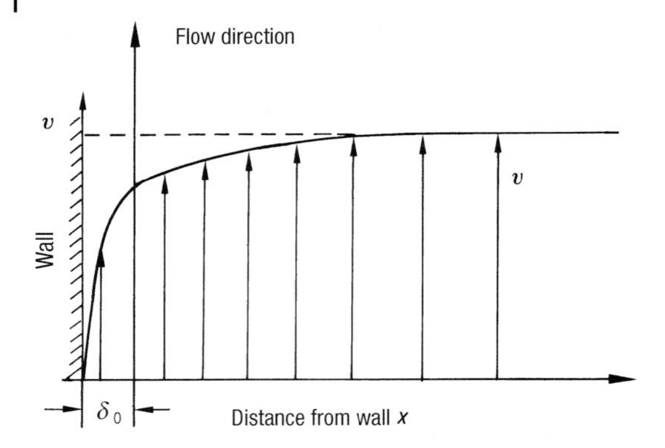

Fig. 5.3a Fall off in liquid velocity near an immobile surface in a turbulent boundary layer

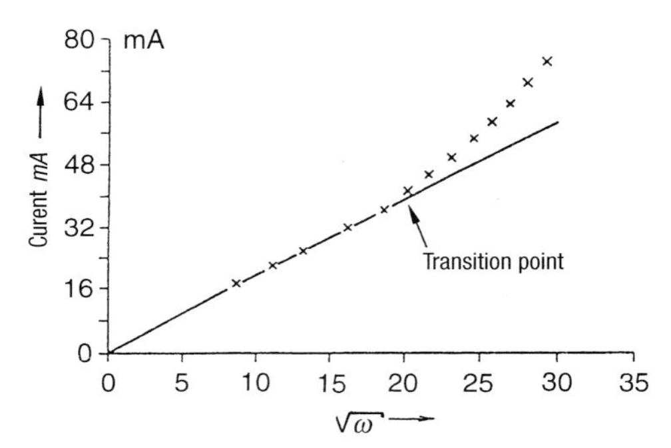

Fig. 5.3b Evolution of the limiting current at a rotating ring-electrode (inner diameter 50 mm; outer diameter 52 mm) during transition from laminar to turbulent flow. The electrochemical reaction is the reduction of $[Fe(CN)_6]^{3-}$ to $[Fe(CN)_6]^{4-}$.

the effects of viscosity. Close to such an interface the decrease in velocity components is almost linear over a distance δ_0 (Fig. 5.3a), and this distance is much smaller than the Nernst diffusion layer δ_N in the case of laminar flow. In fact, to a first approximation, the familiar relationships of laminar flow that we have already met in Section 4.3.5 can be replaced by similar equations for turbulent flow provided δ_N is replaced by $\delta_{N,turb}$ and the diffusion coefficient D by $(D + \varepsilon)$, where ε contains the turbulent contribution to transport and depends on the distance to the electrode surface.

This similarity of underlying equations shows that the current-voltage curves for turbulent flow conditions will show, for example, clear limiting currents, but these currents will generally be larger than those found under laminar conditions, as shown in Fig. 5.3b, where there is an obvious transition point between the two flow regimes.

Measurements under turbulent conditions are particularly suitable for very rapid electrochemical processes, but the use of a rotating disc as in Fig. 5.3 is not really practicable since the required angular velocity for purely turbulent flow is not easily accessible. It is possible to use rotating *ring* electrodes, but at the required rotation speeds, these suffer heavy mechanical loads due to centrifugal forces, and froth formation in the liquid is also troublesome.

These difficulties have been overcome by the use of a hollow tube, in the interior of which is mounted a ring to form the working electrode (Fig. 5.4a). Electrolyte forced down the tube will show turbulent behaviour if $\mathrm{Re} = \upsilon \cdot d/\nu$ (where d is the internal tube diameter) exceeds ca. 2300. In practice, to convert such an arrangement into a cell, we will need a second ring as a reference electrode and a counter-electrode placed as close as possible to the outlet to the tube. Typical values of d are ca. 0.2 cm, so that a flow speed of 400 cm s^{-1} gives a Reynolds number of ca. 8000 in aqueous electrolyte. The metal ring electrodes usually have thicknesses in the region 100–200 μm.

Evaluation of the exchange-current density for a specific electrochemical reaction can be carried out straightforwardly by comparing the experimentally obtained current-voltage relationships in the anodic and cathodic regions with those calculated by computer simulation. By comparison of the experimental data with working curves derived from the computer simulation, values of the exchange-current density can be estimated, and an example is shown in Fig. 5.4b for the redox couple $[Ru(NH_3)_6]^{2+/3+}$ on gold, for which the standard exchange-current density is a remarkable 98 A cm^{-2}. This corresponds to a rate constant k_0 of ~ 1 cms^{-1}, or to ca. 7×10^{20} molecules involved in redox processes on Au per square cm every second. Given that the gold surface will contain ca. 2×10^{15} atoms per unit area, it follows that ca. $3 - 4 \times 10^5$ electron exchange processes take place per surface atom per second for this very fast couple (compare section 4.2.3).

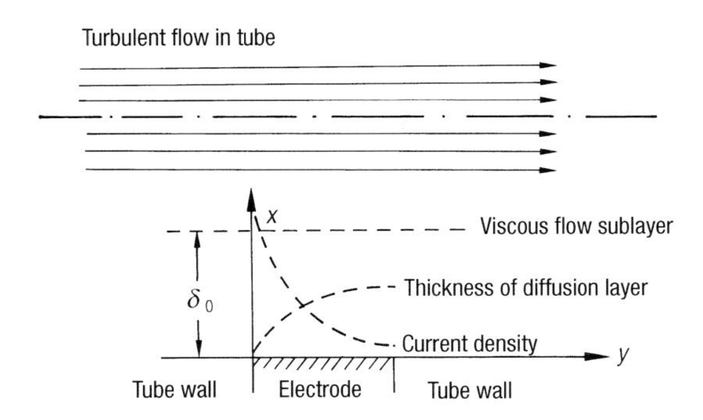

Fig. 5.4a Thickness of the diffusion layer and variation of the electrical current density above a ring-electrode under turbulent conditions (not to scale).

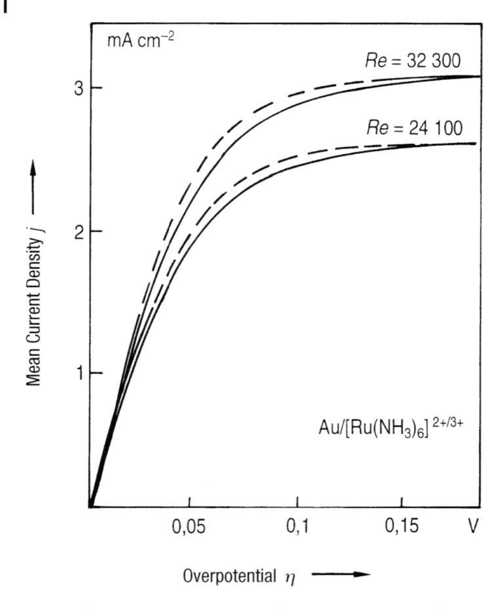

Fig. 5.4b Stationary current density vs. potential curves obtained with a ring electrode in the turbulent flow régime for the system $[Ru(NH_3)_6]^{2+/3+}$ on gold at 20°C. The concentration $c_{ox} = 3.3 \times 10^{-4}$M and $c_{red} = 2.8 \times 10^{-4}$M. For comparison the theoretical curves calculated for infinitely rapid electron transfer, j_{rev} are also shown. From this measurement, the exchange current density for the electron transfer process can be calculated to be $j_0 = 116$ A cm^{-2}.

5.2 Quasi-Stationary Methods

Following the sudden alteration of an electrode parameter such as current, potential or concentration, the system will need, even under favourable conditions, a certain time to reach a new stationary state. Naturally, as the time taken to effect the alteration becomes very long, the system will approach the stationary condition described in Section 5.1 above, but there will be an intermediate regime, for example through imposition of a linear or sinusoidally changing potential that does not fluctuate too rapidly, for which the electrode will respond in some quasi-stationary mode.

5.2.1 Cyclic Voltammetry: Studies of Electrode Films and Electrode Processes – "Electrochemical Spectroscopy"

Cyclic voltammetry involves the imposition of a triangular waveform as the potential on the working electrode, as shown in Fig. 5.5, with the simultaneous measurement of the current. Note that in this and all subsequent figures, the convention is adopted that

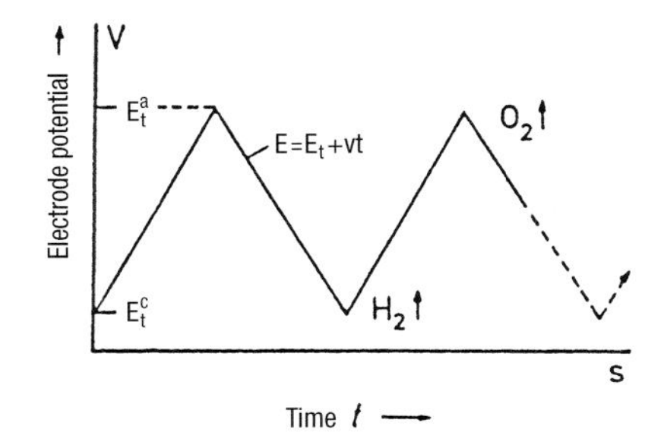

Fig. 5.5 Potential-time behaviour at the working electrode following imposition of a triangular waveform typical of cyclic voltammetry; v is the rate of change of the potential with time, and clearly changes sign at each reversal point.

positive going currents are *anodic* and negative-going currents are *cathodic*. The potential is normally generated with a function generator, and applied to the cell through a potentiostat as described earlier. The current-potential curves can be displayed with an X-Y recorder at slow scan rates, and with a storage oscilloscope or computer terminal if a wider range of scan speeds is required.

Negative and positive turn-round potentials, E_t^c and E_t^a are usually chosen in aqueous electrolyte to lie between the hydrogen and oxygen evolution potentials, which has the advantage that any adsorbed impurities that might otherwise block the electrode process of interest can be removed either by oxidation or reduction. By scanning the potential between these limits, it is possible to obtain highly reproducible current-voltage behaviour (cyclic voltammograms) for solid metals such as platinum, gold and palladium. In general, however, reproducible behaviour in cyclic voltammetry depends on a series of parameters, including electrolyte purity, identity of the electrode material, choice of turn-round potentials and rate of change of the potential. The latter should normally exceed 10 mV s^{-1}; at slower rates perceptible de-activation of the electrode surface is frequently seen.

5.2.1.1 **The Cyclic Voltammogram**

If there are no redox-active couples in the aqueous solution in the potential region between E_t^c and E_t^a, the observed current-potential behaviour in aqueous solution will correspond to the formation and dissolution of chemisorbed hydride and oxide layers on the electrode surface. An example of this is shown in Fig. 5.6 for platinum in 1M KOH, for which the potential scale is referred to a reversible hydrogen electrode *in the same solution* (RHE), for which hydrogen evolution will always take place thermodynamically at 0 V.

If we follow the changes starting at +450 mV and moving the potential in an anodic direction, then between 450 mV and 550 mV, the only current that flows is that required to charge the electrolytic double layer (j_c). If the capacitance of this layer is C_d, then $j_c = C_d \cdot \mathrm{d}E/\mathrm{d}t$; for $C_d = 100\ \mu\mathrm{F\ cm^{-2}}$ and $v = \mathrm{d}E/\mathrm{d}t = 100\ \mathrm{mV\ s^{-1}}$, $j_c = 10\ \mu\mathrm{A\ cm^{-2}}$.

Above 550 mV, oxygen chemisorption begins with the hydroxide discharge process:

$$\mathrm{Pt} + \mathrm{OH^-} \rightarrow \mathrm{Pt} - \mathrm{OH} + e^- \tag{5.9}$$

and above ca. 800 mV further oxidation takes place as:

$$Pt - OH + OH^- \rightarrow Pt - O + H_2O + e^- \quad (5.10)$$

Above ca. 1600 mV, oxygen evolution takes place, and at even higher potentials, a *phase* oxide may form on the platinum surface (i.e. one with thickness significantly larger than a monolayer). As the potential sweep is reversed in direction, any oxygen gas present in the neighbourhood of the electrode is reduced together with the chemisorbed oxide layer, though recent research has demonstrated that reduction of the oxide may give rise to a roughened platinum surface that only slowly re-attains an equilibrium morphology. At lower potentials still, there is a small double-layer region followed by the deposition of hydride as

$$Pt + H_2O + e^- \rightarrow Pt - H + OH^- \quad (5.11)$$

Finally, close to the thermodynamic potential for the H_2/H_2O couple, there is strong evolution of hydrogen which can be re-oxidised once the potential direction is again reversed. One respective cyclic voltammogram in acid solution shows Fig. 5.6a. For the reactions see chapter 6.2.

If the attainment of the equilibrium coverage of the film is rapid compared to the rate of potential change, then the differential capacitance, C^d, of the film, which can be defined as the rate of increase of charge with potential, will not depend on the scan rate. (Note here that C^d is not the same as the double-layer capacitance, C_d defined in the previous section). Since $C^d = dQ/dE = j/(dE/dt) \equiv j/v$, it follows that j will scale with v in a cyclic voltammogram for such films. This is actually observed for the films in Fig. 5.6 up to 0.5 V s^{-1}; above this value j increases more slowly and the peaks in the current become displaced in the direction of the potential change.

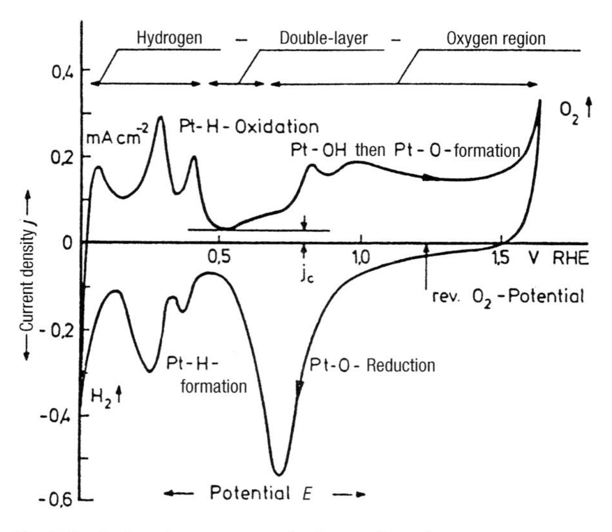

Fig. 5.6 Cyclic voltammogram of polycrystalline platinum in 1 M KOH. The solution was purged with N_2, and the measurement was carried out at 20°C with $v = 100$ mV s^{-1}.

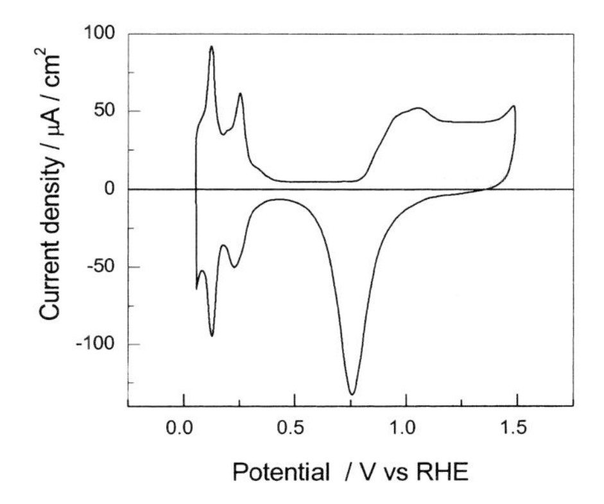

Fig. 5.6a Cyclic voltammogram of smooth polycrystalline platinum in 0.5 M H_2SO_4, nitrogen purging, $\upsilon = 50$ mV s^{-1}, 20°C.

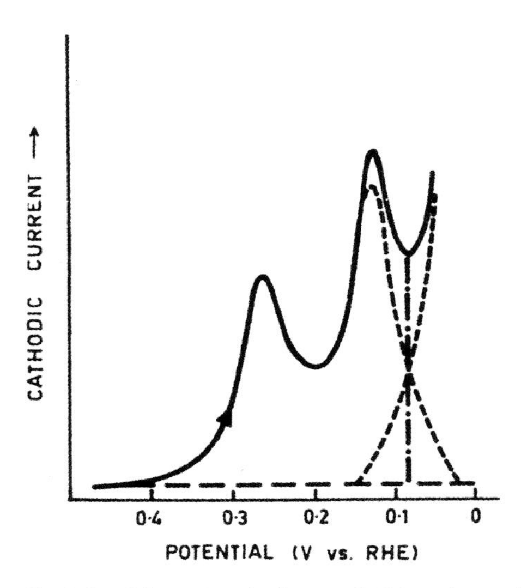

Fig. 5.6b Schematic cathodic part of a CV at polycrystalline platinum in diluted sulphuric acid. Compensation of the currents for hydrogen adsorption and evolution at the end-point potential E_{min} of 0.08 V RHE; solid line: total current, dashed line: double layer charging, dot-dash lines: extrapolated hydrogen adsorption and evolution currents [*see* T. Biegler, D.A.J. Rand, R. Woods, Determination of Real Platinum Area by Hydrogen Adsorption, *J. Electroanal. Chem.*, 29 (1971) 269–277 and M. Breiter, H. Kammermaier, C.N. Knorr, Z. *Elektrochem.*, 60 (1956) 37]. Please note that in this figure the convention is used that the *negative* of the potential is plotted on the *x*-axis and the *cathodic* current is plotted on the (positive) *y*-axis.

In a potential interval ΔE, the integrated charge required to form or remove a film can easily be measured

$$Q_F = \int_{\Delta E} i\mathrm{d}t = \int_{\Delta E} \mathrm{d}Q(E) = \int_{\Delta E} C^{\mathrm{d}}(E)\mathrm{d}E \tag{5.12}$$

and this opens up both the possibility of the determination of the coverage θ of a film or the determination of the true surface area of a roughened electrode, such as that obtained by repeated potential cycling of a metal electrode or by deposition of a metal from solution. This has been much used for platinum, where a complete hydride layer is believed to form, the overall formation charge of which will correspond to the number of exposed platinum atoms on the surface.

5.2.1.2 Determination of Surface Areas, Influence of Electrolyte on CV

For the experimental determination of a true platinum surface via hydrogen adsorption by a *cyclic voltammogram*, two assumptions are made: (i) each platinum atom adsorbs one hydrogen atom and (ii) any charge required to complete the partial monolayer up to a potential E_{min} positive to RHE is compensated by the charge due to the hydrogen evolution current.

Assumption (ii) is certainly not fulfilled near the equilibrium potential. In addition, the ratio of the two current contributions depend on the surface roughness, due to different mass transport conditions. For smooth platinum surfaces, Biegler, Rand and Woods suggested measuring the charge for a partial coverage down to an E_{min} of 0.08 V RHE (see Fig. 5.6b). The corresponding fractional coverage is 0.77, as obtained from isotherms e.g. by Breiter. The most accepted charge for a complete Pt monolayer is 210 μCcm^{-2}. This amount is very close to the charge calculated for Pt(100) surfaces (compare Fig. 5.8 below).

In general, the shape of the cyclic voltammogram will depend only rather weakly on the nature of the electrolyte for a given metal (compare Figs. 5.6 and Fig. 5.6a, the former being a cyclic voltammogram (CV) of platinum in 1M KOH and the latter

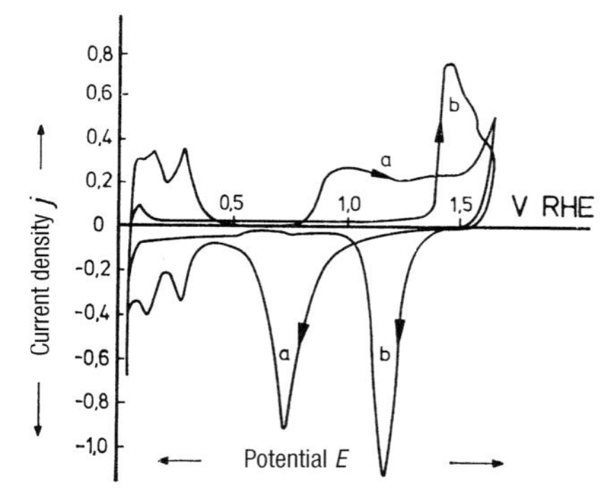

Fig. 5.7 Cyclic voltammograms in 0.5 M H_2SO_4 at room temperature, $\upsilon = 100$ mV s^{-1} with nitrogen purging: (a) platinum; (b) gold.

a CV of platinum in 0.5 M H_2SO_4), but shows a very strong sensitivity to the metal itself as shown in Fig. 5.7, which presents cyclic voltammograms for gold and platinum in 0.5 M H_2SO_4. The platinum CV is clearly not enormously different from that in Fig. 5.6, but the CV for gold shows a much smaller charge in the hydride region, and a correspondingly much larger double-layer region, allowing electrochemical investigations to be conducted over a wide potential range without formation of an oxide or hydride layer as discussed below.

A significant problem in cyclic voltammetry is the possibility of dissolution of the counter electrode. The potentiostat will force an equal and opposite current through the counter electrode as the working electrode, and small amounts of metal at the counter electrode may dissolve and re-deposit on the working electrode. For this reason, a counter electrode of the same material as the working electrode is usually employed during electrochemical investigations on gold, platinum etc.

Another important point is that the cyclic voltammograms of platinum and gold shown above are for the polycrystalline metals; cyclic voltammograms of single-crystal samples show a strong sensitivity to the Miller indices of the face exposed, as shown in Fig. 5.8.

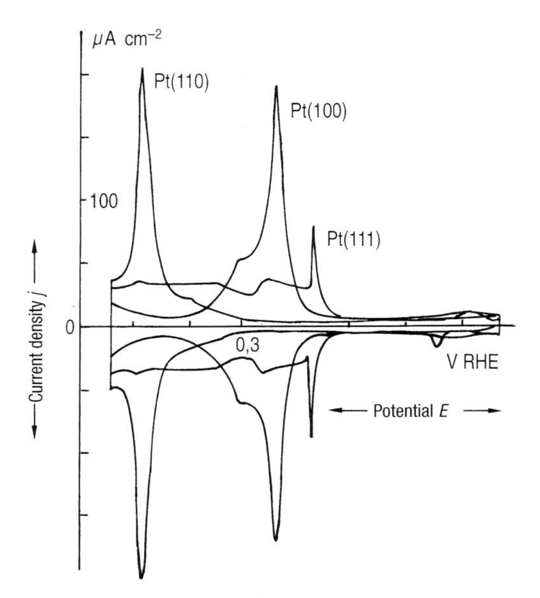

Fig. 5.8 Cyclic voltammograms of platinum single-crystal surfaces in the hydrogen adsorption region in 0.5 M H_2SO_4 under nitrogen purging; $|v| = 50$ mV s^{-1} [from J. Clavilier, A. Rhodes and M.A. Zamakhchari, J. Chim. Phys. **88** (1991) 1291].

5.2.1.3 Cyclic Voltammetry in the Presence of an Electrochemically Active Substance in the Electrolyte

If the electrolyte contains an electrochemically active species, the current due to film formation on the electrode will be overlaid with the current-voltage characteristics of the corresponding electrode reaction. A simple example is that of oxygen reduction on platinum in 0.5 M H_2SO_4: a marked difference between the CV in the absence of O_2 and that during O_2 reduction is apparent over a suitably chosen range of scan speeds, as shown in Fig. 5.9. At 30 mV s^{-1}, we can neglect the contribution from film formation on the underlying platinum in comparison with the oxygen reduction current in oxygen sparged solutions, and the resultant CV is close to the quasi-stationary current-potential response for O_2 reduction. At much more rapid scan speeds, the underlying film formation on Pt appears more preponderant, and it is clear that the formation and dissolution of surface hydride films, for example, takes place independently of oxygen reduction.

Cyclic voltammograms of a much more complex form are found for more complex electrochemical processes such as the oxidation of organic substances, especially in cases where the solution is unstirred, and the CV shows a sensitive dependence

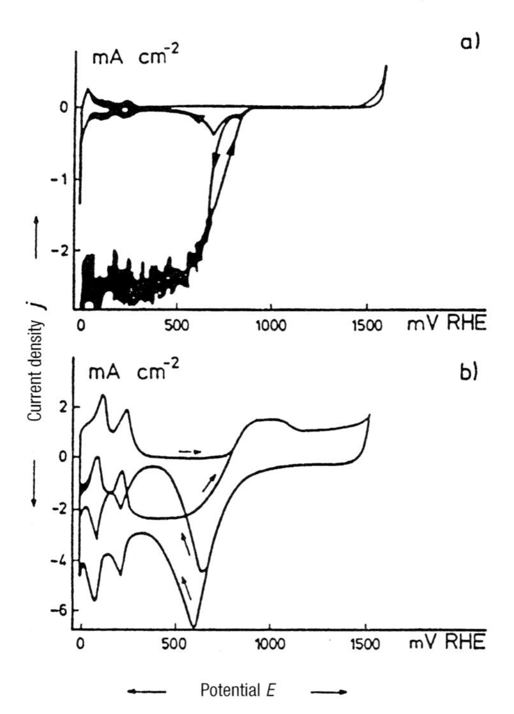

Fig. 5.9 Cyclic voltammograms of platinum in oxygen-sparged 0.5 M H_2SO_4: (a) slow potential sweep, $\upsilon = 30$ mV s^{-1}; upper curve: with nitrogen purging to remove oxygen; lower curve: formation of a quasi-stationary oxygen reduction current in the presence of oxygen sparging; (b) rapid potential sweep, $|\upsilon| = 1$ V s^{-1}; upper curve: with nitrogen purging; lower curve: with oxgyen sparging.

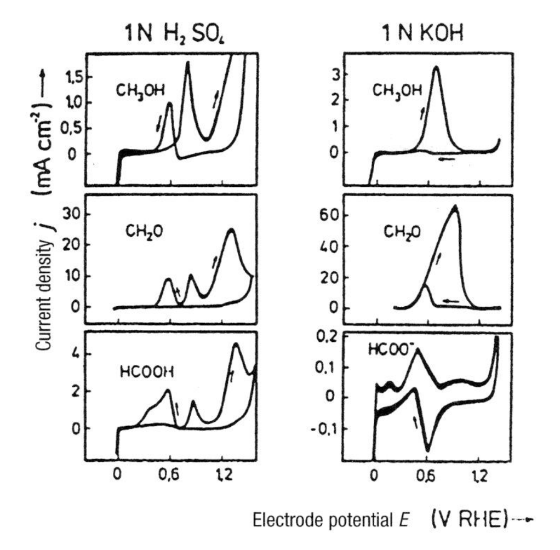

Fig. 5.10 Cyclic voltammograms of 1 M methanol, 1 M formaldehyde and 1 M formic acid at smooth platinum electrodes in 0.5 M H_2SO_4, 1 M KOH, 20°C, all solutions purged with nitrogen to remove oxygen.

on the type of electroactive substance in the electrolyte, the electrolyte itself, and the electrode material, as shown in Fig. 5.10. The existence of multiple peaks often reflects the build up and dissolution of chemisorbed inhibitor layers on the electrode surface; Fig. 5.10 shows, for example, the inhibiting effect of a build up of oxide layer on platinum on the oxidation of small organic molecules. The information obtained from cyclic voltammetry can be substantially extended through simultaneous monitoring of the mass-spectrometer (MS) signal derived from volatile species with the help of the DEMS method or through the dependence of *in situ* FTIR signals on potential (see section 5.4).

5.2.1.4 **The Theory of Cyclic Voltammetry I – Single Potential Sweep in Unstirred Solutions.**

We first consider a simple potential sweep in which the current is determined by the coupling of diffusion to the electrode and an electron transfer process at the surface. In general, this coupling gives rise to a *current maximum*, since after the potential is reached at which electron transfer is initiated, the surface concentration, c^s, of the reacting species falls as the potential rises further, from the bulk concentration, c^0, towards $c^s = 0$. This fall in c^s initially increases the concentration gradient $(c^0 - c^s)/\delta_N$ and hence the flux of material to the surface. However, δ_N is also increasing with time according to the expression $\delta_N = (\pi Dt)^{1/2}$, and at higher potentials, when electron transfer is fast and the current has reached its mass-transport-limited value nFc^0/δ_N, increasing δ_N will cause the current to *decrease* again.

This discussion can be quantified by considering a reversible reaction of the form $S_{red} \rightarrow S_{ox} + ne^-$ taking place during a positive potential sweep. The diffusion of S_{ox} and S_{red} can be described by the usual diffusion equations:

$$\frac{\partial c_{ox}}{\partial t} = D_{ox}\left(\frac{\partial^2 c_{ox}}{\partial x^2}\right) \quad (5.13)$$

$$\frac{\partial c_{red}}{\partial t} = D_{red}\left(\frac{\partial^2 c_{red}}{\partial x^2}\right) \quad (5.14)$$

with the boundary conditions

$$j = -nFD_{ox}\left(\frac{\partial c_{ox}}{\partial x}\right)_0 = nFD_{red}\left(\frac{\partial c_{red}}{\partial x}\right)_0 \quad (5.15)$$

At the beginning of the potential sweep, only S_{red} is present in the solution and we have:

$$t = 0,\ x \geq 0; \quad c_{red} = c^0_{red}; \quad c_{ox} = 0 \quad (5.16)$$

$$t \geq 0, x \rightarrow \infty; \quad c_{red} = c^0_{red}; \quad c_{ox} = 0 \quad (5.17)$$

A further boundary condition is dictated by the rate of the electron transfer process:

Case (a): Fast electron transfer

In this situation, electron transfer at all potentials is sufficiently rapid for the Nernst equation always to hold. If $\upsilon = dE/dt$, and $E = E_t + \upsilon t$, where E_t is the turnaround or starting potential:

$$E = E_t + \upsilon t = E^0 + (RT/nF)\ln\left(\frac{c^s_{ox}}{c^s_{red}}\right) \quad (5.18)$$

which can be written as the boundary condition:

$$x = 0, t > 0; \quad \left(\frac{c^s_{ox}}{c^s_{red}}\right) = \exp\left[\frac{nF(E_t + \upsilon t - E^0)}{RT}\right] \quad (5.19)$$

From these boundary conditions, we obtain, for the current,

$$j = nF\left(\frac{nFD_{red}}{RT}\right)^{\frac{1}{2}} \cdot c^0_{red} \cdot \upsilon^{\frac{1}{2}} \cdot P\left[(E - E^0) \cdot n\right] \quad (5.20)$$

where the function $P[(E - E^0).n)]$ is shown in Fig. 5.11. The maximum value of the current lies at the peak potential, E_p, which is independent of υ, and for a one-electron oxidation step, and at 25°C, lies some 28.5 mV higher that E^0. The corresponding maximum value of P is 0.4463, so that the maximum value of the current is given by the Randles-Sevcik equation:

$$j_{max} = 2.69 \times 10^5 \cdot n^{\frac{3}{2}} \cdot (D_{red})^{\frac{1}{2}} \cdot c^0_{red} \cdot \upsilon^{\frac{1}{2}} \quad (5.21)$$

where j_{max} is in A cm^{-2}, c^0_{red} in mol cm^{-3}, D_{red} in cm^{2}s^{-1} and υ in V s^{-1}. Clearly, the theory predicts a linear increase in j_{max} with $\upsilon^{1/2}$, a result borne out experimentally.

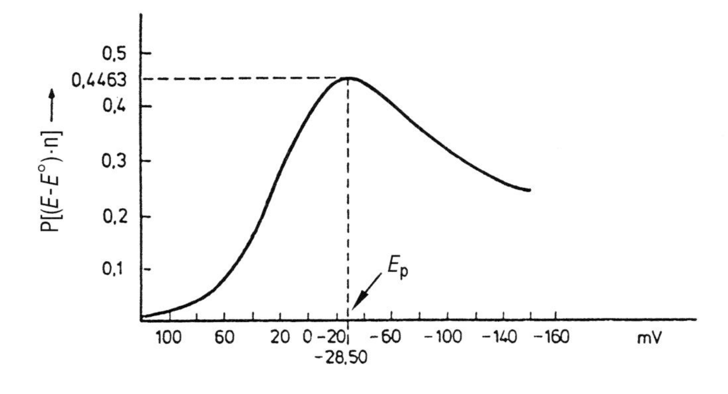

Fig. 5.11 Graphical representation of the function $P[(E - E^0) \cdot n]$ from equation 5.20 corresponding to the case of fast electron transfer and unstirred solution.

Case (b): Slow electron transfer

The coupling of diffusion and electron transfer also leads to a current maximum in this case, but the potential at which the maximum in the anodic current occurs is now no longer independent of scan rate but shifts to positive potentials with increasing scan rate. This is because the concentration of S_{red} at the electrode surface decreases more slowly than in the previous case, so the maximum value of the concentration gradient is reached at more positive potentials. Quantitatively, we have, at any time t, the flux of material arriving at the surface must be the same as the rate of consumption by the electrode reaction:

$$t > 0, x = 0; \quad D_{red}\left(\frac{\partial c_{red}}{\partial x}\right)_0 = c^s_{red} \cdot k_0^+ \cdot \exp\left[\frac{(1-\beta)nF(E_t + \upsilon t)}{RT}\right] \tag{5.22}$$

where β is the asymmetry parameter defined on page 163. This gives, as solution for the current:

$$j = \left(\frac{\pi\beta nF\upsilon D_{red}}{RT}\right)^{\frac{1}{2}} \cdot nFc^0_{red} \cdot Q\left[\frac{(1-\beta)nF\upsilon t}{RT}\right] \tag{5.23}$$

where the function Q shows a similar behaviour to P, with a maximum value of 0.282. At 25°C, this leads to an expression for the current maximum of the form:

$$j_{max} = 3.01 \times 10^5 \cdot n^{\frac{3}{2}}(\beta D_{red}\upsilon)^{\frac{1}{2}} \cdot c^0_{red} \tag{5.24}$$

As before, the anodic current density maximum increases linearly with $\upsilon^{1/2}$, and for $n = 1$ and $\beta = 0.5$, *the peak moves by some 30 mV more positive for each increase in υ by a factor 10.*

In the case of a cathodic reaction as $S_{ox} + ne^- \rightarrow S_{red}$, similar equations are obtained.

5.2.1.5 The Theory of Cyclic Voltammetry II: Cyclic Potential Scan

If, in a solution initially containing S_{red} only, the potential sweep of a voltammogram starts at low potentials and moves in the positive potential direction, a peak due to the anodic reaction $S_{red} \rightarrow S_{ox} + e^-$ will be observed (Fig. 5.12). When, following this peak, the direction of the sweep is reversed, a peak will appear in the current corresponding to the cathodic reduction of the S_{ox} formed during the anodic potential sweep.

However, the baseline for the cathodic sweep is no longer well defined though it is easily established that for fast electron transfer and equal diffusion coefficients for reduced and oxidised species, the peak heights should be exactly equal. If the zero-current line is used as the baseline, the ratio of cathodic to anodic current maxima measured with respect to this line is not unity, though its value depends on the difference between the turn-round potential, E_t, and the reversible electrochemical potential for the couple, E_{rev}.

(a) Fast electron transfer

For the case of a fast electron transfer process, the ratio i_p^-/i_p^+ is actually solely a function of $|E_t - E_{rev}|$ as in Table 5.1. The reversible potential is exactly in the middle of the two peaks.

Table 5.1

$\|E_t - E_{rev}\|$/mV	i_p^-/i_p^+
100	0.5727
200	0.6894
300	0.7439
400	0.7771
500	0.8001

For a fast electron transfer process, the two peaks lie some 57.0 mV apart provided the turn-round potential lies well away from E_{rev}. As $|E_t - E_{rev}|$ decreases, the peak separation ΔE_p *increases*, as shown in Table 5.2.

Table 5.2

$\|E_t - E_{rev}\|$/mV	ΔE_p
71.5	60.5
121.5	59.2
171.5	58.3
271.5	57.8
∞	57.0

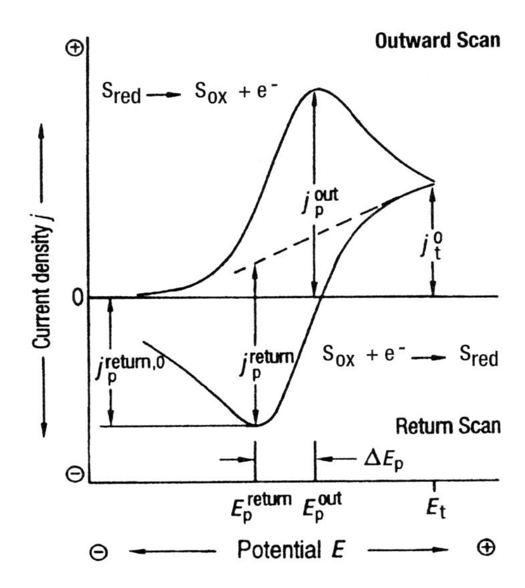

Fig. 5.12 Current-potential curve for a reversible redox system in unstirred solution for a single triangular potential excursion; for fast electron transfer, the peaks are separated by 57 mV.

(b) Slow electron transfer

In the case of a slow electron transfer, equation (5.18) does not hold anymore. According to equation (5.23), peak positions will move, which results in increased peak distances above 57 mV. From the difference between reverse peak positions, an estimate of the rate constant of the electrochemical reaction can be obtained.

A distinction between the regions of validity of equations (5.20) and (5.23), i.e. between fast and slow electron transfer, can be made by considering the value of the dimensionless Matsuda number Λ, given by

$$\Lambda = \frac{k_0}{D^{\frac{1}{2}}(nF/RT)^{\frac{1}{2}}.v^{\frac{1}{2}}} \tag{5.25}$$

where k_0 is the heterogeneous electron-transfer rate constant, defined on page 163, and D is the mean diffusion coefficient. For values of β lying between 0.3 and 0.7, it can be shown that for

$$\Lambda \geq 15 \text{ or } k_0 \geq 0.3v^{\frac{1}{2}} \tag{5.26}$$

then case (a) holds (Nernstian equilibrium maintained) whereas for

$$\Lambda \leq 10^{-3} \quad \text{or} \quad k_0 \leq 2 \times 10^{-5} v^{\frac{1}{2}} \tag{5.27}$$

then case (b) holds with essentially irreversible electron transfer. The intermediate regime, with

$$10^{-3} < \Lambda < 15 \tag{5.28}$$

corresponds to the quasi-reversible case. In this case, the current maximum shifts with scan speed, and in the region defined by equation (5.28), the separation of the anodic and cathodic current peaks can be correlated with Matsuda's number as in Table 5.3.

Table 5.3

Λ	ΔE_p	Λ	ΔE_p
36	61	1.8	84
12	63	1.3	92
11	64	0.9	105
9	65	0.6	121
7	66	0.44	141
5.3	68	0.18	212
3.5	72		

From the peak separation, which should lie in the region 80 to 140 mV for optimum accuracy, an estimate of k_0 can be made from the Matsuda number using equation (5.25). This peak separation can be attained through a suitable choice of the sweep rate, v, though great care is necessary in practice since larger values of v lead to larger currents and to the increasing necessity of avoiding ohmic loss effects in the electrolyte. This type analysis has been carried out for the $[Fe(CN)_6]^{4-/3-}$ couple at a platinum electrode, as shown in Fig. 5.13, giving values of the peak separation of 123 mV, 106 mV and 91 mV at scan rates of 164 Vs^{-1}, 103 Vs^{-1} and 50 Vs^{-1} respectively. From the calculated values of Λ, an exchange current under standard conditions of 17 ± 1 A cm^{-2} is obtained, slightly higher than that given in Table 4.1.

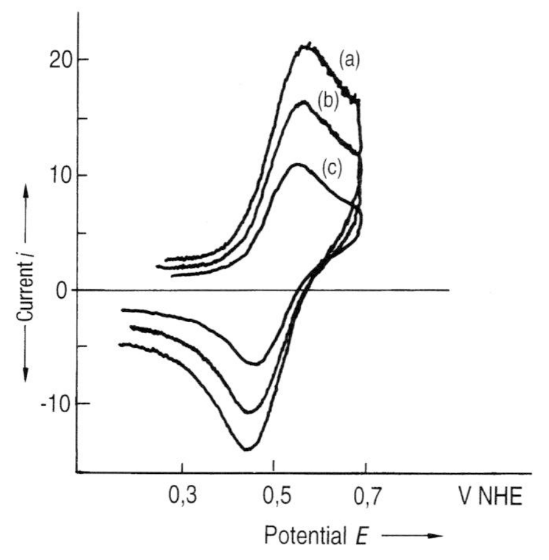

Fig. 5.13 Cyclic voltammograms of the $[Fe(CN)_6]^{3-/4-}$ couple at platinum. The concentration of the couple is 0.005 M and the background electrolyte is 0.5 M K_2SO_4, $T = 21\,°C$:
(a) $|v| = 164$ V s^{-1}, $\Delta E_p = 123$ mV; (b) $|v| = 103$ V s^{-1}, $\Delta E_p = 106$ mV; (c) $|v| = 50$ V s^{-1}, $\Delta E_p = 91$ mV.

5.2.1.6 **Multiple Cyclic Potential Scans and Diagnostic Analysis via Single Cyclic Potential Scans**

In practice, the use of *stationary* cyclic voltammograms is preferred to the analysis of a single potential excursion, since small differences between the first and subsequent scans are frequently found. The simplest example is that of a reversible one-electron couple, and an example is shown in Fig. 5.14 for the oxidation of 9,10 Diphenyl anthracene (DPA) in acetonitrile/0.1 M $LiClO_4$. It is evident that, as expected from Table 5.1, $i_p^-/i_p^+ < 1$, and that both current peaks are increasing with scan speed. The peak separation is clearly independent of υ, and close to 57.0 mV, the exact value being influenced by the turn-round potential as indicated in Table 5.2.

If an irreversible chemical reaction follows electron transfer, then the size of the reverse peak will be affected as indicated above. If the chemical reaction is fast, the reverse peak may disappear altogether, whereas a less rapid reaction simply leads to a diminution of the peak, which can be itself reversed by increasing the rate of the potential sweep. This behaviour is shown in Fig. 5.15, in which the system of the previous figure is altered by addition of a radical scavenger, *N*-*t*-butyl hydroxylamine hydrochloride.

If several electrons can be transferred one at a time, then provided the successive values of E_{rev} are more than ca. 150 mV apart, these electron transfers will be seen as separate peaks in the cyclic voltammogram, as shown in Fig. 5.16. In this experiment, DPA is oxidised first to DPA^+ and then to DPA^{2+}; interestingly, the reduction of the di-cation is not seen clearly, suggesting that this species is disappearing from solution, possibly as a result of reaction with adventitious nucleophiles.

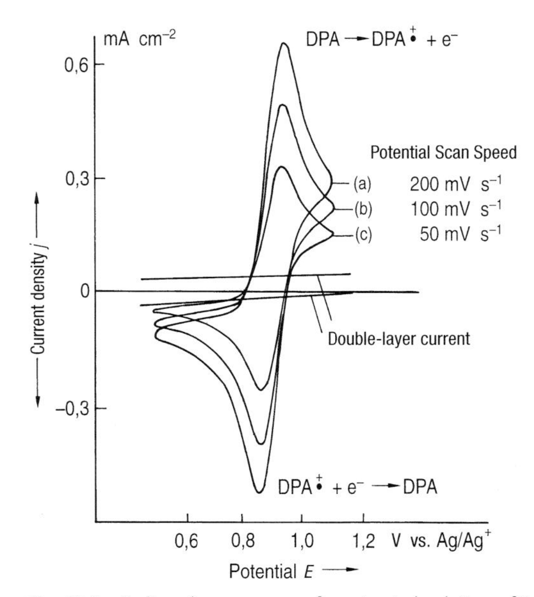

Fig. 5.14 Cyclic voltammogram of a saturated solution of 9,10- diphenylanthracene in acetonitrile at polished platinum. The reference electrode is a silver wire in acetonitrile/0.1 M $LiClO_4$/0.01 M Ag^+.

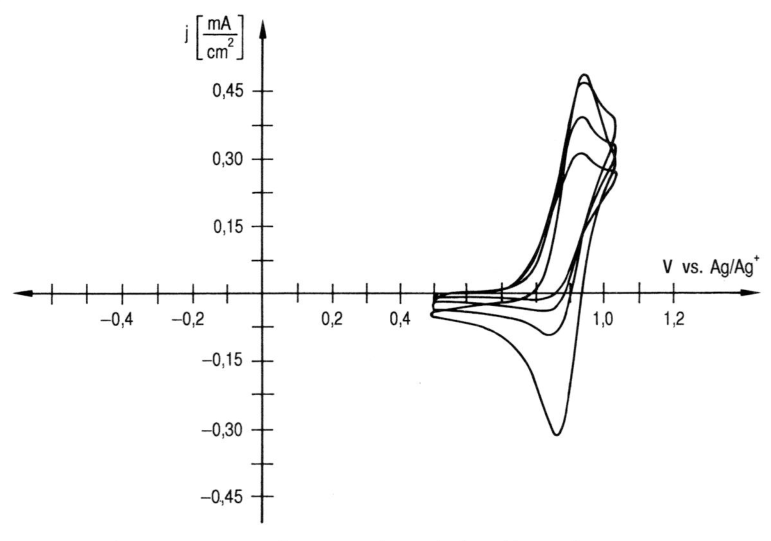

Fig. 5.15 The same system as for Fig. 5.14 but with the addition of a radical scavenger, N-t-butylhydroxylamino hydrochloride.

Quantitative investigation of cyclic voltammograms leads to a series of criteria in which the relative magnitudes of the peak currents, the potential difference between the peaks and the dependence of both these on sweep rate can be used to identify specific cases. Such quantitative investigations require us carefully to identify appro-

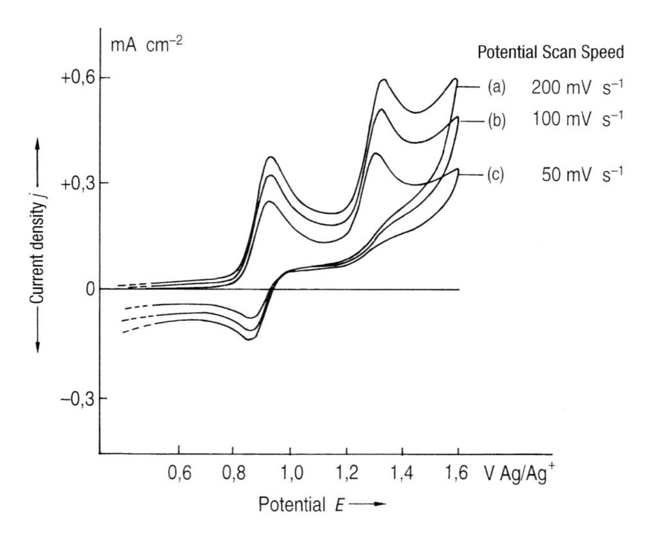

Fig. 5.16 Positive region of cyclic voltammograms for the same system as Fig. 5.14, but with a more positive turn-round potential (the negative turn round potential remains − 0.4 V vs. Ag/Ag^+).

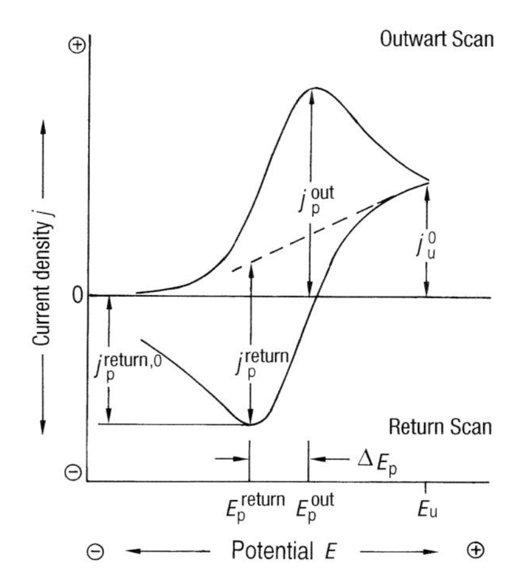

Fig. 5.17 Diagram showing the parameters of importance as diagnostic criteria in the analysis of a cyclic voltammogram.

priate baselines for the current measurements, as shown in Fig. 5.17. For the forward curve, the zero-current line can be used as the baseline, but for the reverse curve, a baseline joining the current values at the turn-round potentials is more suitable. An empirical formula by Nicholson can also be used:

$$\frac{i_{p,back}}{i_{p,forw}} = \frac{i^0_{p,back}}{i^0_{p,forw}} + \frac{0.485 i^0_t}{i_{p,forw}} + 0.086 \tag{5.29}$$

Measurement of $i_{p,back}$ and $i_{p,forw}$ leads to the diagnostic Table 5.4.

The criteria in Table 5.4 have been further subdivided in the literature, incorporating, for example, intermediate rates of electron transfer and additional combinations of chemical and electrochemical steps. However, these subdivisions are not always simply distinguishable, and further information is usually necessary, which may be derived, for example, from the modelled fitting of the data to the conjectured pro-

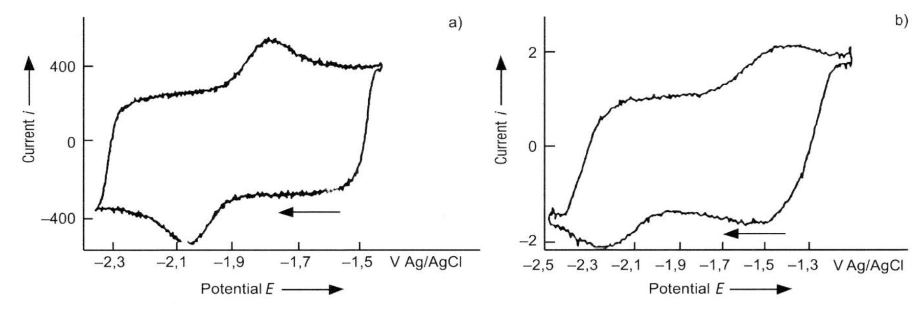

Fig. 5.17a Cyclic voltammograms for the reduction of anthracene in acetonitrile with 0.4 M $TBAPF_6$ as background electrolyte, 25°C; (a) $\upsilon = 120,000\ Vs^{-1}$, Pt wire of diameter 1 μm, $c = 0.008$ M; (b) $\upsilon = 10^6 Vs^{-1}$, Au wire of diameter 2.5 μm, $c = 0.005$ M.

Table 5.4

Mechanism	**Shift of E_p with υ**	**ΔE_p/mV**	$\frac{i_{p,forw}}{\upsilon^{\frac{1}{2}}}$	$\frac{i_{p,back}}{i_{p,forw}}$	**Remarks**
Fast-electron Transfer	None	$57/n$	Constant	1	Diffusion-controlled
Slow-electron Transfer	Yes	$57/n$ increases with υ	Constant	1 ($\beta = 0.5$)	Electron-transfer controlled
Irreversible electron-transfer	Yes	No return peak	Constant		
Charge transfer followed by a reversible chemical step	Yes	No particulars	Constant	1 for small υ but decreases with higher υ	EC-mechanism
Charge transfer followed by an irreversible chemical step	Yes	Return peak may disappear at small υ	Constant	< 1 for small υ but tends to increase with υ and tend to 1	EC-mechanism
An irreversible chemical step precedes electron transfer	Shifts in opposite direction	No particulars	Decreases with υ	1 for small υ but increases as υ rises	CE-mechanism

cess by adjusting the values of the underlying parameters (diffusion coefficients, rate constants, etc), or by data from additional experiments.

5.2.1.7 Cyclic Voltammograms at Ultra-microelectrodes with Sweep Speeds up to 10^6 Vs^{-1} – Fast Scan Voltammetry

The requirements for the measurement of cyclic voltammograms under very rapid potential sweep without having to make extensive corrections are that the iR drop between the working electrode and the tip of the Luggin capillary connexion to the reference electrode should be very small, and that the capacitive current associated with the working electrode should be as small a component of the total current as possible. This latter arises because the Faradaic current increases with $\upsilon^{1/2}$ but the capacitive current is directly proportional to the sweep rate υ.

These requirements can be fairly easily satisfied by the use of ultra-microelectrodes (section 4.3.8) with diameters, for example ca. 1 – 10 μm (compare Fig. 4.21(a)). Electrodes of a suitable design with good electrical screening have been described by Tschuncky and Heinze [P. Tschunky and J. Heinze; Anal. Chem. 67 (1995) 4020], and Fig. 5.17a shows CVs obtained for the reduction of anthracene in acetonitrile with platinum wire of diameter 2 μm and gold wire with diameter 5 μm at sweep rates of 120,000 and 10^6 Vs^{-1} respectively. Owing to the favourable experimental con-

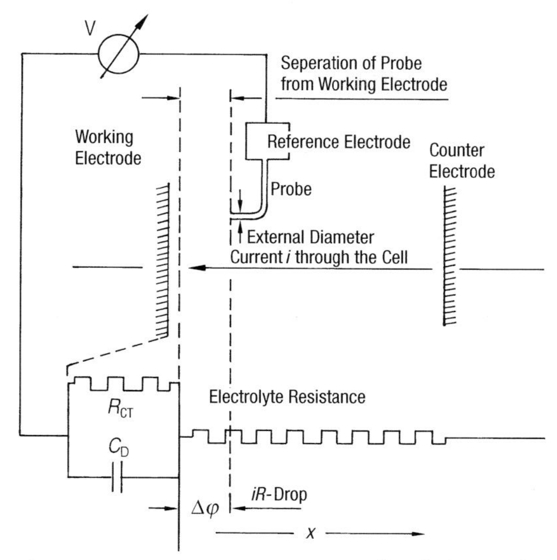

Fig. 5.18 Diagrammatic representation of a cell showing the origin of resistance terms.

ditions pertaining for microelectrodes, it is possible to obtain a measurable distance between current peaks for anodic and cathodic sweeps (cf. Figs. 5.13 and 5.17a) even for very fast charge transfer, allowing the determination of the rate constants for extremely fast electron transfer reactions (cf. equation 5.25 and Table 5.3).

5.2.1.8 Correction for *iR* drop

We have seen already that overpotential data can be in serious error if the potential drop between working and reference electrodes is neglected. To quantify this, the electrochemical cell must be modelled by an appropriate equivalent circuit, and a simplified form is shown in Fig. 5.18. The iR correction, $\Delta\varphi$, is then the potential drop between electrode and capillary tip (note that there is no potential drop between capillary tip and the reference electrode itself since only minute ($< 10^{-12}$ A) currents flow in the reference electrode circuit, as described on page 252. It is obviously essential that the electrolyte conductivity should not vary with capillary position, that the current flow be uniform and that no substantial electrolyte concentration profile be set up in the diffusion layer, which can best be achieved with highly conductive concentrated electrolyte. In addition, for controlled experiments, gas bubbles and thick electrode films should be avoided.

The value of $\Delta\varphi$ can be calculated straightforwardly if the current distribution is uniform. If the electrode area is A, the current i, the electrolyte conductivity κ and the distance from electrode to capillary tip is d, then

$$\Delta\varphi = i \cdot \frac{\mathrm{d}}{\kappa A} \equiv j \cdot \frac{\mathrm{d}}{\kappa} \equiv j \cdot R_{\Omega} \tag{5.30}$$

where j is the current density. For d = 0.1 cm and $\kappa = 0.1\ \Omega^{-1}\mathrm{cm}^{-1}$, the value of $R_\Omega = 1$ Ohm cm^2 and for a current density of 0.1 A cm^{-2}, $\Delta\varphi = 0.1$V, a considerable error. Unfortunately, this error cannot be minimised beyond a certain limit by reducing d: if the tip is placed within 1–3 capillary diameters of the surface, the current distribution becomes inhomogeneous owing to screening of the electrode.

Experimental determination of the iR drop is always possible. Earlier work involved the systematic variation of tip-electrode distance and extrapolated the resultant measurement to zero separation, but nowadays the currrent-interrupt method is most commonly employed. This exploits the fact that on interrupting the current, the iR contribution to the potential drops immediately to zero whereas the electrode potential falls only relatively slowly (in the order ms) owing to the large double-layer capacitance. The potential change can be measured with a storage oscilloscope, or by utilising an appropriate electronic circuit within the potentiostat; iR compensation can be built into the potential control by a variety of means, yielding automatically corrected values of the electrode potential.

In addition to the requirements for accurate work in stationary measurements, it is also the case that iR errors can substantially distort the appearance of the cyclic voltammogram, particularly in poorly conducting electrolytes, both by altering the *position* of the peak and by altering the *shape*. This latter effect arises from the fact that if E is the true electrode potential, and E_d the apparent potential from the function generator applied to the potentiostat, then $E = E_d - \Delta\varphi$, and

$$\frac{\mathrm{d}E}{\mathrm{d}t} = \frac{\mathrm{d}E_d}{\mathrm{d}t} - \frac{\mathrm{d}\Delta\varphi}{\mathrm{d}t} = \frac{\mathrm{d}E_d}{\mathrm{d}t} - R_\Omega \frac{\mathrm{d}j}{\mathrm{d}t} \tag{5.31}$$

from which we see that during the rising part of the peak, the actual electrode potential is increasing *less* rapidly than the function generator read-out would suggest, with the reverse true for the decreasing side of the peak. This combination of peak distortion and shift is particularly inimical to the measurement of accurate electrochemical rate constants, and very great care is needed to ensure that an apparent peak shift is due to a finite rate constant and is not a consequence of iR drop.

5.2.2
AC Measurements

The component steps of an electrochemical process, such as mass transport, chemical and adsorption steps, electron transfer etc., all contribute to the total potential drop across the cell, and for a dc current can be represented by resistances such as R_E (electrolyte) or R_ct (electron transfer). However, if an ac current flows, it becomes necessary to distinguish purely ohmic resistances, such as R_E, from non-ohmic, complex and normally frequency-dependent resistances (often termed impedances). By careful investigation of the frequency dependence of such impedances, it is possible to separate the contributions to the overall electrochemical process, and such studies have been greatly facilitated by the availability of modern measurement equipment. These studies include not only investigations of electrode kinetics and adsorption rates, but more complex problems, such as corrosion processes, battery properties, ageing of sensors and the properties of porous electrode structures. However, the complexity of some of the processes investigated inevitably leads to equivalent model circuits of equal complexity, with the result that interpretational errors become increasingly probable.

Our treatment of the ac response of electrochemical cells will follow the pathway already described above, with initially a general treatment, which we then specialise to the two simpler cases of purely diffusion controlled and purely electrochemically controlled reactions. Finally, in section 5.2.2.7 we will tackle the more difficult problem of *ac* impedance in the presence of adsorption. We will represent all periodic functions, such as potential or current, with expressions of the form:

$$A_0 e^{i\omega t} = A_0(\cos \omega t + \boldsymbol{i} \sin \omega t) \tag{5.32}$$

where $\boldsymbol{i}$ in this section is $\sqrt{-1}$ and ω is 2π times the frequency of the ac signal. As above, we will use j to represent current *density*.

5.2.2.1 **Influence of Transport Processes on the ac Impedance of an Electrochemical Cell**

The application of an *ac* potential to the cell forces the processes at the electrode surface also to oscillate with the applied frequency; owing to the rapidity of the changing potential, in fact, a quasi-stationary state is set up, at least for frequencies in excess of 1 Hz. In particular, if a reversible couple is present at the electrode surface, the concentrations of both *Red* and *Ox* will also oscillate not only at the surface but away from the electrode, though clearly the further from the electrode, the more damped such concentration oscillations will be.

Consider the electrochemical process

$$\text{Ox} + e^{-}\text{Red}$$

The current density flowing will be a function, quite generally, of the potential drop E across the interface and the concentrations of Ox and Red, c_{O}^{s} and c_{R}^{s}, at the electrode surface:

$$j = j(E, c_{\text{O}}^{\text{s}}, c_{\text{R}}^{\text{s}}).$$

Small changes in j can be related to small changes in the other variables provided these changes are all small enough to be linearised:

$$\delta j = (\partial j/\partial E)_{c_{\text{O}}^{\text{s}}, c_{\text{R}}^{\text{s}}} \delta E + (\partial j/\partial c_{\text{O}}^{\text{s}})_{c_{\text{R}}^{\text{s}}, E} \delta c_{\text{O}}^{\text{s}} + (\partial j/\partial c_{\text{R}}^{\text{s}})_{c_{\text{O}}^{\text{s}}, E} \delta c_{\text{R}}^{\text{s}} \tag{5.33}$$

Clearly progress is possible only if we can express the partial derivatives in this expression in terms of known functions. We shall assume, as usual, that the concentration of supporting electrolyte is sufficient to ensure that all the potential drop at the interface is accommodated in the Helmholtz layer, so that the transport of Ox and Red takes place only by diffusion. The equation of motion for either Ox or Red will then take the form

$$\frac{\partial c}{\partial t} = D\frac{\partial^2 c}{\partial x^2} \tag{5.34}$$

where x is the distance from the surface (assumed to be at $x = 0$) and the flux of c in the *positive* x-direction is $J = -D(\partial c/\partial x)$.

In the presence of a sinusoidal perturbation, the concentrations of Ox and Red at any point x can be written:

$$\begin{aligned} c_{\mathrm{O}} &= \bar{c}_{\mathrm{O}} + \tilde{c}_{\mathrm{O}} e^{i\omega t} \\ c_{\mathrm{R}} &= \bar{c}_{\mathrm{R}} + \tilde{c}_{\mathrm{R}} e^{i\omega t} \end{aligned} \tag{5.35}$$

where the concentrations $\bar{c}_{\mathrm{O}}$, $\bar{c}_{\mathrm{R}}$ correspond to those expected at the steady state potential $\bar{E}$ and are *not* time dependent. The small changes in concentration and potential are written with an explicit time-dependence, $e^{i\omega t}$, and have amplitudes $\tilde{c}_{\mathrm{O}}$, $\tilde{c}_{\mathrm{R}}$ and $\tilde{E}$ where it should be noted that the first two of these may be complex, corresponding to a phase shift from $\tilde{E}$. Putting these expressions into the diffusion equation (5.34), we obtain, for Ox:

$$i\omega\tilde{c}_{\mathrm{O}} e^{i\omega t} = D_{\mathrm{O}}\left(\frac{\mathrm{d}^2\bar{c}_{\mathrm{O}}}{\mathrm{d}x^2} + e^{i\omega t}\frac{\mathrm{d}^2\tilde{c}_{\mathrm{O}}}{\mathrm{d}x^2}\right) \tag{5.36}$$

We now make the simplification that we are only measuring AC components in our cell, so the first term in the bracket on the right hand side, which is essentially time independent, can be neglected, and we can simply equate coefficients of $e^{i\omega t}$:

$$D_{\mathrm{O}}\left(\frac{\mathrm{d}^2\tilde{c}_{\mathrm{O}}}{\mathrm{d}x^2}\right) = i\omega\tilde{c}_{\mathrm{O}} \tag{5.37}$$

Assuming that $\tilde{c}_{\mathrm{O}} \to 0$ as $x \to \infty$, and at $x = 0$, $\tilde{c}_{\mathrm{O}} = \tilde{c}_{\mathrm{O}}^{\mathrm{s}}$, then the diffusion equation can be solved to give:

$$\tilde{c}_{\mathrm{O}} = \tilde{c}_{\mathrm{O}}^{\mathrm{s}} \exp\left\{-(i\omega/D_{\mathrm{O}})^{\frac{1}{2}}x\right\} \tag{5.38}$$

and

$$\tilde{c}_{\mathrm{R}} = \tilde{c}_{\mathrm{R}}^{\mathrm{s}} \exp\left\{-(i\omega/D_{\mathrm{R}})^{\frac{1}{2}}x\right\}$$

These formulae demonstrate the strong damping of the *ac* components of O and R away from the electrode surface as shown in Fig. 5.19. The sinusoidal component of the *flux* of $\tilde{c}_{\mathrm{O}}$ at the surface, $\tilde{J}_{\mathrm{O}}$, will clearly be related to the current, since it is the passage of current that leads to the flux of the components of the redox couple. In fact from Fick's first law:

$$\tilde{J}_{\mathrm{O}} = -D_{\mathrm{O}}\left(\frac{\partial\tilde{c}_{\mathrm{O}}}{\partial x}\right)_{x=0} = +D_{\mathrm{O}}(i\omega/D_{\mathrm{O}})^{\frac{1}{2}}\tilde{c}_{\mathrm{O}}^{\mathrm{s}} \tag{5.39}$$

and

$$\tilde{J}_{\mathrm{R}} = -D_{\mathrm{R}}\left(\frac{\partial\tilde{c}_{\mathrm{R}}}{\partial x}\right)_{x=0} = +D_{\mathrm{R}}(i\omega/D_{\mathrm{R}})^{\frac{1}{2}}\tilde{c}_{\mathrm{R}}^{\mathrm{s}}$$

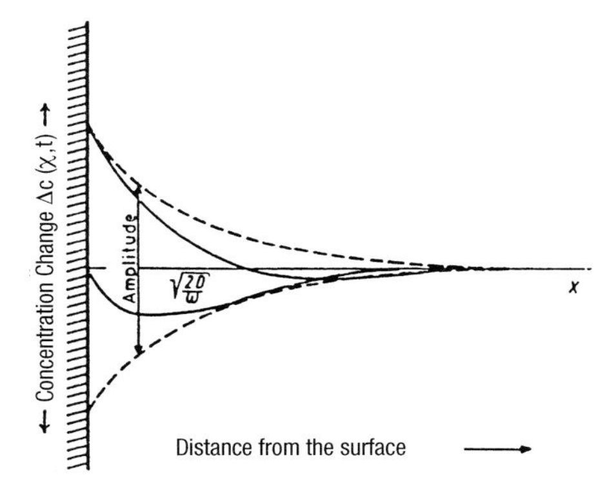

Fig. 5.19 Concentration change, $Re\{\tilde{c}e^{i\omega t}\}$, as a function of the distance x from the electrode for a sinusoidal *ac* current and for a diffusion-controlled reaction. The concentration is shown for two different times, and the dashed line shows at each position, *x*, the maximum amplitude for $Re\{\tilde{c}e^{i\omega t}\}$.

The sinusoidal current flowing is $\tilde{j} = F\tilde{J}_{\mathrm{O}} = -F\tilde{J}_{\mathrm{R}}$ for a one-electron reaction. From equation (5.33) above, writing $\tilde{j}$ for δj etc., and substituting for $\delta c_{\mathrm{O}}^{\mathrm{s}}$, $\delta c_{\mathrm{R}}^{\mathrm{s}}$ the values of $\tilde{c}_{\mathrm{O}}^{\mathrm{s}}$, $\tilde{c}_{\mathrm{R}}^{\mathrm{s}}$, we have:

$$\tilde{j}\left\{1 - \left(\frac{1}{F(i\omega D_{\mathrm{O}})^{\frac{1}{2}}}\right)\cdot\left(\frac{\partial j}{\partial c_{\mathrm{O}}^{\mathrm{s}}}\right)_{E,c_{\mathrm{R}}^{\mathrm{s}}} + \left(\frac{1}{F(i\omega D_{\mathrm{R}})^{\frac{1}{2}}}\right)\cdot\left(\frac{\partial j}{\partial c_{\mathrm{R}}^{\mathrm{s}}}\right)_{E,c_{\mathrm{O}}^{\mathrm{s}}}\right\} = \tilde{E}\left(\frac{\partial j}{\partial E}\right)_{c_{\mathrm{O}}^{\mathrm{s}},c_{\mathrm{R}}^{\mathrm{s}}} \tag{5.40}$$

To proceed further we need to find explicit expressions for the derivatives in equation (5.40), and this we can do from the Butler-Volmer equation, which can be written in the concentration-dependent form:

$$j(\eta) = j_0\left\{\left(\frac{c_{\mathrm{R}}^{\mathrm{s}}}{c_{\mathrm{R}}^{0}}\right)\exp\left[\frac{(1-\beta)F\eta}{RT}\right] - \left(\frac{c_{\mathrm{Ox}}^{\mathrm{s}}}{c_{\mathrm{Ox}}^{0}}\right)\exp\left[-\frac{\beta F\eta}{RT}\right]\right\} \tag{5.41}$$

where, as above, $\eta = E - E^0$. Furthermore, for the analysis to be meaningful, the *dc* changes in concentration of Ox and Red at the electrode surface must be very slow compared to the changes induced by the ac potential, and this implies that η must be very small, else changes induced in the concentrations of electro-active species at the electrode surface by virtue of the underlying dc electrochemical processes will prevent any steady-state being established. Assuming that these conditions hold, and defining the charge-transfer resistance, $R_{\mathrm{CT}} = \left(1/\left(\partial j/\partial E\right)_{c_{\mathrm{O}},c_{\mathrm{R}}}\right)$, and

$$\sigma \equiv \frac{1}{F}\left\{\left(\frac{1}{(2D_{\mathrm{R}})^{\frac{1}{2}}}\right)\cdot\left(\frac{\partial j}{\partial c_{\mathrm{R}}^{\mathrm{s}}}\right)_{E,c_{\mathrm{O}}^{\mathrm{s}}} - \left(\frac{1}{(2D_{\mathrm{O}})^{\frac{1}{2}}}\right)\cdot\left(\frac{\partial j}{\partial c_{\mathrm{O}}^{\mathrm{s}}}\right)_{E,c_{\mathrm{R}}^{\mathrm{s}}}\right\}$$

and equating the measured AC impedance Z to $\tilde{E}/\tilde{j}$, we see that

$$Z = R_{\mathrm{CT}}(1 + \sqrt{2}\sigma/(i\omega)^{\frac{1}{2}}) = R_{\mathrm{CT}} + (1-i)\cdot R_{\mathrm{CT}}\cdot\sigma/\omega^{\frac{1}{2}} \tag{5.42}$$

remembering that $(1/i)^{\frac{1}{2}} = (1-i)/\sqrt{2}$. The actual evaluation of σ and R_{CT} will evidently depend on the potential at which the AC impedance is measured; as indicated above, if the analysis above is to be meaningful, then the impedance must be measured at or very close to the rest potential in order to ensure that the concentrations of reactant and product in the absence of a sinusoidal perturbation are indeed reasonably constant with time. If the DC potential is fixed at the rest potential, and assuming that the bulk concentrations of Ox and Red are equal, and at $\eta = 0$, $c_O^s = c_R^s = c^0$, it is straightforward to show from (5.41) that

$$R_{CT}\sigma = \left(\frac{RT}{\sqrt{2}F^2c^0}\right) \cdot \left\{\frac{1}{D_R^{\frac{1}{2}}} + \frac{1}{D_O^{\frac{1}{2}}}\right\} \tag{5.43}$$

The term $(1-i).R_{CT}.\sigma/\omega^{\frac{1}{2}}$ is called the *Warburg impedance*, Z_W; like all complex numbers it has both magnitude and phase, and since the latter is given by

$$\varphi = \tan^{-1}(\text{Imaginary Part/Real Part})$$

and the real part of Z_W is $R_{CT} \cdot \sigma/\omega^{\frac{1}{2}}$ and the imaginary part is $-R_{CT} \cdot \sigma/\omega^{\frac{1}{2}}$, the ratio of the two is clearly frequency *independent* and the phase angle is $-45°$. It is also evident that as the frequency increases, the magnitude of Z_W *decreases*, since at higher frequencies, owing to the rapid reversals of the reaction direction at the electrode surface, changes in concentration of Ox and Red are no longer propagated into the solution, and the overpotential associated with diffusion therefore vanishes. We can see this more clearly if we separate Z into its real and imaginary components:

$$Z(\omega) = \text{Re}(Z(\omega)) + i \cdot \text{Im}(Z(\omega)) = R_{CT}(1 + \sigma/\omega^{\frac{1}{2}}) - iR_{CT}\sigma/\omega^{\frac{1}{2}} \tag{5.44}$$

From this, it is evident that as $\omega \to \infty$, $\text{Re}(Z) \to R_{CT}$ and $\text{Im}(Z) \to 0$, so at sufficiently high frequency we can explore the electron-transfer kinetics alone through R_{CT}. Finally, if the electrode is held at a *dc* potential corresponding to $\eta = 0$, and the overpotential is purely derived from the *ac* component, $\tilde{E}$, which is such that $\tilde{E} << RT/F$, then we can relate the charge-transfer resistance to the exchange current; in fact, from section 4.2.2

$$\frac{1}{R_{CT}} = \left(\frac{\partial j}{\partial E}\right)_{c_O^s, c_R^s} = \frac{Fj_0}{RT} \tag{5.45}$$

To continue the discussion, we will now introduce the equivalent circuits for electrodes in electrochemical cells.

5.2.2.2 Equivalent Circuit for an Electrode – Diffusion Limited Reaction

Our picture hitherto of the circuit elements describing the flow of *ac* current through an electrode comprises an electron-transfer resistance, R_{CT} and a Warburg Impedance, Z_W, in series. In practice, we must also include the capacitance of the double layer, C_D, in parallel with R_{CT} and Z_W, and the electrolyte resistance, R_E. This gives the complete circuit diagram of Fig. 5.20.

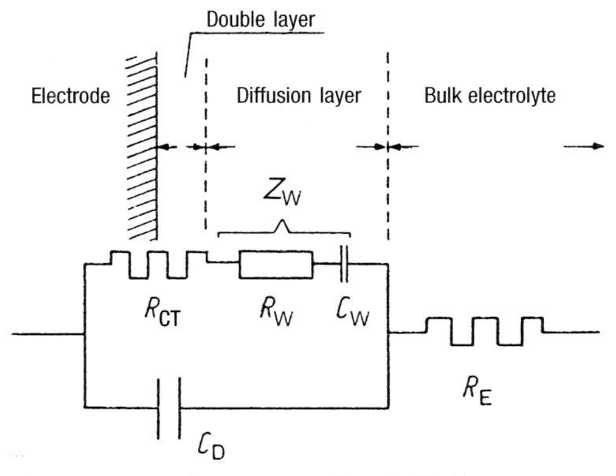

Fig. 5.20 Equivalent ac circuit for a half-cell (see text). R_{CT} is the electron-transfer resistance, $Z_W = R_W + 1/i\omega C_W$ is the Warburg impedance, R_E is the electrolyte resistance and C_D the double-layer capacitance.

During the initial period of development of the impedance technique, the extraction of the values of the various components of the circuit was carried out in a stepwise fashion, by first determining the value of R_E using platinised electrodes, and then the value of C_D by the use of a cell without a redox couple. Finally, the value of $R_{CT} + |Z_W|$ could be measured and plotted as a function of $\omega^{-1/2}$, giving a so-called Randles plot (Fig. 5.20a). Extrapolation to high frequency then allowed R_{CT} to be determined directly, from which the value of j_0 could be calculated (equation 5.45).

Modern instrumentation and software allow the measurement of the *ac* impedance of a cell and the subsequent extraction of circuit elements to be carried out directly with only the form of the equivalent circuit being needed. An instrument of this type is normally termed a Frequency Response Analyser.

If the electron transfer process is coupled to homogeneous or heterogeneous chemical reactions, then the Randles plot is found characteristically to deviate from linearity. These deviations have been observed, for example, for the deposition of silver ions, and are, in this case, associated with the subsequent migration of silver surface atoms from open surface sites to step sites.

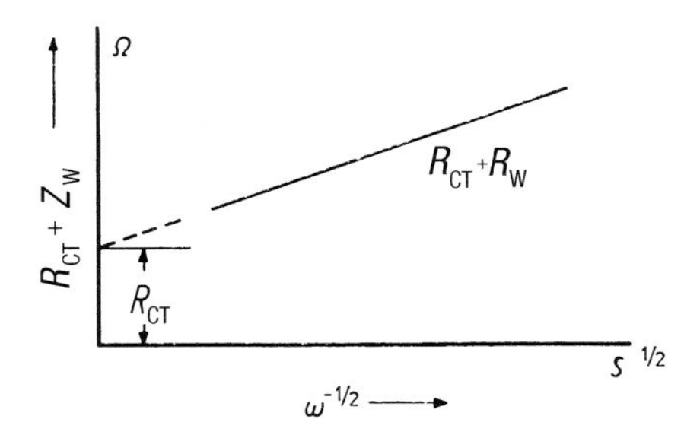

Fig. 5.20a Randles plot of transfer resistance + Warburg impedance.

5.2.2.3 AC Impedance of an Electrode where the Electron-Transfer Process is Rate-limiting

In this second limiting case, in which we assume that the Warburg impedance becomes very small compared to R_{CT}, the equivalent circuit simplifies to that of Fig. 5.21, with the interfacial impedance now being that of a parallel RC circuit, Z_p. The overall impedance, Z, of this circuit can be found using standard methods:

$$Z = Z_p + R_E = \frac{1}{1/R_{CT} + i\omega C_D} + R_E \tag{5.46}$$

and after some manipulation, we find

$$Z = R_E + \frac{R_{CT} - i\omega R_{CT}^2 C_D}{1 + \omega^2 R_{CT}^2 C_D^2} \tag{5.47}$$

from which it is evident that

$$\mathrm{Re}(Z) = R_E + \frac{R_{CT}}{1 + \omega^2 R_{CT}^2 C_D^2}; \quad |\mathrm{Im}(Z)| = \frac{\omega R_{CT}^2 C_D}{1 + \omega^2 R_{CT}^2 C_D^2} \tag{5.48}$$

where Re(Z) is the real part of Z, and Im(Z) the imaginary part. In particular, we obtain the following limiting values from (5.47):

$$\omega \to 0: \quad \mathrm{Re}(Z) \to R_E + R_{CT}; \quad \mathrm{Im}(Z) \to 0 \tag{5.49}$$

$$\omega \to \infty: \quad \mathrm{Re}(Z) \to R_E; \quad \mathrm{Im}(Z) \to 0 \tag{5.50}$$

$$\omega = 1/(R_{CT}C_D): \quad \mathrm{Re}(Z) = R_E + R_{CT}/2; \quad \mathrm{Im}(Z) = R_{CT}/2 \tag{5.51}$$

These limiting values suggest that the most effective way of displaying equations (5.46–5.48) is through an 'impedance plot' or 'Nyquist plot', in which the imaginary part of the impedance, $|\mathrm{Im}(Z)|$, is plotted against Re(Z) for the range of frequencies explored. Carrying out this plot for the equivalent circuit of Fig. 5.21 gives rise to a simple semicircle, as shown in Fig. 5.22; again experimentally, a frequency response analyser would be used. In practice the semicircle must be extrapolated at the high frequency end; in addition, measurements at very low frequencies are extremely time-consuming, connected with the problems of stationary state. Semicircles at high frequency may show the effects of stray inductances.

Interestingly, the full Randles circuit may also be plotted as an impedance plot: in this case, the semicircle at higher frequencies is accompanied by a linear region at lower frequency (compare Fig. 5.24): the Warburg region, whose gradient is unity, corresponding to the 45° phase of Z_W mentioned earlier.

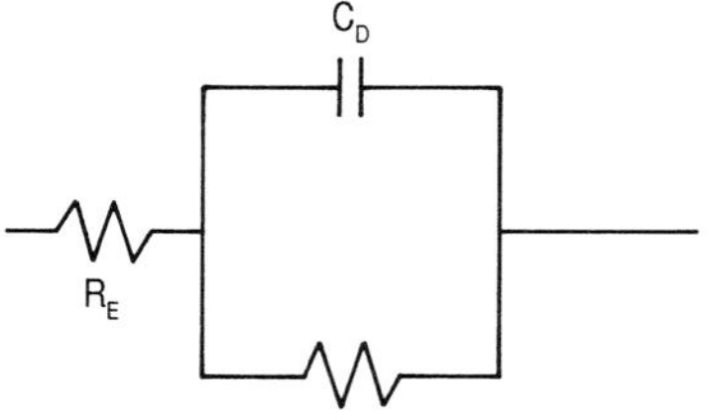

Fig. 5.21 Equivalent ac circuit for the case when the Warburg impedance can be neglected, ie when the impedance of the cell is dominated by the electron-transfer resistance.

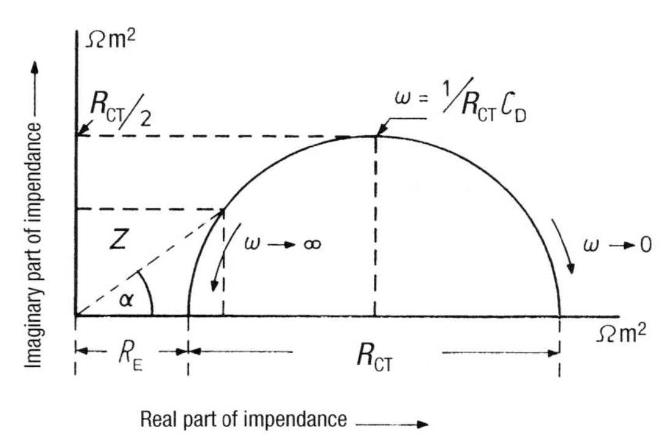

Fig. 5.22 Nyquist plot for the equivalent circuit of Fig. 5.21. For any arbitrarily chosen frequency, ω, the vector Z represents the complex impedance.

5.2.2.4 **Logarithmic or Bode Plot Representations**

The impedance plot has the frequency only as an implicit variable, and a better representation of the frequency variation of Re(Z) and Im(Z) can be found through 'Bode' plots of $\log|Z|$ and phase-angle, α, vs. $\log(\omega)$. For a frequency-independent resistance, a plot of $\log|Z|$ vs. $\log(\omega)$ would be a horizontal straight line, whose phase angle would be zero for all frequencies. For the equivalent circuit of Fig. 5.21, the corresponding Bode plot is shown in Fig. 5.23. Inspection of equations 5.46 to 5.48 shows that at low frequency, $\alpha \to 0$, and $|Z| \to R_{CT} + R_E$, whereas at high frequencies, $\alpha \to 0$, and $|Z| \to R_E$. In the mid-frequency region, a linear section of slope -1 is found; extrapolation of this to $\log(\omega) = 0$ gives $\log(1/C_D)$.

The phase angle goes through a maximum, corresponding to an angular frequency

$$\omega(\alpha_{max}) = (1 + R_{CT}/R_E)^{1/2}/(R_{CT}C_D) \quad (5.52)$$

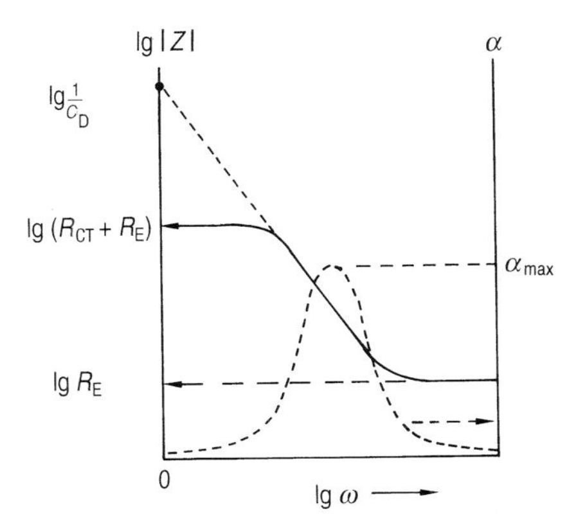

Fig. 5.23 Bode plot for the equivalent circuit of Fig. 5.21.

5.2.2.5 Electrode Reactions under Mixed Control

If mixed control operates, then the two components can be separated: at high frequency the diffusion limitation vanishes, and the formation of at least the beginning of a semicircle is normally seen in the impedance plot, as in Fig. 5.22. At lower frequencies, diffusion limitation will become more and more significant, and a Warburg line will form, as shown in Fig. 5.24. The low-frequency part of the semicircle may be extrapolated, resulting in an estimation of the value for $R_E + R_{CT}$. The intercept of the Warburg line on the real part of the impedance axis, R_N, is a function of electrolyte concentration, diffusion coefficient and double-layer capacity. For rapid electron transfer, the semicircle component of the impedance plot may be very small, and the Warburg line will then cut the x-axis at ca. $R = R_E$.

For $\omega \rightarrow 0$, the Warburg line must curve back towards the real axis in principle, since the *dc* resistance of the cell is finite. Analysis suggests that for a simple quasi-reversible case, $R_{CT} \rightarrow 0$, the Warburg line curves back to form part of a second semicircle at very low frequencies. This behaviour arises because the correct boundary condition is not $\tilde{c}_O \rightarrow 0$ as $x \rightarrow \infty$ as in (5.37) but is more correctly written for the case of convective diffusion: $\tilde{c}_O \rightarrow 0$ as $x \rightarrow \delta_N$, the thickness of the Nernst boundary layer. This very low frequency semicircle finally intersects the x-axis as $\omega \rightarrow 0$ at an intercept R_N (Nernst resistance) from the extrapolated Warburg line intersection, R_E, where R_N has the value:

$$R_N = \frac{2RT\delta_N}{n^2F^2Dc^0} \tag{5.53}$$

In an unstirred solution, R_N, the Nernst resistance, will not be well-defined, but if control of convection is also maintained, through rotation of a disc-electrode, so that δ_N and R_N are well-defined, this second semicircle itself starts to shrink with increasing rotation rate, as shown experimentally in Fig. 5.25 for the system $[Fe(CN)_6]^{4-/3-}$ on platinum (where the first semi-circle in this case cannot be discerned). For this system, the Warburg line cuts the x-axis at $R = R_E = 1.8\ \Omega cm^2$.

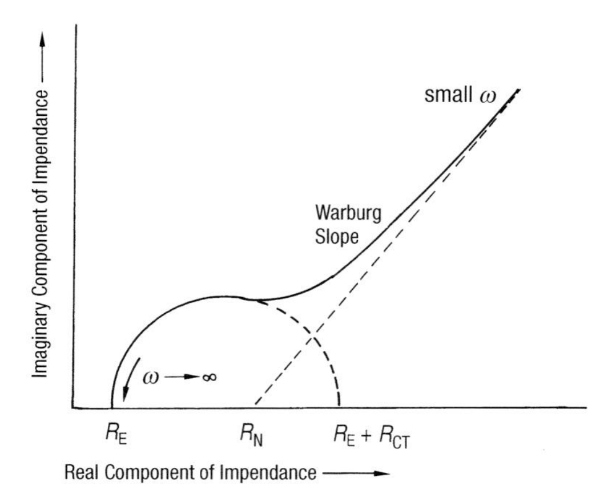

Fig. 5.24 Nyquist plot for the equivalent circuit of Fig. 5.20, showing the effects of mixed control.

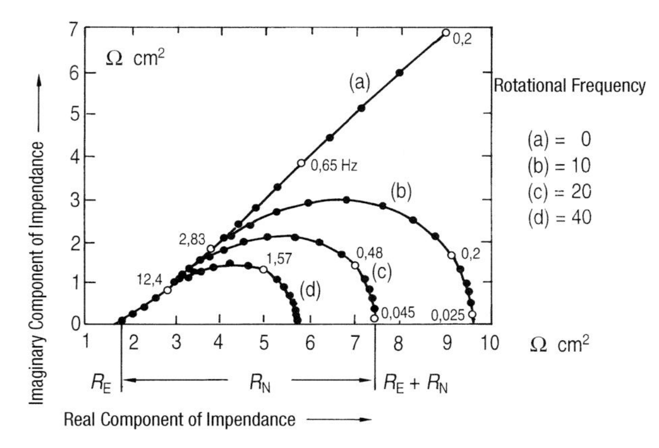

Fig. 5.25 Nyquist plot for the case of convective diffusion at a rotating-disc electrode for the system: $2 \cdot 10^{-3}$ M $[Fe(CN)_6]^{3-/4-}$ in 0.5 M Na_2SO_4.

The values of δ_N calculated from curves (b) – (d) are close to those predicted from the Levich equation.

5.2.2.6 **Methanol Oxidation as an Example of the Investigation of a Complex Electrode Reaction with *ac* Techniques**

If a thin-layer model methanol fuel cell (as shown in Fig. 9.27) with a Nafion membrane as the proton-conducting electrolyte, and with the oxygen cathode replaced by a hydrogen cathode, is studied by *ac* techniques, the following components of the impedance must be allowed for: the ohmic resistance of the membrane, the potential dependent ohmic electron transfer resistance of the methanol anode (that of the hydrogen electrode can be neglected), the frequency dependent capacity of the anode/electrolyte double-layer as well as the frequency-dependent resistance of the mass transport components. The resultant impedance plot (or, commonly, *spectrum*) is shown in Fig. 5.26a with a 10 mV *ac* measurement amplitude, and frequencies varying from 65 kHz to 3 mHz; it is clear that there are *three* semicircles.

The diameter of the first semicircle (the highest frequency one) shrinks with increasing temperature. Clearly this semicircle is determined by the resistance of the membrane and the double-layer capacity. The middle semicircle changes drastically with anode potential (see Fig. 5.26b), and is evidently conveying information about the kinetics of the methanol reaction. This semicircle appears to be a mixture of linear (at the high-frequency end) and circular segments (compare Fig. 5.25), which is the behaviour expected for the interaction of charge transfer and diffusion in pores [H. Moreira and R. DeLevie, J. Electroanal. Chem. 29 (1971) 353].

The low-frequency semicircle finally is strongly dependent on the methanol stoichiometry of the fuel-cell feed (Fig. 5.26c). The stoichiometic factor is the ratio of methanol feed rate to the cell to the stoichiometrically needed amount. Clearly this semicircle depends on a mass-transport effect. The slower, for the same methanol concentration, the feed rate is, the smaller the stoichiometric factor. For the same

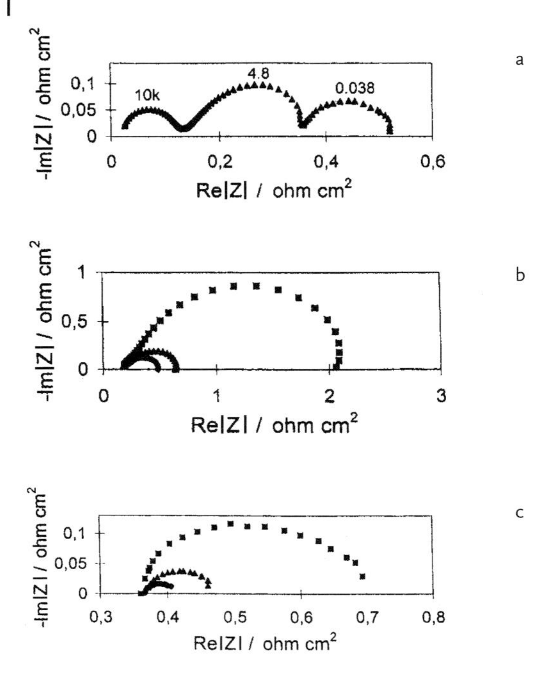

Fig. 5.26 Nyquist plot for a porous PtRu DMFC-Anode including a Nafion membrane as electrolyte with a 1 M methanol feed: (a) Impedance spectrum between 65 kHz and 3 mHz at ca. 250 mV RHE; (b) Potential dependence of the middle frequency region of the spectrum in (a) for anode potentials of 200 mV (■), 250 mV (▲), 300 mV (•); (c) low-frequency region of (a) for different stoichiometric factors for inflow of methanol with the same current density: stoichiometric factors of 1.6 (■); 2.0 (▲) and 5.0 (•) [J.T. Muller and P.M. Urban, J. Power Sources 75 (1998) 139].

current loading on the cell, the slower this feed, the more the CO_2 product accumulates at the anode, blocking the catalytically active sites. To avoide this effect, we must work at sufficiently high stoichiometric factors (e.g. 5-10) for the CO_2 to be fully removed from the anode; at high stoichiometric factors, the low frequency behaviour in this region becomes dominated by methanol oxidation kinetics and the diameter of the low-frequency part can clearly be seen to decrease with increase in current density (Fig. 5.27, part a).

A special feature is that now the Nyquist plot also extends into the fourth quadrant; that is, there is an inductive component to the impedance. This has been traced by Harrington and Conway [D.A. Harrington and B.E. Conway, Electrochim. Acta 32 (1987) 1703] to the presence of adsorbed intermediates in the reaction pathway, and this is further explored in section 5.2.2.7. One reaction pathway for methanol oxidation does indeed lead to the formation of adsorbed carbon monoxide. For the oxidation of methanol in a thin-layer cell, with a stoichiometric factor of 10 or more, an

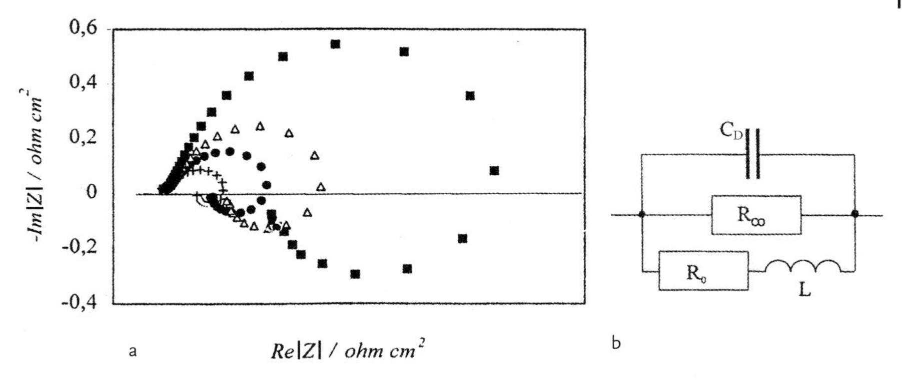

Fig. 5.27 Anode impedance for methanol oxidation as in Fig. 5.26 but with a stoichiometric factor of 10 to ensure that the behaviour represents purely the kinetics of methanol oxidation: (a) Impedance spectrum for different current densities: 500 mAcm^{-2} (+); 300 mAcm^{-2} (•); 200 mAcm^{-2} (Δ); 100 mAcm^{-2} (■); (b) equivalent circuit with ohmic resistances R_∞ and R_0, Inductance L and capacity C_D [J.T. Muller, P.M. Urban, and W.F. Holderich, J. Power Sources 84 (1999) 157].

equivalent electrical circuit of the type shown in Fig. 5.27, part b, can be observed. For the Faradaic component of the impedance, this circuit yields $Z_{\text{faraday}} = \{1/R_\infty + 1/[R_0(1 + i\omega t)]\}^{-1}$.

5.2.2.7 Theoretical Treatment of *ac* Impedance in the Presence of Adsorption

To understand the inductive behaviour during adsorption, we can extend the analysis given above. Consider the case where a species A is oxidised by a two-electron process with an adsorbed intermediate B formed before the product P is generated:

$$\text{AB}_{\text{ads}} + e^-;\ \text{current } j_1$$
$$\text{B}_{\text{ads}}\text{P} + e^-;\ \text{current } j_2$$

Let θ be the coverage of B_{ads}; then since j_1 is a function only of E, c_A and θ, we can write, by analogy with above

$$\tilde{j}_1 = \left(\frac{\partial j_1}{\partial E}\right)_{c_{A,\theta}} \tilde{E} + \left(\frac{\partial j_1}{\partial c_A}\right)_{E,\theta} \tilde{c}_A + \left(\frac{\partial j_1}{\partial \theta}\right)_{E,c_A} \tilde{\theta}$$

where we have replaced the differential quantities that appeared in the earlier equation (5.33) such as δE by the small sinusoidal amplitudes such as $\tilde{E}$. Solving the diffusion equation as before gives

$$\tilde{j}_1 = F(i\omega D_A)^{\frac{1}{2}}\tilde{c}_A$$

and eliminating $\tilde{c}_A$ between these two equations gives:

$$\alpha_1\tilde{j}_1 = \left(\frac{\partial j_1}{\partial E}\right)_{c_{A,\theta}} \tilde{E} + \left(\frac{\partial j_1}{\partial \theta}\right)_{E,c_A} \tilde{\theta}$$

where

$$\alpha_1 = 1 - \frac{1}{F(i\omega D_A)^{\frac{1}{2}}}\left(\frac{\partial j_1}{\partial c_A}\right)_{E,\theta}$$

In a similar way, we can obtain an expression for the current $\tilde{j}_2$ as:

$$\alpha_2\tilde{j}_2 = \left(\frac{\partial j_2}{\partial E}\right)_{c_{P,\theta}} \tilde{E} + \left(\frac{\partial j_2}{\partial \theta}\right)_{E,c_P} \tilde{\theta}$$

where

$$\alpha_2 = 1 + \frac{1}{F(i\omega D_P)^{\frac{1}{2}}}\left(\frac{\partial j_2}{\partial c_P}\right)_{E,\theta}$$

Now it is clear that there must be a relationship between $\left(\frac{\mathrm{d}\theta}{\mathrm{d}t}\right)$ and $\tilde{j}_1$, $\tilde{j}_2$ since passage of current, j_1, will lead to an increase in the coverage θ of B_{ads}, and passage of the second current leads to a decrease in θ. In time δt, $j_1\delta t/F$, moles of A arrive at the surface per unit area and are converted to B_{ads}; if ξ moles per unit area of B_{ads} correspond to unit coverage ($\theta = 1$), then the increase in coverage of B_{ads}, $\delta\theta$, will be $j_1\delta t/\xi F$. Similarly the *decrease* in coverage of B_{ads} on passage of j_2 current for δt time is evidently $j_2\delta t/\xi F$ and it is clear that $\frac{\mathrm{d}\theta}{\mathrm{d}t} = (j_1 - j_2)/\xi F$. However, we know, by analogy with equation (5.35) above, that $\theta = \bar{\theta} + \tilde{\theta}e^{i\omega t}$ and so $\frac{\mathrm{d}\theta}{\mathrm{d}t} = i\omega\tilde{\theta}e^{i\omega t}$. Replacing j_1 and j_2 by their values $\bar{j}_1 + \tilde{j}_1 e^{i\omega t}$, $\bar{j}_2 + \tilde{j}_2 e^{i\omega t}$ and equating coefficients of $e^{i\omega t}$ as above, we see that $i\omega\tilde{\theta} = (\tilde{j}_1 - \tilde{j}_2)/\xi F$, and substituting for the values of the sinusoidal partial currents we find:

$$i\omega\tilde{\theta}\xi F = \left\{\left(\frac{1}{\alpha_1}\right)\left(\frac{\partial j_1}{\partial E}\right)_{c_A,\theta} - \left(\frac{1}{\alpha_2}\right)\left(\frac{\partial j_2}{\partial E}\right)_{c_P,\theta}\right\}\tilde{E} + \\ + \left\{\left(\frac{1}{\alpha_1}\right)\left(\frac{\partial j_1}{\partial \theta}\right)_{c_A,E} - \left(\frac{1}{\alpha_2}\right)\left(\frac{\partial j_2}{\partial \theta}\right)_{c_P,E}\right\}\tilde{\theta}$$

The total AC current $\tilde{j} = \tilde{j}_1 + \tilde{j}_2$ and straightforward, if rather tedious algebra will give rather complex equivalent circuit. If we are working at frequencies high enough for $\alpha_1 \approx 1$ and $\alpha_2 \approx 1$ (normally a few tens of Hz) or if we are using controlled transport through a rotating ring or other hydrodynamic system, where again the transport coefficients α_1, α_2 can be put equal to unity, we can simplify the above expressions to give

$$\tilde{j} = \left(\frac{1}{R_\infty}\right)\tilde{E} + \left(\frac{1}{R_0(1 + i\omega\tau)}\right)\tilde{E}$$

where

$$R_{\infty} = \left\{ \left(\frac{\partial j_1}{\partial E} \right)_{\theta} + \left(\frac{\partial j_2}{\partial E} \right)_{\theta} \right\}^{-1}$$

$$R_0^{-1} = \left\{ \left(\frac{\partial j_1}{\partial \theta} \right)_{E} + \left(\frac{\partial j_2}{\partial \theta} \right)_{E} \right\} \cdot \left\{ \left(\frac{\partial j_1}{\partial E} \right)_{\theta} - \left(\frac{\partial j_2}{\partial E} \right)_{\theta} \right\} \cdot \left\{ \left(\frac{\partial j_2}{\partial \theta} \right)_{E} - \left(\frac{\partial j_1}{\partial \theta} \right)_{E} \right\}^{-1}$$

$$\tau R_0 = \xi F \left\{ \left(\frac{\partial j_2}{\partial \theta} \right)_{E} - \left(\frac{\partial j_1}{\partial \theta} \right)_{E} \right\}^{-1}$$

corresponding to an equivalent circuit with a resistor R_{∞} in parallel with a series combination of a resistor R_0 and an inductance $L \equiv \tau R_0$(compare Fig. 5.27, part b). This type of approach has been used, for example, to study the mechanism of chlorine evolution on platinum; Hillman and co-workers were able to show [Electrochimica Acta 37 (1992) 2715] that the only mechanism consistent with their AC data had the simple form:

$$Cl^{-} \dot{Cl}_{ads} + e^{-}$$

$$2\dot{Cl}_{ads} \rightarrow Cl_2$$

rather than a Heyrovsky type involving a process such as $\dot{Cl}_{ads} + Cl^{-} \rightarrow Cl_2 + e^{-}$ or processes involving such species as Cl^{+}_{ads} or OH_{ads}. The approach can obviously be extended to electrochemical reactions of arbitrary complexity, though the equivalent circuits rapidly become exceedingly complex, and the number of parameters needing to be fitted also becomes sufficiently large for the technique to cease to yield unequivocal information. Furthermore, quite different underlying mechanisms may give rise to essentially identical equivalent circuits, and different equivalent circuits can give similar impedance spectra over a limited frequency range. In general, the modern approach is to use alternative techniques to establish at least the outlines of the mechanism before attempting to provide a definitive interpretation of the *ac* data.

As a final point, it will be seen from the above that there is an inherent assumption that in the absence of any AC perturbation, the concentrations of O and R are assumed to be constant, or at least to vary on a timescale slow compared to the AC perturbation itself. Another way of putting this is that the Fourier components of any variation in the concentrations of O and R due to an underlying potential sweep or to long-term diffusional changes must be of lower frequency than the impressed AC potential. If this is not the case, then the two timescales will overlap, making analysis exceedingly difficult.

5.3
Electrochemical Methods for the Study of Electrode Films

A large number of electrochemical reactions proceed through the formation of adsorbed reactants or intermediates, and these adsorption processes can radically effect the course of the reaction, leading in extreme cases to the complete inhibition of any electrochemical process. The coverage, θ, of an adsorbed species is, therefore, an important parameter in the understanding of reaction mechanism. Unfortunately, whilst electrochemical measurements can lead, particularly through the determination of the number of electrons per adsorbed molecule needed for complete oxidation or conversion, to some feel for the structure of adsorbates in simple cases, it is rare indeed that a completely unambiguous understanding of any electrochemical process involving adsorbed species will emerge from such measurements alone. As we shall see in Section 5.4, however, the development of new *in situ* spectroscopic techniques has subtantially improved the possibility of direct assignment of adsorbate structure.

The quantitative measurement of adsorbate coverage electrochemically can be carried out in two ways:

(a) The amount of charge required wholly to oxidise or reduce the layer may be measured and the surface coverage calculated from Faraday's law or from comparison with a reference adsorbate on the same surface;

(b) The capacitance of the double layer can be measured and compared to that found for the "bare" surface under identical conditions.

5.3.1 Measurement of Charge Passed

If the amount of charge passed for complete conversion of a surface film is Q_B, and the number of electrons per adsorbed molecule required is n, then it is clear that

$$Q_B = nF\Gamma \tag{5.54}$$

where Γ is the surface coverage in moles. If the maximum surface coverage for the species is Γ_{max}, corresponding to a charge Q_B^{max}, then the coverage, θ, is given by

$$\theta = \frac{Q_B}{Q_B^{max}} \tag{5.55}$$

Note that this is an *operational* definition, and Q_B^{max} may not correspond to complete coverage of the surface. To relate θ as measured in (5.55) with the true surface area, at least on platinum, it is common to measure the charge passed in oxidation of the hydride region of the cyclic voltammogram. The charge, integrated between 80 and 450 mV *vs* RHE and denoted by Q_H, takes the value 210 μCcm^{-2} for smooth polycrystalline platinum as discussed above in section 5.2.1.2, and is largely independent of the sweep rate. Furthermore, if a partial electrode film is already present, the hydride layer will only form on the unoccupied part of the surface, and if the hydride oxidation charge is now Q_H^F, then

$$\theta = \frac{Q_H - Q_H^F}{Q_H} \tag{5.56}$$

This type of measurement is shown in fig. 5.28, which shows the cyclic voltammogram for platinum in sulphuric acid in the absence (dotted line) and presence (continuous line) of methanol. The region shown shaded with slanting lines can be used to estimate $Q_H - Q_H^F$ and hence to calculate θ from (5.56) and the region shaded with horizontal lines can be used to calculate θ from (5.55) and to estimate the value of n.

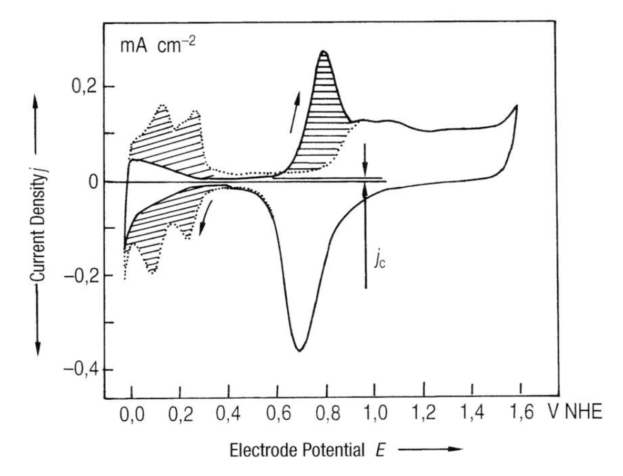

Fig. 5.28 Cyclic voltammogram ($v = 100\ mVs^{-1}$) for the determination of electrode coverage. *Dotted line*: measurement in the base electrolyte of 0.5 M H_2SO_4; *Continuous line*: measurement following adsorption of methanol at 0.5 V vs. NHE from a 0.25 M solution of methanol in 0.5 M H_2SO_4 (admitted by the method of rapid electrolyte exchange (see text).

Experimental details are reasonably straightforward. At the beginning of the experiment, the surface must be fully cleaned and prepared in a reproducible manner, and the electrode then maintained under potential control at all times during the deposition of the film. The determination of the value of Q_B can, in principle, be carried out in the same solution from which the film is adsorbed, but this carries the obvious danger of charge transfer to the solution itself. To avoid this, whilst at the same time retaining potential control, necessitates a cell design capable of replacing the adsorbate solution with one corresponding only to the background electrolyte. This presents no problems if the adsorption process is slow and irreversible, but the rate at which electrolyte can be exchanged in simple cells is limited, and if very short adsorption times are to be measured, more attention has to be paid to cell design.

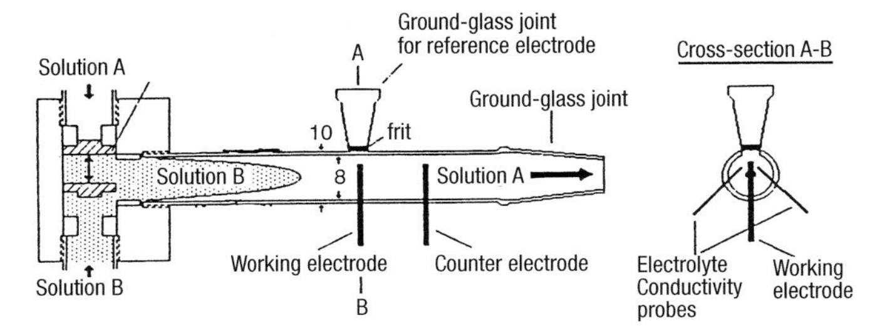

Fig. 5.29 Flow-through cell for rapid electrolyte exchange; measurements shown are in mm [from G. Janoske and C.H. Hamann, Ber. Buns. Phys. Chem. **95** (1991) 1187].

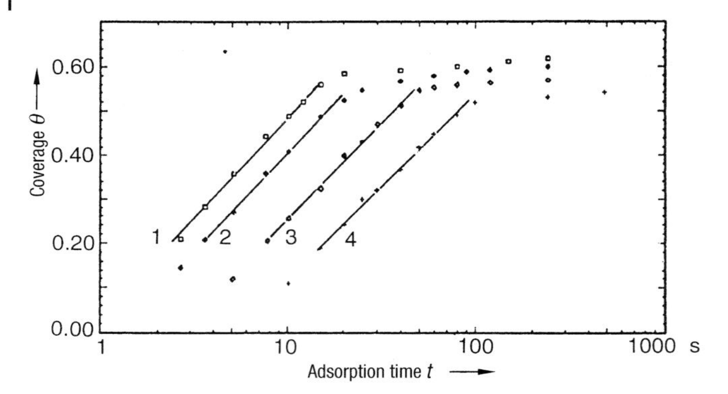

Fig. 5.30 Coverage, θ, as a function of adsorption time for the adsorption of different concentrations of phenol in 0.05 M H_2SO_4 on platinum. The adsorption potential in all cases was 0.3 V vs. NHE and the concentrations are: 1: 10^{-4}M; 2: 5×10^{-5}M; 3: 2.5×10^{-5}M; 4: 10^{-5}M [from G. Janoske and C.H. Hamann, loc. cit.].

One type of fast-electrolyte-exchange cell is shown in Fig. 5.29, in which an electromagnetic valve is used to control first the entry of the adsorbate solution and then the background electrolyte. The replacement front between the two solutions is also shown diagrammatically, and the adsorption time must be measured not from the time the valve is reversed but from the arrival of the front at the electrode; this can be detected, for example, by a change in electrolyte conductivity, and such equipment allows adsorption times of as low as 0.5 s to be measured. Such rapid times are characteristic, for example, of the adsorption of aromatic ring systems onto platinum, and some results for the adsorption of phenol are shown in Fig. 5.30. In this case, saturation coverages of ~ 0.6 (from Q_H) are attained in times of 10–100 s, giving an adsorption rate constant of $\sim 3000\ dm^3 mol^{-1} s^{-1}$.

5.3.2
Capacitance Measurements

If C_0 is the capacitance of the electrolyte double layer in the absence of film formation, and C the capacitance measured after an adsorbed film has formed, then in general $C < C_0$, since the presence of an adsorbed layer will increase the thickness of the double layer. Assuming that the decrease, $C - C_0$, is proportional to the coverage of adsorbed species, and that C_{min} is the measured capacitance for the case of maximum coverage, then

$$\theta = (C_0 - C)/(C_0 - C_{min}) \tag{5.57}$$

Experimentally, again the method is straightforward: the electrode surface must, as before, be cleaned and prepared in a reproducible state, and the capacitance measured by ac methods as described above. A powerful variant is to use *ac* cyclic voltammetry, in which the capacitance of the surface is measured continuously by the superimposition of a small *ac* potential on the linear potential ramp of the cyclic voltammogram as the electrode is scanned; this allows the possibility of following film formation in real time during a potential sweep. This is frequently termed tensammetry, and was a method used to identify the presence of adsorbed surfactants in polarographic experiments (see below).

5.4
Spectroelectrochemical and other Non-classical Methods

5.4.1
Introduction

In these techniques, electromagnetic radiation of a multiplicity of wavelengths is used as a probe of the molecular interactions taking place at or near the electrode surface, including the formation and composition of electrochemically generated films. The spectral regions used vary from synchrotron radiation in the X-ray region, giving structural information through EXAFS (Extended-X-ray-absorption fine structure analysis) and diffraction data, to microwave spectra used to study radical scavengers in order to establish their identity, and hence infer the existence of radical intermediates in electrochemical reactions. The most important spectro-electrochemical method uses infra-red light (of wavelength 2.5 – 100 μm corresponding to wavenumbers in the range 100 – 4000 cm^{-1}), which allows the identification of molecular species formed at or near the electrode, and also is sufficiently sensitive to permit the observation of surface-bound species. In this latter case, the spectral shifts observed as

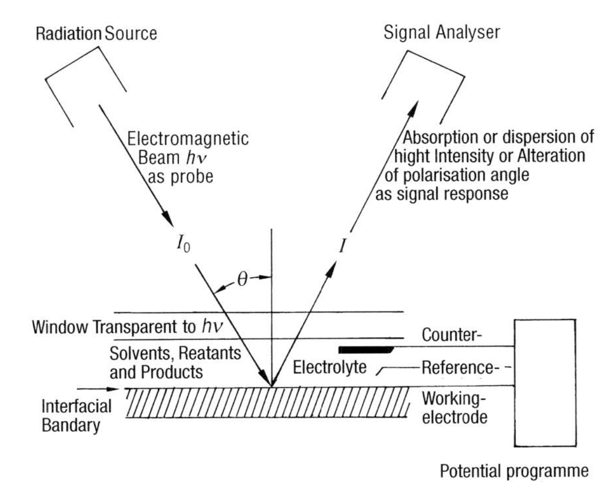

Fig. 5.31 Basic experimental set-up for the use of electromagnetic radiation as an in-situ probe of an electrochemical reaction. I_0 is the intensity of the incident and I the intensity of the reflected rays, and θ the angle of incidence.

compared to solution species give a good indication of the strength of electrode-adsorbate interaction.

The study of electrode films has been particularly facilitated by these techniques, with information now available on the structure, composition, reactivity and molecular orientation that had previously only been possible to infer indirectly from electrochemical measurements. A powerful method for the general study of electrode films has been the technique of ellipsometry, which has normally been used in the wavelength region from 300 – 800 nm. In this technique, the alteration in polarisation of linearly polarised light incident on the electrode surface (usually at an angle of 45°) is measured, usually over a wide wavelength range, to establish the optical properties and thickness of films varying in thickness from nanometres to microns. The sensitivity of the technique is such that, for example, the thickness of the oxide layer on platinum can be followed from its initial monolayer formation at ca. 1.0 V vs. RHE to a multilayer oxide above 1.6 V.

Critical to the success of these methods has been the ability to apply them *in situ*; that is, the techniques can be applied to the electrode under normal operating conditions: potentiostatic, galvanostatic etc. The general operational scheme then has radiation of the chosen frequency incident on the electrode surface, with the reflected beam being analysed, as shown in Fig. 5.31. Naturally for infra-red studies, absorption by the commonly-used polar solvents presents severe problems for the design of the experiment if in-situ studies are required, and much earlier work concentrated on Raman spectroscopy, where absorption of the incident probe beam is not a serious problem. However, the Raman effect is normally very weak, and only for a small number of metals could a sufficiently enhanced signal be obtained to allow the technique to probe adsorbed species at the surface. Even in these cases, the enhancement could only be effected by creating a roughened surface, the behaviour of which may not represent well the "normal" electrode.

Although the most valuable methods for electrochemical use have been those capable of application in situ, it is also the case that ex-situ methods, in which the electrode is removed from the cell under carefully controlled conditions and then examined *in vacuo*, have also proved of great value. The possibility of using electrons as probes in a high vacuum system allows us to use some very powerful structural probes that are not available at the liquid-solid interface, and these have been used to great advantage in, for example, the study of under-potential deposition, where the metal atoms are sufficiently immobile to give us confidence that major rearrangements will not take place during the electrode transfer process. Clearly, however, the use of these techniques to study, for example, the corrosion layer on iron is fraught with difficulties, since this layer is known to be a hydrous oxide, and its dehydration on exposure of the electrode to vacuum is a serious problem.

A word of warning needs to be given at this point, before detailed consideration of individual techniques. It is imperative to realise that the sensitivity of a given technique very much depends on the type of species present. It is possible, as we shall see below, for infra-red reflection to detect a species easily when oriented perpendicular to the surface, but to be completely blind to that same species when it lies flat. It is highly dangerous to rely overmuch on the results of a single spectro-electrochemical experiment; several different probes are usually necessary, and the results must also be reconciled to those derived from classical electrochemical measurements before complete faith in the interpretation can be realised.

5.4.2
Infra-Red Spectroelectrochemistry

5.4.2.1 Basics

The energy levels of a harmonic oscillator are quantised, with the energy E given by

$$E = nh\nu \tag{5.58}$$

where n can have values 1,2... and the vibrational frequency ν is given within the harmonic oscillator model for a diatomic molecule with atomic masses m_1 and m_2 and force constant k, by:

$$\nu = \frac{1}{2\pi}\left(\frac{k}{\mu}\right)^{\frac{1}{2}} \tag{5.59}$$

where the effective mass, $\mu = m_1 m_2/(m_1 + m_2)$, from which can be seen the fact that isotopically different molecules will have different vibrational frequencies. For $^{12}C^{16}O$, $k = 1902$ Nm^{-1}, giving $\nu = 0.643 \times 10^{14}$ s^{-1} with a corresponding wavelength of $\lambda = 4.67 \times 10^{-6}$ m ($\lambda \cdot \nu = 3 \cdot 10^{10}$ cm s^{-1}) and an inverse wavelength, or wavenumber, of 2143 cm^{-1}. For the CO molecule, the excitation energy for the transition from the ground state to the first vibrationally excited state is $\sim 4.26 \times 10^{-20}$ J, about a factor of ten larger than the thermal energy $k_B T$ at room temperature. It follows that CO is overwhelmingly present in the ground state under ambient conditions.

For polyatomic molecules, the motions of all the atoms are coupled together and it is no longer possible usually to envisage the vibrations in terms of single bond stretches or bending motions. The reason for this is that the motion of atoms in one bond will excite motion in neighbouring bonds, and the collective molecular motion is a linear combination of individual bond stretches and bends. In general, the motion of individual bonds does not correspond to a fixed energy, since energy can flow from one vibrating bond to another, but it is always possible to find linear combinations of bond bending and stretching motions that correspond to fixed vibrational energies, these linear combinations being termed normal modes. The energies of these normal modes are given by formulae similar to those of (5.51) and (5.52), and there are 3N-6 such modes for a non-linear molecule (3N-5 for a linear molecule).

Not all modes are active towards the absorption of IR radiation: quite generally it is possible to show that unless excitation of the normal mode leads to a change in dipole moment in the molecule, it will not be active in absorption. A simple example is the fact that homonuclear diatomic molecules will show no IR absorption, nor will the totally symmetric modes of molecules such as CO_2, CH_4 etc. However, in addition to this selection rule, there is a second operative under conditions when species are adsorbed on metallic electrode surfaces, the so-called surface selection rule. This arises because IR radiation incident with the electric field vector parallel to the metal surface (s-polarised light) undergoes a phase change of 180° on reflection. Adding the incident and reflected electric fields at the surface will, therefore, give a null result, essentially ensuring that the intensity of the electric field parallel to the surface is zero. Only the component of the electric field perpendicular to the surface remains finite, so that only molecules having IR-active modes with a nett component in this direction will show IR absorption. Molecules lying flat on the surface therefore absorb no IR radiation (at least in first order), and conversely, s-polarised light will not be absorbed by species on or close to the surface.

5.4.2.2 Spectroelectrochemical Cells

The most commonly used type of cell is shown in Fig. 5.31a. In order to minimise absorption of the incident IR by the types of solvent used in electrochemical experiments, particularly water which has broad strong absorption bands in the mid-IR region, the electrolyte is trapped in a thin layer between the IR-transparent window (usually CaF_2 or ZnSe) and the working electrode. This latter is normally formed from a well-polished disc of diameter 0.5 – 2 cm, which should be large enough to reflect all the incoming elliptical cross-section IR beam. The angle of incidence is fixed between 50° and 80° and must not vary during the experiment. The main disadvantage of the cell design is the long and narrow current path between working electrode and reference/counter electrodes. This is a general problem with external reflection geometry, and even with highly conductive electrolytes, a considerable iR-drop is observed.

This disadvantage can be circumvented by the use of internal reflection geometry, as illustrated in Fig. 5.31b. The incident IR beam is now multiply internally reflected by

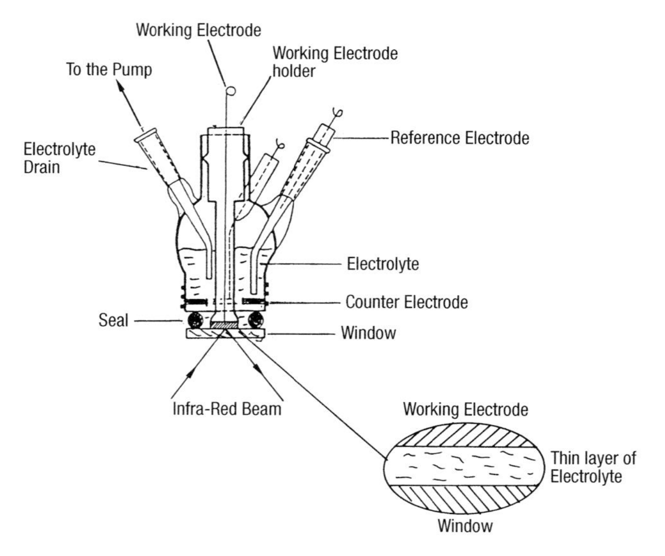

Fig. 5.31a Thin-layer IR-spectroelectrochemical cell for use in the external reflection-mode.

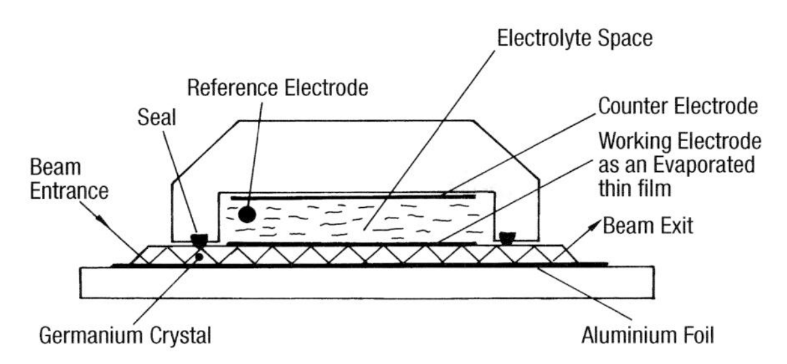

Fig. 5.31b IR-spectroelectrochemical cell for use in the internal reflection mode.

passage through a plane-parallel IR-transparent material, such as a germanium crystal. On each internal reflection, the beam penetrates a fraction of a wavelength into the neighbouring medium, which in the arrangement shown is actually a very thin layer of iron or other metal, acting as an optically transparent electrode. Provided the metal layer is thin enough, the IR will also penetrate into the electrolyte above the electrode to a depth of a fraction of a micron, sampling both the electrode surface and near-surface region. In addition, the multiple reflections give a substantial sensitivity gain on the single external reflection mode. However, the highly resistive electrolyte layer of the external reflection configuration has been replaced in this configuration by a resistive *electrode*, and the use of germanium is itself problematic since it is a semiconductor whose transparency is itself a function of electrode potential.

5.4.2.3 Examples of Different Methodologies in IR and Some Results

In the simplest case, the light from a polychromatic source would be polarised and focussed onto the electrode. Following reflection, the light could be analysed by a monochromator and the intensity determined as a function of wavelength. As indicated above, however, such a process is not really practicable, since the strong absorption by the solvent and the dispersion by the optical elements means that at most a fraction of a percent change in intensity on formation of a monolayer would be expected, and to detect so small a change in a dispersion instrument can only be done with phase-sensitive detection. In this technique, a modulation in intensity is imposed in some fashion on the beam and this intensity modulation is measured at the detector, allowing small changes to be extracted from a large constant background. This is the basis of the EMIRS (Electrochemically modulated IR spectroscopy) technique, in which the potential of the electrode is modulated between a reference potential E_R and a working potential E_M with a frequency of ca. 10 Hz, whilst the monochromator is slowly (< 1 cm^{-1}s^{-1}) scanned over the wavenumber region of interest. The instrument measures directly the differential intensity $\Delta I = I_M - I_R$ and this can be displayed as a spectrum. To remove the effects of the variation of intensity with frequency arising from the emitter, the spectra are normally presented in the form of $\Delta I/I_R$, or equivalently as $\Delta R/R$, where R is the reflectivity of the cell, as shown in the examples below.

Evidently, the electrochemical system being investigated must be reversible and rapid if this technique is to give meaningful results. For the case that an IR-active species is formed at E_M, then the intensity of the reflected light will decrease at this potential at frequencies corresponding to infra-red absorption by that species when compared to the reference potential; the converse is true if the IR-active species disappears at E_M. Clearly the primary advantage of the technique is the elimination of any IR absorptions from species that do not change their concentration or coverage at the working potential; furthermore, at least in principle noise may be further reduced by scanning the frequency region a number of times and adding and averaging the results. However, such a process takes a considerable time, and raises the possibility of artefacts arising due to electrode and electrochemical instabilities.

Subtractively Normalized In situ FTIR Spectrometry (SNIFTIRS)

To circumvent the above problems, advantage has been taken of the development of Fourier-Transform (FT) techniques in Infra-Red spectroscopy, which has allowed the acquisition of complete spectra in less than one second. FTIR spectrometers operate with polychromatic light that at no point is analysed by a monochromator. Instead, before the beam is incident on the electrode, it is modulated by a time-dependent interference pattern. The reflected intensity is then measured as a function of time, and by use of Fourier-Transform techniques the complete spectrum can be recovered. The interference pattern is impressed on the light by the use of a Michelson interferometer, in which the incident polychromatic beam is split by a beam splitter, the two separate beams then being reflected respectively by a fixed mirror and a moving mirror before being recombined. The purpose of the moving mirror is to give a time-dependent path-length difference; clearly, if the path length difference for a ray of wavelength λ is $\lambda/2$, or any odd multiple of $\lambda/2$, then on recombining the two beams, light of that wavelength will be extinguished. It follows that the intensity of any wavelength component of the light will show a sinusoidal variation with time as the moving mirror traverses its scan, and the beam will be the superposition of such sine waves corresponding to all wavelengths. Fourier transforming this intensity then allows us to recover the intensities of all the components separately, and this latter process can now be carried out extremely rapidly with modern high-speed maths co-processors in computer-driven spectrometers.

Owing to the great rapidity with which a single spectrum can be obtained, it is normal to co-add and average a number of spectra at each potential, and then directly to subtract such added spectra obtained at the reference potential from those at the working potential; this again allows us to plot $(I_M - I_R)/I_R = \Delta I/I$ vs. wavenumber.

Fig. 5.32 shows as an example, spectra corresponding to a Pt(111)-electrode in aqueous formic acid solution. The reference spectrum is first obtained at a potential of 50 mV *vs.* RHE, since at this potential chemisorption of formic acid takes place extremely slowly on the electrode. Spectra are then measured at a succession of potentials, and the resultant normalised difference spectra, $(I_M - I_R)/I_R$, show negative bands at 2341, 2057 and 1826 cm^{-1}. These bands correspond respectively to dissolved CO_2, and to linearly bonded CO and bridge-bonded CO on the platinum surface, the latter arising from the dissociative chemisorption of HCOOH. If the frequency of the absorptions is compared to that of CO in the gas-phase, 2143 cm^{-1}, there is evidently a substantial lowering of the absorption frequency. Further analysis of the data of Fig. 5.32 is shown in Fig. 5.33; the lower part of this figure shows the cyclic voltammogram of Pt(111) in the potential region 0.05 to 1.1 V *vs.* RHE; a large current is seen, and the features associated with the CV of Pt(111) in pure $HClO_4$ (dotted curve) are almost completely suppressed in formic acid solution. The upper part of Fig. 5.33 shows the integrated peak intensities for linear and bridge-bonded CO and for solution-phase CO_2 as a function of electrode potential, beginning at 50 mV *vs.* RHE. The formation of CO_2 as an end-product clearly sets in well before the removal of CO from

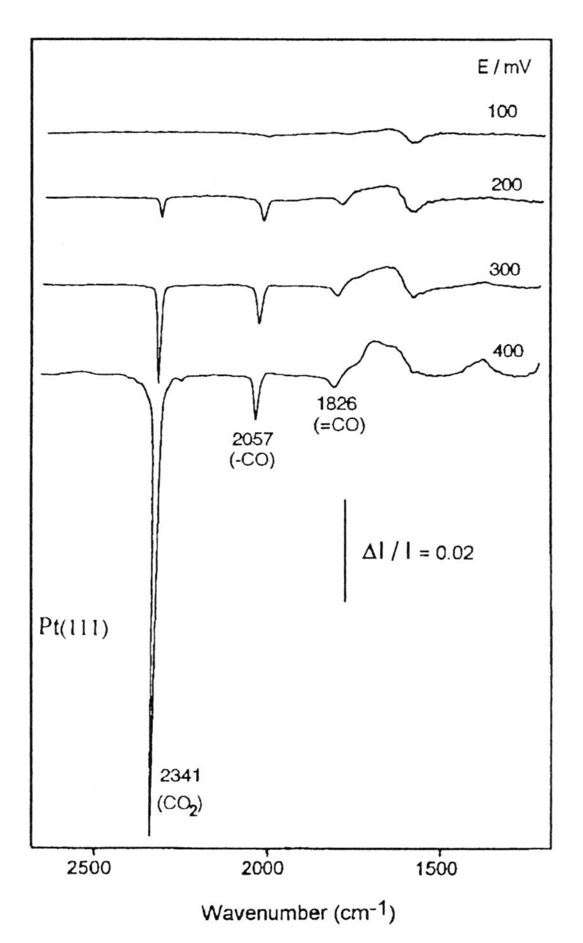

Fig. 5.32 In situ subtractively normalised FTIR spectra (SNIFTIRS) of a Pt(111)-electrode in 1.0 M HCOOH/0.1 M $HClO_4$ taken in a thin-layer cell of the type shown in Fig. 5.31a. The reference spectrum was obtained at 50 mV *vs.* RHE, and spectra were then measured at increasing potentials as shown. For each potential, 250 scans were co-added, with a resolution of 8 cm^{-1}.

the surface, suggesting a parallel reaction mechanism in which CO_2 is formed by one mechanism and CO_{ads} by another.

If one were to carry out the experiment in reverse, by first measuring a reference spectrum at e.g. 400 mV vs. RHE, and then measuring the spectrum at a working potential of 50 mV *vs.* RHE, the CO adsorbed at 400 mV is not lost from the surface on lowering the potential, and it might be expected that under these circumstances no

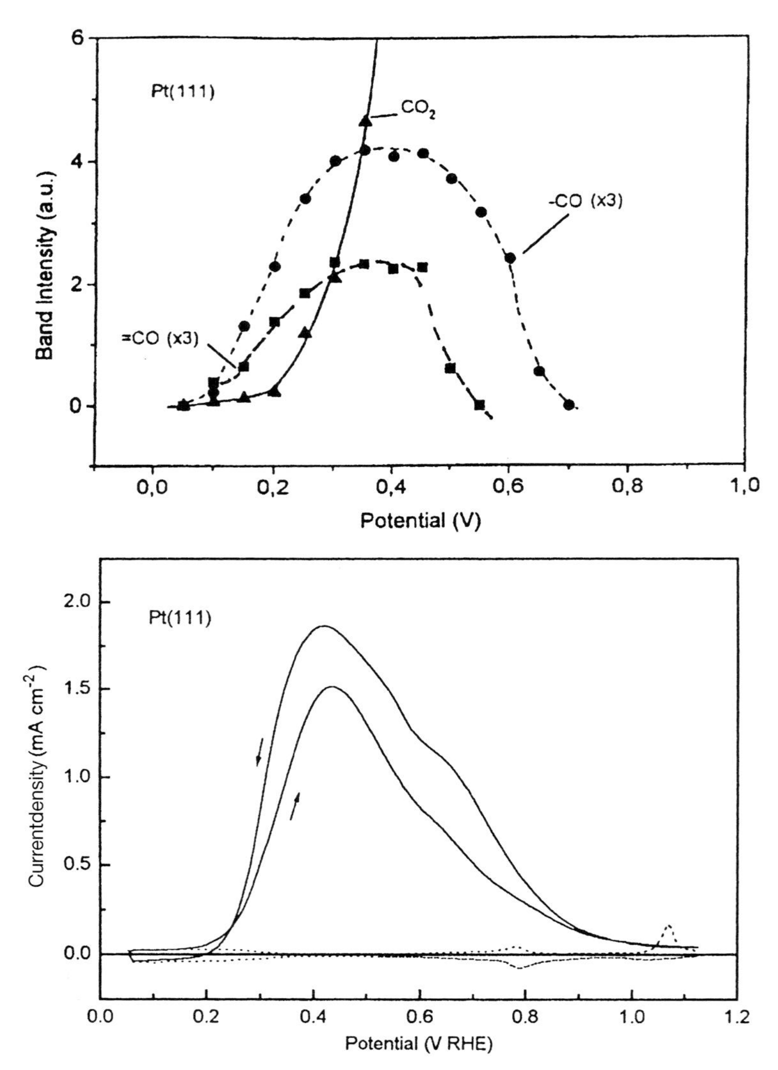

Fig. 5.33 (Lower part) Cyclic voltammogram of Pt(111)-electrode in 0.1 M HCOOH/0.1 M $HClO_4$ at a scan rate of 50 mV s^{-1}; (upper part) plot of integrated band intensities for adsorbed CO (linearly and bridge bonded) and for solution-phase CO_2 in the course of cyclic voltammograms on the Pt(111)-electrode.

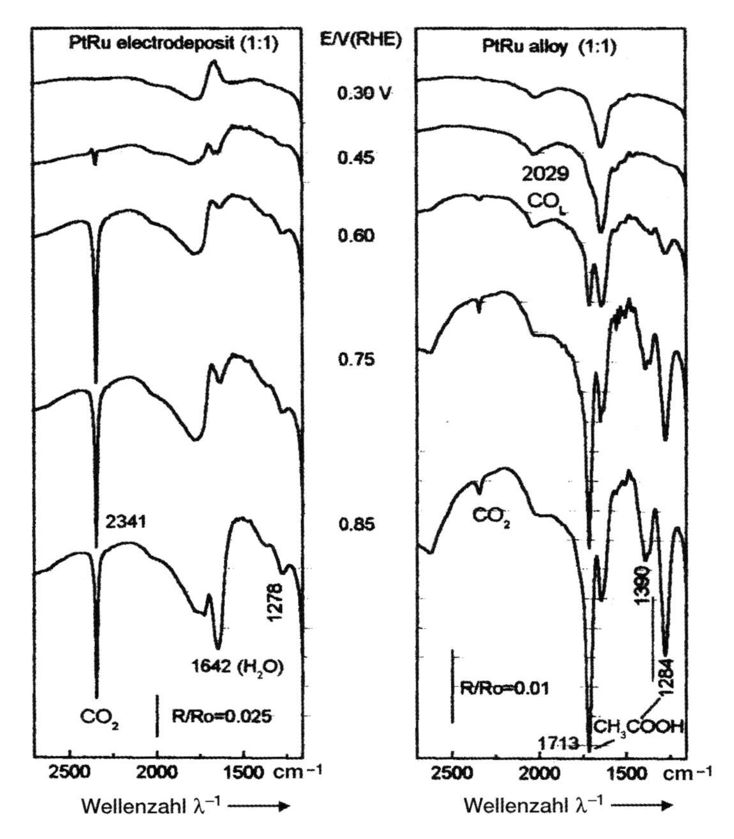

Fig. 5.33a In situ FTIR spectra for the oxidation of 0.1M C_2H_5OH in 0.1 M $HClO_4$ at different potentials *vs.* RHE at 20°C; the left-hand figure shows a (1:1) Pt-Ru electrodeposit on gold; the right-hand figure shows a smooth (1:1) Pt-Ru alloy. The large differences in the relative band intensities for CO_2 (2341 cm^{-1}) and acetic acid (1284, 1390 and 1717 cm^{-1}). The band near 1700 cm^{-1} originates from both acetic acid and acetaldehyde [T. Iwasita, J. Braz. Chem. Soc. 13 (2002) 401].

spectrum would be seen. In fact, through the interaction of the molecular dipole of the CO with the electric field in the Helmholtz layer, there is a shift in absorption frequency with electrode potential, higher frequency absorption being found at more anodic potentials, and a *bipolar signal* for the linearly bonded CO results.

One striking feature of the CV in Fig. 5.33 is the marked reversibility of the current-potential variation. On account of the relative concentrations of reactants, the maximum in the current-voltage plot cannot be sensibly dependent on mass transport effects (as we saw above in Fig. 5.11), and it has been suggested that the fall off in current above 500 mV is predominantly due to the adsorption of water molecules. The water molecules become more strongly adsorbed at higher potentials and hinder the approach of the more weakly adsorbed HCOOH molecules.

Another example of FTIR measurements in the area of electrocatalysis is the investigation of the reaction mechanism for the oxidation of ethanol in acid solution on a Pt-Ru surface. Fig. 5-33a shows in situ FTIR spectra obtained through stepwise increase in the electrode potential (measured vs. RHE) measured in one case for a smooth 1:1 Pt-Ru alloy and in the other case for a porous Pt-Ru layer consisting of small clusters of the alloy deposited on a gold underlayer. A striking feature here is the extraordinary alteration in the relative intensities of the signals derived from CO_2 on the one hand and acetic acid/acetaldehyde on the other. It is clear that CO_2 dominates for the porous catalyst, suggesting that in this case, in contrast to the smooth Pt-Ru catalyst, the primary reaction mechanism involves scission of the C-C bond, and that the porous catalyst has a much higher activity for this process.

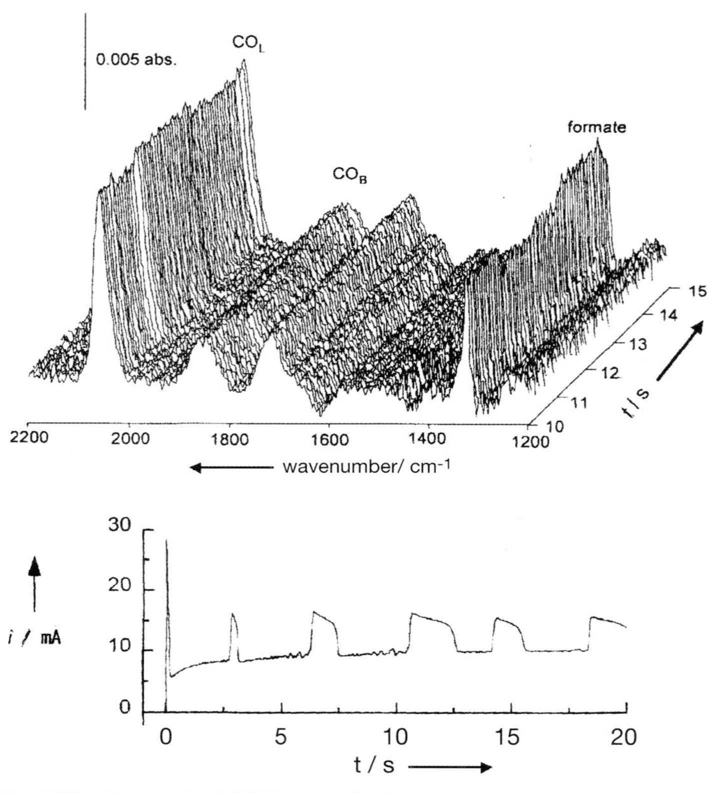

Fig. 5.33b Time resolved SEIRA spectra for the in situ observation of the reaction dynamics of current oscillations during formic acid oxidation at 1.10 V vs. RHE (see the inset). Time IR bands are given for linearly bonded CO (CO_L, ν ca. 2060 cm^{-1}), bridge bonded CO (CO_B, ν ca. 1850 cm^{-1}) and the O-C-O stretching mode of formate (ν ca. 1320 cm^{-1}). Reference spectra at 0.05 V in pure 0.5 M H_2SO_4 [from G. Samjeske, A. Miki, S. Ye, A. Yamakata, H. Okamota, and M. Osawa, J. Phys. Chem. B, 109, 23509–23516 (2005)].

Surface-Enhanced Infrared Absorption Spectroscopy (SEIRAS)

Using attenuated-total-reflection (ATR) spectroscopy as presented in Fig. 5.31b, a relatively high surface sensitivity can be obtained. This technique, therefore, offers the advantage of obtaining a time resolution of different spectra down to 0.1 s.

An example of this is shown in Fig. 5.33b. The figure shows SEIRA spectra, taken during current oscillations (see the inset) associated with formic acid oxidation on a ca. 50 nm film of polycrystalline Pt, deposited on a reflected plane of a Si ATR prism by electroless deposition. The oscillations (compare chapter 6.7) started after stepping from 0.05 to 1.10 V versus RHE in a 1 M HCOOH/0.5 M H_2SO_4 solution.

The in situ observation of the reaction dynamics shows that the formate band intensity (band at ca. 1320 cm^{-1}) decreases first, while the band for linearly bonded CO (ca. 2060 cm^{-1}) is slowly increasing.

5.4.3 Electron-spin Resonance

Electron-spin Resonance is used for the detection and identification of radical intermediates in the neighbourhood of the electrode surface. We consider firstly the basis of the technique and the experimental modifications necessary for electrochemical investigation before considering examples of its use.

5.4.3.1 Basics

A free electron possesses a magnetic dipole moment

$$\mu = g\mu_B m_S \tag{5.60}$$

where m_S is the spin quantum number, g is a number close to 2 for a free electron and μ_B is the Bohr magneton, of magnitude $\frac{e_0 h}{4\pi m_e} = 9.274 \times 10^{-24}$ JT^{-1}. The energy of this dipole in an external magnetic induction, B, is

$$E = -\mu B = +g\mu_B m_S B \tag{5.61}$$

The difference in energy, therefore, between the electron in the spin quantum number state 1/2 and −1/2 is

$$\Delta E = h\nu = g\mu_B B \tag{5.62}$$

and we can see that the degeneracy of the spin is raised by the magnetic flux. Transitions between the two levels can be induced by radiation of frequency ν in the above formula; for a field strength of 0.4 T (Tesla, 1 Tesla = 10^4 Gauss), this corresponds to electromagnetic radiation of frequency ~ 10 GHz, which lies in the microwave region. The basic electron-spin resonance (esr) experiment thus involves placing the sample in a microwave resonator (the source) between the poles of a magnet and adjusting the magnetic flux until absorption of microwaves by the sample takes place.

In the field of chemistry, free spins are usually encountered in the form of unpaired electrons in organic radicals or transition-metal complexes. In such compounds, the

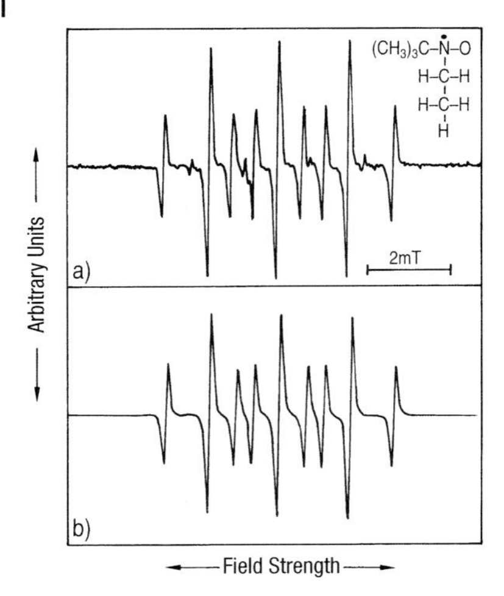

Fig. 5.34 ESR-spectra for the *N*-ethyl-tert-nitrosobutane radical. Upper curve: experiment; lower curve: theoretical simulation [from A. Heinzel, R. Holze and J.U. Blum, Z. Phys. Chem. NF, 160 (1988) 11–23].

simple expression (5.62) will break down both through inter-electronic coupling (which affects the value of g) and coupling to the non-zero spins of the surrounding nuclei, which will give rise to a complex and highly characteristic splitting of the esr signal into absorption peaks appearing at different values of the magnetic field. Both the magnitude and the multiplicity of the splitting contain information on the molecular structure of the radical, often allowing unambiguous identification of the species involved, particularly if isotopic enrichment by ^{13}C is permitted. Given the importance of the magnitude of the coupling constants in esr spectra, it is common to present spectra not in the form of absorption curves but as first-derivative spectra, allowing the more accurate determination of the magnitude of the spectral splitting, and giving rise to the characteristic appearance of esr spectra as exemplified in Fig. 5.34. This shows the esr spectrum of N-ethyl-*tert*-nitrosobutane, $CH_3CH_2N(O)C(CH_3)_3$, and the ninefold splitting pattern can be understood here in terms of the interaction of the unpaired electron with the ^{14}N nucleus (nuclear spin $I = 1$), and with two hydrogen nuclei on the neighbouring carbon to the nitroso group (^1H, $I = 1/2$). The proof of this assertion is not straightforward, and it is now common to model such spectra with estimated values of the coupling constants, which can be adjusted until good agreement with the experimental spectrum is obtained.

5.4.3.2 Electrochemical Electron Spin Resonance

In the classical esr experiment, radicals are introduced into the esr cell which is placed in the microwave resonator. It should be realised that polar solvents and normal glassware also absorb microwaves, and the spectroscopic measurements are therefore carried out in small-volume flat cells made from quartz. With this equipment, it is pos-

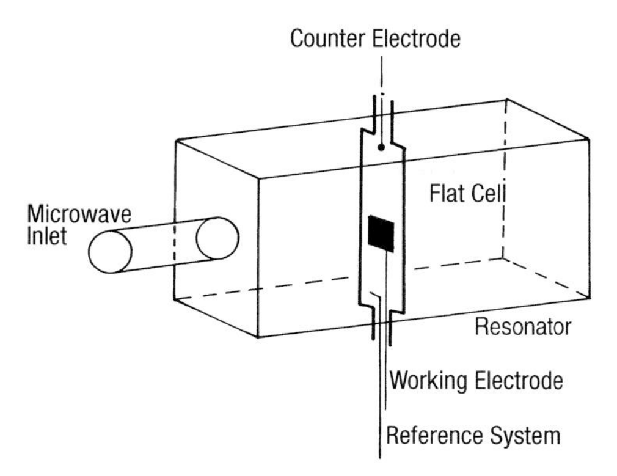

Fig. 5.35 Electrochemical ESR cell [from R. Koopmann and H. Gerischer, Ber. Buns. Phys. Chem. **70** (1966) 118,127; for further developments see, eg, I.B. Goldberg and A.J. Bard, J. Phys. Chem. **75** (1971) 3281].

sible to detect radical concentrations as low as 10^{-9} M. If esr spectroscopy is to be used to study radicals generated electrochemically, then as a first step, the solution from the electrochemical cell could be introduced into the esr cell through a flow system. This is, in essence, an ex-situ measurement and suffers from the disadvantage that many radicals are very short-lived. For the study of short-lived radicals near the electrode surface, the electrochemical cell must be placed within the microwave cavity so that the microwaves can sample the near-electrode volume.

The first esr cells consisted of a quartz capillary, in which, for example, the front surface of a mercury thread formed the working electrode. Subsequently, Gerischer and Koopman described a three-electrode cell, in which a piece of platinum foil placed in the middle of a flat cell served as a working electrode, whilst reference and counter electrodes were placed above and below the resonance cavity as shown in Fig. 5.35.

Owing to the long flat electrolyte path, the cell shows inhomogeneous current flow and poor potential control at the working electrode; later cell designs, for example that of Goldberg and Bard, have improved this, but all such thin flat cells show rapid depletion of electroactive material with time. From the spectroscopic viewpoint as well, such cells are not ideal: the additional cell components disturb the microwave distribution, and calibration measurements carried out with stable paramagnetic ions show that the detection limit, c_{lim} is $10^{-6} - 10^{-5}$ M. From this limit, we can calculate the minimum lifetime of the radical that can be detected. Assuming that the technique is sensitive only to the number of spins and not to their spatial distribution, we can relate the detection limit above to the limiting number of spins by forming the product of c_{lim} and V, the volume of the cell. Typically, V is about 10^{-5} dm^3, so equating the rate of formation of the radicals (i/nF) to the rate of their disappearance (kVc_{lim}, where the lifetime $\tau = \ln 2/k$) we find that for $c_{\text{lim}} \sim 10^{-5}$ M and $i \sim 10$ mA, the limiting lifetime will be of the order milliseconds.

Electrochemical esr is not limited to the detection of radical products that form at a specific potential: the time dependent evolution of a radical product following a po-

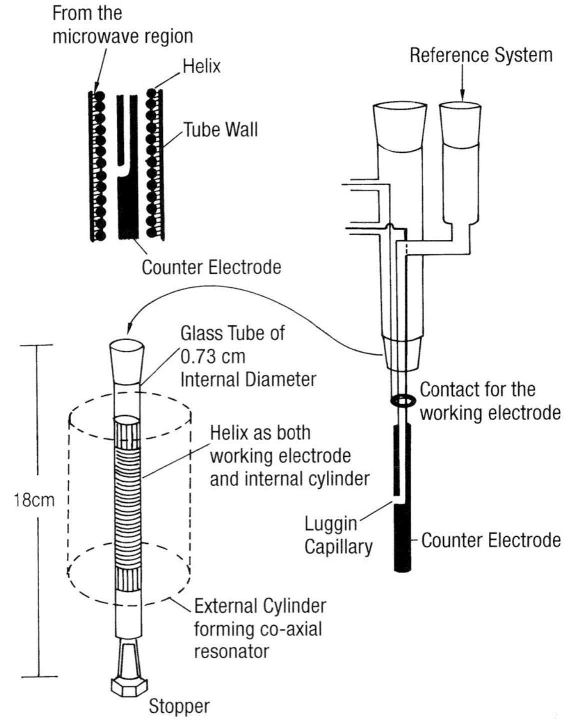

Fig. 5.36 Electrochemical ESR cell designed to sample the greatest possible volume of electrolyte close to the electrode surface (see text).

tential step can be studied, or the radical signal can be determined during potential cycling. It is also possible to follow the decay of the radical after the current is interrupted, allowing the measurement of the radical decay rate constant k defined above. Measurements have also been extended to electrode films, with particularly interesting results obtained from the study of conducting polymers.

The cell designs above are not the optimum possible if we wish to sample the greatest possible volume of electrolyte close to the electrode surface. If this is desirable, then a modified cell can be used, in which the working electrode forms one of the reflecting surfaces of the microwave resonator. This is achieved in practice by having the working electrode in the form of a wire-wound helix, as shown in Fig. 5.36, which is held close to the outer cylindrical wall of the resonator. Inside the helix are placed the counter and reference electrodes, and this geometry allows a reasonable cell volume to be used, reducing substantially the risk of depletion of electroactive material, whilst maximising the near-electrode volume sampled by the microwaves.

If the lifetime of the radicals generated in the experiment is less than the critical level calculated above, it may still be possible to detect them through the use of spin traps, which are chemical compounds that react with the radical generated electrochemically to form a second, more stable radical with a characteristic esr spectrum. Clearly, these spin traps must be stable themselves to electrochemical attack in the potential region being investigated. As an example of the use of spin traps, ethyl radicals have been detected by using tert-nitrosobutane, $(CH_3)_3C–NO$, as the trap during the oxidation of ethanol at a platinum helix electrode in a cell similar to that shown in Fig. 5.36. The radical that forms:

$$(CH_3)_3C-NO + C_2H_5^{\bullet} \rightarrow (CH_3)_3C-N(O)-C_2H_5^{\bullet} \tag{5.63}$$

has already been discussed above (see Fig. 5.34). One attraction of *tert*-nitrosobutane is that it can be generated itself by in situ oxidation of *tert*-butylhydroxylamine

$$(CH_3)_3C-NHOH \rightarrow (CH_3)_3C-NO + 2H^+ + 2e^- \tag{5.64}$$

and this has the advantage that in the investigation of oxidation reactions, it is possible to generate both radical and trap in the near- electrode region, increasing the sensitivity of the technique. This was exploited in the detection of the formyl radical, $CHO^{\bullet}$, during oxidation of methanol at platinum at high potentials (such as 0.95 V vs SCE).

5.4.4 Electrochemical Mass Spectroscopy

5.4.4.1 Basics of Mass Spectroscopy

In the technique of Mass Spectroscopy, atoms, molecules or molecular fragments are identified by the ratios of their masses to the charge z created on the species in the Mass spectrometer. This ratio is usually denoted by the symbol m/e, which has the value 1 for singly charged ^1H (more accurately, 1.007825) and $\sim$ 78 for $^{12}C_6^1H_6^+$. In the classical mass spectrometer, the substance to be identified is evaporated under reduced pressure, and the vapour is passed through an electron beam with an energy of $\sim$ 10 eV, which suffices to ionise some of the molecules in the vapour to ions. These ions are focussed by an electric field in the ionisation chamber and accelerated into the analyser, where they are separated by a magnetic flux B. This flux induces the ions to follow a circular path of radius r; if the accelerating electric field has voltage U, the radius of this circular path is

$$r = \frac{(2mU/e_0)^{\frac{1}{2}}}{B} \tag{5.65}$$

and if the analyser geometry only permits ions of radius r_{crit} to pass, then

$$\frac{m}{e_0} = \frac{B^2 r_{crit}^2}{2U} \tag{5.66}$$

Thus, by varying U or B, ions of differing m/e_0 will arrive at the detector, which is usually a secondary-electron multiplier for optimal sensitivity.

Classical electron-beam ionisation has the property that the particles, once ionised, usually fragment into many fractions. This is particularly true of organic molecules in the medium mass range, where complex spectra result in which the mass peak due to the unfragmented molecular ion (the molecular peak) may be missing completely.

This especially complicates the analysis of mixtures. Other ionisation processes can reduce or even eliminate this disadvantage: field-desorption, for example, in which adsorbed material is ionised under careful control from a surface by a strong inhomogeneous ($\sim 10^6$ V cm^{-1}) electric field, and appears predominantly as the molecular ion, allows the relatively easy assignment of the spectra of mixtures. Other possibilities for ionisation include photoionisation, ionisation through chemical reaction, through collisions with fast atoms or simply through contact of the material with a hot metal band.

An important characteristic in mass spectroscopy is the mass resolution, $m/\Delta m$. A resolution of 1000 means that ions of mass 1000 can just be separated from those of mass 1001, and a resolution of this magnitude is normally sufficient to identify ions of medium mass in simple cases. However, in more complex cases, since different combinations of C,H,O and N can yield ions of closely similar masses (isobaric ions, such as N_2^+: 28.006148; CO^+: 27.994915; $C_2H_4^+$: 28.031300...), a higher resolution is necessary for unambiguous identification. Such higher resolution can be obtained by coupling an additional focussing electric field with the magnetic field of the analyser, which can give resolutions exceeding 50,000. Alternatively, a so-called quadrupole analyser can be used, which consists of four rods arranged with their cross-sections forming a cross. Between opposite pairs of these rods potentials of the form $U + V \cos \omega t$ are applied; it can be shown that for a given choice of parameters U, V and ω, only ions within a very narrow range of values of e/m are transmitted lengthwise in the central part of the cross, the remainder suffering wide amplitude oscillatory motions leading to their loss from the collector. The resolution of the quadrupole system increases with the length of the rods, with several meters being necessary for resolutions in excess of 10,000. However, even if such resolutions are not essential, the quadrupole analyser forms an extremely cost effective alternative.

5.4.4.2 **The Coupling of the Electrochemical Experiment to the Mass Spectrometer**

The critical factor is the sensitivity of the complete experiment: if the charge passed at the electrode is Q, the current yield for the detected species a, and the number of electrons per molecule produced is n, then the current signal at the mass spectrometer detector, Q_{MS}, has the value

$$Q_{MS} = K \cdot (aQ/n) \tag{5.67}$$

where the transfer factor, K, contains the losses associated with desorption and mass transfer in the electrochemical cell, material transfer to the spectrometer and losses in the spectrometer itself. K can be measured, and values of 10^{-8} are attainable. Since Q_{MS} values of 10^{-14} are measurable, this implies that values of $\sim 10^{-6}$ C for Q will lead to detectable signal, which for an electrode of 1 cm^2 corresponds to $\sim 1\%$ of a monolayer.

5.4.4.2.1 Ex-situ Measurements

The first, rather unsuccessful measurements used the field-desorption principle, in which the emitter serves first as the electrode and then as the ion source. This produces electrodes of a rather unfavourable form from the electrochemical perspective, and a considerable improvement can be effected by using thermal desorption, in which the electrode can be of arbitrary geometry. The experimental set-up is shown in detail in Fig. 5.37: the electrochemical cell is shown in the lower part of the figure, and the working electrode is attached to a motorised drive that allows it to be translated from the cell (Position I) to a vacuum lock (Position II) where it can be examined through a viewing window. Once evacuated, the electrode can be further translated to position III, at the entrance to the mass spectrometer, where it can be heated by means of the focussed beam of a projector lamp, with the temperature of the electrode continuously monitored by a thermocouple.

It is evident that the measurements will only be valid if it can be shown that the removal of the electrode from the electrolyte and its subsequent evaporation in vacuo do not destroy the surface adsorbed layer. Evidence for this is provided in Fig. 5.38. The upper part shows a cyclic voltammogram obtained for a platinum electrode after exposure to CO-saturated 0.05 M H_2SO_4 and replacement of the electrolyte with pure

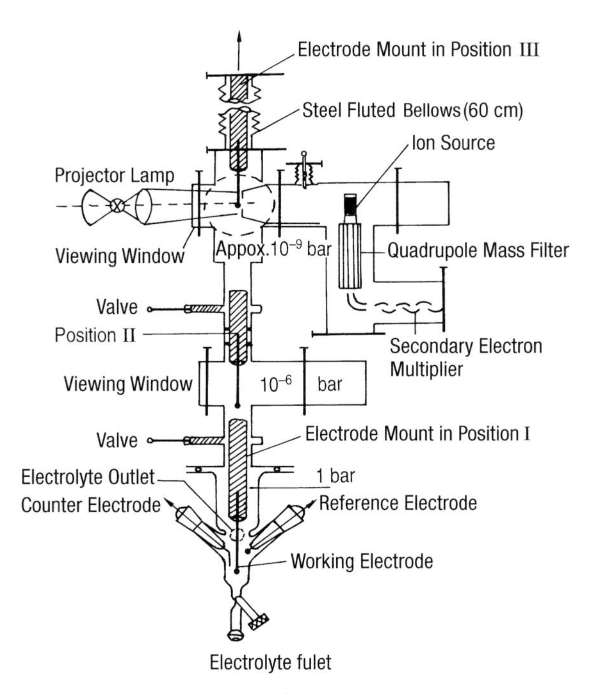

Fig. 5.37 Experimental set-up for electrochemical thermal desorption mass spectrometry (ECTDMS) [from S. Wilhelm, W. Vielstich, H.W. Buschmann and T. Iwasita: J. Electroanal. Chem. **229** (1987) 377; B. Bittins, E. Cattaneo, P. Kenigshoven and W. Vielstich in "Electroanalytical Chemistry", vol. **17**, 1991, ed. A.J. Bard, Marcel Dekker, New York].

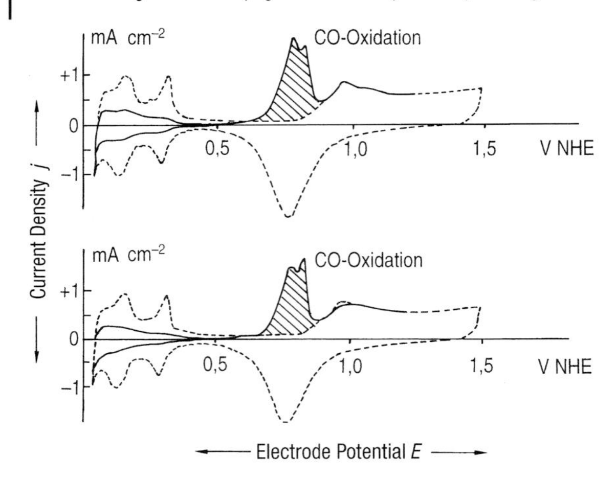

Fig. 5.38 Experimental proof that an adsorbed CO layer is not affected by exposure to high vacuum [see text: from S. Wilhelm et al., loc. cit.].

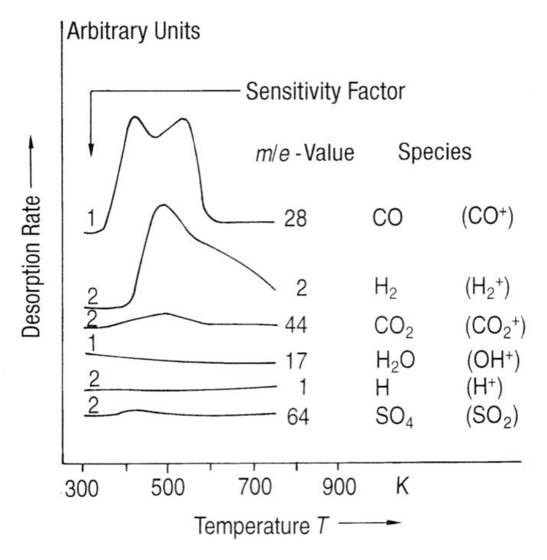

Fig. 5.39 Example of an ECTDMS-investigation, with a heating rate of 5 K s^{-1} [see text: from S. Wilhelm et al., loc. cit.].

0.05 M H_2SO_4. In the lower part of the figure, the same experiment is carried out save that, before the cyclic voltammogram is measured, the electrode is transferred to position III for 10 minutes, and then restored to the cell. Comparison shows that, within the bounds of experimental reproducibility, the two cyclic voltammograms are identical.

An application of the technique is shown in Fig. 5.39, which reproduces the mass spectrum obtained from a platinum electrode exposed to a solution of methanol in 0.05 M H_2SO_4. The intensity of the mass signals for species thermally desorbed from the surface clearly demonstrate the presence primarily of CO and H_2, and it has been suggested that these may arise jointly from an adsorbed species such as ≡C-OH. Other species, such as sulphate anions, can be seen in smaller quantities, and experiments in pure electrolytes show that water and anions can be transported into the mass spectrometer on the electrode surface.

5.4.4.2.2 In-Situ Measurements at Porous Surfaces

The direct coupling of the electrode processes to the mass spectrometer can be achieved by using a porous working electrode both as a wall of the electrochemical cell and as the inlet system to the spectrometer. In this way, volatile material from the electrode surface can be transferred rapidly to the ion source of the spectrometer, and the corresponding mass signal can be measured and correlated with the time axis of the potential variation of the electrode, giving on-line information on volatile products or intermediates of the electrode reaction.

Clearly the joining of the cell to the spectrometer is all important, and this is illustrated in Fig. 5.40. Central is a steel frit to allow the electrode to withstand the pressure difference of ca. 1 bar, upon which is placed a hydrophobic microporous PTFE membrane. On this, in turn, is placed a porous metal or carbon layer that acts as the working electrode, and the current collector is a length of platinum wire pressed onto this layer. The electrode is cleaned by multiple cycling in background electrolyte before admission of the adsorbate solution, and the on-line capability of the experiment allows this cleaning process to be monitored closely.

The counter and reference electrodes are arranged in the bottom of the cell, relatively close to the working electrode of about 1 cm^2, resulting in a working cell volume of 1 to 2 cm^3 of electrolyte. This allows, in turn, a rapid electrolyte exchange in the course of flow-cell experiments. In this latter case, the course of the electrolyte flushing can easily be followed and controlled by the simultaneous observation of the mass signals; it is normally found that flushing with 50 to 100 times the cell volume of fresh solution is necessary.

It is also possible to replace the porous electrode layer by a compact or rotating electrode placed in close proximity to the PTFE membrane (see next section below).

If the intention is to receive mass signals with a time resolution comparable to voltammetric sweeps at the normal rate of near 50 mV s^{-1}, *differential pumping between cell and ion source* is necessary (Fig. 5.40a). Using two turbo-molecular pumps, the total gas volume in the analsyis chamber can be exchanged in the order of 10^{-2} s (Differential Electrochemical MS, DEMS).

One type of experiment is shown in Fig. 5.41, which reproduces once again the oxidation of CO adsorbed from CO-saturated sulphuric acid solution. After changing to CO-free solution, the on-line mass spectrometry clearly shows that the main product is indeed CO_2, and the uncomplicated electrochemistry allows the transfer factor K defined above to be measured with confidence.

A second example shows some further possibilities with the DEMS method. In this case, in order to improve the differentiation of the products of oxidation of ethanol on porous platinum, deuterium-substituted material, C_2D_5OD, is used. In this way, it is possible to detect not only CO_2 ($m/e = 44$) but to distinguish it from deuterated acetaldehyde CD_3CHO ($m/e = 47$), a process not possible for the undeuterated product

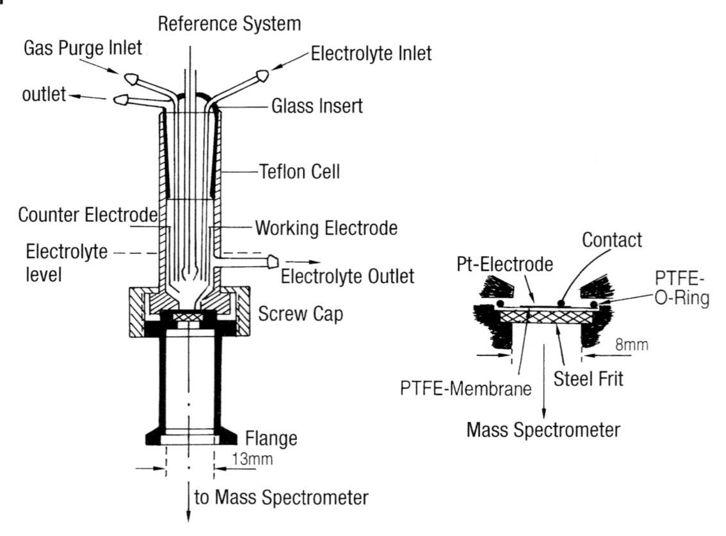

Fig. 5.40 Electrochemical cell with porous working electrode acting as the window of a mass spectrometer [see text: from B. Bittins et al., loc. cit.].

($m/e = 44$). It can be seen from Fig. 5.41a that CO_2 is, in fact, formed only in a relatively narrow potential region, whilst acetaldehyde is the primary reaction product and is formed across the entire potential range studied, closely tracking the anodic current (upper figure).

The increasing interest in methanol fuel cells (see chapters 6 and 9) has motivated experiments to prove that different pathways exist for the oxidation of molecules like

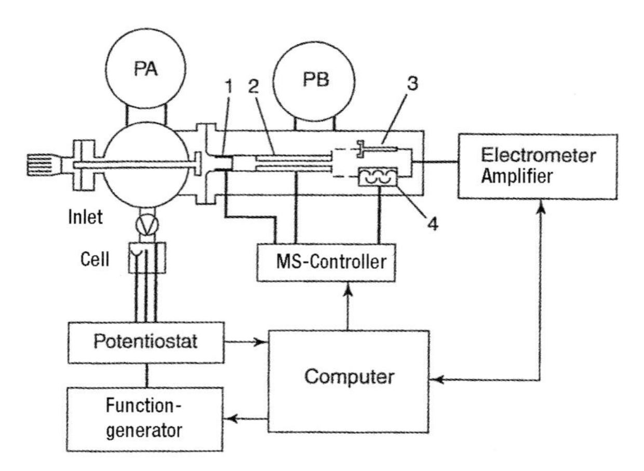

Fig. 5.40a Experimental setup for DEMS with ionization chamber (1), quadrupol mass filter (2), turbo-pumps PA and PB, and Faraday cup (3) or Secondary Electron Multiplier (4) for ion detection.

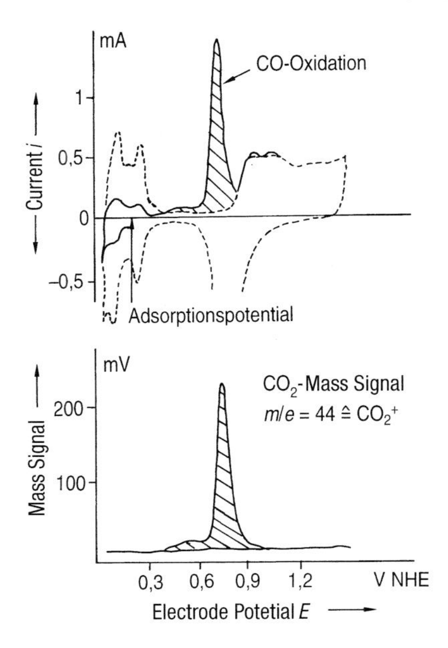

Fig. 5.41 Proof of CO_2 as the main product of oxidation of adsorbed CO during linear sweep voltammetry with a scan speed of 12.5 mV s^{-1} [from O. Wolter, Ch. Giordano, J. Heitbaum and W. Vielstich, Proc.Symp. Electrocatalysis, Vol.82-2, p. 235, El.Chem.Soc. Pennington 1982].

methanol and formic acid. With the help of labelled formic acid, $^{13}COOH$, in a DEMS experiment, the *dual pathway of formic acid oxidation* can be demonstrated:

(a) $HCOOH \rightarrow CO_2 + 2H^+ + 2e^-$

(b) $HCOOH \rightarrow CO_{ad} + H_2O;\ H_2O \rightarrow OH_{ad} + H^+ + e^-$,
and $CO_{ad} + OH_{ad} \rightarrow CO_2 + H^+ + e^-$

The existence of these two pathways in demonstrated in Fig. 5.41b.

After cycling the Pt DEMS electrode in the background electrolyte 0.5 M H_2SO_4, the solution was changed to 0.04 M $H^{13}COOH$/0.5 M H_2SO_4 under potential control at 0.2 V *vs.* RHE for 3 min. During this time ^{13}CO is adsorbed at the Pt surface. Then was followed an exchange of solution to $H^{12}COOH$/0.5 M H_2SO_4 and a potential scan was started in the anodic direction (upper part of figure). The mass signal for the oxidation of the adsorbed ^{13}CO, $^{13}CO_2$ ($m/e = 45$), begins *only above* 0.5 V, while the signal for $^{12}CO_2$ ($m/e = 44$), formed from bulk $H^{12}COOH$, begins much earlier (lower part of the figure). This experiment shows that before the oxidation of CO, formed as ^{13}CO or as ^{12}CO, a current is obtained. The current can be only due to a second pathway like the above *direct* pathway (a).

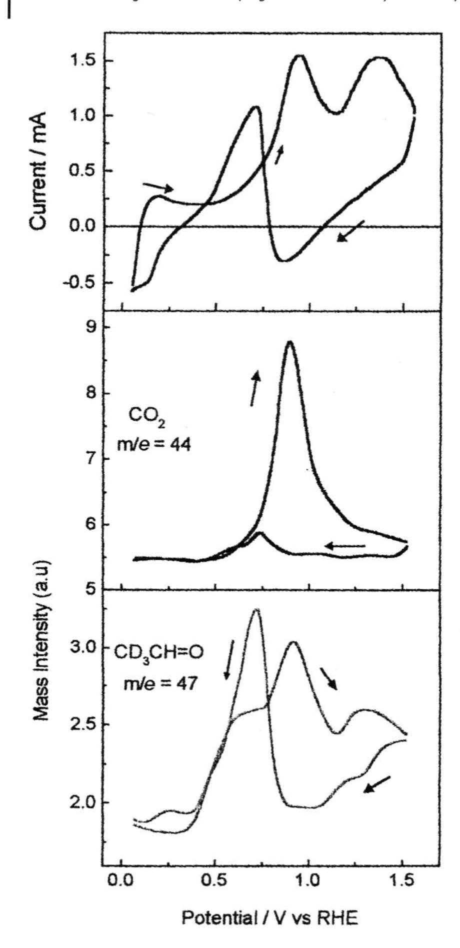

Fig. 5.41a DEMS product analysis for the oxidation of ethanol; *upper figure*: CV of porous platinum in 0.01 M C_2D_5OD + 0,05 M H_2SO_4, 20 mV/s; central figure: corresponding evolution of the mass signal (MSCV) for CO_2, $m/e = 44$; *lower figure*: MSCV for deuterated acetaldehyde CD_3CHO, $m/e = 47$.

5.4.4.2.3 In-situ Measurements at Smooth Surfaces

The online measurement of volatile substances is not limited to porous gas-permeable electrodes. For solid compact electrode materials it is preferable to use rotating electrodes to obtain sufficient mass transport. An example is illustrated in Fig. 5.42, which shows a cylindrically formed electrode rotating close to the window of the mass spectrometer, which is formed from a membrane and frit. This arrangement also permits the study of electrochemical reactions accompanied by strong gas evolution, where investigation on porous electrodes would run the risk of particle dissolution.

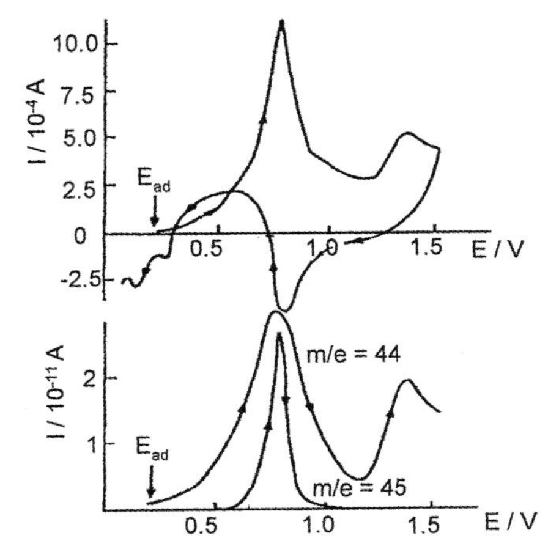

Fig. 5.41b Formic acid oxidation at porous Pt. Upper figure: CV after adsorption at 200 mV; lower figure: potential scan up to 1.5 V with on-line MS for $m/e = 44$ and 45, after adsorption of CO out of a $H^{13}COOH$ solution and solution exchange to $H^{12}COOH$. Scan rate 12.5 mV s^{-1}.

For those cases where it is not possible to machine the electrode into a cylinder, such as a single-crystal substrate, a planar rotating disk can be used as the electrode. This is illustrated in Fig. 5.43, which shows a meniscus of electrolyte adhering to the disc electrode directly above the membrane window of the mass spectrometer; the meniscus is stabilised by the rapid rotation. Some early measurements on a polycrystalline platinum disk, rotating at 50 Hz, are shown in Fig. 5.43a; in this case, the distance between the working electrode and the PTFE membrane was 0.4 – 0.5 mm.

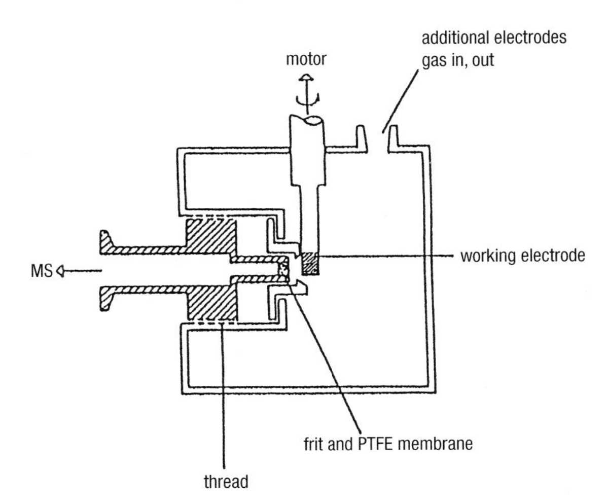

Fig. 5.42 Cylindrical electrode rotating close to the entrance window of the mass spectrometer.

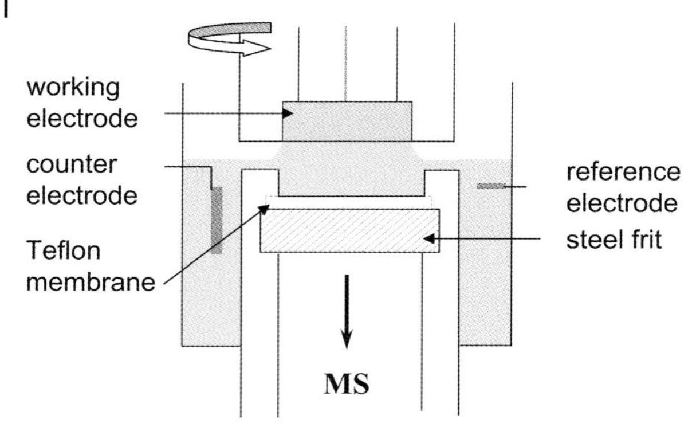

Fig. 5.43 Planar rotating disk electrode with meniscus contact to the electrolyte.

The CV in Fig. 5.43a is different from the CV for HCOOH in Fig. 5.41b. This is mainly caused by the different activities of smooth and porous Pt surfaces. The current density (surface area in both cases near 1 cm^2) is almost the same due to the much better mass transfer at the rotating electrode. And the ion current also is in the same order of magnitude in both cases.

Another attractive possibility for measurements on single-crystal electrodes can be realised by the use of a *dual thin-layer flow-through cell*. The principle is shown in Fig. 5.44. The cell consists of two separate compartments, the electrochemical com-

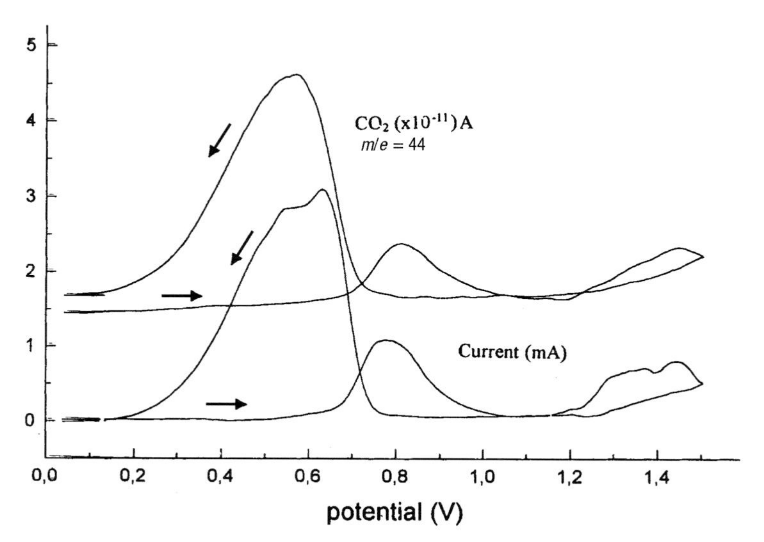

Fig. 5.43a Oxidation of formic acid on a rotating platinum disk electrode over a meniscus contact; 0.1 M HCOOH/ 0.5 M H_2SO_4. Rotation frequency 50 s^{-1}, scan speed 10 mVs^{-1}. On-line MS with $m/e = 44$ (for $^{12}CO_2$).

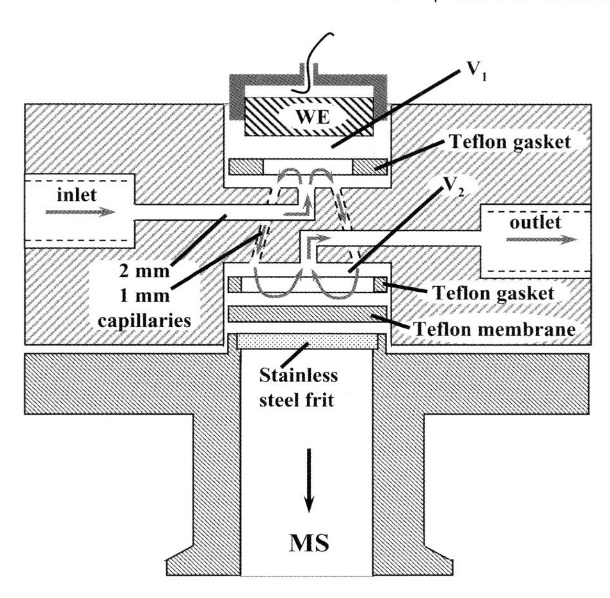

Fig. 5.44 Dual thin layer cell. Electrolyte flow from volume (V_1) at the working electrode WE via 4 channels to volume (V_2) and the MS entrance. On the basis of a model from H. Baltruschat and Z. Jusys.

partment V1 with a flow inlet near the working electrode, and the MS compartment V2 with the electrolyte outlet near the porous membrane at the MS inlet. Both parts are connected e.g. by four 1 mm capillaries. Suitable are working electrode diameters of ca. 1 cm and a liquid flow of 5 microliters per second.

The volumes V1 and V2 have to be so small, that time to move an electrolyte volume from the working electrode to the MS entrance is in the order of one second only.

It has been shown that the dual flow through cell can be used for the study at *thin-film porous electrodes* also. This is of help to avoid mass transfer limitations. In addition, Behm and coworkers have demonstrated that using a thin layer flow cell combinations of internal and external (ATR) FTIR with DEMS are possible.

5.4.5 Additional Methods of Importance

There are numerous other methods that have found considerable application in electrochemistry, particularly in the detection and identification of adsorbed layers on electrodes. Some of the most important of these methods are summarised below.

5.4.5.1 Radiotracer Methods

In this technique, the electrode is exposed for a pre-determined time to a solution spiked with an appropriate radioactive isotope, such as ^{14}C (a weak β-emitter). After removal and rinsing, the intensity of emission, I, of the electrode surface is determined by a suitable counter. For the determination of the coverage, θ, the emission intensity of the clean surface, I_c, and the covered surface, I_{max}, must also be determined,

$$\theta = \frac{I - I_c}{I_{max} - I_c} \qquad (5.68)$$

This method has also been adapted for in situ studies: the counter window is covered with the thinnest possible metal foil which forms the electrode. Immersion of the window and electrode into the adsorbate solution can then be used to monitor in situ the adsorption process. Clearly the method can be extended to incorporate rapid electrolyte flow methods as described earlier in this chapter, as well as the use of continuous potential control.

5.4.5.2 **Microbalance Methods**

The direct weighing of an electrode is an obvious method for the determination of coverage by an adsorbate, and has been used both in the gas phase and in the measurement of adsorption from solution. The experiment requires an adsorbate of low solubility and a volatile solvent, and relies on the difference in weight of the *dry adsorbing material*, such as a metal powder, before and after adsorption.

This type of gravimetry is extremely difficult using normal smooth electrode surfaces, but an indirect and elegant attack on weight changes at electrodes in situ is now possible by means of the quartz microbalance. The idea derives from the dependence of the resonance frequency of a piezo-crystal on changes in external mass present on one of the faces of the crystal. By depositing a thin metal layer electrode on one of the quartz crystal surfaces, the adsorption of species can be detected as alterations in the resonance frequency of the quartz. Increases in weight give rise to *decreases* in frequency, so *desorption* can also be followed by measuring the corresponding *increase* in frequency. However, the interpretation of the frequency response of an *electrochemical* quartz microbalance (EQMB) is complicated by other effects, such as changes in density and viscosity of the solvent near the electrode, and the loss of adsorbed water molecules as these are displaced by more strongly bonded adsorbates.

Quartz crystals possess through the piezo-electric effect a resonance frequency

$$f_0 = \frac{1}{2\mathrm{d}} (\mu_c/\rho_c)^{\frac{1}{2}} \qquad (5.69)$$

where d is the crystal thickness, μ_c the shear modulus of quartz (2.95×10^{11} $\mathrm{g cm^{-1} s^{-2}}$), and ρ_c the density of quartz (2.65 $\mathrm{g cm^{-3}}$). For normal wafer thicknesses of the order of 0.2 – 0.3 mm, the resonance vibration region usually lies between 5 and 10 MHz and is excited by an *ac* potential field across gold contacts.

The relationship between frequency and mass changes is given by the Sauerbrey equation [G. Sauerbrey, Z. Phys. **155** (1959) 206]

$$\Delta f = -\{2f_0^2/(\mu_c\rho_c)^{\frac{1}{2}}\}\Delta m \qquad (5.70)$$

where Δm is the mass change per square cm. This simple dependence is only valid if the mass changes lead to no more than a 2% change in the resonance frequency.

The detection limits for the technique are in the region of ng per cm^2, which corresponds approximately to the change expected for the formation of a single upd-layer of metal. It is important that the frequency change can be measured simultaneously with the cyclic voltammogram or other electrochemical study, and the EQMB has proved especially suitable for the study of metal deposition, intercalation in oxide films, redox properties of conducting polymer films and the oxidation of organic molecules.

5.4.6
Scanning Microscope Techniques

5.4.6.1 Scanning Tunnelling Microscopy (STM)

STM was discovered in 1982 by Binning and Rohrer, and is based on the existence of a so-called tunnelling current, i_t, between two metals placed very close together, but not in electrical contact, in a vacuum. If a potential, V_b, is maintained between the two metals, and the distance apart, d, is of atomic dimensions, the tunnelling current is given by

$$i_t \sim CV_b e^{-2\kappa d} \tag{5.71}$$

where C is proportional to the product of the density of states at the Fermi level of the two metals and κ is a function of the electronic work functions of the two metals, the form of the function corresponding to the height of the potential barrier through which the electron must tunnel. The tunnelling process itself is essentially quantum-mechanical in origin, and arises because the wave functions of the electrons in the two metals show a small but finite spatial extension into the region between them. The tunnelling current is extremely sensitive to d, decreasing by about a factor of ~ 3 for every additional Å separation.

The central idea of the microscope is to exploit this sensitivity to the metal separation by forming a fine metallic tip and rastering this tip over the electrode surface. The control of the tip is through a set of piezoelectric elements (which are formed from crystals whose dimensions alter slightly as a function of the presence of an electric field), allowing the tip to be moved both laterally and vertically. As the tip is moved laterally over the surface, and encounters areas of different height, the tunnelling current alters; in the most common instrumental arrangement, this alteration is incorporated into a feedback circuit such that the vertical position of the tip is also altered so as to restore the tunnelling current to its pre-set value. Alternatively, instead of the 'constant current mode' the height of the tip over the electrode surface can be maintained constant ('constant height mode').

The experimental arrangement is shown in Fig. 5.45. The vertical movements or current changes of the tip then constitute a depth profile which can be combined with the lateral rastering to generate a three-dimensional map of the surface of extremely high resolution: vertical displacements of less then 0.1Å are detectable as are

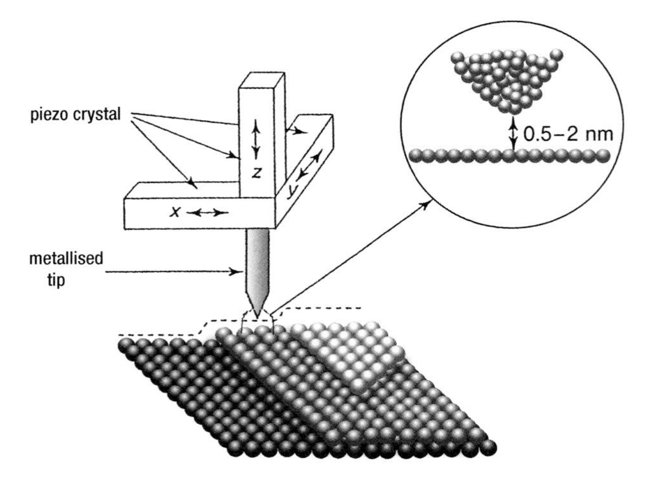

Fig. 5.45 Principles of the STM experiment [L. Kibler and D.M. Kolb: "Structrure Sensitive Methods: AFM, STM" in "Handbook of Fuel Cells" eds. W. Vielstich, H. Gasteiger and A. Lamm, Wiley-UK, Chichester, 2003].

lateral displacements of less than 1Å. Typically scans of the tip over the surface constitute movements of between nanometers and microns, whilst the atomic dimensions of the surface are probed through the piezo-elements (with conversion factors of about 4 nm V^{-1}. This technique has opened up the possibility, for the first time, of studying and monitoring surfaces on the atomic scale, and has been used extensively to investigate solid interfaces.

In practice, the manufacture of the tips is the most demanding technical problem, and is normally effected by electrochemical etching of Pt/Ir (80:20) or W wires; in fact, it is very difficult to generate tip radii of much below micron size, and the efficacy of the technique relies on the presence of single adventitious atoms, or groups of atoms, on the tip through which the tunnelling current dominates. Remarkably, the technique has found application in electrolyte solutions, in spite of the expected screening of the tip-electrode potential, and such processes as metal deposition and dissolution and the formation and dissolution of oxide layers have now been extensively studied. Technical difficulties that have been encountered on the transition to electrolytes have primarily been associated with electrochemical activity of the tip, giving rise to measured currents that contain contributions from both tunnelling and faradaic components. The latter can be minimised, however, either by poising the tip near its rest potential or by coating all but the very end of the tip with electrochemically inactive varnish.

A good example of the type of resolution and detail obtainable in STM is shown in Figs. 5.46 and 5.46a. The first figure shows the cyclic voltammogram of single-crystal Au(111) in 0.1 M H_2SO_4 at 10 mVs^{-1}, showing two sharp peaks associated with the adsorption of anions and a broad peak extending from ca. 0.2–0.6 V associated with the structural transition of the originally reconstructed Au(111) surface into the unreconstructed (1×1) surface. The sharp peak at ca. 0.7 V is observed to shift

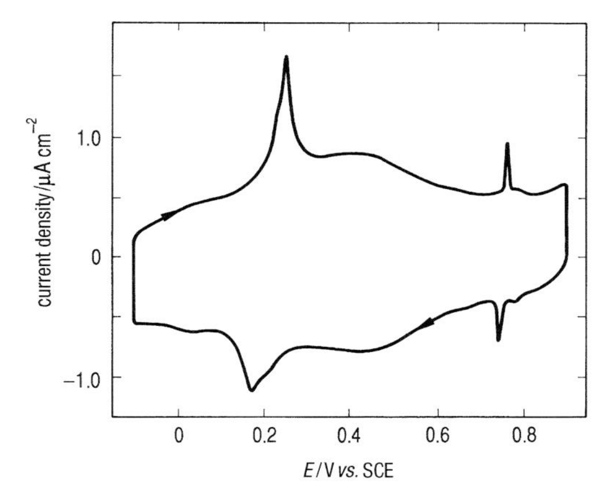

Fig. 5.46 Cyclic voltammogram of Au(111) in deaerated 0.1 M H_2SO_4 with a scan speed of 10 mV s^{-1} [from O. Magnussen, R.J. Behm, Farad.Disc. Chem. Soc. **94** (1992) 330].

by 50 mV per decade change in anion concentration, and is associated with substantial changes in the STM image, as shown in Fig. 5.46a. An analysis of the data shows that a lower potentials, HSO_4^- is present in a mobile and disordered form on the surface, with a rapid and reversible formation of an ordered and much more compact adlayer of these anions at higher potentials. This ordering has been found to occur initially in islands, which then spread and join up to form a continuous structure, the details of which are given in Fig. 5.46b. This figure also compares the structure to that found in solid $H_2SO_4 \cdot H_2O$. Interestingly, this latter structure is stabilised by hydrogen bond-

Fig. 5.46a High resolution STM image of Au(111) in 1M H_2SO_4 (It = 25 nA; area sampled 210 × 120 52) recorded on the same area directly before and after formation of the ordered anion adlayer. In the upper part, at a cell potential of 0.68 V, the Au(111) substrate lattice is visible. After raising the potential to 0.69 V, the superstructure is observed. [from O. Magnussen et al., loc. cit.]

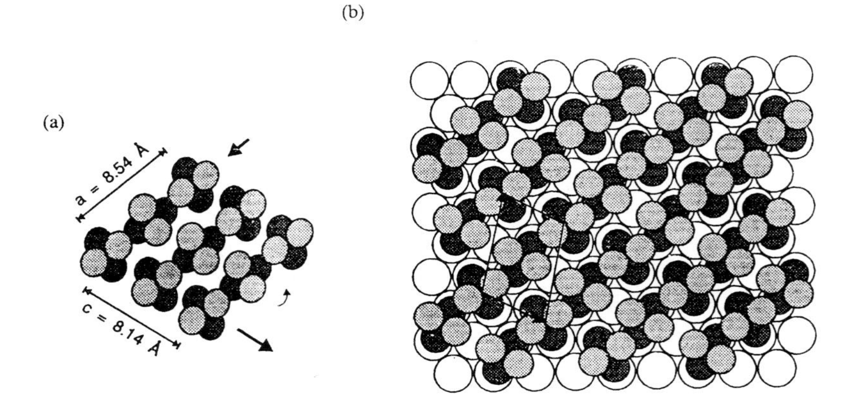

Fig. 5.46b (a) Structure of solid H_2SO_4 projected along the b axis (after R.W.G. Wyckoff: "Crystal Structures", Wiley, New York, vol. **2**, 1963), with the arrows indicating in which way the structure is modified to match the anion adlayer parameters; (b) expanded model of the superstructure showing the arrangement of the anions in the adlayer. Oxygen atoms at the corners of the sulphate tetrahedra pointing towards the surface (towards the solution) are represented by dark (light) circles.

ing, and it seems likely that the energetics of hydrogen-bond formation are comparable to the difference in energy between the different possible adlayer configurations.

5.4.6.2 **Variants on STM**

The invention of STM has led rapidly to the development of a large number of variants, all of which rely on similar technology but exploit different physical effects. Perhaps the most significant electrochemically is Atomic Force Microscopy (AFM), the principles of which are shown in Fig. 5.47. In place of the tunnelling current, the AFM uses the physical contact force between tip and surface as a monitor, the contact force itself arising from electrostatic forces, van der Waals exchange terms and/or incipient chemical binding effects. With the AFM it is also possible to investigate non-conducting surfaces. The tips themselves are normally constructed from Si, SiO_2 or Si_3N_4, for which methods of micro-machining are available, and they are integrated into a micro-fabricated cantilever device which acts as a weak spring and can bend under the action of the tip-surface forces. This deflection is detected and monitored optically by reflection of a laser beam from the back of the cantilever to a position-sensitive detector (see Fig. 5.47) and the lateral dimensions of the cantilever are in the order of 100 μm with a thickness of about 1 μm.

Considering van der Waals forces alone, for a sphere of radius r at a distance s from a surface, integration over all pairwise atomic van der Waals forces gives:

$$\mathbf{F}_{\mathrm{VDW}} = \frac{H\mathbf{r}}{6s^2} \tag{5.72}$$

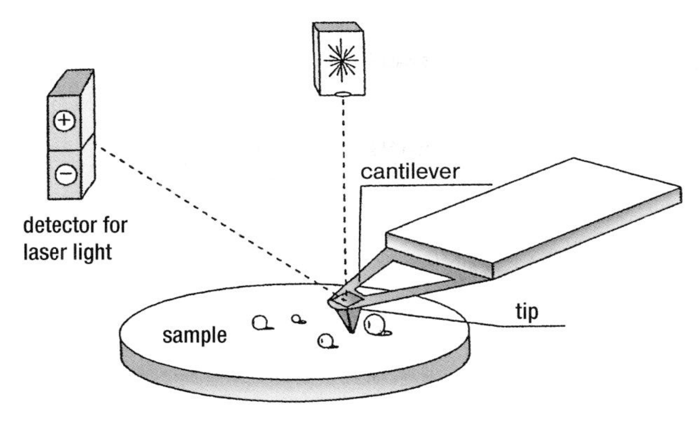

Fig. 5.47 Schematic diagram showing the principle of AFM [N.J. DiNardo, Nanoscale Characterisation of Surfaces and Interfaces, VCH, New York, 1994].

where H is the Hamacker constant, dependent on the materials of tip and surface. The primary data from AFM measurements are lines of constant force, which, provided the surface is homogeneous, can be related directly to the surface topography.

A scanning microscope that is also able to raster across thin insulating surface layers is the *Scanning Kelvin Microscope* (SKM). This device permits measurement of the work function of the surface with a lateral resolution of $2 - 5\ \mu m$ by means of the measurement of the Kelvin capacitance. Through clever use of harmonics, it is possible to measure the surface topography and the work function simultaneously.

A further variant of the STM, replacing the tunnelling current with a faradaic current, is the *Scanning Electrochemical Microscope* (SECM). It uses an ultra-microelectrode (or even the original tip of the STM is completely suitable), which can be moved over the surface of a conducting, or even non-conducting surface, at constant height. The measured current results from electroactive species in solution and is a sensitive function of changes in structure and composition of the underlying surface.

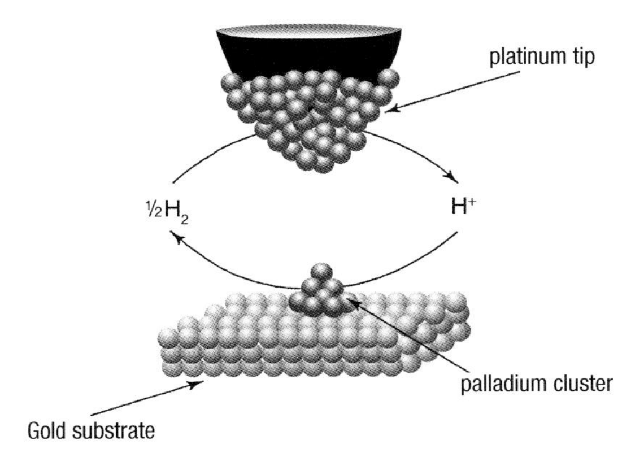

Fig. 5.48 Principle of the measurement of the catalytic activity of a Pd-nanocluster on a Au(111) substrate in an SECM experiment.

One example of the use of SECM is shown in Fig. 5.48. Hydrogen molecules generated on the surface of Pd-nanoclusters electrochemically by reduction of protons in solution can diffuse to the platinum tip of the SECM and there can be re-oxidised to protons. In principle, this method allows us to measure the catalytic activity of these nanoparticles directly. Interestingly, it is also possible to explore the microstructure associated with corrosion effects using SECM. The lateral resolution of the technique is determined by the diameter of the microelectrode used, being a few microns for ultra-microelectrodes and for STM tips as low as 50 –100 nm.

5.5
Preparation of Nanostructures, Combination of STM and UHV-Transfer

5.5.1
Use of an STM-tip in SECM Experiments for the Preparation of Definite Nanostructure

An STM tip arranged in an SECM experiment (i.e. where the tip is held sufficiently far from the surface that no tunnelling current can flow) can for instance be used to prepare cobalt-metal clusters of defined size on a Au(111) substrate from 10^{-3}M Co^{2+}/ 0.5 M H_2SO_4 solution. The tip can be prepared by etching a 0.25 mm gold wire pre-covered with Apiezon wax in 32 % HCl, from which a un-insulated tip of area 10 μm is obtained with a corresponding double-layer charging current smaller than 100 pA at 10 mVs^{-1}. Onto this nanoelectrode *a layer of cobalt metal* of defined thickness is electrodeposited from 10^{-3} M Co^{2+} in 0.5 M H_2SO_4. Following this, the concentration of Co^{2+} in solution close to the underlying gold substrate can be suddenly increased by a positive 800 mV pulse on the tip. Since the potential of the Au(111) surface is held constant, one has in the region close to the STM-tip, a strong electrodeposition of cobalt on the Au(111) surface, as can be seen in Fig. 5.49 (a). The smaller the distance of the tip from the Au(111) surface, the smaller the diameter of the clusters of Co formed, since small tip-surface distances give rise to higher concentration of Co^{2+} near the centre of the nano-electrode. From the known diffusion coefficient for Co^{2+}, the concentration distribution following the potential pulse on the tip can be calculated, and from this the diameter and height of the cluster (see Fig. 5.49 (b) and (c)).

5.5.2
Combination of STM and UHV Transfer

Until recently, UHV systems with STM offered little in the way of direct transfer from the electrochemical cell to the UHV chamber under controlled conditions. However, recently Omicron have offered an electrochemical "mini"-cell as an insert into their UHV equipment. Fig. 5.50 shows a mini-cell used, with a device to permit electrolyte exchange under potential control. The required solution can be introduced through the end of a 20 cm long glass tube of 19 mm diameter, fused to which is a second glass tube whose end contains a platinum wire to act as a counter electrode and a platinum platelet to carry the sample as well as a connexion to the reference electrode. Contact to the electrolyte is through a meniscus, with the working electrode being held above the electrolyte as shown in Fig. 5.50. With this arrangement, it is possible to obtain CVs and current-time transient measurements on a sample pre-prepared in UHV without it encountering the laboratory atmosphere. The STM measurements on a Pt(111) electrode with Ru clusters prepared in the UHV part of the equipment clearly shows the magnitudes and structures of the Ru clusters as can be seem in Fig. 5.51, and with this technique the activity of particular electrodes can be correlated with definite electrode structures (see Section 6.4.6 and Fig. 6.11).

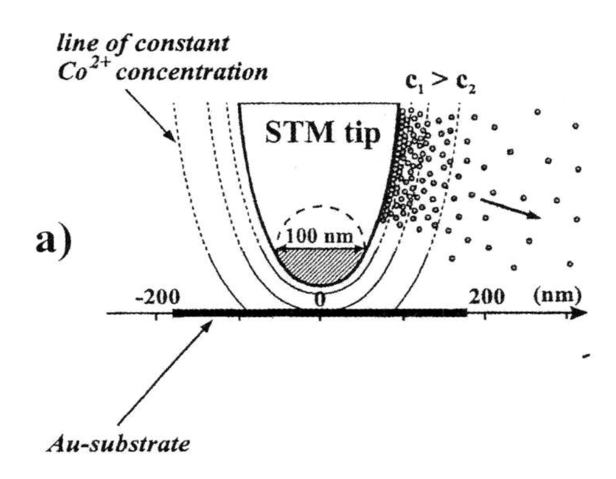

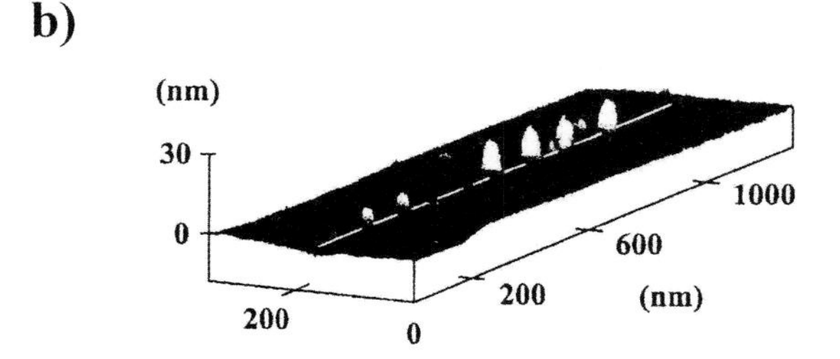

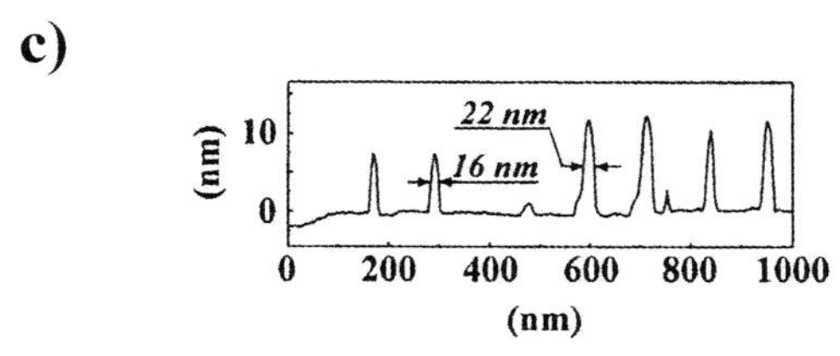

Fig. 5.49 Preparation of Co-clusters on a Au(111) substrate from Co^{2+} solution with the help of an STM-tip arranged in an SECM experimental configuration (*cf* Fig. 5.48). (a) Co^{2+} concentration distribution on anodic dissolution of the Co-layer from the STM-tip, starting concentration of 10^{-3} M Co^{2+} in 0.5 M H_2SO_4. (b) STM image showing the Co-clusters laid down by the technique described in (a); (c) height profile along the white line of (b) [W. Schindler, D. Hofmann and J. Kirchner, J. Electrochem. Soc. **148** (2001) C124].

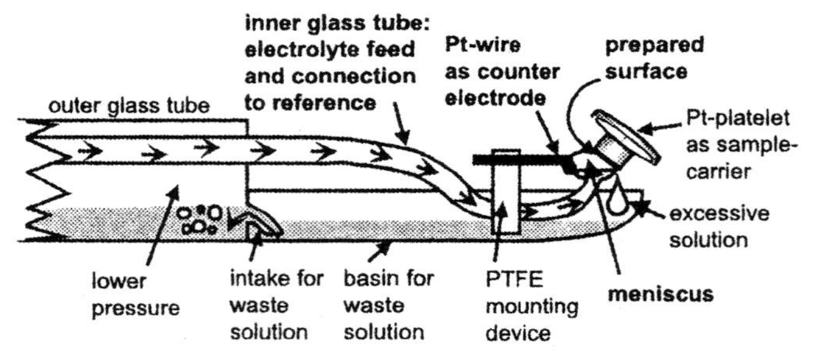

Fig. 5.50 Miniature flow cell for horizontal approach, applicable for many load-lock systems, with possibility of solution exchange and meniscus position for single crystal electrodes. From H. Hoster et al., Phys. Chem. Chem. Phys., 3(2001)337.

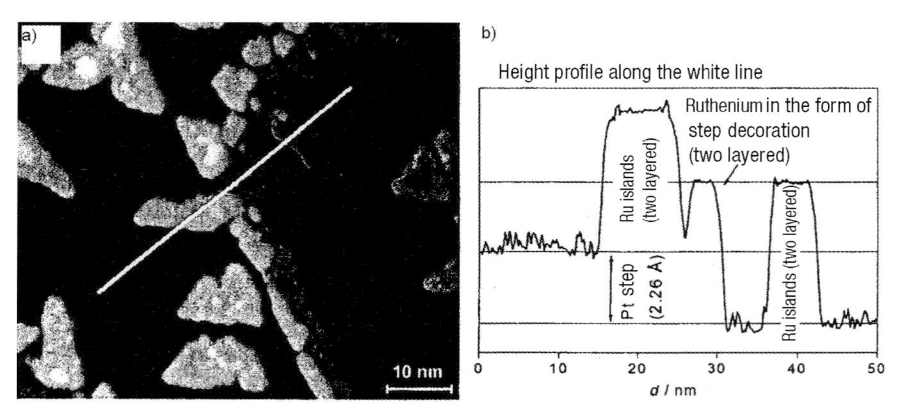

Fig. 5.51 Pt(111) surface with Ru-clusters introduced at 400 K in UHV. (a) STM picture taken at higher cluster density (at times in the form of double monolayers) on a step on the (111) surface; (b) height profile of steps and clusters along the white line in (a) [H. Hoster, T. Iwasita, H. Baumgartner and W. Vielstich, PCCP **3** (2001) 337].

5.6 Optical Methods

These constitute some of the oldest methods of studying electrode surfaces and reactions, and still find application today. The most straightforward is optical absorption spectroscopy, which is frequently carried out in cells with optically transparent electrodes, such as degenerately doped tin oxide. Such materials show both very high conductivity and very little absorption in the visible region of the spectrum, and studies of this type have been extensively carried out to determine the route taken in complex organic, or especially inorganic electrochemical reactions, since the latter species often show suitable chromophores in the 400 – 700 nm spectral region.

The technique has now been extended into the infra-red region, where its usefulness in disentangling organic electrochemical processes is extremely valuable.

One of the most interesting applications of this technique is to the study of electrochemically generated species that are unstable and all but impossible to isolate and study separately. Good examples of this can be found in recent studies of metal clusters stabilised in solution by surface chemisorption of thiolate, phosphine or similar ligands. It is possible, for example, to synthesise clusters of the form $Au_{38}(PhC_2S)_{24}$ where PhC_2SH is the ligand 2-phenylethanethiol and the 38 gold atoms form a nanoparticle of with dimensions of the order one nanometre. Such clusters often show multiple electron-transfer peaks in cyclic voltammetry, as shown in Fig. 5.52a, where up to four electrons can be removed from the cluster without irreversible decomposition. Optical absorption spectra of the neutral species, together with $[Au_{38}(PhC_2S)_{24}]^+$ and $[Au_{38}(PhC_2S)_{24}]^{2+}$ are shown in Fig. 5.52b. These species are stable in CH_2Cl_2 solution but are exceedingly difficult to isolate.

Direct investigation of surface film formation can be carried out most simply through reflection spectroscopy, which exploits the alteration in intensity of light reflected from the electrode surface associated with the formation of an optically absorbing film. To increase sensitivity, it is common to employ multiple reflection techniques; however, even for this arrangement, the sensitivity is poor if the film absorbs only weakly, and such techniques have, by and large, been superseded by ellipsometry.

5.6.1 Ellipsometry

In this technique, the surface is illuminated by light of a fixed, linear polarisation, the direction of polarisation normally being fixed at 45° to the electrode surface. The polarisation state of the resultant reflected light is then analysed. It can be shown that such light will normally be elliptically polarised (that is, the resultant electric field vector of the reflected light would be seen to vary in direction if observed at a fixed point, with the tip of the vector describing an ellipse). This ellipse can be defined by three parameters: the length of the major axis (corresponding to the intensity of the reflected light), the angle of the major axis of the ellipse to the plane of reflection (the azimuthal angle), and the eccentricity (actually the inverse tangent of the ratio of minor and major axes). These latter two angles are influenced by the status of the surface, and can be measured with considerable precision; they are normally converted into two further angles, Δ and Ψ, corresponding to the difference in the change in phase of the light reflected from the surface for light rays polarised parallel and perpendicular to the plane of reflection, and to the arc tangent of the ratio of the magnitudes of the corresponding reflection coefficients. The precision with which Δ and Ψ can be measured allows changes corresponding to fractions of a monolayer to be determined. Such sensitivity also has been exploited in studying the growth of oxide films on metals, and of conducting polymer films, the latter proving particularly useful in elucidating the early processes of growth following potential step initiation.

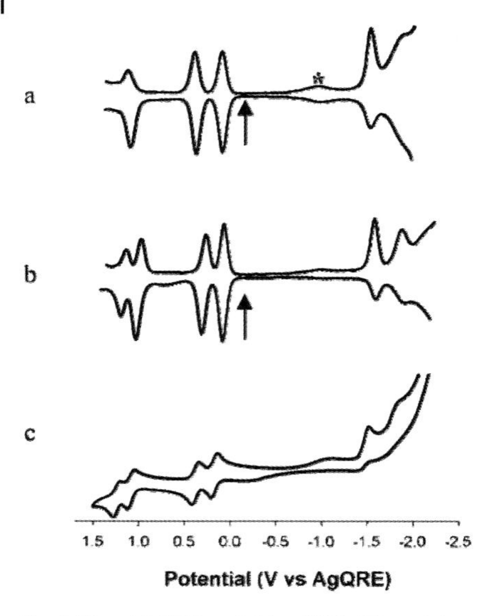

Fig. 5.52a (a) 25°C and (b) −70°C differential pulse voltammograms (DPVs; see section 10.2.1.2) at 0.02 V/s, and (c) −70°C cyclic voltammogram (0.1 V/s) of $Au_{38}(PhC_2S)_{24}$ in 0.1 M Bu_2NPF_6 in degassed CH_2Cl_2 at 0.4 mm diameter Pt working electrode, with a Ag wire quasi-reference electrode (AgQRE, which is commonly used in aprotic solvents), and Pt wire counter electrode. Arrows indicate the rest potentials for each solution, and * indicates the current corresponding to incompletely removed O_2, which varied from experiment to experiment [Dongil Lee, Robert L. Donkers, Gangli Wang, Amanda S. Harper, and Royce W. Murray, *J. Am. Chem. Soc.*, **126** (2004) 6193].

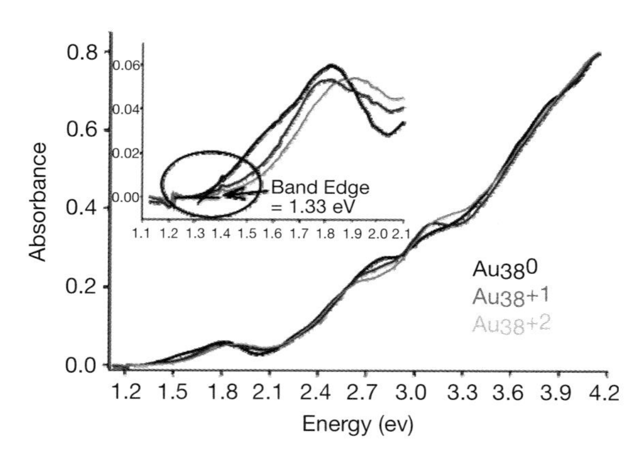

Fig. 5.52b UV-vis spectra (25°C) of (a) $Au_{38}(PhC_2S)_{24}$ (dark line), (b) $[Au_{38}(PhC_2S)_{24}]^{+}$(intermediate line), and (c) $[Au_{38}(PhC_2S)_{24}]^{2+}$ (light-shaded line) in degassed CH_2Cl_2 solution. The three spectra are of the same solution; the 1+ and 2+ charge states were generated by electrolysis in the spectroelectrochemical cell. The insert shows clearly that these small stabilised clusters have a clear optical absorption onset, and hence a clear gap between the highest filled and lowest unfilled molecular orbitals. [*loc.cit.*]

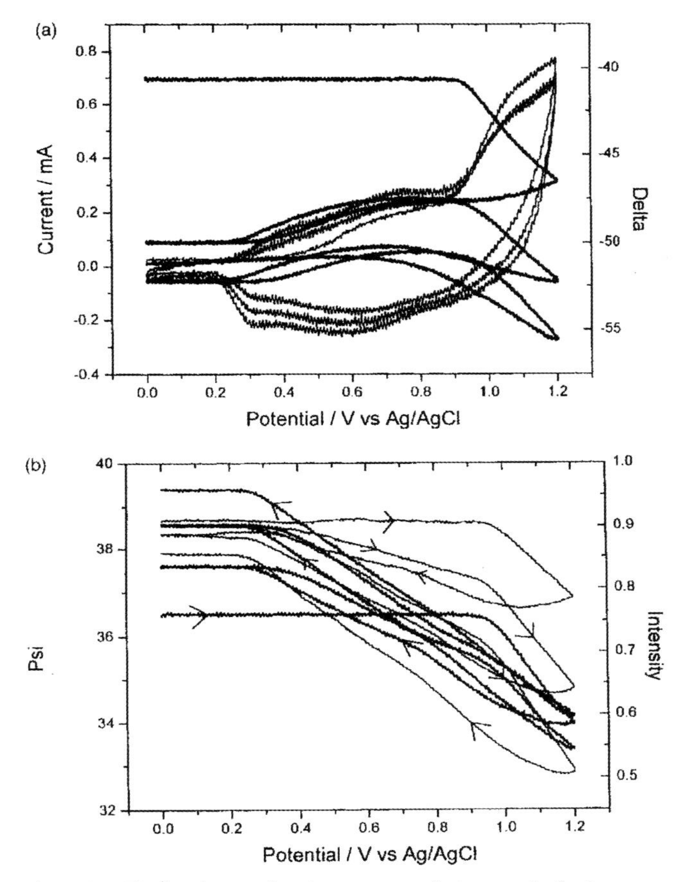

Fig. 5.53a The first three cyclic voltammograms during growth of poly-[Pd(OMeSalen)] showing the variation in Δ, Ψ and intensity. Growth was from a 1 mM solution of Pd(OMeSalen) in 0.1 M $Bu_4N^+BF_4^-$ in acetonitrile. The platinum electrode had an area of 0.64 cm^2 [P.A. Christensen and A. Hamnett, Electrochimica Acta **45** (2000) 2443].

A key issue in ellipsometry is the fact that homogeneous films are characterised by three parameters: the real (n) and imaginary (k) parts of the complex refractive index, $\hat{n} = n - ik$ (where k is related to the optical absorption coefficient α by the expression $\alpha = 4\pi k/\lambda$), and the film thickness, L. However, normal ellipsometric measurements give only two experimental angles, Δ and Ψ. There are commonly two ways in which additional information can be acquired: either by simultaneously measuring the intensity of reflected light or some other optical property of the film at a single wavelength, or by measuring the values of Δ and Ψ at a number of different wavelengths and seeking models that fit these data assuming a constant thickness. This latter approach, termed spectroscopic ellipsometry, has become much more important in recent years as the requirement for the same physical structure of the film, together with

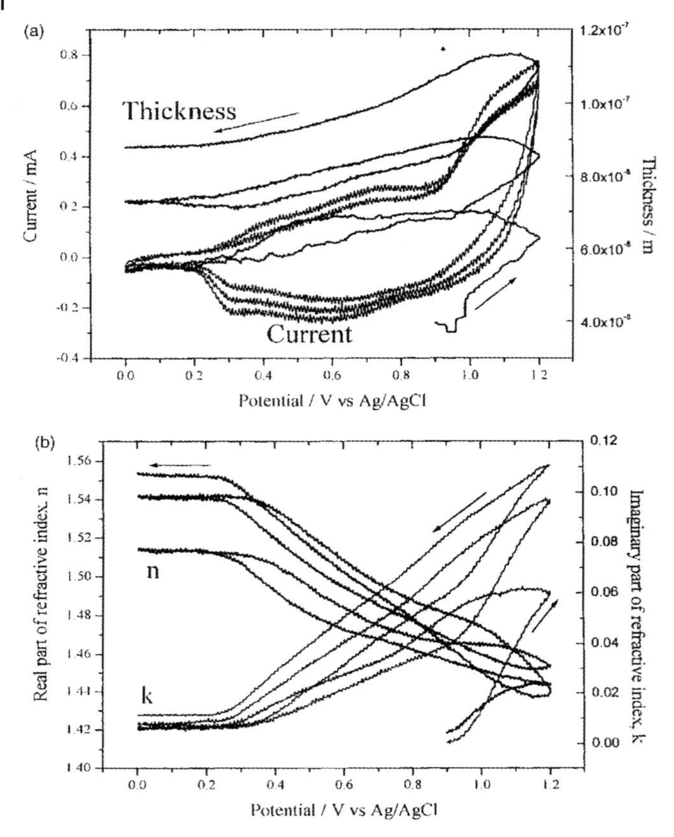

Fig. 5.53b The variation in (a) thickness and (b) real and imaginary components of the refractive index, n and k, for the first three growth cycles of Fig. 5.53a assuming the film to be homogeneous. [*loc. cit.*]

simple models for the variation of n and k with wavelength impose severe restrictions on the possible solutions.

A simple example of the use of modern ellipsometry is shown in Figs. 5.53 and 5.54. As discussed in more detail in section 6.4.5, it is now well known that many simple heterocyclic molecules and metal complexes based on heterocyclic ligands can be polymerised electrochemically, usually by oxidation of the monomer at an electrode surface. These polymeric films have important electrochromic and catalytic properties; their refractive indices change very markedly on potential cycling and the polymers based on metal complexes often retain the catalytic activity of the monomer whilst being immobilised on the electrode. Fig. 5.53a shows the ellipsometric angles Δ and Ψ and the reflected intensity varying as the important metal catalyst Pd(OMeSalen) is polymerised as a film on a Pt electrode by cycling the potential between 0 and 1.2 V from a 0.001 M solution in acetonitrile with 0.1 M ammonium tetrafluoroborate as supporting electrolyte. Fig. 5.53b shows the calculated values of thickness, n and k. It is evident that even at very short times, there is quite a thick film on the surface, but this film initially has substantial amounts of solvent associated with it; subsequent cycles show the film both growing in thickness and consolidating.

Fig. 5.54 shows the variation in n, k and thickness for a film grown over a large number of potential cycles and then removed and immersed and cycled in electrolyte alone. The cyclic voltammogram shows the passage of a large current at more positive potentials, with the total charge passed suggesting that the film is quite compact in the un-oxidised form. On oxidation, the film expands substantially as both solvent and counter-ions enter the structure, k rises and the fall in n is associated with the fact that there is an intense electronic absorption in the oxidised film in the near infra-red. Analysis shows that such an absorption will cause n to fall to very low values

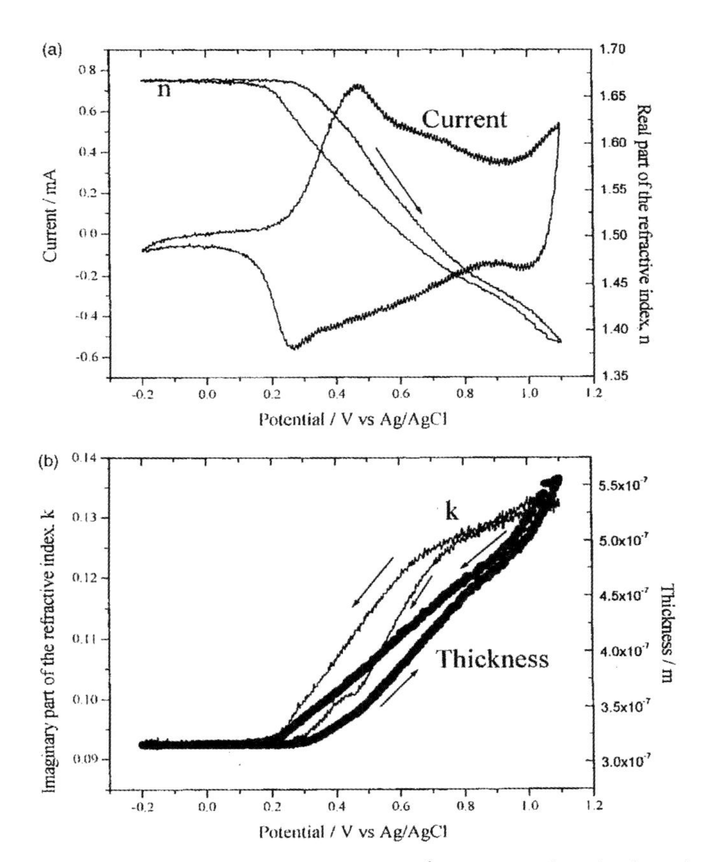

Fig. 5.54 The cyclic voltammogram of a 3000 Å film of poly-[Pd(OMeSalen)] and the variation in thickness and optical properties as the film is cycled. In this model, confirmed by data at a number of different wavelengths, the film is homogeneous when in the reduced state but on oxidation shows a strong variation in acetonitrile content from the electrode surface, where very little solvent penetrates, to the outer part of the film, which becomes very diffuse. The values of n and k given are for the innermost part of the film. The values for the outermost part of the film are almost the same as the solvent. [*loc. cit.*]

in the visible part of the spectrum. More detailed analysis shows that in the oxidised form the film is not homogeneous; the innermost layers next to the electrode remain quite compact but the outer part of the film becomes highly solvated.

5.6.2
XAS, SXS and XANES

Synchrotron radiation is obtained from storage rings in which electrons accelerated by strong magnetic fields to very high velocity emit electromagnetic radiation in a continuous distribution from infra-red to hard X-ray wavelengths. X-ray absorption spec-

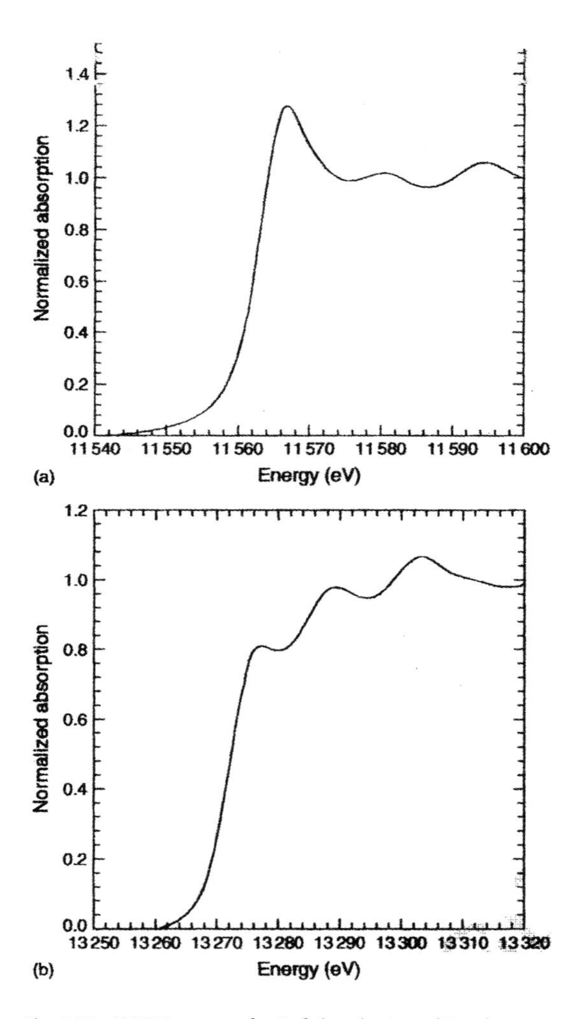

Fig. 5.55 XANES spectra for Pt foil at the L_3 and L_2 edges.

troscopy (XAS) has been used to study two-dimensional structures on surfaces, a technique closely related to surface X-ray scattering (SXS).

Successful investigations have been reported for such electrochemical phenomena as metal deposition, surface oxide formation, and surface water and anion adsorption. A special advantage is that XAS is element specific, which permits the investigation of composite materials such as alloy clusters on electrode surfaces. In addition, XAS may be used to probe the d-states of platinum catalysts. For this study, the near edge part of the spectrum, the part near the very steep edge in the spectrum, is suitable (X-ray Absorption Near-Edge Structure, XANES). The XANES spectra L_2 (11,564 eV) and L_3 (13,273 eV) for a platinum foil are shown in Fig. 5.55. The analysis of the spectra allows the determination of the d-band occupancy as a function of particle size and electrode potential. Another related technique, Extended X-ray Absorption Fine Structure (EXAFS) has been used as a local structural probe, as described in section 2.5.3.2.

References for Chapter Five

The potentiostat is described in detail in:

D.T. Sawyer and J.L. Roberts: "Experimental Electrochemistry for Chemists", John Wiley and Sons, 1974.

P.T. Kissinger and W.R. Heinemann eds. "Laboratory Techniques in Electroanalytical Chemistry", Dekker, New York, 1984.

R. Kalvoda: "Operational Amplifiers in Chemical Instrumentation", Ellis Horwood, Chichester, 1975.

E. Gileadi, E. Kirowa-Eisner and J. Penciner: "Interfacial Electrochemistry: An Experimental Approach", Addison-Wesley, Reading, MA, 1975.

and specific means of achieving *iR* elimination described in D. Britz, J. Electrochem. Soc. **88** (1978) 309-352.

Rotating Disc and Ring-Disc Electrodes are described in:

A. Frumkin and G. Tedoradse, *Z. Elektrochem.* 62 (1958) 251.

V.G. Levich: "Physicochemical Hydrodynamics", Prentice Hall, Englewood Cliffs, NJ, 1962.

D. Jahn and W. Vielstich, "Rates of Electrode Processes by the Rotating Disc Method", *J. Electrochem. Soc.* 109 (1962) 849.

A.C. Riddiford, in "Advances in Electrochemistry and Electrochemical Engineering", volume 4, ed. P. Delahay and C.W. Tobias, J. Wiley, New York, 1966.

W.J. Albery and M.L. Hitchman, "Ring-Disc Electrodes", Clarendon Press, Oxford, 1971.

Yu.V. Pleskov and V. Yu. Filinovsky: "The Rotating Disc Electrode" Consultants Bureau, New York, 1976.

Turbulent Hydrodynamics are described in:

F. Barz, Ch. Bernstein and W, Vielstich: "On the Investigation of Electrochemical Reactions by the Application of Turbulent Hydrodynamics" in "Advances in Electrochemistry and Electrochemical Engineering", vol. 13, ed. H. Gerischer, John Wiley, New York, 1984.

C.H. Hamann, H. Schöner and W. Vielstich: Ber. Buns. Phys. Chem. **77** (1973) 484.

General Treatments of Cyclic Voltammetry are given in:

H. Matsuda and Y. Ayabe, Z. Elektrochem. Ber. Bunsenges. Phys. Chem., 59 (1955) 494.

A.J. Bard and L.R. Faulkner, "Electrochemical Methods: Fundamentals and Applications", John Wiley, New York, 2001.

P.A. Christensen and A. Hamnett, "Techniques and Mechanisms in Electrochemistry", Blackie and Son, Edinburgh, 1995.

Vielstich, W., *Cyclic Voltammetry*, in Handbook of Fuel Cells, Wiley-UK, Chichester 2003, Vol. 2, chapter 14.

P.T. Kissinger and W.R. Heinemann eds. "Laboratory Techniques in Electroanalytical Chemistry", Dekker, New York, 1984.

J.O'M. Bockris and S.U.M. Khan: "Surface Electrochemistry", Plenum Press, New York, 1993.

V.D. Parker "*Linear Sweep and Cyclic Voltammetry*" in "Comprehensive Chemical Kinetics", vol. 26, eds. C.H. Bamford and R.G. Compton, Elsevier, Oxford, 1986.

R.M. Wightman and D.O. Wipf: "Voltammetry at Ultramicroelectrodes" in "Electroanalytical Chemistry" ed. A.J. Bard, Marcel Dekker, New York, 1989.

R.S. Nicholson and I. Shain: Anal. Chem. **36** (1964) 706.

R.S. Nicholson: Anal. Chem. **37** (1965) 1361.

Further discussions of AC methods can be found in:

A.J. Bard and L.R. Faulkner, "Electrochemical Methods: Fundamentals and Applications", John Wiley, New York, 2001.

P.A. Christensen and A. Hamnett, "Techniques and Mechanisms in Electrochemistry", Blackie and Son, Edinburgh, 1995.

D.D. MacDonald, "Transient Techniques in Electrochemistry", Plenum Press, New York, 1977.

I. Epelboin, C. Gabrielli, M. Keddam and H. Takenouti in "Comprehensive Treatise of Electrochemistry", eds. J.O'M Bockris, B.E. Conway, E. Yeager and R.E. White, vol. **4**, p.151, Plenum Press, New York, 1981.

J.R. MacDonald (ed.): "Impedance Spectroscopy", Wiley, New York, 1987.

M. Sluyters-Rehbach and J.H. Sluyters: "Alternating Current and Pulse Methods" in "Comprehensive Chemical Kinetics", eds. C.H. Bamford and R.G. Compton, Vol. **26**, p. 203, Elsevier, Oxford, 1986.

S. Krause, "Impedance Methods" in "Encyclopaedia of Electrochemistry", vol. 3, Wiley-VCH, Weinheim, 2003.

For further discussions on adsorption see references in the previous chapter. See also:

M.W. Breiter: "Adsorption of Organic Species on Platinum Metal Electrodes", in Modern Aspects of Electrochemistry, **10** (1975) eds. J.O'M. Bockris and B.E. Conway, Plenum Press, New York.

J. Lipkowski and P.N. Ross (eds.): "Adsorption of Molecules at Electrode Surfaces", VCH Publishers Inc., New York, 1992.

General overviews of Methods in Spectroelectrochemistry can be found in:

P.A. Christensen and A. Hamnett, "Techniques and Mechanisms in Electrochemistry", Blackie and Son, Edinburgh, 1995.

R.J. Gale (ed.): "Spectroelectrochemistry: Theory and Practice", Plenum Press, New York, 1988.

R.G. Compton and A. Hamnett (eds.): "Comprehensive Chemical Kinetics", Elsevier, Amsterdam, 1989, vol. 29.

G. Gutierrez and C. Melendres (eds.): "Spectroscopic and diffraction techniques in interfacial electrochemistry", Proceedings of NATO ASI 1988, Kluwer, Dordrecht, 1990.

R. Varma and J.R. Selman (eds.): "Techniques for characterisation of Electrodes and Electrochemical Processes", Wiley, New York, 1991.

H.D. Abruna (ed.): "Electrochemical Interfaces: modern techniques for in- situ intercace characterisation", VCH, New York, 1991.

Further Treatments of IR-spectroscopy in electrochemistry can be found in:

A. Bewick and S. Pons, in "Advances in Infra-Red and Raman Spectroscopy", eds. R.J.H. Clark and R.E. Hester, vol. **12**, 1985, Wiley, Chichester.

P.A. Christensen and A. Hamnett in : "Comprehensive Chemical Kinetics", R.G. Compton and A. Hamnett (eds.), 1989, vol. **29**, Elsevier, Amsterdam.

T. Iwasita and F.C. Nart in "Advances in Electrochemical Science and Engineering", vol. **4**, 1995, eds. C. Tobias and H. Gerischer, VCH, Weinheim.

In addition to Chapters in most of the general texts cited above, further Treatments of Electrochemical ESR can be found in:

B. Kastening, in "Comprehensive Treatise of Electrochemistry", vol. **8**, 1984, eds. R.E. White, J. O'M. BOckris, B.E. Conway and E. Yeager, Plenum Press, New York.

A. Heinzel, R. Holze, C.H.Hamann, J.K. Blum, Z.Phys.Chem.N.F. 160(1988)1–23, Electrochim.Acta 34 (1989) 657.

Mass Spectroscopy

B. Bittins, E. Cattaneo, P. Königshoven and W. Vielstich: "*New developments in Electrochemical Mass Spectroscopy*" in "Electroanalytical Chemistry", ed. A.J. Bard, vol. 17 (1991) p. 181.

H. Baltruschat, "Differential Electrochemical Mass Spectrometry", in *Interfecial electrochemistry*, A. Wieckowski Ed., Marcel Dekker Inc., New York 1999, pp. 577–597.

Radiotracer Methods:

V.A. Kazarinov and V.N. Andreev in "Comprehensive Treatise of Electrochemistry", vol. **9**, 1985, eds. R.E. White, J. O'M. BOckris, B.E. Conway and E. Yeager, Plenum Press, New York.

G. Horanyi: Electrochim. Acta **25** (1980) 43.

Quartz Microbalance:

D.A. Buttry in "Electroanalytical Chemistry", vol. **17**, 1991 ed. A.J. Bard, Marcel Dekker, New York.

S. Bruckenstein and M. Shay, Electrochim. Acta **30** (1985) 1295.

Scanning Tunnelling Microscopy and other scanning probe techniques:

R. Sonnenfeld, J. Schneir and P.K. Hansma in "Modern Aspects of Electrochemistry", vol. **21**, 1990, eds. R.E. White, J. O'M. Bockris and B.E. Conway.

R.J. Behm, R. Garcia and H. Rohrer in "Scanning Tunnelling Microscopy and Related Methods", NATO ASI Series E: Applied Sciences, vol. **184**, Kluwer Academic Publishers, Dordrecht.

P.A. Christensen, Chem. Soc. Rev. **21** (1992) 207.

K. Itaya, S. Sugawara and K. Higaki: J. Phys. Chem. **92** (1988) 6714.

L. Kibler, D.M. Kolb, "*Structure Sensitive Methods: AFM, STM*", in "Handbook of Fuel Cells", Wiley-UK, Chichester 2003, Vol. 2, chapter 19.

Zhou, S.F., Bard, A.J., *Scanning Electrochemical Microscopy*, J. Amer. Chem. Soc. **116** (1994) 393.

N.J. DiNardo, "*Nanoscale Characterisation of Surfaces and Interfaces*", VCH, New York, 1994.

Ellipsometry:

R. Greef in "Comprehensive Treatise of Electrochemistry", vol. **8**, 1984, eds. R.E. White, J. O'M. BOckris, B.E. Conway and E. Yeager, Plenum Press, New York.

S. Gottesfeld in "Electroanalytical Chemistry", vol. **15**, 1989, ed. A.J. Bard, Marcel Dekker, New York.

R.H. Müller in "Adv. Electrochem. Electrochem. Eng." vol. **9**, 1973, ed. H. Gerischer.

A. Hamnett in J. Chem. Soc. Faraday Trans. **89** (1993) 1593.

6
Electrocatalysis and Reaction Mechanisms

This chapter is primarily concerned with the basis of electrocatalysis and the elucidation of reaction mechanisms in a small number of practically important electrochemical processes. Examples of different approaches will be used, including the influence of adsorbed intermediates on the hydrogen electrode; and the use of the rotating disc electrode to characterise two *parallel* pathways in the oxygen reduction reaction. Further investigation of mechanisms will show that whilst inorganic electrochemical reactions often involve only a change of valency, electro-organic processes are almost invariably complex, involving the generation and subsequent multiple-pathway reactions of radicals and radical ions. Finally an introduction to dynamic instabilities and oscillations of currents and potentials is given.

Techniques for the investigation of mechanisms have grown rapidly in number and molecular specificity in recent years, and include now not only such "standard" approaches as cyclic voltammetry and ac impedance measurements, but also spectroscopic and mass spectrometric methods. It cannot be over-emphasised how important it is to study any electrode process by as wide a variety of techniques as possible; only by this means can a reliable picture of the reaction be built up.

6.1
On Electrocatalysis

Experimentally, it is often observed that an electrochemical reaction at the interface between the electrolyte and particular electrode materials either does not take place close to the thermodynamic potential of the system, or takes place only at a very small rate. However, if other metals or modified surfaces are used the rate of reaction can be greatly increased. As in the more general case of a heterogeneous chemical reaction, one terms the activating surface a *catalyst*, provided that the surface itself is not consumed or irreversibly altered during the electrochemical process. The action of the catalyst may be due to structural or chemical modifications of the electrode surface and may also derive from additions to the electrolyte. Structural effects may be associated with changes in the electronic state at the surface, e.g. a change in the d-band occupation, or by variation in the geometric nature (crystal planes, clusters, alloys, surface defects). In the field of *electro*catalysis it is hardly possible to speak of a *non-*

Electrochemistry. Carl H. Hamann, Andrew Hamnett, Wolf Vielstich

ISBN: 978-3-527-31069-2

catalyzed reaction, because such a pathway would have to exist in the absence of an electrode surface. Electrocatalysis is based only on *relative* data; a zero-activity electrode does not exist.

Electrochemical reactions are associated with the transfer of electric charge through the electrode/electrolyte interface. The charge carriers can be ions or electrons. In the case of *ions*, the surface changes continuously, either by the deposition of ions or by the dissolution of electrode material. Additives to the electrolyte, which are not consumed, can increase the rate of ion transfer; for instance, small amounts of organic or inorganic species can accelerate the anodic dissolution of metals. In the case of *electrons* as charge carriers, the catalyst provides one of the reactants for the process, the electrons, which are either consumed or generated in the nett reaction. The electrode surface remains unaffected after the reaction has reached a steady state.

In contrast to heterogeneous catalysis, the driving force of an electrochemical reaction is not only controlled by parameters like concentration, pressure and temperature, but also by the potential drop accommodated at the interface, which, in turn, affects the charge transfer through the interface. This potential drop is characterised by the so-called electrode potential which can be altered in an electrochemical cell by an external voltage. A change in potential leads to a change in electronic structure, e.g., to a change in electronic work function.

Recent studies have shown that the rate of a heterogeneous *chemical* reaction can also be influenced by the change in potential (work function) of the catalyst, and by concomitant changes in the coverage by adsorbed ions. This phenomenon is called *non-faradaic electrochemical modification of chemical activity*, NEMCA, and is discussed in chapter 8.7.1. The closest connection between electrocatalysis and catalysis of a heterogeneous chemical reaction exists in the role of adsorption and chemisorption of reactants or intermediates on the rate of the process. As an example, molecular hydrogen adsorbed at a metal surface, can be split into reactive hydrogen atoms.

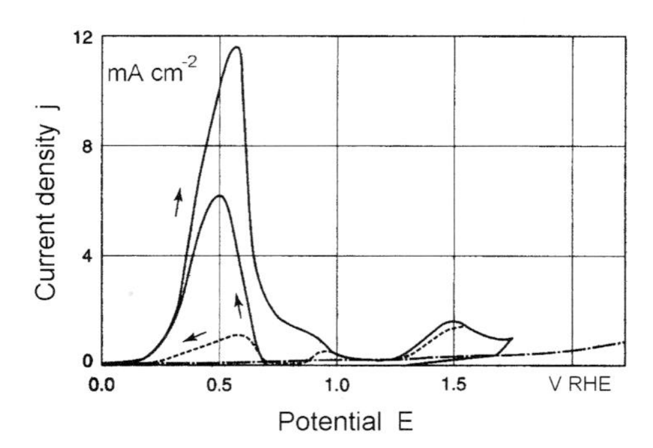

Fig. 6.1 (a) CV for formic acid oxidation at smooth polycrystalline platinum; 1M HCOOH/1M H_2SO_4, 100mVs^{-1}. (dotted line); (b) CV with addition of 5×10^{-5} M lead acetate to the electrolyte (full lines). From E. Schwarzer, W. Vielstich, *Der Einfluss von Blei auf die anodische Oxidation der Ameisensaeure an Platin*, Proc. 3rd International Fuel Cell Symposium, Bruessel, SERAI, Presses Academic Europeennes, Vol. III (1969), pp. 220–229.

The necessary energy for breaking the bond is given by the heat of adsorption. This first step in the catalysis of hydrogen reaction takes place as a heterogeneous reaction at the gas/metal interface as well as an electrocatalytic reaction at the electrolyte/electrode interface. The electrochemical oxidation of hydrogen and other electrocatalytic reactions will be discussed below in this chapter.

A basic concept in the field of electrocatalysis is the formation of suitable clusters, which hinder the adsorption of reaction intermediates. Platinum is known to be a catalyst for the oxidation of formic acid to CO_2. The rate of oxidation at a smooth Pt surface is still relatively low, as shown in the cyclic voltammogram of Fig. 6.1(a). The activity of the surface is increased by two orders of magnitude by adding lead acetate to the electrolyte in a concentration larger than 10^{-5} M (see Fig. 6.1(b)). Even above the thermodynamic Pb^{2+} deposition potential, Pb clusters are formed on the Pt surface. This effect is known as *underpotential deposition* (compare section 4.6.3). Obviously, the partial coverage with Pb clusters hinders the adsorption of less active intermediates like CO in the case of formic acid.

More details will be discussed below in sections 6.4 and 6.5. The role of morphology of electrocatalytic surfaces is dicussed in section 8.1.3.

6.2 The Hydrogen Electrode

In aqueous acid, the hydrogen reaction has the nett stoicheiometry:

$$H_2 + 2H_2O \rightleftharpoons 2H_3O^+ + 2e^- \quad (E^0 = 0.0V) \tag{6.1}$$

and in alkaline solution

$$H_2 + 2OH^- \rightleftharpoons 2H_2O + 2e^- \quad (E^0 = -0.828V \text{ vs. NHE}) \tag{6.2}$$

In neutral solution, (6.1) predominates in the anodic direction, since (6.2) requires a finite concentration of OH^- ions; in a similar way, (6.2) is predominant in the cathodic direction.

On many metals, H_2 is primarily adsorbed in the atomic form; the energy required to break the H-H bond coming from the enthalpy of adsorption. Very many mechanistic investigations have been carried out into the hydrogen reaction mechanism, and these investigations have allowed us to partition the overall reaction, in aqueous acid for example, into a number of steps as follows:

1. Mass transport of dissolved H_2 to the surface and physisorption of H_2

$$H_{2,aq} \rightarrow H_{2,ads} \tag{6.3}$$

2. Chemisorption of hydrogen as atoms

$$H_{2,ads} \rightarrow 2H_{ads} \text{ (Tafel Reaction)} \tag{6.4}$$

3. Ionisation and hydration. There are two possibilities here, depending on whether oxidation of H_{ads} or $H_{2,\,ads}$ takes place. In the first case

$$H_{ads} + H_2O \rightarrow H_3O^+ + e^- \text{ (Volmer Reaction)} \tag{6.5}$$

where the overall reaction takes place twice, as in Case A of section 4.2.4. An alternative is

$$H_{2,ads} + H_2O \rightarrow [H_{ads} \cdot H_3O]^+ + e^- \rightarrow H_{ads} + H_3O^+ + e^- \tag{6.6}$$

("Heyrovsky Reaction")

followed by

$$H_{ads} + H_2O \rightarrow H_3O^+ + e^-$$

corresponding to Case B of section 4.2.4.

4. Transport of the H_3O^+ ions away from the electrode surface

The reactions are reversible, the reverse process leading to hydrogen evolution. The mechanism:

$$2H_3O^+ + 2e^- \rightarrow 2H_{ads} + 2H_2O; \quad 2H_{ads} \rightarrow H_{2,ads} \tag{6.6a}$$

is termed the Volmer-Tafel mechanism, whereas

$$H_3O^+ + e^- \rightarrow H_{ads}; \quad H_{ads} + H^+ + e^- \rightarrow H_{2,ads} \tag{6.6b}$$

is termed the Volmer-Heyrovsky mechanism.

Both hydrogen oxidation and evolution have high exchange current densities on suitable electrocatalysts. Practical electrodes in fuel cells can exhibit current densities of 0.5A cm^{-2} at overpotentials below 100 mV.

6.2.1
Influence of Adsorbed Intermediates on *i-V* Curves

Depending on the experimental conditions, such as the condition of the catalyst surface, the pH, the current density and the electrolyte composition, different reaction pathways are possible on the hydrogen electrode, and we will restrict our attention to the Volmer-Tafel mechanism in acid solution. This mechanism is pertinent to hydrogen oxidation on platinum for which the equilibrium atomic hydrogen coverage, θ_H^0, derived from cyclic voltammetry (Section 5.2.1.), exceeds 0.9 and also for which H_2 molecules are only very weakly adsorbed. (This should be contrasted with gold, for which $\theta_H^0 \ll 1$, and mercury, for which $\theta_H^0 \approx 0$).

If the Tafel reaction is rate-limiting, then θ_H becomes current dependent, and we can treat the coverage as leading to a reaction overpotential. By analogy with Section 4.4.1, we can write, for the overpotential, in the case of the anodic reaction, with $\nu = -2$ and $n = 2$

$$\eta_r = -(RT/F)\ln(\theta_H/\theta_H^0) \tag{6.7}$$

The current density can be written, for the Tafel reaction, as

$$j_T = j^+ + j^- = k_T^+(1 - \theta_H)^2 c^s_{H_2,aq} - k_T^- \theta_H^2 \tag{6.8}$$

where is the concentration of $c^s_{H_{2,aq}}$ at the electrode surface, and the squared factors arise from the fact that *two* spaces are needed for H_2 to chemisorb and *two* H_{ads} are required to evolve H_2.

The exchange current density is given by

$$j_{0,T} = k_T^+(1 - \theta_H)^2 c^0_{H_2,aq} = k_T^- \theta^0_{H_2} \tag{6.9}$$

where $c^0_{H_{2,aq}}$ is the bulk hydrogen concentration. From (6.8) and (6.9) we have

$$j_T = j_{0,T}\left[\left(\frac{1-\theta_H}{1-\theta_H^0}\right)^2 \cdot \frac{c^s_{H_2,aq}}{c^0_{H_2,aq}} - \left(\frac{\theta_H}{\theta_H^0}\right)^2\right] \tag{6.10}$$

For high anodic overpotentials, we would expect that $\theta_H \to 0$, and provided the transport of H_2 to the surface is sufficiently rapid, we obtain an anodic limiting current of the form

$$j^+_{lim,T} = j_{0,T}(1 - \theta_H^0)^{-2} \tag{6.11}$$

though this is not attained in practice on Pt since it is impossible to transport H_2 in solution sufficiently rapidly.

For the cathodic limit, the maximum current density would be expected when $\theta_H \to 1$ and

$$j^-_{lim,T} = -j_{0,T}(\theta_H^0)^{-2} \tag{6.12}$$

Again, however, this limiting behaviour is not observed on platinum; in fact, over the entire accessible overpotential region, provided the transport of H_2 away from the surface can be maintained sufficiently rapid, *there is a linear relationship between overpotential and logarithmic current*. Given the apparently large value θ_H^0 of as derived from cyclic voltammetry, this is extremely difficult to understand, especially within the context of (6.10). The experimental behaviour is, in fact, much more naturally accounted for by assuming $\theta_H^0 \approx 0$ even though this is manifestly not consistent with the cyclic voltammetric data. One possible resolution of this puzzle has been suggested by spectroscopic work by Nichols and Bewick (J. Electroanal. Chem. **243** (1988) 445). These authors suggested that there were *three* distinct forms of adsorbed hydrogen on the polycrystalline platinum surface. A "strongly- bonded" form, present below ca. 250 mV, which is multiply bonded, a "weakly-bonded" form, present below ca. 120 mV, which may be doubly or multiply bonded and a third form, present only in the hydrogen evolution region, which is apparently singly bonded. It would appear that this third form is the only form that is *kinetically* active, and its coverage at equilibrium is indeed very low, as required by the kinetic data.

If both the Tafel and Volmer reactions affect the rate of reaction, then the expression becomes more complex, though the derivation remains straightforward. For the Volmer reaction, we have:

$$j_V = F\theta_H k_0^+ \exp\left\{\frac{(1-\beta_V)FE}{RT}\right\} - F(1-\theta_H)c^s_{H^+}k_0^- \exp\left\{-\frac{\beta_V FE}{RT}\right\} \tag{6.13}$$

and if this reaction is rate-limiting alone, so that $\theta_H \approx \theta_H^0$, then we expect, in the cathodic region, a Tafel slope of ca. 120 mV provided $\beta_V \approx 0.5$. This is, in fact, found for a number of metals including mercury, copper, silver and iron.

It can be seen that the adsorption of hydrogen atoms plays a decisive role in the reaction, particularly in the anodic direction, since unless the adsorption enthalpy is sufficient, dissociative chemisorption cannot take place. Of course, the adsorption enthalpy should not be too high, else it will prove kinetically difficult to remove the adsorbed hydrogen atoms as H^+, and as with other examples of catalysis, we expect, and find, a "volcano" curve.

6.2.2
Influence of the pH-value of the Solution and the Catalyst Surface

The overall equations (6.1) and (6.2) lead us to expect a minimum in the exchange-current density at neutral pH, and this is indeed the case, as shown in Fig. 6.2. If the Volmer reaction is assumed rate-limiting, we see that in acid solution we can write from (4.38)

$$\ln j_{0,V} = \ln j^0_{0,V} + (1-\beta_V)\ln c^0_{H^+} + \beta_V \ln\theta^0_H \tag{6.14}$$

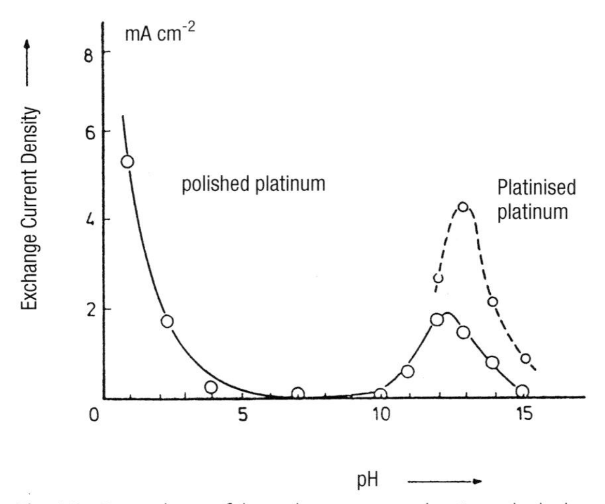

Fig. 6.2 Dependence of the exchange current density at the hydrogen electrode on the pH at 20°C [from S. Ernst and C.H. Hamann: J. Electroanal. Chem. **60** (1975) 97].

and under the assumption that is θ_H^0-invariant, the expectation is that $\ln(j_{0,V})$ would decrease linearly with pH, as is observed. The same type of argument can be used to show that the exchange current would be expected also to decrease with decreasing OH^- concentration in alkaline solution, as is again observed. Rather surprisingly, the exchange current density is observed to decrease strongly in very alkaline solutions, for reasons that are not clear, but may be due to influence of OH^- ions on the strength of the Pt-H bond, as well as the decreasing solubility of H_2 in strongly alkaline media.

A clear example of the influence of the state of the catalyst surface on the hydrogen oxidation reaction is found for Pt in the whole of the pH region at potentials above 0.8 V vs. the reversible hydrogen potential for that pH. Above 0.8 V, coverage of the Pt surface by chemisorbed oxygen becomes dominant, and this prevents hydrogen chemisorption. Similar displacement processes are found for certain other strongly chemisorbed ligands, such as I^- and phosphate.

6.2.3
Hydrogen Oxidation at Platinum and Chemisorbed Oxygen

In section 5.2.1.1 the formation of OH- and O-layers on polycrystalline platinum during a cyclic voltammogram was described. As shown in the Figs. 5.6 and 5.7a for alkaline and acid solution, the respective charges are well defined, reproducible and somewhat different for the two pH‘s. In alkaline media the formation is extended between 550 and 1600 mV, in acid only between 800 and 1450 mV. And a strong increase in the cathodic reduction is observed only below 1000 mV RHE with a maximum in both cases near 750 mV.

In order to study the influence of oxygen covered surfaces on the rate of hydrogen oxidation, cyclic voltammetry can again be used to give a first survey. Fig. 6.3 shows the fast kinetics of hydrogen oxidation at smooth platinum with a steep increase of the current already apparent at the thermodynamic equilibrium potential. The current eventually attains a stationary value, limited by the convective diffusion caused by the hydrogen bubbles near the electrode (the relative strong noise in this current is also due to the hydrogen bubbles). Near ca. 750 mV, a strong decrease in the hydrogen oxidation current starts, due evidently to the formation of a layer of chemisorbed oxygen. On the cathodic sweep, a hysteresis in the CV is observed, depending on the rate of the potential change.

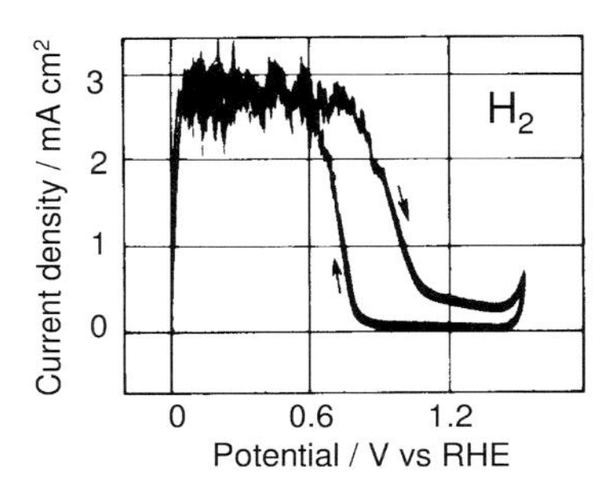

Fig. 6.3 Cyclic voltammogram of a smooth platinum electrode immersed in a hydrogen-saturated solution of 0.5M H_2SO_4 at 25°C. The sweep rate is 50 mVs^{-1}. As the potential rises to 0.8 V, it can be seen that the hydrogen oxidation current has fallen from the diffusion limiting value to close to zero.

The coverage of the platinum surface by strongly bound oxygen moieties also hinders other electrode reactions, like the reduction of oxygen itself (see Fig. 5.9) or the oxidation of CO (Fig. 6.17).

6.3 The Oxygen Electrode

The processes at the oxygen electrode are, like those of the hydrogen electrode, of great technical importance in fuel cells and electrolysers, as well as in the corrosion of metals. Characteristic of the oxygen electrode reactions are the following properties:

1. Large overpotentials, in excess of 400 mV are observed, even for very low current densities (ca. 1 $mAcm^{-2}$) for both cathodic and anodic reactions, as a consequence of the very low (< 1 $nAcm^{-2}$) exchange-current densities. This restriction can be overcome, but only by the development of extremely active electrocatalysts.
2. For the cathodic reaction on such metals as silver and platinum *two parallel reaction pathways* are observed: the *direct* reduction of O_2 to H_2O (in acid solution) or OH^- ions, and an indirect pathway with H_2O_2 (in acid solution) or HO_2^- as intermediates. These latter can attain significant concentration in the electrolyte.

The reaction paths show a strong pH dependence and have the following forms:

In *acid* solution:

direct reduction:

$$O_2 + 4H^+ + 4e^- \rightarrow 2H_2O \qquad E^0 = 1.23V \tag{6.15}$$

indirect reduction

$$O_2 + 2H^+ + 2e^- \rightarrow H_2O_2 \qquad E^0 = 0.682V \tag{6.16}$$

$$H_2O_2 + 2H^+ + 2e^- \rightarrow 2H_2O \qquad E^0 = 1.77V \tag{6.17}$$

In *alkaline* solution

direct reduction

$$O_2 + 2H_2O + 4e^- \rightarrow 4OH^- \qquad E^0 = 0.401V \tag{6.18}$$

indirect reduction

$$O_2 + H_2O + 2e^- \rightarrow HO_2^- + OH^- \qquad E^0 = -0.076V \tag{6.19}$$

$$HO_2^- + H_2O + 2e^- \rightarrow 3OH^- \qquad E^- = 0.88V \tag{6.20}$$

In order for the oxygen reduction reaction to take place, the oxygen bond must be weakened, which in turn implies that a strong interaction with the surface of the catalyst will be necessary. It has been suggested that if adsorption of the O_2 molecule is *end-on*, this does not weaken the O-O bond sufficiently, and the indirect mechanism is favoured, whereas bidentate bonding of the O_2 to a surface metal atom, or bridge bonding between two surface metal atoms will favour the direct process, since in this way the O-O bond is substantially weakened. The necessity for bridge bonding will imply that the *direct* reduction pathway is more sensitive to poisoning than the indirect one, an effect commonly observed experimentally.

The very small exchange-current density leads to very poor reproducibility for the rest potential, which in any case is only slowly attained. In very clean acidic solutions, the rest potential usually lies between those of (6.15) and (6.16), mostly near to 1.1 V.

6.3.1 Investigation of the Oxygen Reduction Reaction with Rotating Ring-Disc Electrode

The rotating ring-disc electrode (RRDE) has proved ideal for the study of the oxygen reduction reaction, since the intermediate, H_2O_2, is itself electroactive, and can be re-oxidised at a Pt ring to O_2. If the total current at the disc can be written as the sum of the direct and indirect currents, then the latter can be obtained from the ring-current using the expression

$$i_R = N \cdot i_{indirect} \tag{6.21}$$

An example of this is shown in Fig. 6.4a, where the reduction of O_2 on a Pt disc in very high purity 0.05 M H_2SO_4 is seen to give a (small) ring current. Given $N = 0.43$ for this RRDE, then in the region 100–500 mV, the limiting current region, the ratio $i_{indirect}/i_{direct} \approx 0.01$, showing that the four-electron pathway dominates for pure Pt in this potential region. Interestingly, as the data show, the ratio is larger in the Tafel

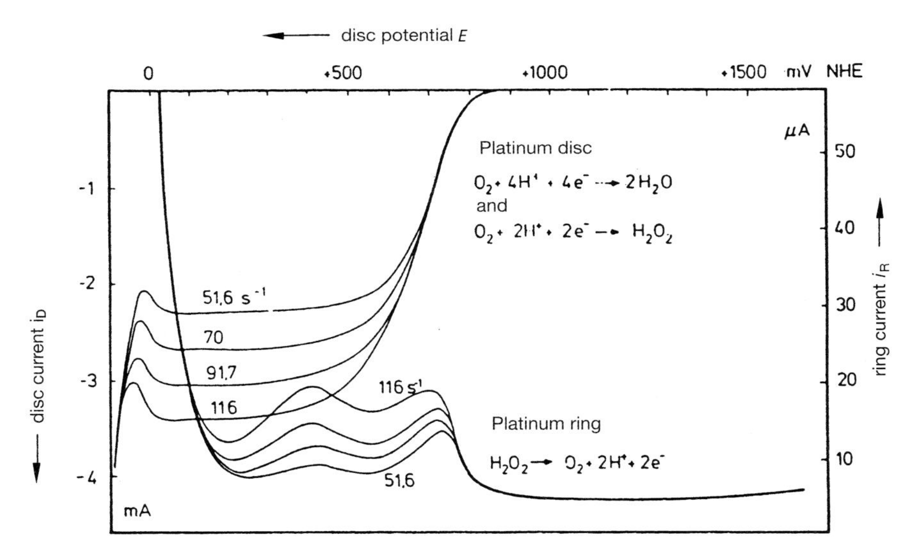

Fig. 6.4a Current-potential curves for oxygen reduction (on the disc) and hydrogen peroxide (on the ring) at a rotating ring-disc electrode in 0.05M H_2SO_4 with rotation frequency as the parameter. The ring potential was set at 1200 mV vs. NHE (see text) and the collection efficiency was measured to be 0.43, a result confirmed by calculation.

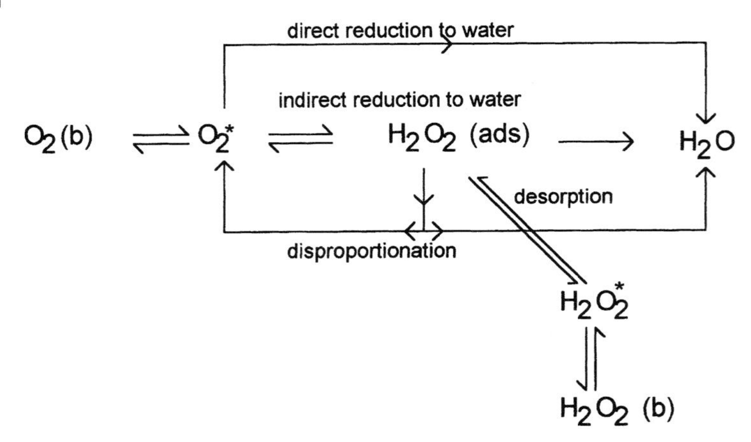

Fig. 6.4b Detailed mechanistic scheme for the reduction of O_2 (from C.A. Paliteiro, D.Phil. Thesis, Oxford University, 1984). In the figure, O_2 (b) and H_2O_2 (b) refer to these species in the bulk of the solution, and O_2* and H_2O_2* are the same species close to the electrode surface.

region (900–700 mV), and shows some underlying fluctuations at lower potentials. This larger ratio in the Tafel region is also found in alkaline solutions, and it has been suggested that it has its origin in the effects of residual chemisorbed oxygen on the Pt surface. The strong rise in ring current near 0 V is caused by the evolution of hydrogen on the disc, which is then re-oxidised on the ring.

A more complete analysis of the parallel processes can be carried out according to the scheme of Fig. 6.4b, in which desorption of H_2O_2 and its disproportionation at the surface are also considered. As has been shown elsewhere, it is best to consider the separate fluxes of O_2 and H_2O_2 at the surface, and the resultant expressions, though complex, still allow a separation of two from four-electron pathways.

Parallel mechanisms are also observed on silver, and on gold and carbon, though in the latter cases, the two-electron route is predominant. On mercury, only the indirect route is observed, as discussed in Section 10.2 below; two polarographic waves are, in fact seen, with the first corresponding to reduction of O_2 to H_2O_2 and the second corresponding to onset of the reduction of H_2O_2 to H_2O.

6.4 Methanol Oxidation

The methanol oxidation reaction has been the subject of a large number of studies over many years in the past. Early work indicated a complex reaction mechanism. The cyclic voltammograms of Fig. 6.5 demonstrate that on a platinum electrode the anodic current starts at large overpotentials in both alkaline and acid media. The thermodynamic potential E^0 is about 20 mV positive to hydrogen. However, the current peaks do not

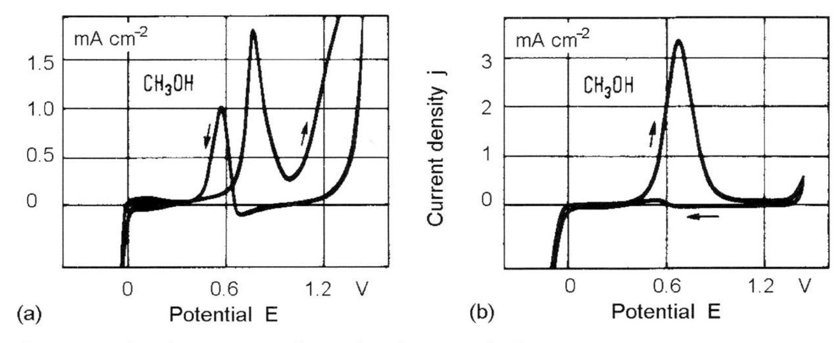

Fig. 6.5 Cyclic voltammograms for methanol at smooth platinum at room temperature; (a) in 1N H_2SO_4, (b) in 1N KOH.

follow the behavior of the simple cyclic voltammograms of Figs. 5.11 and 5.12 demonstrating that methanol oxidation reaction is not determined by the simple coupling of mass diffusion and electron transfer processes. Indeed, above 700 mV the coverage of the platinum surface with adsorbed O and OH species steadily increases (see Figs. 5.6 and 5.7), and the electron transfer process involves several parallel pathways.

In acid solution water is consumed with the formation of CO_2

$$CH_3OH + H_2O \rightarrow CO_2 + 6H^+ + 6e^- \tag{6.22a}$$

For alkaline solutions one has the formation of carbonate

$$CH_3OH + 8OH^- \rightarrow CO_3^{2-} + 6H_2O + 6e^- \tag{6.22b}$$

For practical reasons, we shall discuss below only the reaction in acidic media. In acid electrolytes, several metal catalysts have been proposed for the reaction, most of them based on modifications of Pt with some other metal; currently promotion of methanol oxidation by ruthenium on platinum is the most effective. In addition to CO_2, formed by the complete oxidation of methanol, non-negligible yields of formaldehyde and formic acid are encountered on many electrodes, depending among other factors on concentration, potential, electrode roughness and temperature.

Thermodynamic data for methanol, carbon monoxide and other small organic molecules are given in Table 6.1. From this table it is apparent that all of these substances could be considered good candidates for operation in a fuel cell, their thermodynamic open circuit potentials being of the same order as that of hydrogen. However, whilst hydrogen oxidation on platinum occurs at relatively high rates near the reversible potential, the electro-oxidation of small organic molecules has severe kinetic limitations, arising from the blocking of the Pt surface by strongly adsorbed intermediates. These intermediates prevent the facile breaking of C-H and O-H bonds, but in spite of this, no better catalyst than Pt is known at present for this purpose.

Table 6.1 Thermodynamic data on the oxidation of CO, methanol and related substances at 25°C and 1 atm.

Theoretical cell reaction	ΔH° kcal/mole	ΔS° cal/mole	ΔG° kcal/mole	n	E_c° volt
$H_2 + {}^1/_2O_2 \rightarrow H_2O$ (liq)	– 68.14	– 39.0	– 56.69	2	1.23
$C + {}^1/_2O_2 \rightarrow CO$	– 26.4	+ 21.4	– 32.81	2	0.71
$CO + {}^1/_2O_2 \rightarrow CO_2$	– 67.62	– 20.9	– 61.45	2	1.33
$CH_3OH + {}^3/_2O_2 \rightarrow CO_2 + 2H_2O$	– 173.8	– 23.5	– 166.8	6	1.21
$CH_2O + O_2 \rightarrow CO_2 + H_2O$	– 134,28	– 32.1	– 124.7	4	1.35
$HCOOH + {}^1/_2O_2 \rightarrow CO_2 + H_2O$	– 64,66	+ 11.8	– 68.2	2	1.48

6.4.1 Parallel Pathways of Methanol Oxidation in Acid Electrolyte

The thermodynamic equilibrium potential for complete oxidation of methanol to CO_2, lies close to the equilibrium potential of hydrogen, but kinetic inhibition strongly shifts the potential by several hundreds of mV at pure platinum. The total oxidation seems to occur through several parallel reactions:

$$\text{I} \quad CH_3OH \rightarrow CO_{ad} \rightarrow CO_2 \tag{6.23}$$

$$\text{II} \quad CH_3OH \rightarrow (\text{reactive intermediate}) \rightarrow CO_2$$

A catalyst for methanol oxidation should be able to (a) dissociate the C-H bond and (b) facilitate the reaction of the resulting residue with some O-containing species to form CO_2. On a pure Pt electrode, which is known to be the best catalyst for breaking the C-H bond, these two processes, both necessary for complete oxidation, occur in different potential regions:

- Process (a), involving the dissociative adsorption of methanol molecules, requires several neighbouring unoccupied places at the surface and, due to the fact that methanol is not able to displace adsorbed H atoms, adsorption can only begin at potentials where enough Pt sites become free from H, i.e. near 0.2 V vs RHE for a polycrystalline Pt electrode.
- The second process (b) requires dissociation of water, which is the oxygen donor of the reaction. On pure Pt electrode, sufficient interaction of water with the catalyst surface is only possible at potentials above 0.4–0.45 V vs RHE.

Thus, on pure Pt, complete methanol oxidation to CO_2 cannot begin below about 0.45 V.

6.4.2 Methanol Adsorption

It has been suggested that methanol adsorption takes place in several steps, forming different species due to dissociation of the molecule:

$$\begin{aligned}
&CH_3OH \rightarrow \underset{x}{CH_2OH} + H^+ + e^- && (1)\\
&\underset{x}{CH_2OH} \rightarrow \underset{xx}{CHOH} + H^+ + e^- && (2)\\
&\underset{xx}{CHOH} \rightarrow \underset{xxx}{OH} + H^+ + e^- && (3)\\
&\underset{xxx}{COH} \rightarrow \underset{x}{CO} + H^+ + e^- && (4)
\end{aligned} \qquad (6.24)$$

where x stands for a free (water covered) Pt site. It has also been suggested (but not proved) that formaldehyde and formic acid could be formed from the intermediates CH_2OH and CHOH respectively.

If a cyclic voltammogram is started after contacting a polycrystalline Pt electrode with a methanol containing solution at a potential of 0.05 V or less, methanol adsorption can be observed as soon as hydrogen coverage begins sensibly to decrease. The dissociation process gives rise to a current peak in the H-region (Fig. 6.6), which can be observed only during the first potential scan, i.e. when the surface is free from organic residues. The experiment in this figure was performed using the DEMS technique (see chapter 5).

As illustrated in Fig. 6.6 no signal for mass ($m/e = 44$) corresponding to CO_2 was detected at potentials in the region of the current peak. Since other volatile products

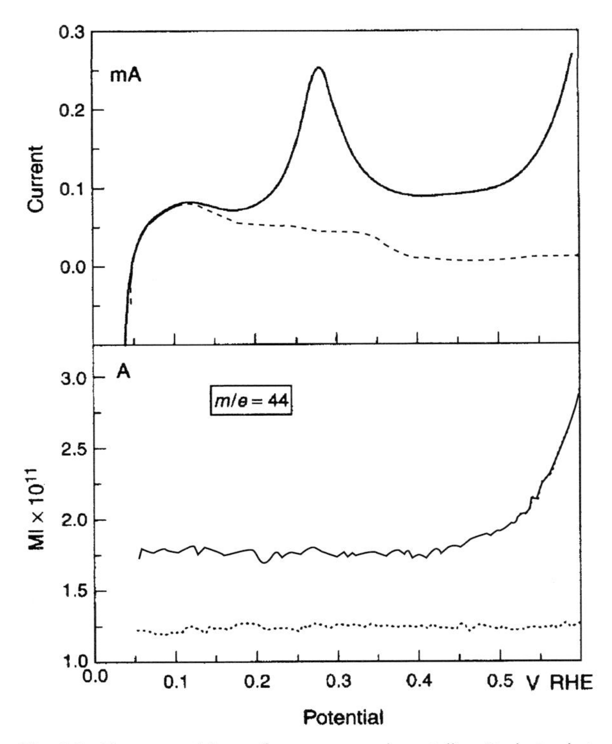

Fig. 6.6 First potential scan for a porous polycrystalline Pt electrode in 0.2M CH_3OH/0.1M $HClO_4$ solution (upper part) and simultaneous recording of mass intensity for CO_2 production (lower part); 10 mV/s. Dashed lines: current and MS signals in supporting electrolyte.

were also absent, it can be concluded that the current peak is related to faradaic processes occurring during methanol adsorption.

6.4.3
Reaction Products and Adsorbed Intermediates of Methanol Oxidation

Establishing the nature of adsorbed species formed during adsorption of small organic molecules is a difficult task. As discussed in chapter 5, simple charge measurements during adsorption of methanol and oxidation of the adsorbed residue can be made. In addition, data from infrared spectroscopy, thermal-desorption MS and DEMS have proved very helpful.

In situ infrared spectra obtained during methanol oxidation at different potentials on Pt(111) (Fig. 6.7) show well characterized bands for linearly and bridge-bonded CO at about 2060 cm^{-1} and 1840 cm^{-1} respectively. The spectra also show bands for the soluble products CO_2 (at 2342 cm^{-1}) and methyl formate at 1710 cm^{-1} and 1230 cm^{-1}. The latter demonstrates the formation of formic acid (see below). A band at 1260 cm^{-1} (not resolved in Fig. 6.7) has been interpreted in terms of the C-OH stretch of hydrogenated species as, e.g. COH or HCOH.

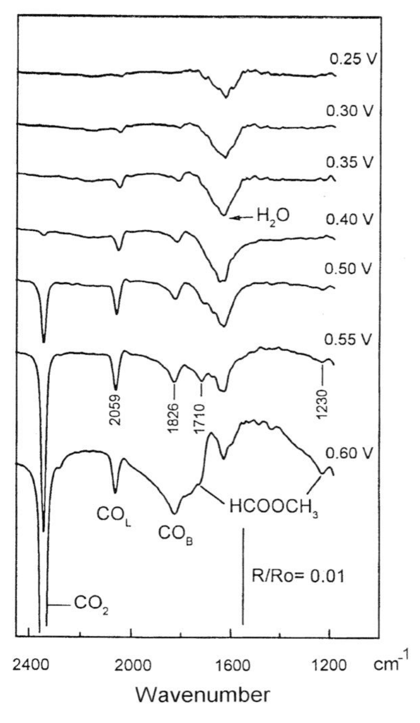

Fig. 6.7 In situ FTIR spectra at Pt(111) in 0.5M CH_3OH/0.1M $HClO_4$ at different potentials, reference spectra at 0.1 V versus RHE.

Another approach to establish the nature of the adsorbate can be made by thermal desorption mass spectrometry, performed on electrodes transferred into UHV. These results confirmed the presence of hydrogenated species in the residues produced upon methanol adsorption. Moreover, these data have shown that the ratio between the amount of CO and other (hydrogenated) species depends on methanol concentration.

The oxidation products of CH_3OH from long-term electrolysis at potentials between 0.5 and 0.6 V vs RHE are CO_2, H_2CO, HCOOH and $HCOOCH_3$. The product methyl formate originates in a reaction between HCOOH and CH_3OH:

$$HCOOH + CH_3OH \rightarrow HCOOCH_3 + H_2O \tag{6.25}$$

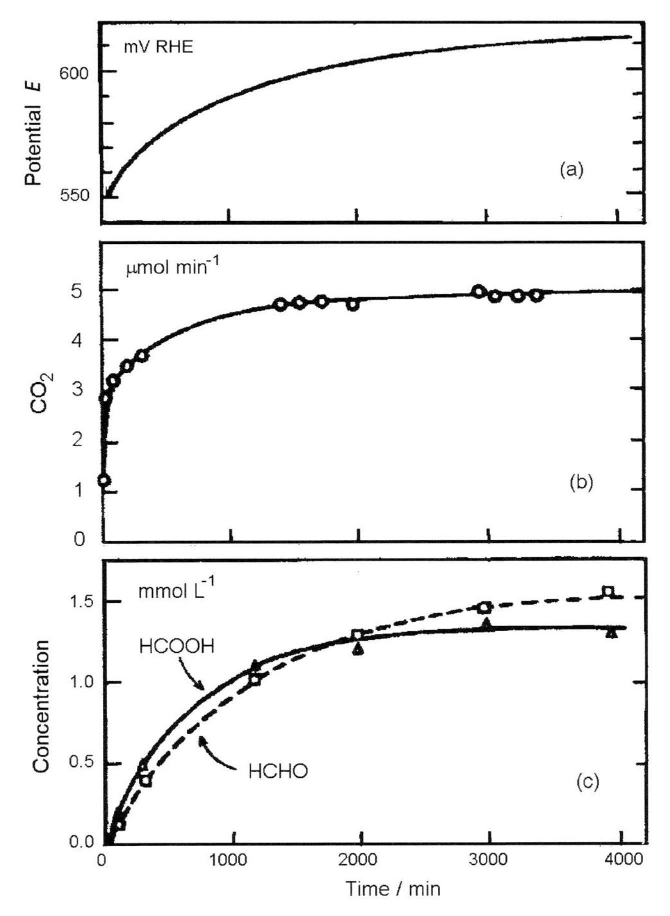

Fig. 6.8 Parallel pathways of methanol oxidation, load 25 mA at 25°C; 5 cm^2 platinized Pt (RF690), 1M CH_3OH/1M H_2SO_4; (a) change in potential, (b) formation of CO_2 via gas chromatography, (c) Concentration of HCHO via colorimetry and HCOOH via iodometric titration.

The yields of different products depend on methanol concentration, temperature, electrode roughness and electrolysis time. Carbon dioxide formation was found to be favoured at high temperature and on rough electrodes. Fig. 6.8 shows, for a 5 cm^2 platinized Pt electrode (RF 690) and a constant current of 25 mA, the time dependence of the electrode potential, of the rate of CO_2 formation and of the product concentrations in solution. After ca. 1 h the current efficiency for CO_2 is practically 100 %, the efficiencies for HCOOH and H_2CO declining close to zero.

6.4.4
Effects of Surface Structure and Adsorbed Anions

The complexities involved in methanol adsorption and oxidation are reflected in the strong sensitivity of the reaction to surface structure and adsorbed anions. For the single crystal Pt(111) surface, methanol oxidation shows a strong inhibiting effect in the presence of sulphate ions (Fig. 6.9); for polycrystalline Pt this effect is less pronounced. The infrared spectra taken in Fig. 6.10 are helpful in understanding the above findings. After the application of a single potential step from 0.05 V to 0.6 V the spectra were followed as function of time up to 5 min. Much more CO_2 is observed in $HClO_4$ and no features for adsorbed CO are found in H_2SO_4. This fact and the low current obtained with H_2SO_4 reinforce the conclusion that adsorbed sul-

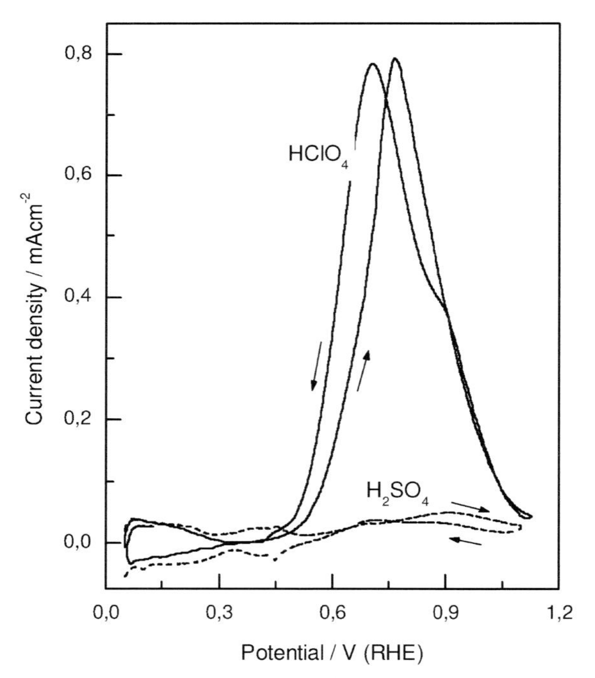

Fig. 6.9 Cyclic voltammograms of methanol oxidation at 50 mVs^{-1} in 0.1M $HClO_4$ and 0.5M H_2SO_4 at a Pt(111) surface.

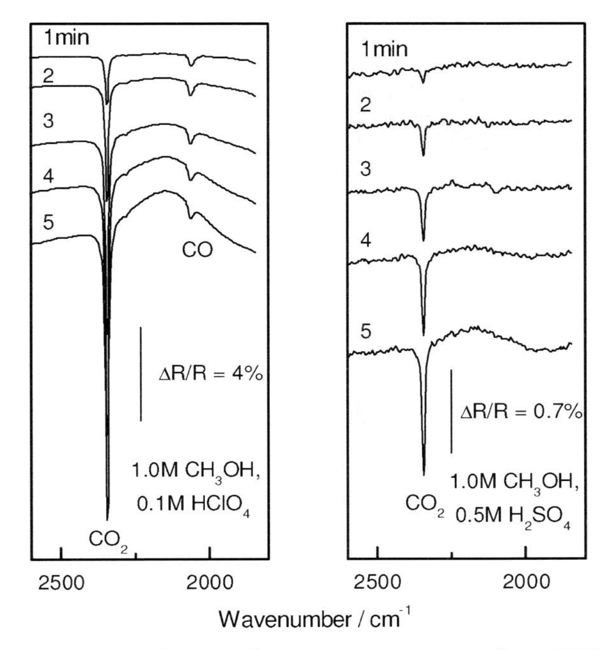

Fig. 6.10 Development of in situ IR spectra with time for a Pt(111) electrode, after applying a single potential step from 0.05 V to 0.6 V in 1.0M $CH_3OH/HClO_4$ (left) and 0.5M H_2SO_4 (right). Reference spectrum taken at 0.05 V vs. RHE.

phate ions hinder the adsorption of methanol molecules, thus the formation of CO is the *rate-determining step*. On the other hand, in $HClO_4$ the CO_2 signals (from the solution in the narrow gap of the IR cell, see Fig. 5.31a) grow with time, the values for the CO_{ad} coverage being stationary. In addition, the oxidation current and CO_2 formed both increase with potential. Obviously, in this situation, methanol adsorption to form CO *is not a rate-determining step*.

At this point, it is interesting to discuss whether CO_{ad} is a poison or simply an intermediate of methanol oxidation. For example, there is no doubt that in the case of hydrogen oxidation with traces of CO in the gas, CO will be a poison of the reaction. But with methanol, in both base electrolytes, CO_{ad} does not act as a poison. On the contrary, in H_2SO_4 at Pt(111) (and also at Pt(110)) CO_{ad} *formation is inhibited*. In conclusion, we can state that during methanol oxidation at platinum CO is not a "poison", it should be considered a *slowly oxidizable reaction intermediate*.

6.4.5 On the Mechanism of Methanol Oxidation

We can distinguish two partial processes, as already indicated above: adsorption of methanol molecules and oxidation of adsorbed residues. We can now try to infer which of these two processes in general includes the rate-determining step of the reaction. Owing to the fact that the state and coverage of the different Pt surfaces both depend on potential, and admitting the necessity of oxygen donor species at the surface for oxidation of the adsorbed residues, both these effects can only take place simultaneously at potentials above 0.5 V or higher on pure Pt.

Theoretical studies of the rate of CO oxidation at Pt also emphasise the importance of (a) the CO_{ad} mobility on the surface (due to the fact that CO forms islands on the surface) and (b) the necessity of forming OH_{ads} via water dissociation (instead of simply using H_2O_{ad} as the O-donor). This CO mobility is an important additional consideration, since formation of OH_{ads} at lower potentials is likely to be restricted to relatively few sites, and CO mobility provides an obvious and important mechanism for low-potential oxidation. Studies also indicate that the $CO_{ads} + OH_{ads}$ reaction at platinum is intrinsically fast, but the actual rate of reaction will depend sensitively on OH_{ads} coverage.

6.4.6 Catalyst Promoters for Methanol Oxidation

Several binary and ternary catalysts have been proposed for methanol oxidation, most of them based on modifications of Pt with some other metal. Given the comments above, it is clear that the limiting process at low potentials is the availability of OH_{ads} species on the surface, and the promoter metal must fulfil the requirement of forming O-containing surface species at low potentials. This seems to be a necessary condition for increasing the catalytic activity of Pt towards methanol oxidation; among other promoters, Ru, Sn, Bi and Mo have all been suggested. There are, of course, several practical factors limiting the choice of the metal. Many O-adsorbing metals can produce negative effects, e.g., inhibit methanol adsorption or may be not sufficiently stable for long term use, as required for a fuel cell. At present, there is a general consensus that PtRu offers the most promising results.

When discussing the reason for the enhancement effect of PtRu, the *bi-functional* mechanism is often invoked. This term was suggested to give emphasis on the joint activities of both metals, Pt being the one adsorbing and dissociating methanol and Ru, the one oxidizing the adsorbed residues. In a simplified manner, with the aim of illustration, the bi-functional mechanism can be written as follows:

$$\begin{aligned} &CH_3OH + Pt \rightarrow Pt(CO)_{ad} + 4H^+ + 4e^- \\ &Ru + H_2O \rightarrow Ru(OH)_{ad} + H^+ + e^- \\ &Pt(CO)_{ad} + Ru(OH)_{ad} \rightarrow CO_2 + H^+ + e^- \end{aligned} \tag{6.26}$$

To judge the capability of PtRu materials, results are usually presented in the form of current-time curves at constant potential. For smooth surfaces, currents obtained after 20 minutes are plotted as function of Ru:Pt composition in Fig. 6.11.

As compared with pure Pt(111), the currents for the alloys are several orders of magnitude higher. Alloys were sputter cleaned and carefully heated in UHV. The

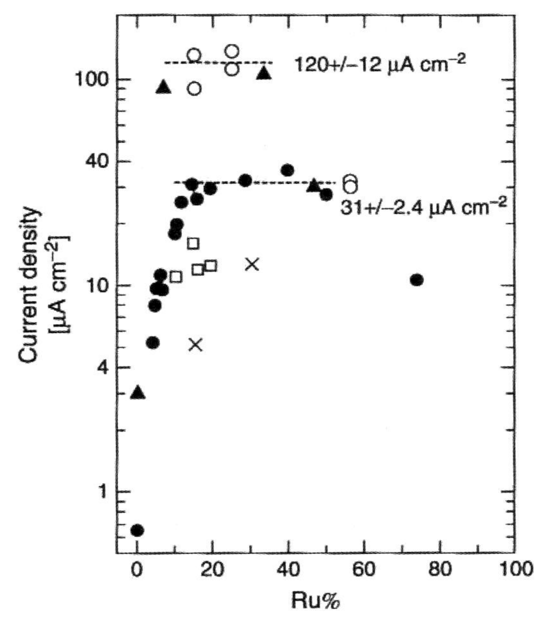

Fig. 6.11 Plot of current densities for methanol oxidation from amperometric curves at 0.5 V after 300 s as function of PtRu surface composition; (•) Ru/Pt(111) formed by spontaneous deposition, (□) Ru modified Pt(111) via Ru adsorption and simultaneous reduction by hydrogen, (x) Ru coverage by UHV vapor deposition, (o) and (▲) UHV prepared PtRu alloys; data obtained after 20 min. ▲-data obtained in 0.5M CH_3OH/0.5M H_2SO_4, all other data with 0.5M CH_3OH/0.1M $HClO_4$.

reproducibility of this method for pre-treating the alloys derives from both the fact that the resultant surfaces are smooth and that they present the same composition as the bulk. Two main points can be extracted from this plot: i) PtRu alloys are better catalysts than Pt(111)/Ru (adsorbed Ru atoms on Pt(111)) and ii) both materials present a wide maximum (between ca. 10–40 % Ru for the alloy and ca. 15–50 % for the Pt(111)/Ru electrodes, at room temperature).

In terms of the bi-functional mechanism, these maxima indicate that a Ru percentage of ca. 10–45 % on the surface is enough to provide an efficient oxidation of adsorbed methanol residues. For compositions within the maxima, the quantity of Ru is not a limiting factor for the reaction. The upper limit of Ru coverage, on the other hand, is given by the necessity of having enough Pt sites for adsorbing and dissociating methanol. This limit is somewhat higher for adsorbed Ru than for the alloy (observe the wider maximum). This and the higher currents observed on the alloys may be related to the fact that the latter present a more homogeneous distribution of Ru atoms than the Pt(111)/Ru electrode.

It is noteworthy that with increasing temperature, the maximum of the j – Ru % plot for the alloy is shifted to somewhat higher Ru % values. This effect probably originates in the fact that at higher temperatures (e.g. 60°C), Ru becomes active for adsorbing methanol.

6.5 Carbon Monoxide Oxidation at Platinum Surfaces

Carbon monoxide is one of the most studied species in surface science. In electrochemical experiments CO is often used as a probe molecule for vibrational spectroscopy and scanning tunnelling microscopy. Such studies can be used to compare the responses of the metal in the electrochemical cell with those in the gas/solid interface. This link is of valuable help for understanding the physicochemical properties of the metal in the electrochemical environment and the behavior of adsorbed species on electrode surfaces.

The interest of CO in fuel cells arises from the search for a catalyst for methanol oxidation, which should be able both to break C-H bonds and to oxidise the adsorbed intermediate CO_{ads} species to CO_2. Additionally, the H_2/O_2 fuel cell requires the development of CO-tolerant anodes, i.e., catalysts oxidizing CO traces present in H_2 samples produced in the reforming process of alcohols or hydrocarbons.

In the present section, some aspects of carbon monoxide oxidation that are relevant to the electrocatalysis of fuel cell reactions will be described. We restrict our discussion to acid media only.

6.5.1 Identification of Surface Structures for CO Adsorbed on Pt(111)

Data from scanning tunnelling microscopy (STM) and infrared spectroscopy (IR) for CO adsorbed on Pt(111) were obtained by Villegas and Weaver. Their results are shown in Figs. 6.12 (STM) and 6.13 (in situ FTIR). A change of potential from +0.05 to +0.5 V vs. RHE induces a phase transition in the adlayer structure, as shown by the ball models in Fig. 6.12. The unit cells at 0.05 V and 0.4 V vs. RHE are respectively (2×2)-3CO and $(\sqrt{19} \times \sqrt{19})23.4°$–13CO. The transition is accompanied by changes in the infrared spectra as shown in Fig. 6.13. At 0.05 V, IR bands correspond to terminal (2073 cm^{-1}) and three-fold bonded CO (1773 cm^{-1}); while at 0.4 V, in addition to terminal CO bridge bonded CO (1850 cm^{-1}) is observed. These structures were observed in the presence of CO in the bulk of the solution and the oxidation of carbon monoxide under such conditions occurs simultaneously with the transition of the CO adlayer structure.

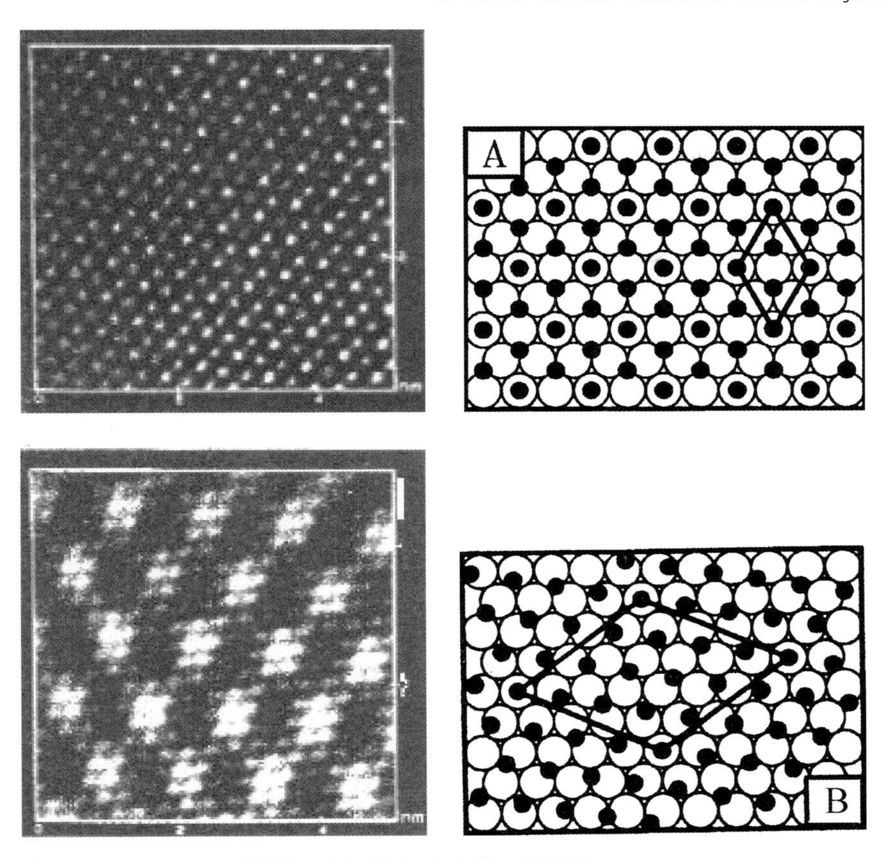

Fig. 6.12 STM images and ball models of adsorbed CO at Pt(111), taken in the presence of dissolved CO. (a) (2 × 2)-3CO structure at 0.05 V, (b) (19 × 19)-13CO structure at 0.4 V RHE.

6.5.2
Oxidation of CO in the Presence of Dissolved CO

The oxidative behaviour of carbon monoxide at Pt depends on several experimental parameters, including the adsorption potential, the degree of coverage and the presence of CO in the bulk of the electrolyte, all of which strongly affect the rate of reaction.

An interesting feature of CO electro-oxidation on Pt electrodes is that much higher overpotentials are required for stripping a CO adlayer in pure supporting electrolyte than for oxidizing CO molecules present in the bulk of the solution. This result is due to repulsion between neighboring CO molecules at the surface, which causes a pronounced decrease of adsorption energy as the coverage with CO_{ad} reaches saturation.

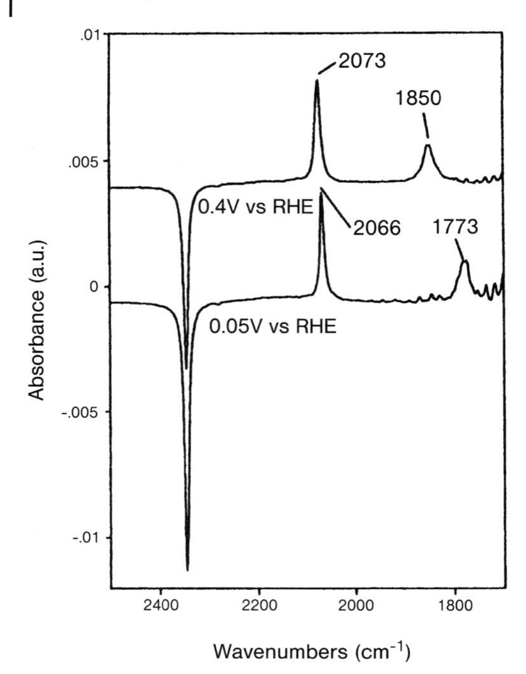

Fig. 6.13 In situ FTIR spectra of the CO adlayer, formed on Pt(111) in CO-saturated 0.1M $HClO_4$ at 0.05 V and at 0.4 V RHE, see band assignment in the text, band for CO_2 at 2341 cm^{-1}. I. Villegas, M.J. Weaver, *J. Chem. Phys., B* **101** (1994) 1648.

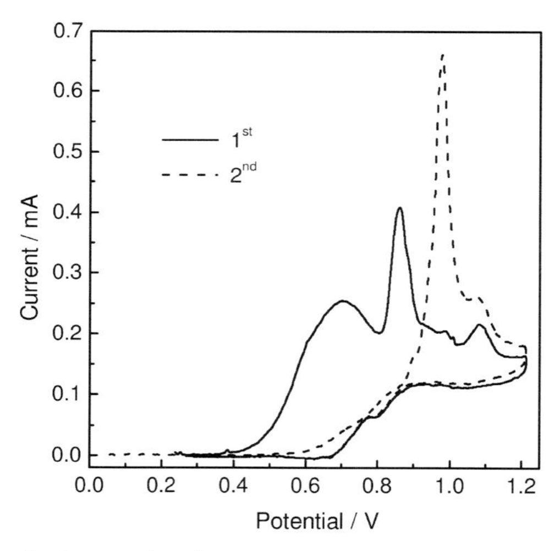

Fig. 6.14 Cyclic voltammograms at Pt(111) in CO-saturated 0.1M $HClO_4$ solution. Scan rate 50 mVs^{-1}. CO admission at 0.05 V, comparison of first and second cycle.

In Fig. 6.14 we show the first and the second cyclic voltammogram measured on a Pt(111) electrode in CO saturated 0.1 M $HClO_4$ solution, after admitting CO in the solution with the electrode *polarized at 0.05 V*. Clear differences in the current for the positive-going sweeps can be observed. Thus, for the first scan, the current starts growing at about 0.35 V and a wide peak at around 0.7 V, followed by a sharp peak at 0.85 V are observed. In contrast, the second sweep exhibits an almost featureless region of low current between 0.5 V and 0.85 V, followed by one sharp peak at about 1.0 V. The current observed at low potentials during the first potential scan, is related to oxidation of weakly bonded CO molecules, present in the CO saturated system. The inhibition of current during the second and subsequent scans is only produced when the switching potentials are higher than ca. 0.6 V. This phenomenon has been attributed to the formation, at high potentials of clusters of strongly H-bonded water, which block defect sites and steps at the Pt surface, i.e., active sites for CO oxidation.

When CO is admitted in the cell at potentials above ca. 0.37 V, the first scan already shows an inhibition of the current at low potentials as in the second scan.

6.5.3 The Oxidation of Carbon Monoxide: Langmuir-Hinshelwood Mechanism

Oxidation of carbon monoxide molecules in a CO *adlayer* requires interaction with adsorbed water as a reaction partner. A reaction mechanism can be written as follows:

$$H_2O + * \rightleftharpoons OH_{ads} + H^+ + e^- \tag{6.27}$$

$$CO_{ads} + OH_{ads} \rightarrow CO_2 + H^+ + e^- + 2* \tag{6.28}$$

where * stands for a free site on the surface, i.e. the oxidation of a CO *adlayer* proceeds through a Langmuir-Hinshelwood mechanism. It was suggested that the reaction takes place mainly at steps (or defect) sites and given the fact that on platinum electrodes CO is adsorbed in the form of islands, oxidation of CO requires a monolayer of adsorbed CO requires a diffusion of CO molecules to the active sites. This step has to be considered when discussing the reaction kinetics since it cannot be discarded as a possible rate-determining step.

As stated in 6.5.2, larger currents are observed at low potentials in the presence of CO in the *bulk* of the solution, than for oxidizing a CO *adlayer*. It is expected, under the former conditions, that only a few Pt sites become available for water adsorption and a Langmuir-Hinshelwood mechanism could be, therefore, inhibited. It was, thus necessary to check what kind of mechanism was operating in CO saturated solutions. For this purpose, the potential step method described in section 5.1.2 was used.

In the experiment shown in Fig. 6.15, after saturating the Pt surface at 0.05 V with pure carbon monoxide, the CO concentration in the solution was systematically changed. For this purpose, gas mixtures with different CO/N_2 ratios were used. With the electrode positioned on the surface in a meniscus configuration, a potential step from 0.05 V to 0.5V was applied and the current response monitored over a couple of minutes. In Fig. 6.15 the current is plotted as a function of $t^{1/2}$ (see section 5.1.2).

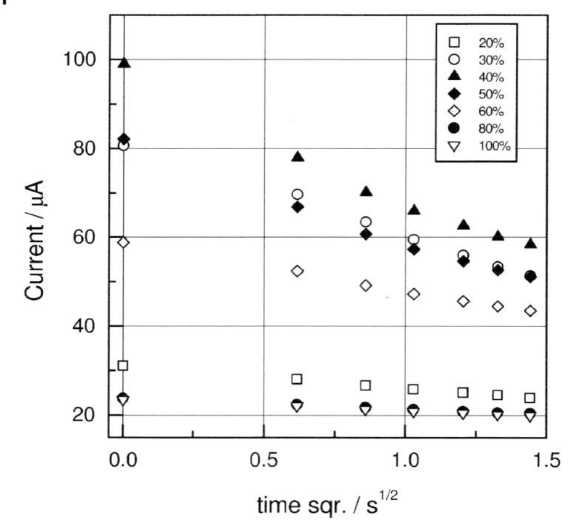

Fig. 6.15 Dependence of the current at 0.5 V on the square root of time, for bulk CO oxidation at Pt(111) in 0.1M $HClO_4$ solution. CO concentration in the gas mixture (CO/N_2) used to saturate the solution are as indicated in the inset.

The current, at short times, decays with $t^{1/2}$ as to be expected by the combined control of electrochemical reaction and diffusion. The current values extrapolated to $t = 0$, are plotted in Fig. 6.16 as a function of the CO concentration in the gas mixture. The results show that for a concentration of CO of 40 % in the gas mixture a maximum rate of oxidation of 126 $\mu A/cm^2$ is attained.

Saturated CO layers are in equilibrium with CO in the solution, as demonstrated by isotope exchange experiments. Therefore, in spite of the fact that, at the beginning of

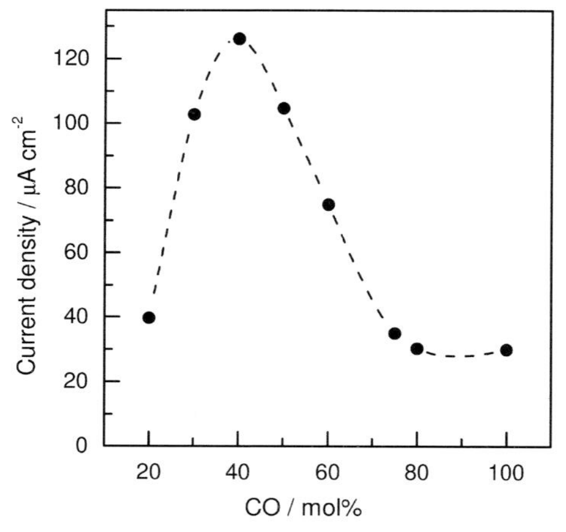

Fig. 6.16 Plot of the initial current density values as extrapolated from the plot in Fig. 6.15 for $t = 0$, as a function of the CO concentration in the CO/N_2–gas mixture used to saturate the solution.

the experiment, the surface is saturated by bubbling pure CO in the solution, some CO molecules desorb during the subsequent step as the CO/N_2 mixture replaces the pure gas. Thus, high but slightly different degrees of coverage can be expected for each concentration of CO in the solution. The increasing part of the curve in Fig. 6.16 can be explained as due to the rapid lowering of the desorption energy with θ_{CO} as the bulk concentration increases. Beyond the maximum, competition with water for adsorption sites to form OH_{ad} hinders the oxidation reaction. The curve obtained implies a *L-H mechanism* as in the case of a CO adlayer oxidation.

6.5.4
CO Oxidation at Higher Overpotentials, Influence of Mass Transfer and Oxygen Coverage

In order to obtain more information about the interaction of reaction kinetics, mass transfer and oxygen coverage of platinum surfaces for *CO oxidation above 0.5 V*, the use of a rotating electrode is suitable. Fig. 6.17 shows cyclic voltammograms for a polycrystalline Pt disc electrode at a rotation of 1000 rpm and up to different positive potentials (solid lines). The sharp peak near 0.9 V is obviously caused by anodic stripping of the CO covered surface after adsorption at 0.05 V and starting the cyclic voltammogram. Following this pseudo-capacity effect, the rate is mainly determined by mass transfer in solution and by the oxygen coverage of the surface. The difference in the current for the lines taken with 400 and 1000 rpm is obvious. In addition with more anodic potentials applied, the oxygen coverage formed hinders CO oxidation, until finally oxygen evolution starts. After an anodic scan to 1.75 V the oxygen layer is sufficiently extensive that for the cathodic scan in all cases the current due to CO oxidation has about the same very low value.

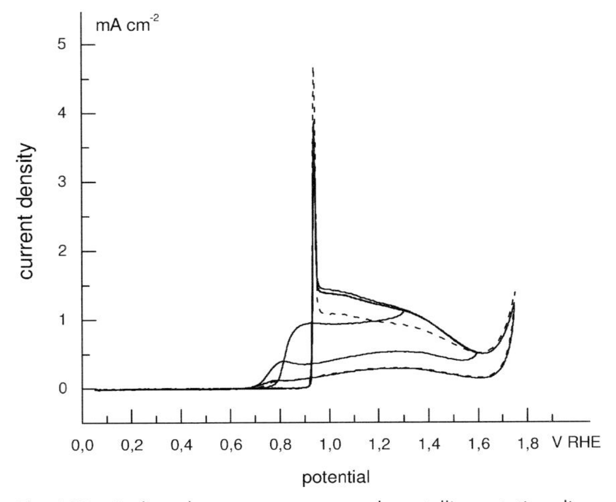

Fig. 6.17 Cyclic voltammograms at a polycrystalline rotating disc electrode, between 0.05 V and different positive potentials vs. RHE; CO-saturated 0.5M H_2SO_4, scan rate 50 mVs^{-1}, 1000 rpm. For comparison one CV (dotted line) is shown for a rotation speed of 400 rpm.

6.6
Conversion of Chemical Energy of Ethanol into Electricity

The energy of a chemical reaction depends only on the difference in energy of the reacting species before and after the complete reaction. Different pathways are possible, e.g. purely chemical or purely electrochemical routes. In the case of ethanol the chemical reaction with oxygen can be used in propulsion motors for automobiles. For the electrochemical route, the complete oxidation at the anode to CO_2 is

$$C_2H_5OH + 3H_2O \rightarrow 2CO_2 + 12H^+ + 12e^- \quad (6.29)$$

Besides ethanol, in acid solution, only water has to be present additionally at the interface of the anode in acid solution. The standard free energy of the reaction with oxygen at the cathode is $\Delta G^\circ = -1325$ kJmol^{-1}, i.e. under standard conditions 8.00 kWh per kg can be obtained. From this it follows that for the cell reaction $C_2H_5OH + 3O_2 \rightarrow 2CO_2 + 3H_2O$ the thermodynamic standard cell voltage $E_c^0 = 1.145$ V, and the standard potential for ethanol is $E^0 = +0.084$ V vs. RHE.

Using platinum as a basic catalyst for ethanol oxidation, the observed open circuit potential at room temperature, is $E_o > 0.4$ V, i.e., the overvoltage is similar to that of methanol. The main problem in electrocatalysis of ethanol oxidation is the scission of the C-C bond; therefore, low yields of CO_2 are observed, in particular at smooth Pt electrodes. The partial oxidation to acetaldehyde and acetic releases 2 and 4 e^-, respectively:

$$C_2H_5OH \rightarrow CH_3CHO + 2H^+ + 2e^- \quad (6.30)$$

and

$$CH_3CHO + H_2O \rightarrow CH_3COOH + 2H^+ + 2e^- \quad (6.31)$$

According to reactions (6.29 – 6.31) several pathways for the ethanol reaction are possible (see Fig. 6.18).

The cyclic voltammogram for ethanol oxidation at polycrystalline Pt (Fig. 6.19) shows a sluggish increase of current between ca. 0.4 and 0.6 V. The activity at potentials below 0.6 V can be improved by adding to Pt metals like Ru and Sn. For Ru as promoter, the current density measured at a constant potential of 0.5 V is a function of the Ru atomic percentage. This dependence has, in contrast to the case of methanol (see Fig. 6.11), a steep maximum at about 40 %.

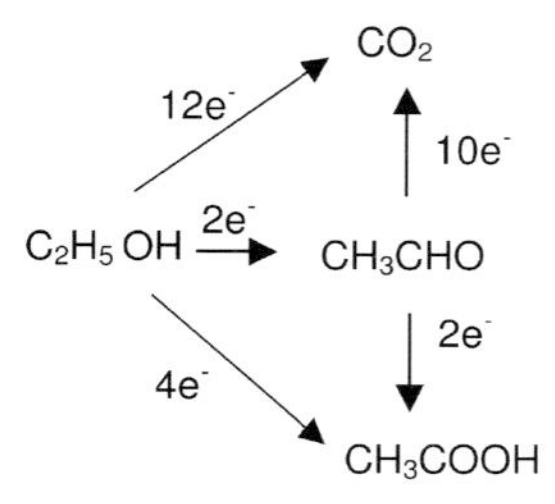

Fig. 6.18 Schematic representation of parallel pathways of ethanol oxidation.

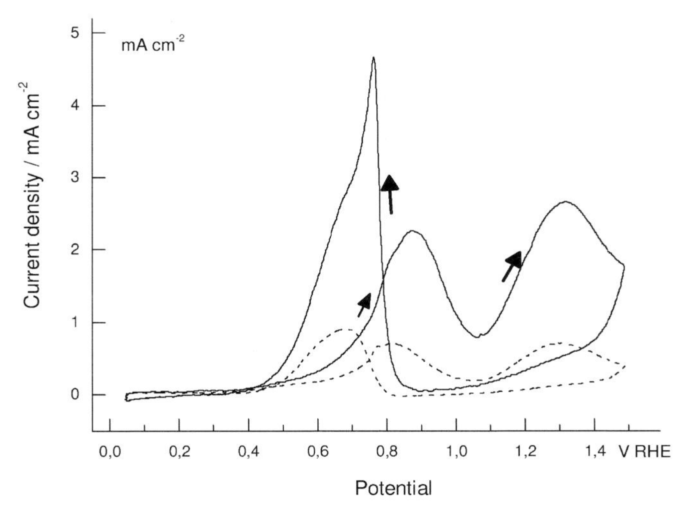

Fig. 6.19 Cyclic voltammogram of ethanol oxidation at smooth polycrystalline Pt for 0.1M (dashed line) and 1M ethanol in 0.1M $HClO_4$; 50 mVs^{-1}, 25°C.

The relative amounts of the different products strongly depend on the Ru atomic percentage of the anode. The yields of soluble products were studied by *in situ* IR spectroscopy. By using so-called "effective" absorption coefficients (values calculated for the thin layer of electrolyte used in the technique), integrated band intensities can be converted to molar amounts Q in mol/cm^2, as shown by Weaver and co-workers.

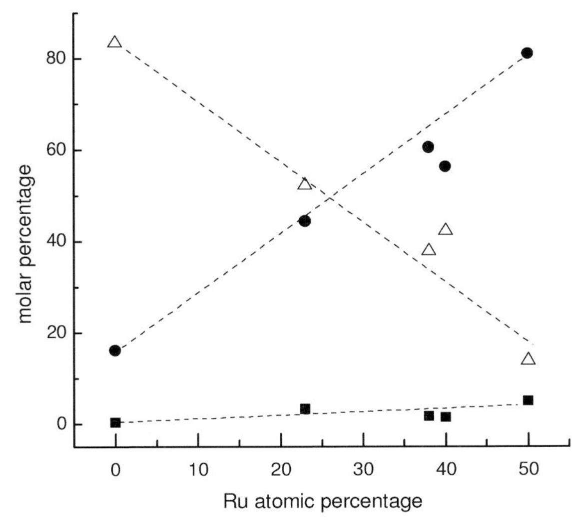

Fig. 6.20 Molar percentages of (■) CO_2, (•) acetic acid and (Δ) acetaldehyde, taken after 3 min of polarization at 0.5 V at different Ru atomic percentages, solution as in Fig. 6.19.

This procedure was used to calculate the data in Fig. 6.20, from spectra measured after 3 min of polarization at 0.5 V. The yields of CO_2, CH_3CHO and CH_3COOH are plotted as function of the percentage of Ru at the electrode. The molar percentages of acetaldehyde are affected by an error of ca. 7 %, but the corresponding errors of acetic acid and CO_2 are near 1 % only.

It is noteworthy that the pathway forming CO_2 is slightly favoured by the increase of Ru at the interface. However, the yield of CO_2 remains by 1–2 orders below the sum of the yields for acetic acid and acetaldehyde. Obviously, PtRu catalysts have a poor activity towards the scission of the C-C bond. On the other hand, the main effect of Ru is reflected in the increase of acetic acid production. The sum of the yields of acetic acid and acetaldehyde is almost constant for the Ru contents studied. This is an indication that acetic acid is formed in a pathway having acetaldehyde as intermediate (Fig. 6.18).

From the data above follows that the number of electrons obtained per molecule of ethanol is rather low, the values of 2–4 being well below the theoretical 12 electrons per molecule. For an acceptable conversion of chemical energy into electricity, not only the number of electrons is of importance, but also the current density at potentials interesting for fuel cell application. The current/potential relationship for PtRu catalysts is still quite low for possible applications.

6.7
Reaction Mechanisms in Electro-organic Chemistry

There are vastly more organic than inorganic compounds, and since almost all organic compounds can be electrochemically oxidised and/or reduced, the field of electro-organic chemistry is potentially vast. The textbooks in this area usually sub-divide the subject into such fields as "electrochemical addition", "electrochemical substitution", "electrochemistry of amines" etc., but we will concentrate here on the fundamentals underlying all such processes. They can be summarised in the statement that although electro-organic chemistry introduces no fundamentally novel synthetic pathways, it does allow the generation of radicals and radical ions simply, rapidly and in high spatial yield.

6.7.1
General Issues

All the considerations developed above such as electron transfer, diffusion and reaction layers, adsorption, electrocatalysis etc. are transferable to the organic sphere. However, from the experimental viewpoint, it is often necessary to use organic co-solvents with water, or even to use nonaqueous solvents alone, if a reasonable concentration of reactants and products is to be maintained. Even if highly polar co-solvents, such as alcohols, are used, this significantly reduces the solubility of the supporting electrolyte salt, and conductivities at least an order of magnitude smaller than those found in simple inorganic aqueous studies are usually encountered. Indeed, if purely non-aqueous solvents are used, the conductivity may be yet further reduced, and very great care is needed properly to incorporate iR losses into the techniques used. This is especially true of cyclic voltammetry, where, as pointed out in Section 5.2.1.8, interpretation of shifting peaks in the CV with increasing scan speed is fraught with difficulty in resistive media.

One way in which solubility of organic materials in aqueous media can be enhanced is through the use of two-phase systems, usually in the form of *emulsions*, in which the substrate is usually very soluble in the organic phase, but electron transfer takes place in the aqueous phase. The aqueous solution remains permanently saturated even near the electrode surface, provided equilibrium between emulsion particles and aqueous phase is fast.

In general, the goal of good selectivity in a particular pathway will depend critically on the appropriate choice of electrode material, solvent, supporting electrolyte, electrode potential and current density.

6.7.2 Classification of Electrode Processes

The primary distinction is between *direct* and *indirect* (including *electro-initiated*) processes. In the first case, there is direct electron transfer to or from the substrate; if the latter is uncharged, then a radical anion or cation will form respectively as, for example:

$$C_6H_6^{\bullet -} \leftarrow e^- + C_6H_6; \quad C_6H_6 - e^- \rightarrow C_6H_6^{\bullet +} \tag{6.32}$$

Quite generally, such "Single Electron Transfers (SET's)" either move the electron from the cathode to the lowest unoccupied molecular orbital (LUMO) of the compound or from the highest occupied MO (HOMO) of the compound to the anode, and the electron transfer will take place at some specific electrode potential corresponding to the energy of that orbital, more specifically to the electron affinity of the LUMO or the ionisation energy of the HOMO. It should, of course be noted that there is no quantitative relationship between such gas-phase concepts as Ionisation Energy and the electrode potential, since the electron ends up not at an infinite distance away in the gas-phase but rather in the metal, and the resultant ion will in any case be strongly solvated.

Indirect processes have at least two separate stages. The first involves the reduction or oxidation of some substance to give a reactive radical anion or cation that then reacts (not necessarily close to the electrode surface) with the substrate M to yield the desired product. In case of a reduction we have:

$$e^- + Y \rightarrow Y^{\bullet -} \tag{6.33}$$

$$Y^{\bullet -} + M \rightarrow M^{\bullet -} + Y \tag{6.34}$$

with Y re-cycled from reaction (6.34) to reaction (6.33). This scheme may have an important advantage if direct reduction of M is kinetically sluggish. Under these circumstances, it may be necessary in the absence of species Y to resort to rather large negative potentials before any appreciable reaction takes place, with such a potential possibly giving rise to highly undesirable side-reactions. However, if Y can be reduced at lower potentials, a much more selective reaction can often be induced. Y may, of course, be attached chemically to the surface, and the process of indirect electrosynthesis then becomes all but indistinguishable from normal heterogeneous catalysis; the process is often referred to for this reason as redox catalysis, and it forms the basis of a large number of recently developed amperometric sensors.

Indirect reduction is nowadays commonly effected by metal catalysts, particularly Ni, Co and Pd complexes which undergo oxidative addition by the carbon-halogen bond in alkyl and aryl halides, and further reduction through the catalytic cycle shown allows the synthesis of di-aryl derivatives such as biphenyl from bromobenzene:

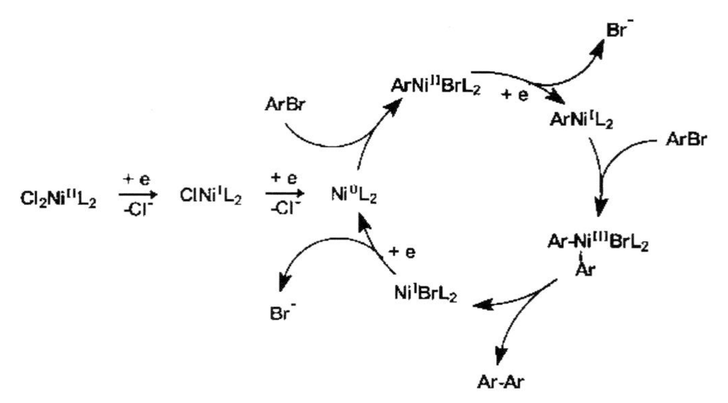

A scheme similar to (6.33/34) may be drawn for indirect *oxidation*, a process that is common with halides, particularly bromide and iodide. One example is the oxidation of primary amines to nitriles, which is usually carried out in undivided cells with closely spaced Pt anodes and cathodes:

At anode $2Br^- \rightarrow Br_2 + 2e^-$

At cathode $2\ MeOH + 2e^- \rightarrow 2MeO^- + H_2$

In solution $RCH_2NH_2 + Br_2 + MeO^- \rightarrow RCH_2NHBr + MeOH + Br^-$

$RCH_2NHBr + MeO^- \rightarrow RCH{=}NH + MeOH + Br^-$

Subsequently $RCH{=}NH + Br_2 + 2MeO^- \rightarrow RCN + 2MeOH + 2Br^-$

Electro-initiated processes are those in which $Y^{\bullet-}$ in the scheme above is not electrochemically but *chemically* re-generated, and in the ideal case the electron transfer from or to the electrode is only needed to initiate the reaction. An example is the synthesis of phenol from benzene in the presence of H_2O_2 and Fe^{3+}, which has the mechanism:

$$Fe^{3+} + e^- \rightarrow Fe^{2+} \quad (6.35)$$

$$Fe^{2+} + H_2O_2 \rightarrow Fe^{3+} + OH^{\bullet} + OH^- \quad (6.36)$$

$$C_6H_6 + OH^{\bullet} \rightarrow C_6H_5(H)(OH)^{\bullet} \quad (6.37)$$

$$C_6H_5(H)(OH)^{\bullet} + Fe^{3+} \rightarrow [C_6H_5(H)(OH)]^+ + Fe^{2+} \quad (6.38)$$

$$[C_6H_5(H)(OH)]^+ \rightarrow C_6H_5OH + H^+ \quad (6.39)$$

This is evidently closely related to the propagation of an unbranched chain reaction.

6.7.3 Oxidation Processes: Potentials, Intermediates and End Products

Oxidation potentials depend qualitatively on the ionisation energy of the HOMO. For saturated chain hydrocarbons, these IEs are large and lead to very high oxidation potentials (> 3.0 V vs. SHE) that are not attainable save in very specialised solvent systems such as supercritical CO_2 for which ultra-microelectrodes must be used. However, the oxidation potentials of unsaturated or aromatic hydrocarbons are much lower, and a collection of such data can be found in standard compilations. Such data must be used with caution, since experimentally it is derived from half-wave potentials, which are not thermodynamic in origin but kinetic, and dependent on electrode material and electrolyte system. In addition, potential scales in different solvents are often difficult to interconvert, as discussed in Section 3.1.13, and it is usually advisable to carry out some preliminary investigations in real systems in order to establish the synthesis potential.

Once oxidation (or reduction) has taken place, the resultant radical is likely to be highly reactive and short-lived and, for *direct* processes, this implies that reactions will generally take place within a few nm of the electrode surface. The reactivity of these radicals can lead to a multiplicity of pathways, as shown in Fig. 6.21, and the dominant pathway taken will depend on reaction conditions. Thus, for the oxidation of toluene in glacial ethanoic (acetic) acid/ethanoate solution, the charge stabilisation reaction (f) in Fig. 6.21 dominates:

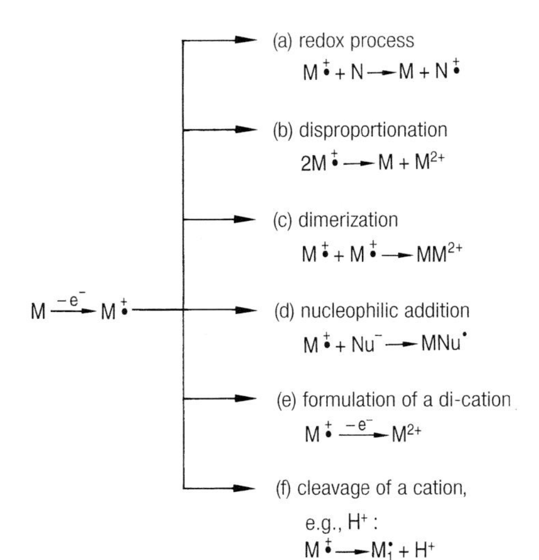

Fig. 6.21 Further reactions of anodically generated radical cations: Nu^- is any adventitious nucleophile.

$$C_6H_5CH_3 \xrightarrow{-e^-} [C_6H_5CH_3]^{+} \longrightarrow C_6H_5CH_2^{\bullet} + H^+ \qquad (6.40)$$

and since the electron in a neutral radical species such as that formed in (6.40) is usually more easily removed than the parent species, rapid further electron transfer is possible as:

$$C_6H_5CH_2^{\bullet} \xrightarrow{-e^-} [C_6H_5CH_2]^{+} \xrightarrow[-H^+]{HOAc} C_6H_5CH_2OAc \qquad (6.41)$$

and the final product is the benzyl ester of ethanoic acid. A second pathway that can be followed in this solvent system corresponds to pathway (d) in Fig. 6.21 (nucleophilic attack on the ring following the initial oxidation process):

$$C_6H_5CH_3 \xrightarrow{-e^-} [C_6H_5CH_3]^{+} \xrightarrow[-H^+]{HOAc} CH_3C_6H_5(H)(OAc) \qquad (6.42)$$

and the neutral radical now formed is also subject to further rapid electron transfer as:

$$CH_3C_6H_5(H)(OAc)^{\bullet} \xrightarrow{-e^-} [CH_3C_6H_5(H)(OAc)]^{+} \qquad (6.43)$$

and the resultant carbo-cation undergoes charge stabilisation by loss of a proton to form the phenoxyester:

$$[CH_3C_6H_5(H)(OAc)]^{+} \xrightarrow{-H^+} CH_3C_6H_4OAc \qquad (6.44)$$

Which route dominates depends sensitively on the nucleophile present and on the concentration of H^+. For weak acids, such as ethanoic acid, the benzyl ester predominates, but for stronger acids, such as 1,1,1-trifluroethanoic acid, which inhibits reaction (6.40), the phenoxyester forms predominantly.

Another difficulty that may be encountered is that of rearrangement of the carbonium ion formed. The oxidation of cyclohexene in methanol illustrates this, with the carbonium ion formed in one pathway showing a Wagner-Meerwein rearrangement:

If non-coordinating solvents, such as alkylammonium salts in saturated chlorinated solvents, are used, a radical-initiated polymerisation process is found, leading to an electroactive polymer of indeterminate composition on the surface of the electrode. This type of process is discussed further in section 6.7.6 below.

6.7.4
Reduction Processes: Potentials, Intermediates and Products

The reduction potential for a compound will depend, again qualitatively, on the electron affinity, and for saturated hydrocarbons such potentials are extremely negative. Unsaturated or aromatic hydrocarbons are more easily reduced, as are compounds containing multiple bonds between carbon and N, O or S. As for oxidation, the resultant radical can follow a number of different pathways, which are shown in Fig. 6.22, and as an example, we investigate the electro-reduction of acrylonitrile ($CH_2{=}CHCN$), which has considerable commercial significance.

The first step leads to the anion:

$$CH_2CHCN + e^- \rightarrow [CH_2{=}CHCN]^{\bullet -} \qquad (6.45)$$

and this can be followed by electrophile attack if protons are present (pathway (d) of Fig. 6.22):

$$[CH_2{=}CHCN]^{\bullet -} + H^+ \rightarrow {}^{\bullet}CH_2 - CH_2 - CN \qquad (6.46)$$

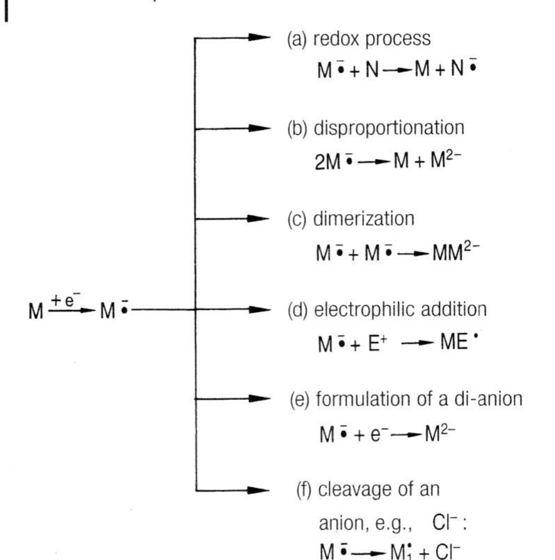

Fig. 6.22 Further reactions of cathodically generated radical anions: E^+ is any adventitious electrophile.

Alternatively, radical nucleophilic attack on a second substrate molecule may take place:

$$[CH_2{=}CHCN]^{\bullet -} + CH_2{=}CHCN \rightarrow NC{-}CH^{-}{-}CH_2{-}CH_2{-}CH^{\bullet}{-}CN \quad (6.47)$$

and a third possibility is a second electron transfer following (6.45) as:

$$[CH_2{=}CHCN]^{\bullet -} + e^- \rightarrow [CH_2{=}CHCN]^{2-} \quad (6.48)$$

Further reactions are described in more detail below (Section 8.5.2), the final product under appropriately controlled circumstances being adiponitrile, $NC{-}CH_2{-}CH_2{-}CH_2{-}CH_2{-}CN$, a precursor for nylon manufacture.

Reduction of substances with OH groups is frequently followed by charge compensation reactions as

$$ROH + e^- \rightarrow [ROH]^{\bullet -} \rightarrow RO^- + \tfrac{1}{2}H_2 \quad (6.49)$$

If ROH is an aliphatic alcohol, an alcoholate can be formed. If ROH is a carbohydrate, then further reaction with alkylating or acylating reagents can generate ethers or esters of the carbohydrate.

6.7.5
Further Electroorganic Reactions and the Influence of the Electrode Surface

An electrochemical reaction of enormous historic importance is the Kolbe reaction, first investigated by Faraday in 1834, who noted that during the electrolysis of potassium acetate, hydrogen was evolved at the cathode and a mixture of CO_2 and hydrocarbons was formed at the anode. Later Kolbe carried out a systematic study of the reaction, and established the basic mechanism, and Hofer and Moest showed that under rather different conditions, different products could be isolated.

The basic mechanism of the reaction is:

$$RCOO^- \rightarrow RCOO^{\bullet} + e^- \rightarrow R^{\bullet} + CO_2; \quad 2R^{\bullet} \rightarrow R{-}R,$$

The reaction takes place at rather high overpotentials, and for electrodes such as smooth platinum, oxygen evolution is a competing process. In fact, the anodic current density in 1M NaOAc/1M HOAc in water shows two distinct regions for both Pt and Ir anodes; at lower potentials, the reaction is entirely comprised of oxygen evolution and formation of an oxide layer. At higher potentials on both metals, once a phase oxide has formed on the surface, the carboxylate ions are more strongly bound than water and form a monolayer that effectively blocks the oxidation of water allowing preferential oxidation of the carboxylate anion to take place as above. Given the high coverage of carbooxylate, dimerisation of the radical $R^{\bullet}$ is the preferred route. If a competing adsorbate, such as carbonate, is added, then the coverage by carboxylate is diluted and the radical can be further oxidised to a carbo-cation. These are the classic conditions for the Hofer-Moest reaction, and oxidation of acetate in water now gives methanol rather than ethane.

Another specialised electrode material, specifically for the oxidation of alkanols in aqueous alkali, is hydrous nickel oxide, which is deposited as Ni(III) oxide onto nickel plate above 1.13 V vs. SCE. The initial step in oxidation of alkanols is removal of the α-hydrogen atom by the Ni(III) followed by oxidation of the radical so formed as:

$$RCH_2OH \xrightarrow{-H^{\bullet}} R\dot{C}HOH \xrightarrow[-H^+]{-e} R{-}CHO \rightarrow R{-}COOH$$

This nickel-oxide anode was developed on the technical scale for the oxidation of a *D*-glucose derivative to the corresponding carboxylic acid as a step in the synthesis of *L*-ascorbic acid.

6.7.6 Electrochemical Polymerisation

Electrochemically initiated polymerisation is similar conceptually to the familiar anionic or cationic polymerisation processes encountered, for example, in "living" polymers. The radical anion formed electrochemically initiates growth in the solution, and only a very small current is necessary to sustain the reaction. An example is the polymerisation of methylmethacrylate, which can be initiated either by an electrochemically generated anion or even by hydrogen atoms formed at the electrode surface. This type of electrochemical polymerisation can be very facile, and is often an unwanted side reaction, leading in severe cases to complete blocking of the electrode surface and cessation of the electrochemistry.

Polymers formed in this way are entirely analogous to polymers formed by more conventional processes, and the backbone of such polymers is usually a saturated C-C chain. Such chains can be made electroactive by attaching pendant groups with redox centres, but if such a polymer film is plated onto an electrode surface, electron transfer through the film can only take place by distortions of the polymer backbone allowing redox centres on neighbouring monomeric units to come together and exchange an electron. However, if materials such as ethyne (acetylene: CH≡CH) are polymerised, the polymer backbone will possess a set of sequential π-bonds as:

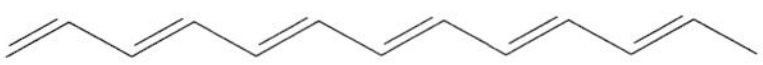

and the overlap between these π-bonds permits charge carriers generated, for example, by oxidation of the backbone, to become mobile. Clearly, at least in principle, if electrons can be transferred between pendant redox centres and such a backbone, then this will greatly facilitate charge transfer to and from an electrode to the polymer film.

Although polyacetylene was the first such conducting polymer to be made, polymerisation was actually by conventional non-electrochemical processes, but shortly after the discovery of polyacetylene, direct electrochemical polymerisation was found to be possible for a wide range of unsaturated heterocyclic compounds. One example is the polymerisation of pyrrole to give a conducting polymer of the basic structure:

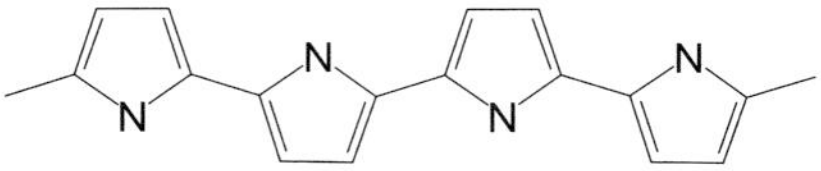

N N N N

Polymerisation is initiated through oxidation of a pyrrole monomer to give a radical cation; this monomer can then either react with a further radical cation, or with a second pyrrole monomer, generating an intermediate that is easily oxidised with loss of protons to form a dimer as:

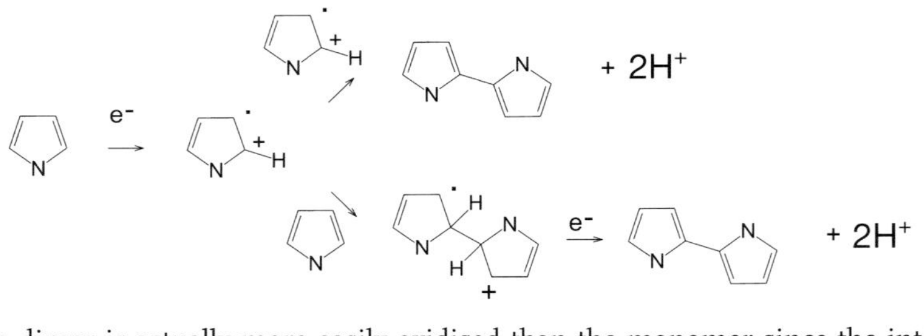

N + H N N + 2H⁺
N e⁻ → N + H
N N H N H + e⁻ → N N + 2H⁺

The dimer is actually more easily oxidised than the monomer since the interactions between the π-systems on the two rings gives rise to a set of orbitals, the highest occupied of which is of higher energy (i.e. more easily oxidised) than the HOMO of the pyrrole monomer. Further growth can then take place both by further oxidation of monomers but also by oxidation of the dimer, the dominant process depending sensitively on the conditions of transport and potential. As polymerisation proceeds, and the chains lengthen, they usually rapidly becomes insoluble, plating out on the electrode surface to form a film. Because the polymeric forms of polypyrrole are more easily oxidised than the monomer, the chains themselves plate out in an oxidised form, and because the backbone is unsaturated, as in polyacetylene, it becomes electronically conducting, allowing charges to be transported from the ends of the growing chains to the electrode surface. This allows extremely thick films to be formed, often very quickly indeed. Given also that the dominant polymerisation mechanism depends sensitively on conditions, it is not surprising that different authors have reported very different film morphologies.

Conduction in polypyrrole and related films has been the subject of intensive investigation, using a wide variety of techniques, including spectroscopic and conventional conductimetric studies. It is well known in the area of solid-state chemistry that for traditional metals, electrons can move freely in the solid lattice with velocities sufficiently rapid that the nuclei do not have time to move to accommodate the presence of the charge. Such behaviour is characteristic of high levels of orbital overlap giving rise to bands of electronic levels of substantial width in energy terms. However, as orbital overlap decreases, the electrons hop less frequently, eventually moving sufficiently slowly that the lattice has the time to adjust its local structure to take account of the presence of the charge. In effect, the electron becomes 'trapped', and its mobility then decreases dramatically, with the conductivity falling to levels more similar to those found in semiconductors. This is found to be the case in such polymers as polypyrrole: the relatively high conductivity seen being a function of the very high density of charge carriers rather than their intrinsic high mobility. The form of these charge carriers is also of considerable interest: a moment's reflection will show that a single positive charge is difficult to accommodate on the chain since it causes a major rearrangement of the bonding, but a *double* charge can be easily accommodated as shown:

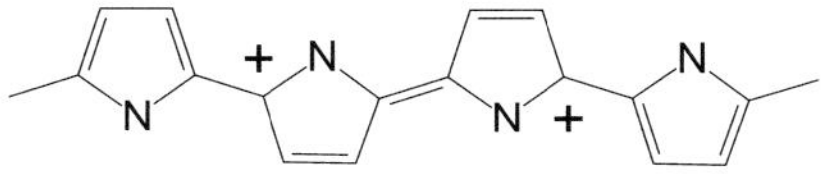

N + N N + N

The actual distance between the positive charges may be larger than that shown: it will be a balance between charge repulsion and the energy required to re-hybridise the bonding in the rings.

The fact that the film can be oxidised to form polymeric chains with substantial electronic conductivity is of considerable potential technical importance; such films have applications not only in "smart windows" and "molecular transistors" but also in bio-electrochemical and bio-electronic devices, analytical devices, including amperometric biosensors, ion-exchange membranes, and charge storage devices,. These latter possibilities arise from the fact that as the chain becomes oxidised, counter-ions from the electrolyte must diffuse into the film to maintain charge compensation. Developments of this technology for both super-capacitors and batteries are described below (Sections 9.3.2.4 and 9.8).

6.8 Oscillations in Electrochemical Systems

Oscillatory phenomena are ubiquitous in electrochemical systems. In fact, it is plausible to say that any electrochemical system might present, under certain conditions, dynamic instabilities in the form of current or potential oscillations. Examples displaying oscillatory kinetics include metal electrodeposition, corrosion and galvanic cells.

Owing to the interplay between electrical and chemical variables, the description of kinetic instabilities in electrochemical systems has to account not only for surface coverage and concentrations of electroactive species, but also for potential drops across the interface, applied voltages and current densities as well as for internal and external ohmic resistances.

The equivalent circuit of Fig. 6.23 shows that the current i from the connection with the potentiostat is passes through an external ohmic resistance R_e, the working electrode with its associated double layer capacitance C_D^W and faradaic impedance Z_F^W, and the electrolyte resistance R_E between working electrode and the tip of the Reference-electrode Luggin capillary. The current i finally traverses the electrolyte towards the counter electrode. Given the high faradaic impedance Z_{Ref} for the reference electrode, practically no current flows into the reference electrode. In this way the voltage U applied to the potentiostat is divided into the potential drops across the two double layers and the ohmic drops on R_e and R_E. Further, we can define $\Delta\varphi_D^W, \Delta\varphi_D^{Ref}$ as the potential difference between the Galvani potentials of the metal φ_{Me} and the solution φ_S, (i.e. $\Delta\varphi_D = \varphi_{Me} - \varphi_s$) for working and reference electrodes respectively. Therefore, defining $\Delta\varphi_D^W - \Delta\varphi_D^{Ref} = \Delta\varphi$, U can be written as

$$U = \Delta\varphi + iR \tag{6.50}$$

Using a simple mathematical approach (see references), it can be shown that under potentiostatic conditions in the potential region of a negative faradaic impedance Z_F^W current oscillations occur if the total resistance $R = R_e + R_E$ exceeds the value of $|Z_F^W|$.

The current oscillations are produced in spite of a fixed voltage U. According to above relation oscillations may occur also without any external resistance R_e if $R_E > |Z_F^W|$.

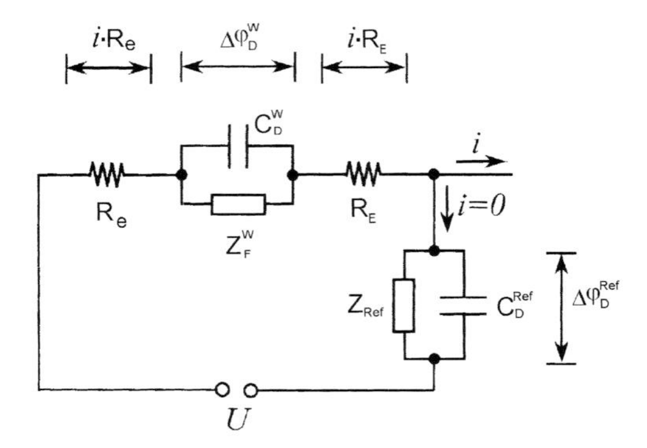

Fig. 6.23 Equivalent circuit of an electrochemical interface under potentiostatic control using a reference electrode Ref. C_D double layer capacity, Z_F faradaic impedance, $\Delta\varphi_D^W$ and $\Delta\varphi_D^{Ref}$ potential drop across the double layer of working and reference electrode, R_e external ohmic resistance and R_E electrolyte resistance between working and reference electrode, U voltage applied.

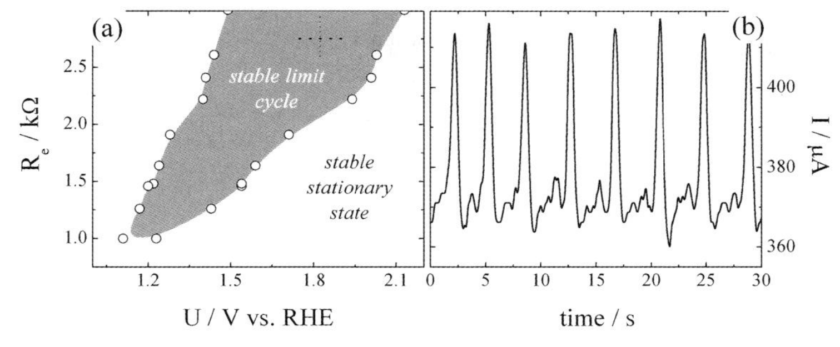

Fig. 6.24 (a) Bifurcation diagram in the R_e vs. U plane, illustrating the regions where current oscillations and stationary stable state (stable limit cycle) are found during the oxidation of methanol (0.68M) on platinum in 0.49M H_2SO_4 solution. (b) Example of current oscillations for the system, obtained at U = 1.82 V and Re = 2.75 kΩ. In this experiment, R_E was negligible in comparison to R_e.

An important point to be considered when thinking about oscillations is the presence of a region in the i/E curve where the current decreases with increasing electrode potential E, in other words, in the presence of a *negative differential resistance* (NDR).

A NDR corresponds to a negative faradaic impedance, $Z_F < 0$, and its occurrence can be associated with different aspects of electrode kinetics. An NDR might occur if: (a) the rate constant decreases with E, (b) there is a coulombic repulsion between electroactive species in the double layer (Frumkin effect) and as a consequence a decrease of the concentration at the reaction plane, and (c) the electroactive area decreases with E, resulting for instance with the adsorption of poisoning reaction products.

The adsorption of species that, at least partially, poison the surface during the reaction is quite common in electrocatalysis. Indeed, many electrocatalytic reactions can display temporal instabilities such as bistability and oscillations, e.g. during the oxidation of hydrogen, carbon monoxide, methanol and formic acid. An example of oscillations during the oxidation of methanol is given in Fig. 6.24. In part (a) the regions where oscillations can be found are shown in a R_e vs. U plane. This type of plot is referred to as a *bifurcation diagram* and illustrates the line delimiting the region where oscillations are stable (*stable limit cycle*). Current oscillations, obtained at applied voltage U = 1.82 V and R_e = 2.75 kΩ, are depicted in part (b), see Fig. 6.24.

Besides the fundamental interest in uncovering the mechanistic aspects underlying the oscillatory dynamics, electrochemical oscillators have been used as a model system to study different aspects of nonlinear dynamics and spatio-temporal pattern formation. In fact, given some peculiarities such as the information exchange along the electrified solid/liquid interface, the research in this area is continually growing.

References for Chapter 6

For a discussion of the hydrogen electrode reactions see:

M. Breiter: "Electrochemical Processes in Fuel Cells", Springer-Verlag, Berlin, 1969.

K.J. Vetter: "Electrochemical Kinetics", Springer Verlag, Berlin, 1961.

P.A. Christensen and A. Hamnett, "Techniques and Mechanisms in Electrochemistry", Blackie and Son, Edinburgh, 1995.

R.J. Nichols and A. Bewick: J. Electroanal. Chem. **243** (1988) 445.

For a discussion of oxygen electrochemistry see:

J. Hoare: "The Electrochemistry of Oxygen", John Wiley, New York, 1968.

V.S. Bagotsky, M.R. Tarasevich and V. Filinovsky: Soviet Electrochemistry, **5** (1969) 1158.

A. Damjanovic and A.T. Ward in "Electrochemistry: The past Thirty Years and the next Thirty Years", Plenum Press, New York, 1977.

K.L. Hsueh, D.T. Chin and S. Srinivasan: J. Electroanal. Chem. **153** (1983) 79.

K. Kinoshita: "Electrochemical Oxygen Technology", New York, 1992.

D.T. Sawyer: "Oxygen Chemistry", OUP, 1991, New York.

D.J. Schiffrin: "Specialist Periodical Reports: Electrochemistry", vol. 8, The Chemical Society, 1983, p. 126.

H. Gerischer and W. Vielstich in "Handbook of Heterogeneous Catalysis", eds. G. Ertl, H. Knözinger and J. Weitkamp, Wiley-VCH, Weinheim, 1997, p.1325.

For discussions of the Mechanism of Methanol Oxidation see:

M. Watanabe, S. Motoo, *J. Electroanal. Chem.*, **69**, 429 (1976).

K. -I. Ota, Y. Nakagawa, M. Takahashi, *J. Electroanal. Chem.*, **179**, (1984).

A. Hamnett, in *Interfacial Electrochemistry* ed. A, Wieckowski, Marcel Dekker, New York, 1999, p. 843.

H. Hoster, T. Iwasita, H. Baumgärtner, W. Vielstich, *Phys. Chem. Chem. Phys.*, **3**, 337 (2001).

T. Iwasita, in *Handbook of Fuel Cells*, Volume 2, Wiley Inc. (2003).

E.A. Batista, G.R.P. Malpass, T. Iwasita, *J. Electroanal. Chem.*, **571**, 273 (2004).

For discussions of the mechanism of CO oxidation see:

I.Villegas and M. J. Weaver, *J. Phys. Chem.* B 101, 10166 (1997).

E.A. Batista, T. Iwasita, W. Vielstich, *J. Phys. Chem.*, B 108 (2004) 14216.

G. Ertl, M. Newmann and K. M. Streit, *Surf. Sci.*, 64, 101 (1977).

M. T. M. Koper, A. P. J Jansen, R. A. van Santen, J. J. Lukkien, and P. A. J. Hibers, *J. Chem. Phys.*, 109, 6051 (1998).

References to Ethanol Oxidation:

C. Lamy, S. Rousseau, E.M. Belgsir, C. Contanceau, J.M. Leger, Electrochim.Acta 22–23 (2004) 3901.

G.A. Camara, T. Iwasita, "Parallel Pathways of Ethanol Oxidation: The Effect of Ethanol Concentration", J. Electronal. Chem. 578 (2005) 315.

Introductory references to Organic Electrochemistry:

M. Baizer and H. Lund (eds.): "Organic Electrochemistry", Marcel Dekker, New York, 1985.

J. Grimshaw, "Electrochemical Reactions and Mechanisms in Organic Chemistry", Elsevier Press, Amsterdam, 2000.

A.J. Fry and W.E. Britton (eds.): "Topics in Organic Electrochemistry", Plenum Press, New York, 1986.

L. Eberson and H. Schäfer: "Organic Electrochemistry" in "Fortschritte der chemischen Forschung", **21** (1971), Springer Verlag, Berlin.

L. Eberson: "Electron Transfer Reactions in Organic Chemistry" in "Advances in Physical Organic Chemistry", vol. **18** (1982), eds. V. Gold and D. Bethell, Academic Press, London.

S. Torii: "Electro-organic Syntheses; Part I (Oxidations); Part II (Reductions)", Verlag Chemie, Weinheim, 1985.

A.J. Bard and H. Lund (eds.): "Encyclopaedia of Electrochemistry of the Elements, Organic Section", vols. XI – XV, Marcel Dekker, New York, 1978 – 1984.

A.F. Diaz and J. Bargon: "Electrochemical Synthesis of Conducting Polymers" in "Handbook of Conducting Polymers", ed. T.A. Skotheim, Marcel Dekker, New York, 1986.

M.E.G. Lyons (ed.): "Electroactive Polymer Electrochemistry", Plenum Press, New York, 1996 (2 vols.)

Oscillating Reactions are considered in:

K. Krischer, "Nonlinear Dynamics in Electrochemical Systems", in D.M. Kolb, R.C. Alkire (Eds.): *Advances in Electrochemical Sciences and Engenering*, Wiley-VCH (Weinheim 2003), vol. 8, 21p. 89.

W. Vielstich, "CO, Formic Acid, and Methanol Oxidation in Acid Electrolytes – Mechanism and Electrocatalysis", in A.J Bard, M. Stratmann, *"Encyclopedia of Electrochemistry"*, Wiley-VCH 2003, vol. 2, pp. 466–511.

H. Valera, K. Krischer, *Catal.Today* 70 (2001) 411.

7
Solid and Molten-salt Ionic Conductors as Electrolytes

Solid ionic conductors and molten salts can serve directly as electrolytes and have a number of important uses. The sodium-sulphur battery uses a sodium-ion-conducting ceramic at ca. 350°C, which serves not only as an electrolyte but also to separate the liquid sodium anode and liquid sulphur cathode regions, and solid oxide-ion conducting ceramics, operating near 1000°C, serve as electrolytes in experimental water electrolysers and fuel cells. Cation-conducting polymers are also used at temperatures below 100°C in chlor-alkali cells and in "solid-polymer" fuel cells, all of which are described in Chapter 9. A classical example of the use of melts in an industrial process is the extraction of aluminium, which is carried out in a cryolite(Na_3AlF_6)/ alumina melt at c. 1000°C, and a more recent example is the use of molten carbonate electrolytes in fuel cells operating near 650°C.

It is clear that the use of melts and solid ionic conductors often requires high temperatures, and cell materials such as quartz, porcelain, ceramics, boron nitride or high-melting metals are needed. In addition, thermoelectric effects can disturb the measured cell potentials, and material balances can be falsified by evaporation or dissolution of metals in the melt; great care is therefore necessary.

7.1
Ionically Conducting Solids

Low temperature ionic conduction in solids has been known for many years but only recently has any serious effort been made to understand the phenomenon in detail.

7.1.1
Origins of Ionic Conductivity in Solids

In principle, a perfect lattice should not be capable of sustaining ionic conductivity, but no lattice is perfect above 0 K, with vacancies forming on lattice sites and interstitial sites becoming occupied, and there is a number of ways in which an ion can move under the influence of an electric field in a real crystal:

Electrochemistry. Carl H. Hamann, Andrew Hamnett, Wolf Vielstich

ISBN: 978-3-527-31069-2

– from vacancy to vacancy
– from interstitial to interstitial
– in a concerted fashion in which an ion moves from an interstitial site to a lattice site expelling the ion on that site to an interstitial site.

An extreme form of the last case is found in so-called *superionic* conductors, such as AgI above 147°C, in which one of the sublattices, in this case the silver-ion sublattice, in effect melts, whilst the iodide lattice remains solid. In this case the distinction between lattice site and interstitial breaks down, and the ion motion is much more reminiscent of a liquid. Such *first-order* transitions are easily observed as discontinuities in conductivity, often of many orders of magnitude, accompanied by strong maxima in the specific heat.

The measurement of ionic conductivity in a solid is similar to that in an ionically conducting liquid (see sections 2.1.2 & 2.1.3). As in the latter case, a double layer will form at the electrode/electrolyte interface, and *ac* methods can be employed. Of course, these measurements can only yield the total contribution of all mobile charge carriers (anions, cations, electrons and holes) to the conductivity, unless they are carried out over a wide range of frequencies or unless additional studies are carried out. As an example, we consider the conductivity of Ag_2S, which has contributions from both Ag^+ ions and electrons. The contribution of the latter can be eliminated by the use of two discs of AgI (solely a Ag^+ conductor) and by using *dc* techniques as shown in Fig. 7.1. The disadvantage of *dc* techniques is, of course, that silver will be deposited at the cathode and will dissolve from the anode, leading to overpotentials at both electrodes. In addition, there will be contact resistances at both the AgI/Ag_2S interfaces as Ag^+ ions migrate across. These can, in principle, be eliminated either by systematically varying the length of the Ag_2S rod in Fig. 7.1 or by measuring the potential drop between two points on the Ag_2S rod with an additional pair of electrodes (the "four-point" technique).

For non-superionic conductors, the ionic conductivity is normally found to increase exponentially with temperature. This can be understood with reference to equations (2.4) to (2.5) in section 2.1.1, in which the conductivity is related to the product of the *concentration* of mobile ions, c_i, and their *mobility* μ_i. Assuming the mobile ions are present in interstitials or lattice vacancies, then the population of such sites will be governed by Boltzmann statistics, with an exponential energy term related to the energy required to generate such an ion. In addition, the place-exchange mechanism proposed for ionic motion will also show an exponential form, with an *activation* energy. For ionic conductivity mediated through defect sites, therefore, we expect to find

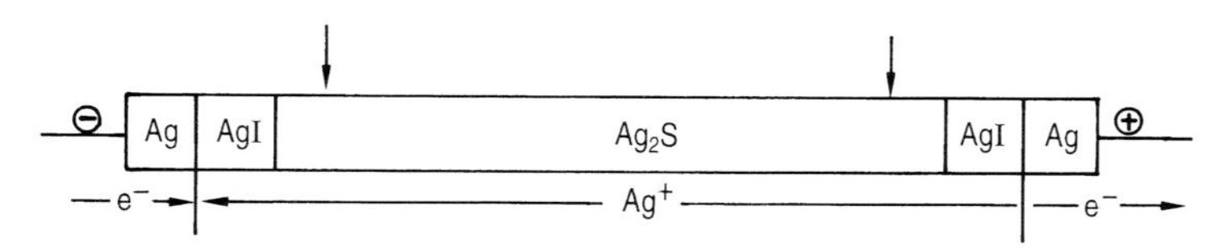

Fig. 7.1 The measurement of the Ag^+-ion conductivity in the mixed ionic/electronic conductor Ag_2S.

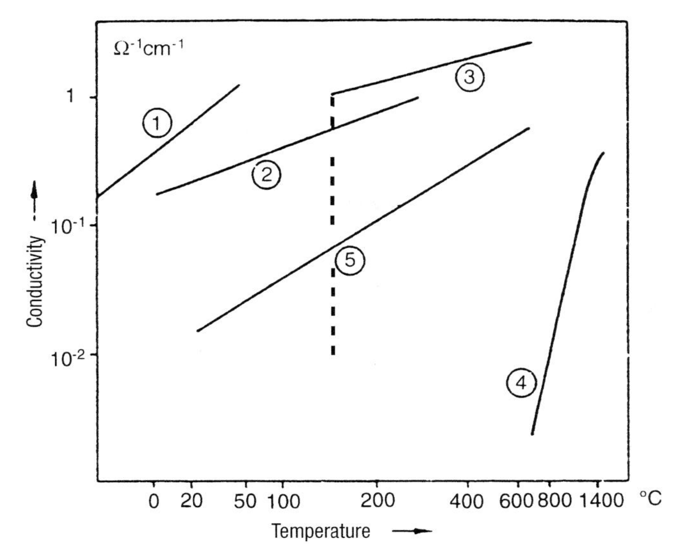

Fig. 7.2 Conductivity of solid electrolytes and of aqueous sulphuric acid as a function of temperature: 1. 6M H_2SO_4; 2. $RbAg_4I_5$; 3. α-AgI; 4. $ZrO_2 \cdot (Y_2O_3)_{0.15}$; 5. $Na_2O \cdot 11Al_2O_3$.

quite high activation energies. The situation is rather different for superionic conductors, since *all* the ions on a sublattice are expected to be mobile and c_i is, therefore, a constant. In addition, the mobility will also have a low activation energy, as is commonly found in liquids (see Fig. 7.2). For those solids where there is a transition at a specific temperature between lattice defect conductivity and superionic conductivity, such as AgI, the overall activation energy will thus fall to a very low value (see Fig. 7.2), though it is usually found that the *entropy* of activation is also rather low in superionic conductors since the ionic motion becomes strongly correlated.

The high activation energy for normal ionic conductivity implies that the resistance of the solid falls rapidly with temperature, quite unlike the case for metals, and this is made use of in the Nernst glower. If a filament of a suitable ceramic, such as $ZrO_2 \cdot 0.15Y_2O_3$ described below, is subjected to a sufficiently high voltage and then warmed with a flame until an appreciable current flows, the ohmic heating associated with that current further heats the filament, whose resistance drops yet further, allowing higher and higher currents to flow until the filament glows bright white, and current limiting becomes essential.

Whilst the above considerations have not involved any particular structural features of the lattice, there is one extremely important class of compounds for which high ionic mobility is commonly found. These materials have *layer-like* lattices, consisting of slabs of material separated by layers within which ions or molecules can diffuse relatively freely. A large number of such systems has now been studied, including pillared clays, graphites, transition-metal oxyhalides etc. Of great technical importance are a set of sodium-ion conducting materials termed sodium β-aluminas, of

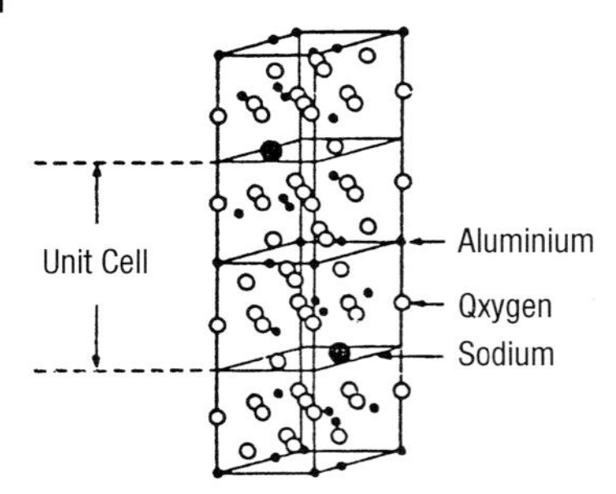

Fig. 7.3 Structure of the sodium-ion conductor, $Na_2O \cdot 11Al_2O_3$.

indeterminate composition $Na_2O \cdot xAl_2O_3$, $5 < x < 11$. The structure is shown in Fig. 7.3, and the conductivity here derives from the fact that the sodium-containing layers are highly disordered, even at room temperature. Above 200°C, the conductivity has risen to values above 0.1 Ω^{-1} cm^{-1}, comparable to those found for aqueous sulphuric acid (see Fig. 7.2).

High ionic conductivity is also found in other materials. The ceramic $ZrO_2 \cdot (15\ \%)Y_2O_3$ is termed *yttria-stabilised zirconia* (*YSZ*), and was developed initially to inhibit a structural transformation that takes place in pure ZrO_2 at 1100°C. This transformation leads to a significant volume change and causes the ceramic to shatter, but admixture of ca. 15 % Y_2O_3 stabilises the cubic phase down to ambient temperatures. It was subsequently discovered that this material has good high-temperature oxide-ion conductivity, with $\sigma > 0.01\ \Omega^{-1}\ cm^{-1}$. Subsequently, high conductivity was found in $Ce_{0.9}Gd_{0.1}O_{1.95}$ and related materials, all of which have large numbers of oxygen vacancies built into the lattice. Finally, good ionic conductivity for Li^+, H^+, Cu^+ and other singly charged cations has also been reported in a wide range of ceramic materials, but mobile multi-valent cations are much rarer.

It should be emphasised that in all these materials, the actual conductivity measured depends critically on the microstructure of the ceramic, owing to the presence of grain-boundary resistances. Furthermore, we have assumed throughout the above treatment that the conductivity arises essentially from a single mobile species, but as indicated for Ag_2S, this may not always be the case, and for the experiment shown in Fig. 7.1, the *transport number* of Ag^+ can be measured by removing the AgI discs and then measuring the weight loss/gain of silver at the two electrodes following the passage of a fixed charge. The very high temperature oxide-ion conductors are sometimes found to show $t_{O^{2-}} < 1$ (this latter being measured by measuring the diffusion coefficient of ^{18}O and calculating the mobility of the oxide ion); this arises because adventitious transition-metal impurities can give rise to an appreciable electronic current at higher temperatures.

7.1.2
Current/Voltage Measurements on Solid Electrodes

Since the thermodynamic relationships derived in Section 3.1 do not depend on the type of electrolyte, we still expect the Nernst equation to be valid, and hence to obtain thermodynamic data from appropriate measurements. As an example, we can consider the oxygen concentration cell:

$$O_2(p_1)|Pt|ZrO_2 \cdot 0.15Y_2O_3|Pt|O_2(p_2) \tag{7.1}$$

The electrochemical equilibrium at both electrodes has the form:

$$\tfrac{1}{2}O_2 + 2e^- \rightleftharpoons O^{2-} \tag{7.2}$$

we find, for each electrode

$$E = E^0 + \frac{RT}{2F}\ln\left\{\frac{(p_{O_2})^{\frac{1}{2}}}{a_{O^{2-}}}\right\} \tag{7.3}$$

where $a_{O^{2-}}$ is the activity of the oxide ion in the ceramic. The *emf* of the cell (7.1) can then be written

$$E = \frac{RT}{4F}\ln\left\{\frac{p_2}{p_1}\right\} \tag{7.4}$$

Above ca. 650°C, this equation is often found to be accurately obeyed, since above this temperature, the reaction rate for oxygen reduction is sufficiently fast for equilibrium to be established. In practice, a thin layer of porous platinum is deposited on both sides of a ceramic disc to form the electrodes, and the arrangement can be modified to measure the partial pressure of oxygen with considerable precision.

If hydrogen replaces oxygen at the anode, we have a fuel cell of the form:

$$H_2|Pt|ZrO_2 \cdot 0.15Y_2O_3|Pt|O_2 \tag{7.5}$$

with the anode equilibrium now

$$H_2 + O^{2-} \rightleftharpoons H_2O + 2e^- \tag{7.6}$$

The overall cell reaction is now

$$H_2 + \tfrac{1}{2}O_2 \rightarrow H_2O \tag{7.7}$$

and placing a load between anode and cathode allows us to generate electrical energy. If instead, the *emf* of this cell is measured under conditions of zero current flow, the free energy, enthalpy and entropy of reaction (7.7) can be obtained above ca. 650°C.

By contrast, *kinetic* measurements with solid electrolytes are much more difficult. It is obviously not possible to control mass transport to the electrode through hydrodynamic means, and there are frequently rather high resistances that build up between electrode and ionic conductor, such that in the worst case all that can be measured is an ohmic response. Perhaps most serious is the fact that the solid/solid phase boundary

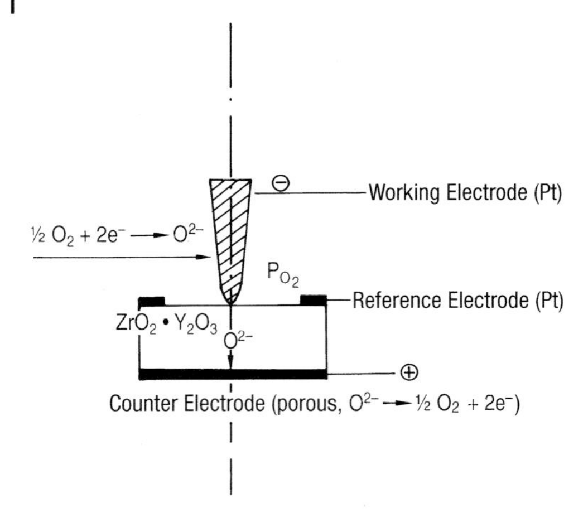

Fig. 7.4 Experimental set-up for the investigation of the oxygen-reduction reaction at an electrode/solid electrolyte interface.

undergoes changes during current flow (including the formation of cavities and dendrites) and impurities accumulating at the boundary cannot be removed by simple electrode activation as in the liquid/solid cell.

These difficulties have severely inhibited high-quality electrode kinetic studies in solid electrolytes, but one example, that of oxygen reduction on Pt/ZrO_2, has been carried out with the experimental set-up shown in Fig. 7.4. The working electrode was a platinum needle, designed to eliminate diffusion within the Pt pores of a conventional porous electrode and to prevent any alteration in the morphology of the interface. Studies between 800°C and 1000°C showed that the rate limiting processes in this region were either the dissociative adsorption of O_2 or the surface diffusion of O^{2-} ions to the Pt/ZrO_2 contact point. Neither electron transfer nor oxide-ion transport in the solid electrolyte were rate limiting at any temperature in this range. Note that the reference electrode in Fig. 7.4 is identical to the counter electrode.

7.2 Solid Polymer Electrolytes (SPE's)

By far the most important class of solid polymer electrolytes are the polymeric perfluorosulphonic acid derivatives, commonly referred to by the Trade Name Nafion®, the first to be marketed, which are singly-charged cation conductors that can be used both as separators and electrolytes. Other polymeric cation and anion conductors are currently being actively developed and are discussed in more detail below.

The basic chemical form of the polymer chains in the perfluorosulphonic acid systems varies between manufacturers to some extent, but is shown schematically in Fig. 7.5. It can be seen to consist of a perfluoro $-CF_2-CF_2-$ backbone with pendant perfluoro side-chains ending in $-SO_3^-$ groups. Charge transport in these materials

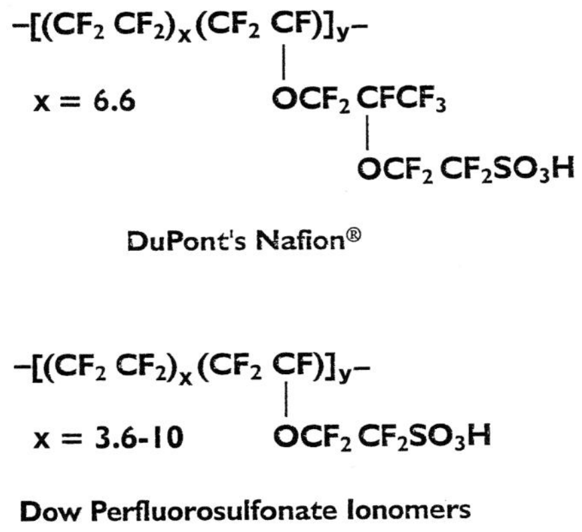

$-[(CF_2\,CF_2)_x(CF_2\,CF)]_y-$

x = 6.6 $OCF_2\,CFCF_3$

$OCF_2\,CF_2SO_3H$

DuPont's Nafion®

$-[(CF_2\,CF_2)_x(CF_2\,CF)]_y-$

x = 3.6-10 $OCF_2\,CF_2SO_3H$

Dow Perfluorosulfonate Ionomers

Fig. 7.5 Molecular structures of Nafion and the Dow perfluorinated membranes.

is usually by Na^+ or H^+ ions that wander from one sulphonic acid group to the next. These ions are normally solvated, and the membranes themselves are strongly hydrated. The internal structure of the membranes has been investigated by a variety of techniques, and is now known to consist of essentially inverse micelles joined through canals, as shown in Fig. 7.6. The inner surfaces of these micelles contain $-SO_3^-$ groups, and the structure is apparently adopted to minimise repulsive interactions between water and the fluorocarbon backbone whilst maximising solvation of the $-SO_3^-$ groups. An examination of the structure reveals that cations can easily transfer from one micelle to the next, whereas anions cannot traverse the canals due to repulsion from the $-SO_3^-$ groups in those canals.

The main advantages of these SPE membranes are their very high conductivity, exceptional stability, and the fact that they can be cast as thin pore-free films. Since the important parameter is the conductance per unit area, κ_A, it is usually this that is quoted, rather than the conductivity, κ. The relationship is simply

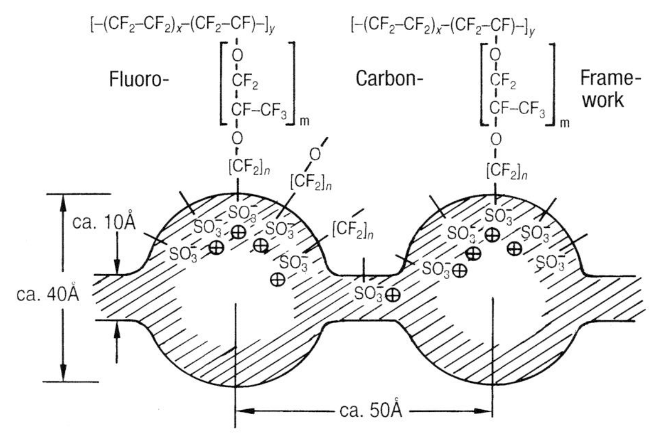

Fig. 7.6 Details of the charge-transport mechanism through the Nafion membrane: $x = 5 - 13$; $y \approx 1000$; $m = 0 - 3$; $n = 2 - 6$.

$$\kappa_A = \kappa/d \tag{7.8}$$

where d is the film thickness, and its units are Ω^{-1} cm^{-2} or Ω^{-1}m^{-2} If the current density is j, then the potential drop across the membrane is simply

$$\Delta E = j/\kappa_A \tag{7.9}$$

For Nafion used in chlor-alkali cells to separate saturated NaCl solutions, the thickness is normally ca. 0.5 mm and the value of $\kappa_A \sim 1\Omega^{-1}$ cm^{-2}, corresponding to a κ_A value of 0.05 Ω^{-1} cm^{-1}, comparable to concentrated aqueous NaCl. The corresponding potential drop across the membrane at a current density of 0.1 Acm^{-2} is 0.1 V. For these films used in the chlor-alkali cells, the transport number of the cation is very close to unity; the contribution of Cl^- conduction amounts at most to a few percent.

7.2.1 Current/Voltage Measurements with SPE's

The boundary between an SPE-membrane and an ionic solution gives rise to a concentration-dependent potential as shown in Section 3.3. As for solid ionic conductors, the number of techniques available to control mass transport is severely limited, but studies have been carried out on the system:

H_2|Pt|proton conducting SPE|Pt|O_2

the reference electrode being attached to the cell with a SPE tongue in contact with aqueous sulphuric acid. The Tafel slopes, asymmetry parameters and exchange current densities for all four reactions (i.e. H_2 oxidation and evolution and O_2 reduction and evolution) were found to be similar to those determined with aqueous sulphuric acid as the electrolyte, suggesting that the mechanisms in all cases were the same in the two systems.

7.2.2 Other Polymeric Membranes

Nafion has many desirable properties, but it is not wholly stable in chemical terms, showing severe problems at low humidities and/or higher temperatures, as well as susceptibility to chemical attack by the H_2O_2 formed during the oxygen reduction at the cathode; this attack can lead to cross-linked sulphonate formation, with degradation of the conductivity. It is also permeable not only to water, but to methanol and ethanol as well. Its use in fuel cells, especially in such devices as the direct methanol fuel cell is, therefore, problematic (see section 9.5.3). There has been a substantial effort over the last two decades, therefore, to develop improved membranes for both fuel cells and water electrolysis, concentrating on modified perfluorinated systems, alternative sulphonated polymers and composites, and on acid-base polymers.

7.2.2.1 Modified Perfluorinated Membranes

To tackle the problems associated with water loss, one approach is to replace the water in the Nafion with a less volatile solvent such as phosphoric acid, but this leads to failure of the anode, possibly through phosphate migration. Other ideas have included impregnation by molten salts, such as tetra-n-butylammonium chloride (m.p. 58°C), gel-based systems (using ionic electrolytes dispersed in polyvinylidine-fluoride (PVDF)-based polymers) and ionic liquids, such as 1-butyl,3-methyl imidazolium triflate (see below).

An alternative approach is to use composites with hygroscopic particulate solids, usually oxides for stability. Initially, such oxides as TiO_2 were used to try to prevent water loss, and later SiO_2, introduced through sol-gel chemistry into the membrane was also found to stabilise the Nafion. More recently, solid inorganic proton conductors have been tried, including zirconium phosphates of the form $Zr(HPO_4)_2 \cdot H_2O$, which show good proton conductivity up to 300°C, heteropolyacids based on the Keggin structure $[PM_{12}O_{40}]^{3+}$ ($M = Mo$, W), or acid salts, of the form $MHXO_4$ ($M = Rb^+$, Cs^+ (to be consistent with NH_4^+); $X = S$, Se, P or As). These materials can be introduced into the Nafion easily by re-casting the polymer from solutions of the ionomer, and some promising results have been obtained.

7.2.2.2 Alternative Sulphonated Polymer Membranes

The requirements for any new polymer are stringent: they include proton and Na^+-ion conductivity, chemical stability, thermal stability, good mechanical properties, low gas permeability, low water drag, fast kinetics for the electrode reactions, and low cost. The two basic types of system that have been extensively investigated are those containing heteroatoms such as fluorine and silicon (the latter in polysiloxanes), and aromatic polymers with phenylene backbones. The earliest systems to be investigated were based on sulphonated polystyrenes

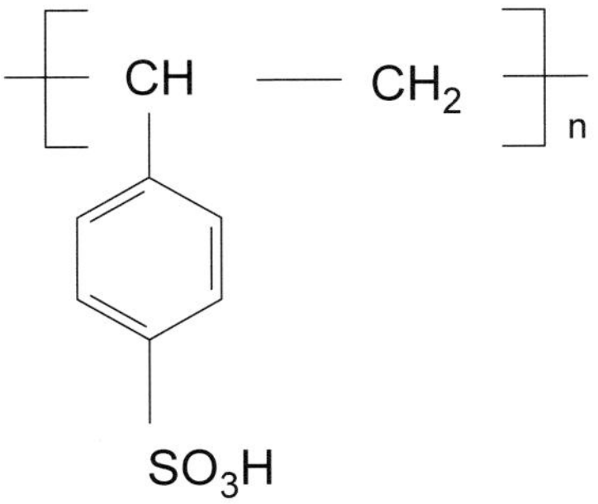

$$\left[-CH(C_6H_4SO_3H)-CH_2- \right]_n$$

but these had poor chemical stability, the tertiary hydrogens being attacked by O_2 and H_2O_2. Partial fluorination, particularly of the backbone, leads to much greater stability. Inorganic polymers have also been investigated, based on the polysiloxane structure:

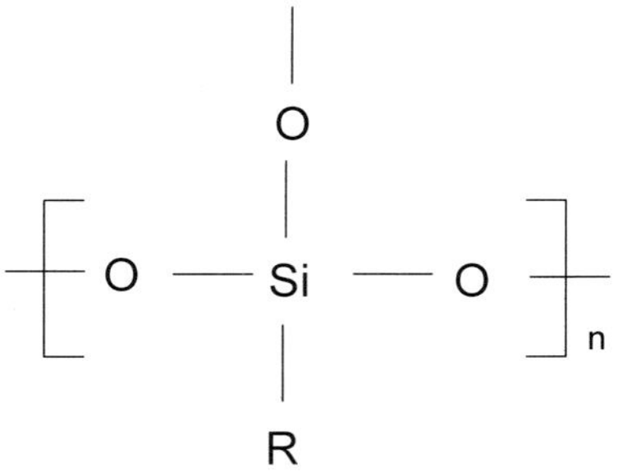

$$\left[-O-\overset{\overset{|}{O}}{\underset{R}{Si}}-O- \right]_n$$

where R is a functionalised organic moiety. The resultant species are referred to as ORMOSILS, and have a variety of applications. For fuel cells, investigations have concentrated on R as a benzylsulphonate species, $-CH_2-C_6H_4-SO_3H$, with the polymers showing good stability.

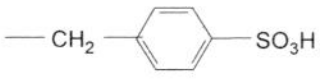

Aromatic hydrocarbons, particularly those based on poly-*p*-phenylene, have the advantages of both cheapness and stability, but must be modified if suitable physical properties are to be obtained, since the parent polymer has a very stiff, rod-like backbone. The simplest modification is to introduce an ether or sulphide linkage into the backbone, to give structures of the type:

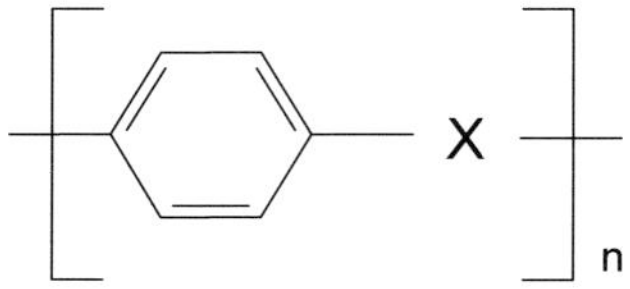

$$\left[-C_6H_4-X- \right]_n$$

where X may be O or S. Such polymers are still very high melting point, however, and more commonly X is $-SO_2-$, $-NHCO-$, $-COO-$ or $CO-$. Particularly important are those in which X alternates between ether and ketone linkages: such polymers have the generic name polyetheretherketones (PEEK) or polyetheretherketoneketones (PEEKK), and proton conductivity is achieved by sulpononating the benzene ring. An important related class are the polybenzimidazoles, of basic structure:

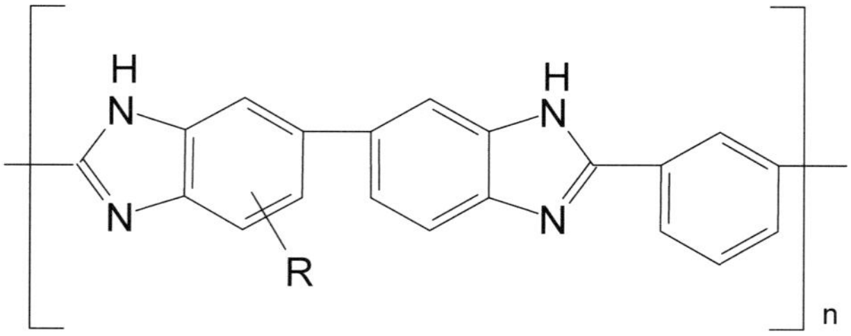

[Structure: repeat unit $[\ldots]_n$ of a polybenzimidazole — two benzimidazole units (each with H–N and N) joined through their benzene rings, substituent R on one benzene ring, linked to a phenylene ring]

The properties of the polymer may be tuned by altering the substituent(s) R on the benzene ring, and proton conductivity can be conferred by having a sulphonate group attached to R. If R is a sulphonated alkyl group, the larger this group, the greater the level of hydration and the more stable the material against water loss, but this has to be tensioned against the increasing instability against H_2O_2.

7.2.2.3 Acid-Base Composite Polymers

The types of polymer listed above can also be made conducting by forming composites of the basic polymer with acids. Most such composites, such as polyethylene oxide with phosphoric acid or sulphuric acid, have rather low conductivities unless the acid content is high, but such a high acid content results in poor mechanical properties and fillers must be used to stiffen the resultant pastes. One successful acid-polymer composite is acid-doped polybenzimidazole (PBI). PBI is soluble in acid and can be cast from such solutions to give a highly thermally stable proton conductor. The conductivity, κ, requires free acid, and at doping levels of 5.7 mol H_3PO_4, the measured conductivity is $4.6 \times 10^{-3}\ \Omega^{-1}\ cm^{-1}$ at room temperature, $4.8 \times 10^{-2}\ \Omega^{-1}\ cm^{-1}$ at 170°C and $7.9 \times 10^{-2}\ \Omega^{-1}\ cm^{-1}$ at 200°C.

7.2.2.4 Anionically Conducting Polymers

These are far less developed as yet than their proton-conducting counterparts. A simple type of example would have a tertiary ammonium group with OH^- ions moving in the polymer lattice as:

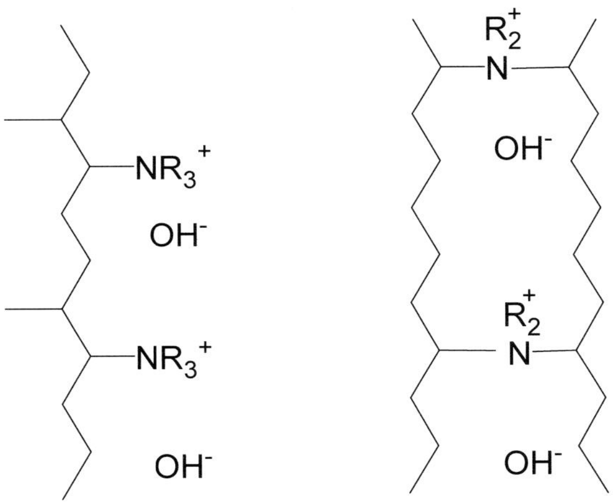

The driving force behind the development of such membranes is that oxygen reduction is kinetically more facile in alkaline electrolytes, and such membranes might, therefore, offer a way of obtaining the same advantages over the current liquid-electrolyte cells as the Nafion membranes offered earlier. Anionic membranes have been developed in the past primarily for use in sensors, as discussed in Chapter 10, and have rather high resistances. Quaternised lattices can give conductivities in the range $10^{-2}\ \Omega^{-1}\ cm^{-1}$, but this is an order of magnitude lower than Nafion, and transport numbers are usually in the order 0.95 for OH^- ions.

7.3 Ionically-conducting Melts

Molten salts confer electrolytic conductivity in a similar manner to aqueous and non-aqueous electrolyte solutions. The analogy can be continued by distinguishing between strong and weak electrolytes, the former being molten salts derived from ionic lattices, such as those exhibited by alkali and alkaline-earth halides, hydroxides, nitrates, carbonates, sulphates etc, which are wholly dissociated on melting, whereas the latter arise from the melting of molecular or semi-ionic lattices, such as $AlCl_3$, and contain in the melt both ions and undissociated molecules. It is also possible to dissolve a salt into a melt to give a molten electrolyte; such melts are of considerable interest in modern battery research, often showing very wide electrochemical windows and exceptionally wide temperature ranges. In recent years, chemists have discovered ionic salts that are even molten at room temperature, and these have led to a completely new type of electrochemistry, with applications in energy conversion, as described below. In terms of industrial applications, however, it is the higher temperature melts that are the most important currently, with aluminium electrolysis in molten Al_2O_3-Na_3AlF_6 (section 8.3.3) and the extraction of highly reactive metals (Mg, Na, Li and the lanthanides) and refractive early transition-metals (Ti, Zr, Nb, Ta, and Mo) the most important processes.

7.3.1 Conductivity

For strong electrolyte melts, the migration of ions can be considered in first order to be through a medium of viscosity η in the same way as for equation (2.1). If we neglect interionic interactions, then the conductivity will be simply proportional to the concentration of charge carriers, and given that the conductivity of 0.1M aqueous NaCl is ca. 0.01 Ω^{-1} cm^{-1}, that the viscosity of the melt is comparable to that of an aqueous solution, and that the concentration of Na^+ ions in the melt is ca. 30M, we expect a conductivity of ca. 3 Ω^{-1} cm^{-1}. However, this estimate takes no account of the difference in temperature between the melt (850°C) and the aqueous solution, and also neglects the obviously powerful interionic forces, which might be expected strongly to inhibit ionic motion. In fact, the conductivity of molten NaCl is 3.75 Ω^{-1} cm^{-1} at 850°C and 4.17 Ω^{-1} cm^{-1} at 1000°C, suggesting that temperature plays little role, and that ionic interactions must be strongly shielded. These facts have led to the conclusion that the mechanism for conductivity in the melt must be rather similar to that in a superionic conductor, with the same rationale for the low temperature dependence being suggested.

For weak electrolyte melts, the temperature dependence of the conductivity is usually larger, since there is a contribution from the dissociation step. In multi-component melts, finally, the conductivity is usually lower than expected as complex formation frequently takes place with the complex ions being larger and therefore less mobile; an example is the dissolution of NaCl into an $AlCl_3$ melt, forming $Na^+AlCl_4^-$.

Table 7.1

Type of Melt	Temperature (°C)	Conductivity (Ω^{-1} cm^{-1})
LiCl	620	5.83
NaCl	850	3.75
$CaCl_2$	800	2.21
KNO_3	400	0.81
AgI	600	2.35
NaOH	400	2.82
Hg_2Cl_2	529	1.00
$HgCl_2$	350	1.1×10^{-4}
$BeCl_2$	472	8.68×10^{-3}

The experimental determination of molten-salt conductivity is not fundamentally different from the conductivity measurements of Chapter 2, save that temperature- and corrosion-resistant materials must be used, and a longer electrolyte pathway is needed to take account of the much higher conductivities normally encountered. Some typical values of the conductivity of melts are given in Table 7.1.

The determination of transport numbers cannot be carried out as in section 2.3.2 since no change in concentration in possible in a melt. Instead, changes in volume must be used or radioactive isotopes employed. Such measurements frequently show high cation transport numbers (e.g. in molten NaCl at 835°C, $t^+ = 0.76$). A counter-example is $PbCl_2$ where the large multivalent cation only moves quite slowly.

7.3.2 Current-Voltage Studies

Reference electrodes in molten electrolytes have considerable problems associated with them, as indicated in Section 3.1.13. Given the very strong interionic interactions, it is very difficult to establish a common reference system in different melts, and redox series can only be defined for single melts.

So far as kinetic measurements are concerned, melts have many similarities to liquid electrolyte solutions, with diffusion coefficients in the range 10^{-6} to 10^{-4} cm^{2}s^{-1}. Even polarographic measurements can be carried out, though it is commoner to use a dropping *silver* electrode above the latter's melting point (961°C). In general, the high temperatures of commonly studied melts means that almost all electrode reactions are diffusion controlled, and an example of the study of an electrode reaction is the carefully studied *molten-carbonate fuel cell:*

$$H_2|M|\text{carbonate melt}|M'|O_2$$

for which the cathode reaction is:

$$\tfrac{1}{2}O_2 + CO_2 + 2e^- \rightarrow CO_3^{2-} \quad (7.10)$$

and the anode reaction is:

$$H_2 + CO_3^{2-} \rightarrow H_2O + CO_2 + 2e^- \quad (7.11)$$

and the metal electrodes M, M′ can be formed from a variety of metals. The carbonate melt is usually a eutectic mixture of Li_2CO_3 (62 mol%) and K_2CO_3 at 650°C, and *ac* impedance and step measurements have normally carried out under flow of a mix of CO_2 and either O_2 or water-saturated H_2.

Results showed quite a strong dependence of the exchange-current density for H_2 oxidation on M, with Pd having the highest value of j_0 (136 mAcm^{-2}), decreasing in the order Pd > Ni > Pt > Ir > Au > Ag, with the last having $j_0 = 13$ mAcm^{-2}. Unfortunately, of the above metals, only Ni is sufficiently resistant in practice.

The mechanism deduced from the data is:

$$H_2 \rightarrow 2H_{ads} \quad (7.12)$$

$$H_{ads} + CO_3^{2-} \rightarrow OH^- + CO_2 + e^- \quad (7.13)$$

$$OH^- + H_{ads} \rightarrow H_2O + e^- \quad (7.14)$$

For the oxygen reduction reaction, values of the exchange-current density on gold are in the region 10 – 40 mAcm^{-2}, many orders of magnitude higher than those found at room temperature in aqueous solutions. The mechanism is *indirect*:

$$\tfrac{1}{2}O_2 + CO_3^{2-} \rightarrow O_2^{2-} + CO_2 \quad (7.15)$$

$$O_2^{2-} + 2e^- \rightarrow 2O^{2-} \quad (7.16)$$

$$2CO_2 + 2O^{2-} \rightarrow 2CO_3^{2-} \quad (7.17)$$

7.3.3
Further Applications of High-temperature Melts

As indicated above, the most important applications of high-temperature melts are in extraction processes. In recent years, new processes have been developed in response to advances in materials science: Uranium can be extracted as a Zn alloy from KCl-NaCl-UCl_4 melts using zinc cathodes, and the important magnetic alloy Fe-Nd made from electrolysis of a LiF-NdF_3 melt using an Fe cathode. Alloys based on transition-metals and rare-earth metals are of great interest owing to their excellent magnetic, hydrogen absorbing or permeating and catalytic properties. They can be formed by reduction of rare-earth cations in melts on transition-metal cathodes; for example, highly uniform films of Ni_2Y can be formed from Ni cathodes and a LiCl-KCl-YCl_3 melt. Related is the interesting steel-coating Al-Mn alloy, which can be electroplated from a $AlCl_3$-NaCl-KCl-$MnCl_2$ melt at 200°C.

Melts containing the H^- ion, formed by dissolving LiH in molten alkali halides, have very interesting applications: the most potentially exciting is the electrolysis of a LiCl-KCl-LiH melt with a Si anode to generate high-purity SiH_4 inexpensively for solar cells.

New materials based on nitrides have been developed for their hardness and wear-resistance, as well as novel magnetic and electronic properties. By using melts containing the N^{3-} ion, for example, electrolysis of a $LiCl\text{-}KCl\text{-}Li_3N$ melt with Ti anodes can form TiN films; the cathode here is a gas electrode at which N_2 gas is reduced to N^{3-}.

7.3.4
Room Temperature Melts

The most common room-temperature melts are those based on the imidazolium cation:

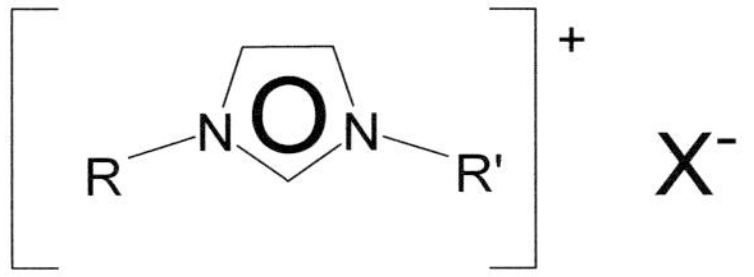

$$\left[R\text{—}N\bigcirc N\text{—}R' \right]^{+} \quad X^{-}$$

Where R and R′ are commonly ethyl and methyl (1-ethyl-3-methylimidazolium) (EMIC) and X^- may be chloride, BF_4^-, PF_6^- etc. Ethyl may be replaced by *n*-butyl and many other species. Other organic cations that have been used include 1-butyl-pyridinium and other common anions include $AlCl_4^-$ and $Al_2Cl_7^-$.

These room temperature melts have wide electrochemical windows, high conductivities, especially compared to organic electrolytes; high reactant solubility, non-flammability and low vapour pressure. The most frequently studied are the chloroaluminate molten salts; the 1:1 EMIC-$AlCl_3$ mix is liquid down to 8°C and the 1:2 mix down to − 98°C. The main equilibria are of the form

$$Cl^- + AlCl_3 \rightleftharpoons AlCl_4^-$$
$$AlCl_4^- + AlCl_3 \rightleftharpoons Al_2Cl_7^-$$
$$Al_2Cl_7^- + AlCl_3 \rightleftharpoons Al_3Cl_{10}^-$$

and these equilibria can be buffered by addition of NaCl:

$$NaCl + Al_2Cl_7^- \rightleftharpoons Na^+ + 2AlCl_4^-$$

In addition to the obvious use of these salts to electroplate very pure Al, they have also been used as solvents to electroplate Al alloys of transition metals such as Co and Ni. The deposition of Nb-Sn alloys from EMIC chloride mixtures has also been reported.

An example of electrochemistry in a room-temperature melt is shown in Fig. 7.7, which illustrates the cyclic voltammogram obtained at a gold disc electrode of $r = 5\ \mu m$, used as ultra-microelectrode, for the reduction of oxygen in a room-temperature ionic liquid at a scan rate of $0.5\ Vs^{-1}$. In this particular ionic liquid, the diffusion coefficient of oxygen is more than one order of magnitude larger than that of superoxide on the return scan. Therefore, while the forward wave (starting at −0.4 V) is determined by stationary radial diffusion, the wave back is strongly peak-shaped, due to a contribution from the effects of planar diffusion. In a single voltammogram both, staedy-state and transient behaviour, are demonstrated.

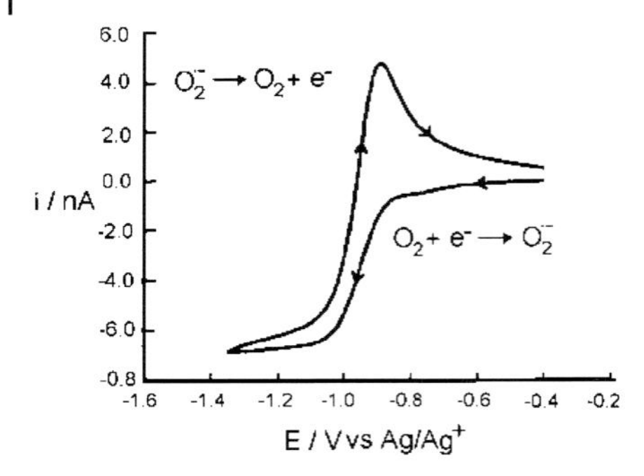

Fig. 7.7 Cyclic voltammogram obtained in a room temperature ionic liquid for the reduction of oxygen, gold disc electrode, $0.5 Vs^{-1}$, $[N_{6222}][N(Tf)_2]$, 308K.

The most important use of these salts is, however, in energy conversion devices, particularly high energy-density batteries (Chapter 9) based on Li. Given the problems associated with the extreme water and air sensitivity of the EMIC-$AlCl_3$ system, interest has turned to BF_4^- and to the fluoride salts, both of which show much greater stability. This type of salt is also the basis for recent work on fuel-cell systems utilising 1-*n*-butyl-3-methylimidzolium BF_4^-, a Pt-based anode and a Ag-based cathode. This cell showed a remarkably high open-circuit potential of 1.0 V and very high overall efficiencies of operation. Power densities of over 1 W cm^{-2} have been reported for these types of device.

References for Chapter 7

Conductivity and Electrochemistry of Solid Electrolytes

P.G. Bruce: "Solid-state Electrochemistry", Cambridge University Press, Cambridge, 1995.

S. Chandra: "Super-ionic Solids – Principles and Applications", North- Holland Publ. Co., Amsterdam, 1981.

E.C. Subbarao (ed.): "Solid Electrolytes and their Applications", Plenum Press, New York, 1980.

H. Rickert: "Electrochemistry of Solids", Springer Verlag, Berlin, 1982.

A.R. West: Ber Bunsen. Phys. Chem. **93** (1989) 1235.

Polymer Electrolytes

A. Eisenberg and H.L. Yeager (eds.): "Perfluorinated Ionomer Membranes", American Chemical Society, Washington D.C., 1982.

F.M. Gray: "Solid Polymer Electrolytes: Fundamentals and Technological Applications", VCH, New York, 1991.

Q.-F. Li, R.-H. He, J.O. Jensen and N.J. Bjerrum, "Approaches and Recent developments of Polymer Electrolyte Membranes for Fuel Cells Operating above 100°C"; Chem. Mater. **15** (2003) 4869.

Fundamental data on chloride melts

J.A. Plambeck, "Encyclopaedia of the Electrochemistry of the Elements", vol. 10, ed. A.J. Bard, Marcel Dekker, New York 1976.

Modern Molten Salt Electrochemistry

Y. Ito and T. Nohira, "Non-conventional Electrolytes for Electrochemical Applications", Electrochimica Acta **45** (2000) 2611.

R.F. de Souza, J.C. Padiha, R.S. Goncalves and J. Dupont, "Room Temperature Dialkylimidazolium Ionic Liquid-Based Fuel Cells", Electrochem. Commun. **5** (2003) 728.

M.C. Buzzeo, R.G. Evans and R.G. Compton, "Non-Haloaluminate Room Temperature Ionic Liquids in Electrochemistry", ChemPhysChem **5** (2004) 1106–1120.

8
Industrial Electrochemical Processes

8.1
Introduction and Fundamentals

The discovery of the ac dynamo in 1866 by Siemens allowed, for the first time, current densities to be generated that were useful for processes on the industrial scale. These processes include, today, the extraction and purification of a wide range of metals, and the electrosynthesis of large numbers of organic and inorganic materials. The scale of this industry can be deduced from the fact that one process, the generation of chlorine, produces 22.5 Mtons of the gas and consumes 7.2×10^{10} kWh of power, and a second process, the extraction of aluminium, generates 13 Mtons of metal and consumes 1.8×10^{11} kWh.

8.1.1
Special Features of Electrochemical Processes

As compared to chemical processes, electrochemical methods have some important advantages:

1) Better *energy yields* (i.e. ratios of weight of product to the energy employed) can be attained than for chemical reactions, even when ohmic losses and overpotentials account for as much as the thermodynamic free energy.
2) Electrochemical reactions can frequently show high selectivity, giving products of high purity, and thereby avoiding cost-intensive purification steps.
3) Reactions take place at normal temperatures and pressures.
4) Current and voltage are easily controlled, and electrochemical reactors are easy to automate.
5) Electrochemical processes are intrinsically environmentally benign, since the electron transfer process to or from the electrodes leaves no by-products.

Electrochemistry. Carl H. Hamann, Andrew Hamnett, Wolf Vielstich

ISBN: 978-3-527-31069-2

However, set against these advantages are two significant *disadvantages*:

1) Electrical energy is much more expensive than the thermal energy used in chemical processes.
2) Electrochemical plant necessitates higher investment costs than chemical plant, since individual cells require only a few volts, but the lower limit of economic operation of electrical transmission lines is ca. 100 V, so electrochemical plants must be constructed of *stacks* of cells rather than the large single reactors appropriate to chemical reactors. In addition, costs of rectification and transmission equipment exceed the costs of heat transfer, and the total plant volume is also greater.

The decision as to whether to use a chemical or electrochemical synthesis will, therefore, depend on balancing the factors above, the most significant being the costs of electrical power. The relevant indicator here is the *specific energy usage, ω*, in units of kWh per tonne of desired product:

$$\omega = 1000 \cdot \frac{\text{Energy required } (\equiv i.t.E_C)}{\text{Total yield of product } (\equiv a.i.t.f)} \tag{8.1}$$

where i is the current in the cell, t the time, E_c the cell voltage, a the current yield ($a = 1$ being the ideal) and f the electrochemical equivalent, defined as:

$$f = \frac{\text{Molecular Weight } (M)}{\text{Number of electrons } (n) \times \text{Faraday Constant } (F)} \tag{8.2}$$

and having units of $kg(kAh)^{-1}$. Simplifying (8.1), we obtain

$$\omega = \frac{1000 E_c}{f.a} \tag{8.3}$$

For chlorine evolution, with $E_C = 3.5$ V and $a = 0.96$, we find $\omega = 2800$ kWh/tonne, which, for an electricity price of £0.05 per unit (kWh), corresponds to a price of c. £14 per 100 kg, neglecting, for the moment the value of the cathode product (NaOH). In a similar way, the specific energy usage for aluminium extraction is 12000 – 16000 kWh/tonne, whereas that for copper purification is only 150 – 300 kWh/tonne.

It is unlikely that the future will bring substantial further decreases in the specific energy usage, ω, for reactions such as chlorine evolution, since we are limited by thermodynamic effects and the technology has, in any case, reached a high level here. Nor are electricity prices likely to fall. On the positive side, however, there has been a large number of new processes developed in the last few years, and the attraction of electrochemical processes in environmental terms is ensuring that they are seen in an increasingly favourable light.

8.1.2
Classical Cell Designs and the Space-Time Yield

There are two types of design encountered in classical electrochemical engineering: *monopolar* and *bipolar*, with parallel-plate electrodes. For *monopolar cells*, the negative charge enters one or more cathodes *connected in parallel*, leaving the cell by corresponding anodes, as shown in Fig. 8.1. In the simplest case, the cell would consist of an open trough in which the electrodes were immersed and suspended from appropriate current collectors. In *bipolar cells*, the anode and cathode of neighbouring cells are connected in series, giving a total stack voltage of NE_C, where N is the number of cells connected, as shown in Fig. 8.2, and N normally lies between 10 and 70.

The cells shown in the two figures have anode and cathode reactions taking place in the same solution, but this may not always be desirable, and where division of the cell into anolyte and catholyte is required, a *membrane* or *diaphragm* must be used to prevent mixing. The installation of such a membrane for monopolar cells is straightforward, but for bipolar cells it is advantageous to use a filter-press arrangement as shown schematically in Fig. 8.3. The electrodes are inserted into plastic corrosion-resistant frames on long bolts, with the diaphragm and spacers inserted between each pair. The disadvantage of this arrangement is the increased maintenance of the cell and the fact that single defect cells cannot easily be bridged.

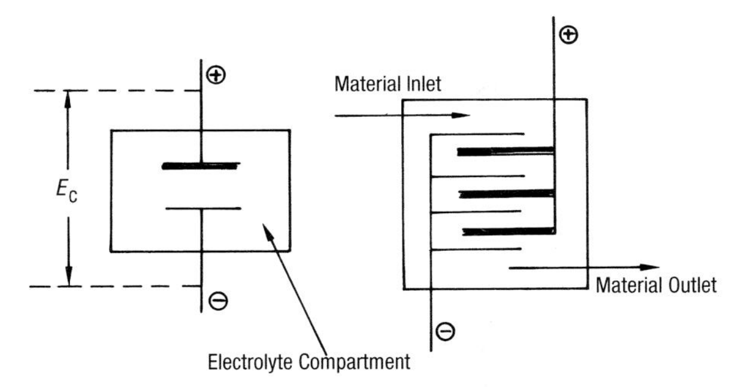

Fig. 8.1 Schematic representation of a monopolar electrolysis cell.

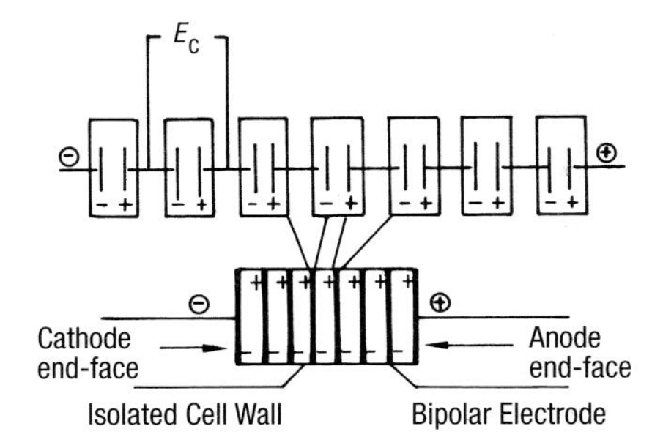

Fig. 8.2 Electrical interconnections for a stack of bipolar cells.

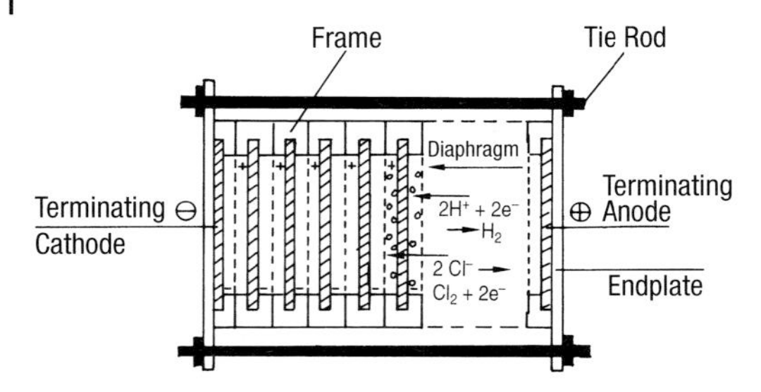

Fig. 8.3 Simplified diagram showing the filter-press principle for the assembly of a stack of bipolar cells.

Introduction and removal of material from the cells is usually carried out by pumping through the working solution, but the corresponding conduits constitute current *shunts* as shown in Fig. 8.4. These shunts increase the specific energy usage and their effect must be minimised by careful attention to overall design.

Primary requirements for the cells themselves are good mass transport and the minimisation of ohmic losses by bringing the anode and cathode as close together as possible. The importance of the latter can hardly be over- emphasised: even for saturated NaCl at 80°C, the conductivity is only 0.6 Ω^{-1} cm^{-1}, and for a current density of 2.5 kA m^{-2} (the technically used unit, corresponding to 250 mA cm^{-2}), there will be an ohmic drop of 0.4 V cm^{-1}. In addition, there will be an associated production of *heat*, particularly in those cells using electrolytes of poor conductivity, and this heat must be removed. The *minimum* electrode separation is ca. 0.1 mm, and for such a capillary system, the continuous pumping of electrolyte is sufficient both to provide mass transport to the electrodes of starting material and to remove the excess heat, certainly up to current densities of 1 kA m^{-2}.

In the case of divided cells, the arrangement of Fig. 8.5, corresponding to the so-called *zero-gap principle*, is used, in which anode and cathode are separated by a diaphragm or ion-exchange membrane (such as Nafion) ca. 1 mm thick, and the electrodes contact the working electrolyte solutions on their reverse sides. This ar-

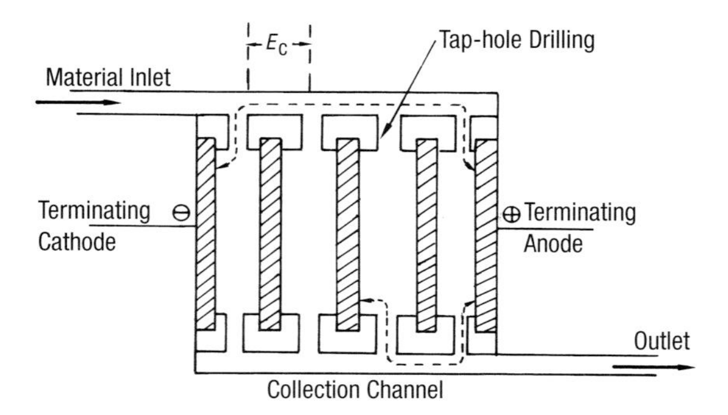

Fig. 8.4 The pumping arrangement for a stack of undivided bipolar cells showing how electrical shunts arise (dashed arrows).

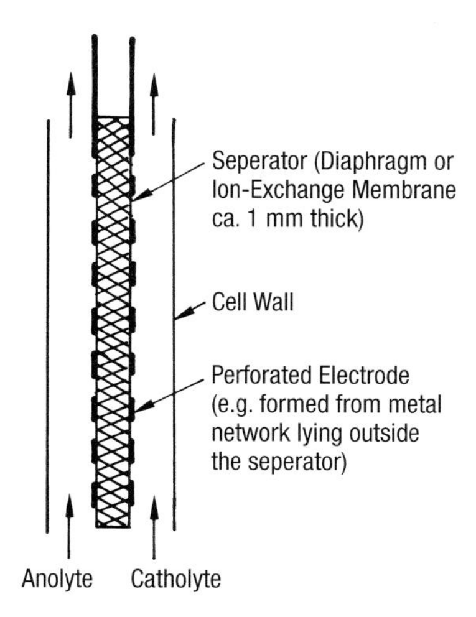

Fig. 8.5 Schematic representation of a divided cell employing the "zero- gap" principle.

rangement is particularly advantageous for gas evolution, since it avoids the formation of gas bubbles between the electrodes, the effect of which would be to increase significantly the electrolyte resistance. Where there is a thin layer (ca. 1 mm) of electrolyte between one or both of the electrodes and the separator, the arrangement is termed a *finite-gap cell.*

A consideration of the technical size of the overall cell stack leads to a second important technical variable, the *space-time-yield*, ρ, which is the yield of the desired product per unit cell volume, V, and operating time, t (in hours). It has the units kg dm^{-3} hr^{-1}, and is given by the formula

$$\rho = 3600 \cdot \frac{M}{nF} \cdot a \cdot j \cdot A_v = 3600 \cdot f \cdot a \cdot j \cdot A_v \tag{8.4}$$

where j is the current *density* (in A cm^{-2}) and A_v is the ratio of the total area of the electrode surfaces to the cell volume (in cm^{-1}). The product $j.A_v$ is sometimes referred to as the current concentration, since it has units of A cm^{-3}. The space-time-yield for any electrochemical synthetic process should, on cost grounds be as high as possible, which implies both a high value of j and a high A_v. Values of A_v for classical reactors of the type described above are ca. 0.1 cm^{-1} for monopolar cells and 1 cm^{-1} for bipolar cells; this contrasts with chemical fixed-bed reactors used in heterogeneous catalytic processes, where values of ca. 20 cm^{-1} are found.

8.1.3
Morphology of Electrocatalysts

Electron transfer takes place, for the most part, at the electrode surface and the interaction of substrate with surface strongly influences the overall reaction rate. For pure metals, we can take the flat polycrystalline surface as the reference point, and the reaction rate can then be increased by increasing the effective surface area. In the

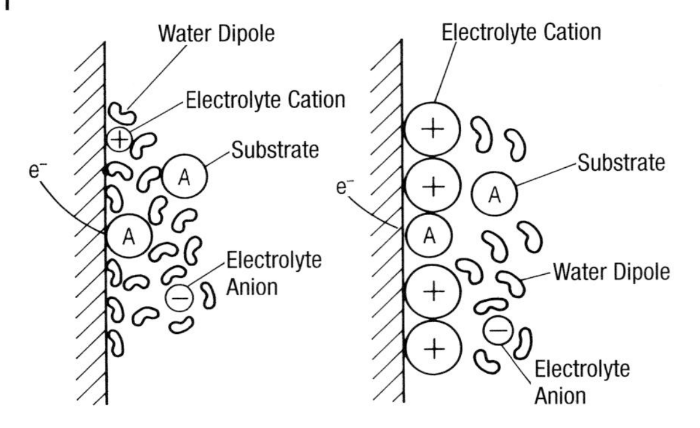

Fig. 8.6 Creation of a hydrophobic layer close to an electrode surface by adsorption of a tetra-alkyl ammonium cation.

simplest case, this can be done by roughening the surface, with the current density (referred to the *true* surface area) remaining constant at a fixed potential, provided mass transport remains sufficiently high. However, in general this roughening process will also lead to an increase in the number of surface dislocations or defects, both of which can act as active sites for electron transfer. A high density of such sites can also be achieved, for example, by cathodic deposition of the metal, or by preparation of Raney-nickel (from powdered NiAl alloy, from which the Al is removed by dissolution into hot KOH).

Metal oxides can also show good electrocatalytic properties, the most familiar case being RuO_2, which can be deposited on TiO_2 as a mixed Ti/Ru oxide layer about 1 μm thick. Carbon is also a most useful substrate for electrochemical engineering, with appropriate catalyst coverage to ensure selectivity. The catalytic importance of metal *alloys* is frequently greater than that of the pure metals, since not only can the electronic properties of one metal be fine-tuned by alloying with another, but *bifunctional* mechanisms become possible. A simple example is the case in which species A only adsorbs on one component of the alloy and species B on another; the surface of the alloy will then contain neighbouring sites at which A and B separately can adsorb, leading to the possible formation of such species as AB^+.

Of great technical importance is the influence of formally non-participating solution species on the course of a particular reaction. In a simple but well-known example, we can consider the formation of A^- from A and its possible reactions to give AA^- or AH, the latter by reaction with a proton arising from the solvent water. This reaction can be steered by appropriate choice of supporting electrolyte cation: if a normal cation, such as K^+ is used, AH is formed, whereas if a large poorly hydrated cation such as $(C_2H_5)_4N^+$ is used, then this will tend to adsorb at the electrolyte surface creating a hydrophobic layer, as shown in Fig. 8.6, which prevents protonation and permits the dimerisation reaction to take place.

8.1.4
The Activation Overpotential

In the interest of low specific energy usage and simultaneous high space-time-yield, it is essential that electrode materials are used that have high exchange currents for the desired reactions and suitable electron transfer factors; a simple example is the use of iron in alkaline solutions for hydrogen evolution, where even at an overpotential of 100 mV, the current density already exceeds 500 mA cm^{-2}. Unfortunately, for not every desired electrochemical reaction can so favourable a material be identified. Indeed, for some reactions, the exchange current densities are so low, even on the best electrocatalysts, that several hundred mV overpotential may be needed for a current density of a few mA cm^{-2} (corresponding to the value at which visible bubble formation in gas evolution commences). An example is oxygen evolution in aqueous solution on platinum, where an overpotential of 0.3–0.4 V is necessary before bubble formation takes place.

The potential difference between the thermodynamic rest potential and the onset of a "visible current" is termed the "activation overpotential" by engineers; of course, the actual overpotential at the electrode will, in general, be larger than this to obtain a usable current, and the relationship between the activation overpotential and the total overpotential for water electrolysis is shown in Fig. 8.7.

In addition to minimising the overpotential, however, the choice of electrode material will also be dictated by the requirement that there is no undesired by-product forming during the electrode reaction. For electrochemical processes (other than water

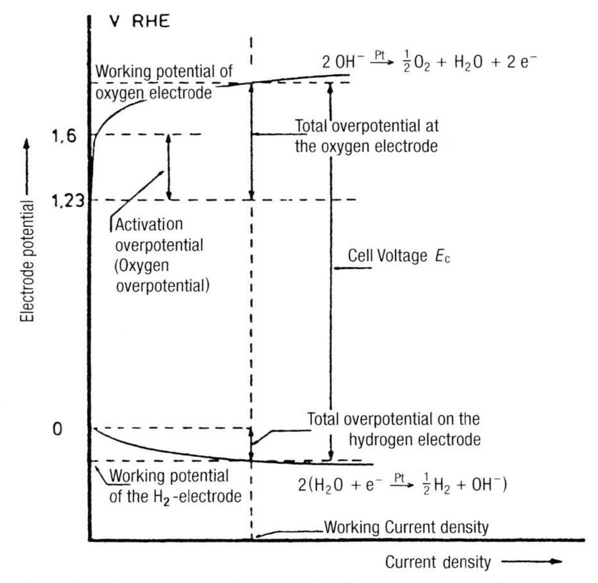

Fig. 8.7 Current-voltage diagram showing an example of activation overpotential for the electrolysis of water at a platinum anode in alkaline solution.

electrolysis itself) in aqueous solution, this requirement usually translates into *maximising* the overpotential for oxygen or hydrogen evolution and minimising it for the desired process, particularly for cell voltages in excess of 1.23 V. Thus, the electrolytic preparation of chlorine from aqueous salt solution, which requires a cell voltage of at least 1.37 V, would not be possible without a large oxygen evolution overpotential at the anode.

8.2 The Electrochemical Preparation of Chlorine and NaOH

Chlorine is the basic raw material for the preparation of a range of important products including chlorinated solvents, bleaches, pesticides and polymers, and it is prepared almost exclusively nowadays by the electrolysis of aqueous NaCl or HCl solutions, the former yielding NaOH and H_2 and the latter H_2 alone as by-products.

8.2.1 Electrode Reactions during the Electrolysis of Aqueous NaCl

The thermodynamic deposition potential of Na metal is −2.71 V vs. NHE, and clearly only a cathode with an extraordinarily high overpotential for hydrogen evolution (whose thermodynamic potential is at −0.41 V in aqueous NaCl) would not preferentially evolve hydrogen. The thermodynamic potential for chlorine evolution is +1.37 V vs. NHE, which lies appreciably above that of oxygen evolution (+0.825 V at pH 7). However, by the use of appropriate anode materials, chlorine gas evolution is favoured: initially carbon anodes were used with iron cathodes. The latter leads to hydrogen evolution with a relatively low overpotential, particularly as the primary cathodic process

$$2H_2O + 2e^- \rightarrow H_2 + 2OH^- \tag{8.5}$$

leads to a strongly alkaline solution in the neighbourhood of the cathode. The carbon anodes evolved little oxygen, but were slowly attacked by the chlorine. More recently, improved catalysts based on a mixture of TiO_2 and RuO_2 on titanium have been found to be far more stable (and are referred to as "Dimensionally Stable Anodes", or DSA's).

The primary products of the electrolysis are, therefore, H_2 and Cl_2, which are collected easily above the cell. The OH^- ions generated at the cathode form sodium hydroxide solution with the Na^+ ions, but the OH^- ions must be kept out of the *anode* compartment, since not only can the Cl_2 react to form hypochlorite:

$$Cl_2 + 2OH^- \rightarrow ClO^- + Cl^- + H_2O \tag{8.6}$$

but at the potential of the anode, further oxidation of the hypochlorite is possible:

$$6ClO^- + 3H_2O \rightarrow 2ClO_3^- + 4Cl^- + 6H^+ + 1.5O_2 + 6e^- \tag{8.7}$$

leading to contamination of the NaOH solution with ClO_3^- and of the Cl_2 with O_2, and to a rapid fall in current yield for the chlorine. Thus, a primary technical requirement for chlorine preparation is the prevention of OH^- migration to the anode, either by the presence of a diaphragm or membrane or by finding a cathode process that does not generate OH^- in the first place.

8.2.2 The Diaphragm Cell

The basic diaphragm cell uses a synthetic plastic or asbestos diaphragm surrounding the *cathode*, which serves to impede OH^- transport. This is re-enforced by a continual flux of the NaCl electrolyte into the cathode compartment, thus equalising the effects of migration out and convection in of the OH^- ions. The cathodes themselves are formed of steel braid and the anodes are formed of hollow expanded titanium braids covered with a layer of RuO_2 on the surface, as shown in Fig. 8.8. It will be seen from this figure that the level of liquid in the cathode compartment is lower than that in the cell, and hydrostatic pressure continually forces NaCl electrolyte into the cathode region, preventing OH^- escape.

Typically, a modern industrial cell operates at 80–95°C, with 6 M NaCl at a current density of 250 mA cm^{-2} and a cell voltage of ca. 3.5 V, generating Cl_2, H_2 and a mixture of ca. 3 M NaCl and 3 M NaOH. Such cells are based on the principle of Fig. 8.8, but

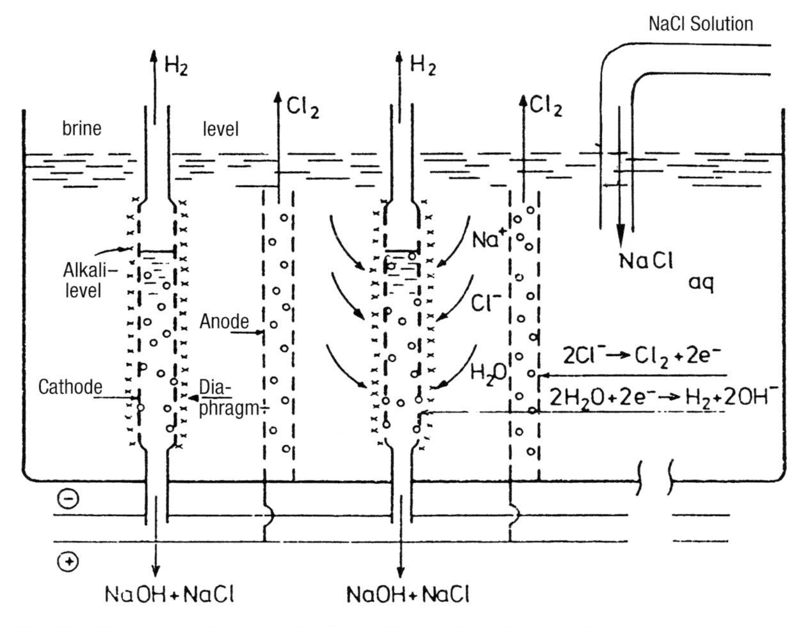

Fig. 8.8 The construction principles for a chlor-alkali diaphragm cell with vertical electrodes.

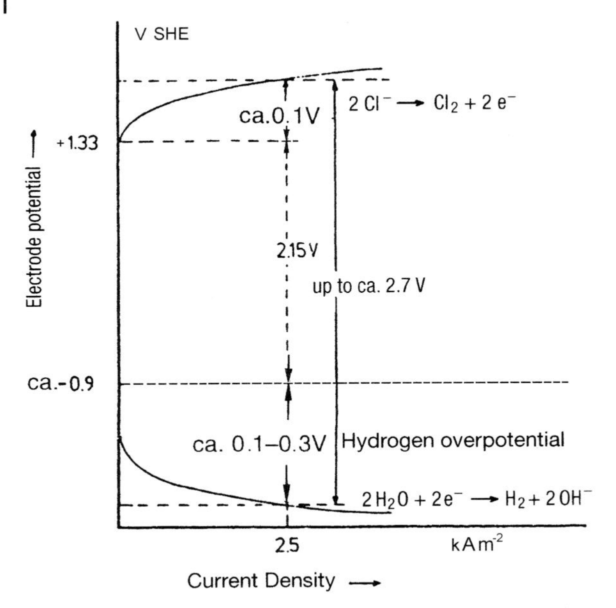

Fig. 8.9 The electrolysis of 6 M NaCl solution between activated titanium anodes and iron cathodes using the diaphragm process; the *iR* drop in the electrolyte itself is not included.

with rectangular anodes and cathodes and an inter-electrode separation of 0.5–0.7 cm. Even with this separation, however, the operating cell voltage is very high, in part because the high hydroxide ion concentration in the cathode region lowers the thermodynamic potential for H_2 evolution to -0.9 V; other contributions to the overpotential are shown in Fig. 8.9, and are primarily due to the formation of bubbles at both cathode and anode which lead to high apparent resistances, giving an electrochemical cell voltage of ca. 3 V. The remaining voltage arises from the ohmic resistance of electrolyte and diaphragm and the connections themselves.

8.2.3
The Amalgam Cell

In the amalgam cell, the formation of hydroxide ions is prevented by using mercury as the cathode and DSA as the anode. As for the diaphragm cell, Cl_2 is evolved at the anode but at the cathode, hydrogen evolution is replaced by the formation of a sodium amalgam

$$Na^+ + xHg + e^- \rightarrow NaHg_x \tag{8.8}$$

At first sight, this is rather surprising, since the overpotential for H_2 evolution on mercury is ca. 1.3 V, implying that H_2 evolution should commence near -1.7 V. This is close to -1.78 V, the potential for formation of a 0.2 % amalgam from 5 M

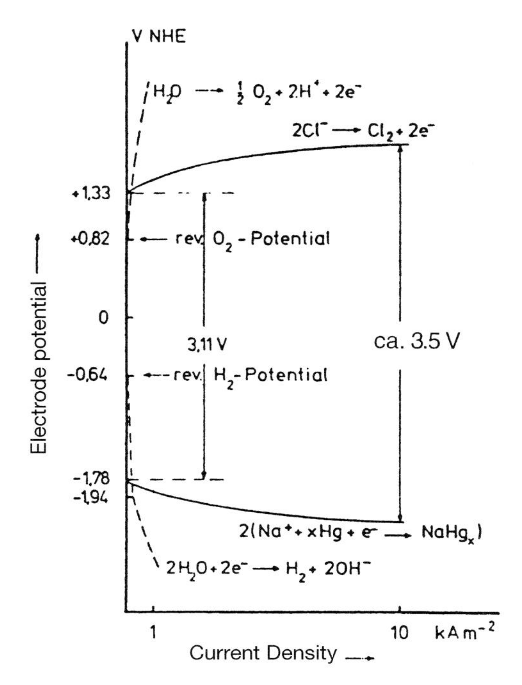

Fig. 8.10 Electrolysis of saturated NaCl solution between a DSA and a mercury cathode; the *iR* drop in the electrolyte is not shown.

NaCl. However, the asymmetry factor is far larger for reaction (8.8) than for H_2 evolution, and another inhibiting factor is that any hydrogen evolution that does take place on the mercury will rapidly shift the pH to higher values, thereby strongly *decreasing* the thermodynamic potential for H_2 evolution. In fact, under steady-state operating conditions in amalgam cells, the pH near the cathode is ca. 11, giving a thermodynamic H_2/H_2O potential of −0.64 V and an activation potential for H_2 evolution of ca. −1.94 V, as shown in Fig. 8.10. The overall cell potential neglecting ohmic losses is ca. 3.1 V for the cell reaction

$$2\text{NaCl} + 2x\text{Hg} \rightarrow \text{Cl}_2 + 2\text{NaHg}_x \tag{8.9}$$

Under working conditions, as shown in Fig. 8.11, pure 6 M NaCl enters the top left hand of the cell at 60°C, together with 0.01 % Na amalgam pumped from below. Mercury and electrolyte flow down the cell, where evolution of Cl_2 and formation of the amalgam takes place, so that at the end of the cell, the NaCl concentration has fallen to ca. 5 M and the amalgam is 0.2 % in concentration (Note that the formation of higher concentrations are increasingly accompanied by H_2 evolution; above 0.7 % amalgam, the mercury becomes very viscous). Current densities of 1–1.5 A cm^{-2} are employed with a total cell voltage of ∼ 4.0 V.

The sodium amalgam and electrolyte are separated by a transverse partition at the right-hand end of the cell, and the electrolyte is enriched with further NaCl and re-

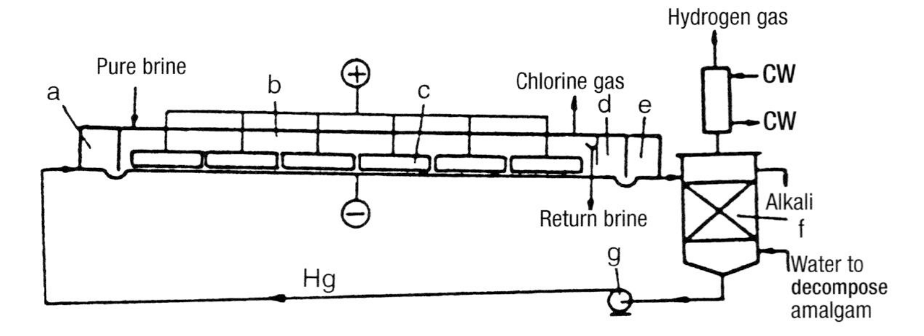

Fig. 8.11 Overall scheme of a chlor-alkali amalgam electrolysis cell: (a) Mercury entry point; (b) Electrolysis cell; (c) Titanium DSA anodes; (d) exit tank; (e) amalgam washing tank; (f) amalgam decomposition tank; (g) mercury pump; CW: cooling water.

introduced into the left-hand end. The mercury amalgam is washed with water and enters the *decomposer*, as shown in Fig. 8.11. The decomposition of the amalgam is achieved by using a ballast of graphite beads of ca. 1 cm diameter; the amalgam trickles down through the beads, with water passing upwards as a counter current and being removed as aqueous alkali at the top of the decomposer. In contrast to mercury, carbon has quite a low overpotential for hydrogen evolution, and a mixed potential is set up in which the cathodic process of hydrogen evolution on the carbon is accompanied by sodium dissolution from the mercury, the nett process being

$$2NaHg_x + 2H_2O \rightarrow 2NaOH + xHg + H_2 \qquad (8.10)$$

The flow rate of the water can be adjusted to give up to 50 % NaOH solution, and the mercury containing a small amount of Na (usually $\sim 0.01\ \%$) is pumped back into the main cell.

It should be noted that the NaCl solution *must* be free of heavy metals, particularly vanadium, chromium and molybdenum, which need to be kept below the ppb level; deposition of these metals onto the mercury leads to a rapid decrease in the hydrogen overpotential and the formation of an explosive mixture of hydrogen and chlorine. Purification is usually through precipitation by addition of carbonate, sulphate or hydroxide.

8.2.4 The Membrane Process

The specific advantages of the diaphragm cell (low cell voltage) and the amalgam cell (NaCl-free alkali) are combined in the cell shown in Fig. 8.12, in which an ion-exchange membrane is used as the cell separator. This membrane should conduct Na^+ ions but be impermeable to Cl^- and OH^- ions. The membrane itself has to satisfy very stringent requirements: it must be stable for long periods in the very aggressive medium, and it must have high transport selectivity and low electrical resistance and

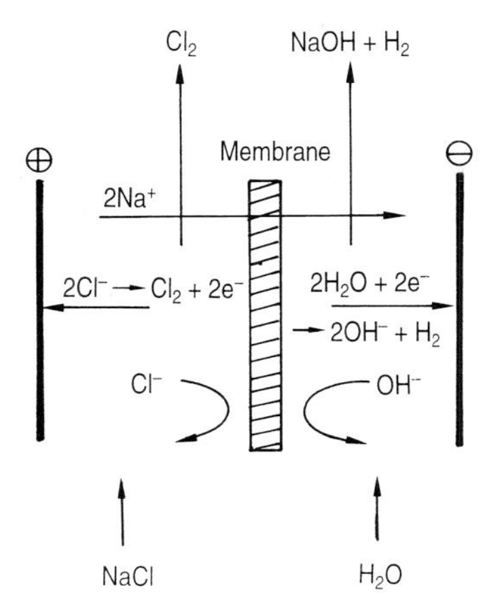

Fig. 8.12 Principle of the membrane process for chlor-alkali electrolysis.

be mechanically robust. Such requirements were only satisfied with the development of perfluoro-sulphonic acid membranes, as described in section 7.2. These membranes have now reached a high state of development, and membrane cells are the preferred type for new chlor-alkali installations.

One possible bipolar arrangement is shown in Fig. 8.13: the anode is again of DSA type and the cathode is Raney Nickel dispersed on an appropriate substrate. The cell voltage for current densities of 3–4 kA m^{-2} is only 3–3.2 V, some 10–20 % lower than

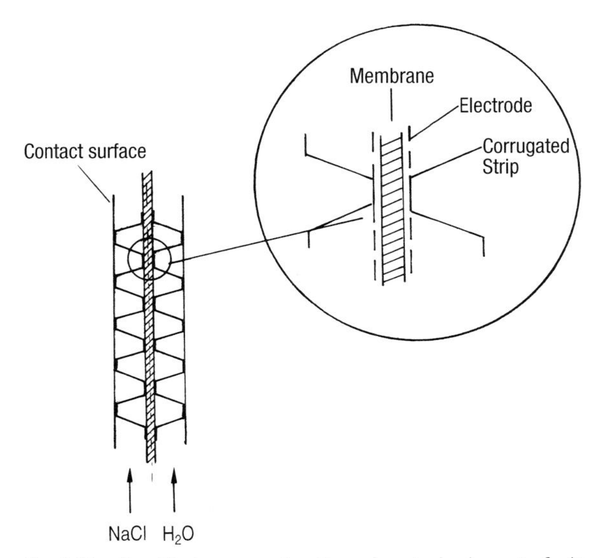

Fig. 8.13 Simplified cross-section through a single element of a bipolar membrane cell in which the electrode/membrane configuration approximates to the zero-gap principle. The electrodes are supported by corrugated bands which are, in turn, supported by contact plates. Compression of the single cells together gives the series connections for the stack.

the diaphragm cell, and the achievable alkali concentration is ca. 33 %. To minimise the membrane resistance, it must be as thin as possible, and thicknesses of 0.1–0.2 mm are used, with the membrane being mechanically strengthened by attachment of a web of stiff threads. The membrane is very sensitive to foreign cations, particularly Ca^{2+} and Mg^{2+}, both of which must be below 0.02 mg dm^{-3}. Purification demands on the electrolyte are even higher than for the amalgam cell, and a second purification step involving ion-exchange columns is generally necessary. The membranes are also not completely impermeable to Cl^- and OH^-, with ca. 100 ppm Cl^- being found in the alkali and reactions 8.6 and 8.7 lead to a reduction in yield of ca. 5 %.

8.2.5
Membrane Processes using an Oxygen Cathode

The essential engineering data for the three chlor-alkali processes are presented in comparative diagrammatic form in Fig. 8.14, from which it is evident that the membrane process will be the preferred option under normal circumstances, and that it may even be worthwhile retro-fitting membrane instead of diaphragms in older diaphragm cells, particularly in view of the probability of further future environmental restrictions on the use of asbestos. Similar concerns are, of course, also present for mercury cells, though it should be noted that only about a gram of mercury is lost per tonne of chlorine, and the total loss of mercury to the environment through the chlor-alkali industry only corresponds to about 0.1 % of the total natural and anthropogenic discharge of mercury.

E_z [V]	3.11	2.15	2.15
U_z [V, ca. - Approx. Value]	4	3.5	3–3.2
Current Density [kAm^{-2}]	10	2.5	3–4

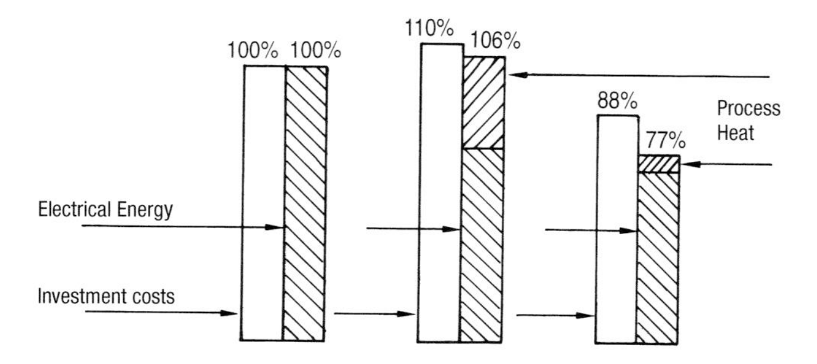

Fig. 8.14 Comparison of key performance indicators for the different chlor-alkali processes; the process heat serves essentially to increase the concentration of the electrolyte.

Given that the primary cost of the chlor-alkali process is the electricity consumption, immense efforts have been made to reduce the cell voltage. In membrane cells, an additional small gain may be possible by utilising the "Membrel©-principle", in which the electrodes are deposited on *both sides* of the membrane surface, as described in section 8.6 below. However, further substantial reduction in the cell voltage seem unlikely unless a different electrode reaction is employed with a reduced free energy. Assuming that the anode reaction remains evolution of Cl_2, this implies alteration in the *cathode* reaction, though with the following restrictions:

1) The cathodic reversible potential must be more positive than −0.9 V
2) The reactants at the cathode must be inexpensive
3) The reaction must have a sufficient exchange-current density

In fact, just about the only cathode reaction that satisfies these criteria is oxygen reduction

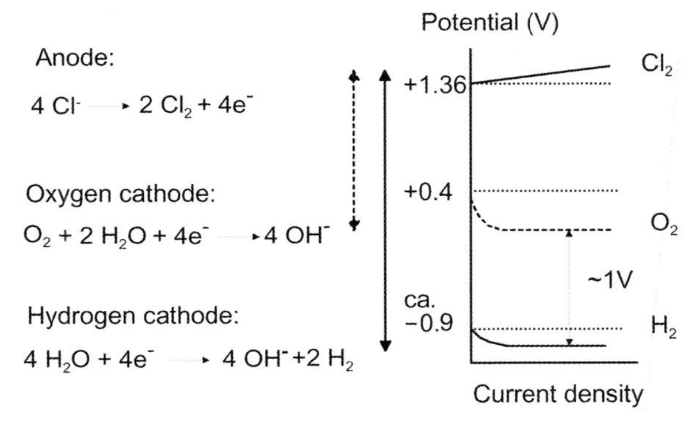

Fig. 8.15 NaCl electrolysis coupled with oxygen reduction or hydrogen evolution.

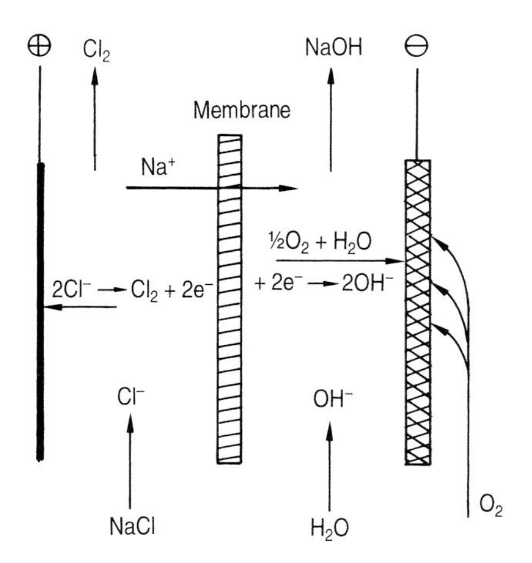

Fig. 8.15a Schematic representation of a membrane process using an oxygen reduction cathode.

$$\frac{1}{2}O_2 + H_2O + 2e^- \rightarrow 2OH^- \tag{8.11}$$

and coupling this to chlorine evolution in alkali gives a nett cell voltage gain of about 1 V, as illustrated in Fig. 8.15. Practical realisation of this idea requires a gas-diffusion electrode, which consists of an electrocatalytically active porous material, usually in the form of finely divided carbon with an appropriate catalyst dispersed onto its surface, which in turn is fabricated into the form of a film using a polymer binder such as PTFE (see section 9.5.1.1 below). The resultant cell is shown in Fig. 8.15a, and its introduction into industrial processes depends critically on the manufacture of gas-diffusion electrodes that are both inexpensive and durable.

8.2.5.1 NaCl-Electrolysis with Oxygen Electrode and Cathode Gap

The primary cause of the lowering of cell voltage on moving from hydrogen evolution to oxygen reduction as the cathode reaction has already been shown in Fig. 8.15 and 8.15a. In practice, great care must be taken to ensure that the gap between cathode and membrane can allow sufficient water to pass through for the reduction of the oxygen and the formation of the electrolyte, and experiments have demonstrated that an apparatus with a carefully calibrated 'finite gap' between membrane and oxygen diffusion electrode can lead to reproducible cell voltages. Furthermore, in this arrangement, peripheral electrolyte can also be utilised.

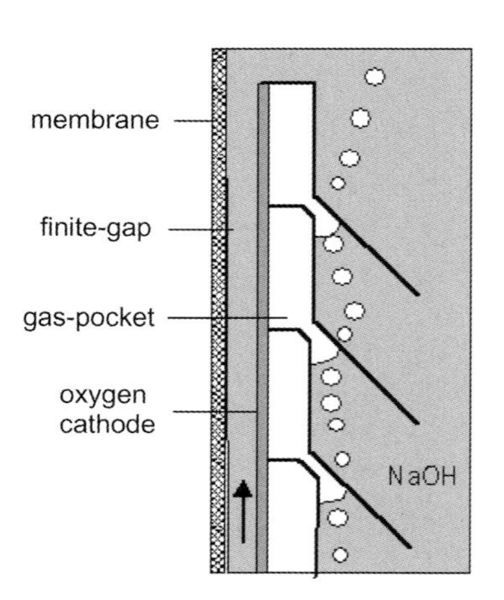

Fig. 8.15b Membrane cell with oxygen cathode, finite-gap and gas-pockets for pressure equalisation between the electrolyte on the anode side and the oxygen inlet. The chlorine anode is not shown.

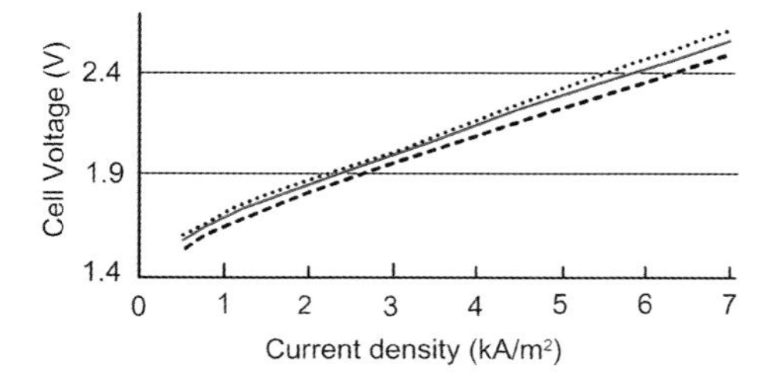

Fig. 8.15c NaCl electrolysis with oxygen cathode: cell voltage vs. current density for three different cells, 32 % NaOH, 90°C.

The main problem with the finite-gap apparatus is the critical dependence on the pressure difference between the electrolyte in the cathode gap and the oxygen pressure on the other side of the porous electrode. If the pressure is too high the air will penetrate through the porous electrode into the thin layer; if it is too low the electrode will flood. To add to the complications, the electrodes are a good metre or more high, so there is a hydraulic pressure difference between top and bottom of the electrode of about 0.1 bar. Ideally, the gas pressure difference should be the same all the way up the electrode and this is achieved through the 'gas-pocket' principle as shown in Fig. 8.15b. Oxygen enters the first pocket of the cell at a pressure appropriate to the lowest part of the cell. Some penetrates to the cathode, but most escapes and rises as bubbles through the alkaline electrolyte on the right-hand side of Fig. 8.15b until it is captured by the entrance to the second pocket. There some penetrates again to the cathode, a the pressure appropriate to that depth, but most again escapes, only to be re-captured at the bottom of the third pocket. It can be seen that under these circumstances the gas will always have a pressure appropriate to the corresponding depth of the cathode. Through the insertion of successive elements in the cell with a height of 120 cm and with a surface area of 2.5 m^2, an energy usage of less than 1400 kWh per tonne of NaOH can be obtained under normal electrolysis conditions (3 $kAcm^{-2}$, 90°C, 32 % NaOH). The measured cell voltage as a function of current density for a four-cell configuration is shown in Fig. 8.15c. A parallel test of oxygen diffusion cells and hydrogen evolution cells has demonstrated impressively that an energy saving of more than 600 kWh/tonne is also possible on the technical scale. The hydraulic-cell concept satisfies the demands of both anode and cathode sides of the apparatus.

8.2.5.2 **HCl-electrolysis with Membrane and Oxygen Cathode**

For the electrolysis of HCl, the reaction product on the cathode side is water:

$$\frac{1}{2}O_2 + 2H^+ + 2e^- \rightarrow H_2O$$

This water can easily be removed from the cathode without the water content of the membrane being perturbed. The cathode can be brought into direct contact with the membrane in a 'zero-gap' configuration and pressure compensation is not necessary. The construction principles for HCl electrolysis with an oxygen cathode can be inferred from Fig. 8.15d. Laboratory tests show the feasibility of HCl electrolysis with membrane and oxygen cathode as well as the possibility to lower the energy usage by about a third compared to classical HCl electrolysis.

Since the cathode can now be placed directly next to the membrane, the critical issue now becomes the thermal regulation of the anode. On the cathode side there is little electrolyte circulation, and only the small quantities of water formed in the reaction are constantly removed. The materials constraints on the equipment mean that the overall temperature must be kept at about 60°C and a continuous thermal equilibration is essential to permit cold electrolyte to be admitted to the cell. To achieve this, a gas bubble jet must be constructed in the anode compartment to mix the electrolyte and permit temperature and electrolyte concentration to be continuously equilibrated.

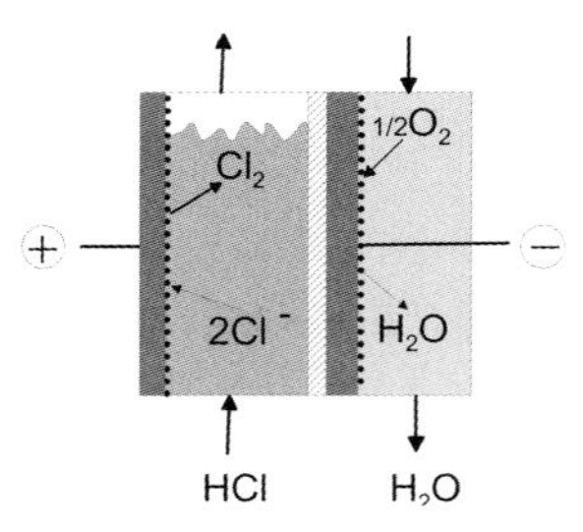

Fig. 8.15d Principle of HCl electrolysis with membrane and zero-gap oxygen cathode.

In continuous operation, the cells are loaded at 4 kAm^{-2}, leading to an energy usage of 1,100 kWh per tonne of chlorine, with a current yield of close to 100 % and a chlorine purity of 99.9 %. Recently, improved cathodes based on Rh instead of Pt have been introduced. This permits a simplification of the equipment since the Rh catalyst does not require a protective potential or gas when the apparatus is switched off; it can remain under a chlorine atmosphere. On restarting, the apparatus produces pure chlorine again immediately.

The preparation of organic chloro-compounds usually involves chlorination of appropriate hydrocarbons as:

$$RH + Cl_2 \rightarrow RCl + HCl \tag{8.12}$$

and it is evident that half the chlorine necessary for the reaction ends up as either HCl gas or aqueous hydrochloric acid. If RCl is an intermediate in the manufacture of fluorocarbons, then all the chlorine is eventually recovered as HCl, but the market for this material is small, and much is re-cycled through the electrolysers to give Cl_2. and H_2 again. The electrolysis is carried out in bipolar cells analogous to those shown in Fig. 8.3, with ca. 2.5 m^2 graphite plates as electrodes and a PVC net as membrane. The cell voltage is now only ca. 2 V per cell for a current density of 3 kAm^{-2} and the electrode separation is ca. 6 mm.

8.3
The Electrochemical Extraction and Purification of Metals

8.3.1
Extraction from Aqueous Solution

Commonly the metal ore is roasted in air to form the oxide which is then dissolved in aqueous sulphuric acid; from this solution the metal is then electrochemically deposited at the cathode and (usually) oxygen is evolved simultaneously from the anode:

$$MeO_x + 2xH^+ \rightarrow Me^{2x+} + xH_2O \tag{8.13}$$

$$Me^{2x+} + 2xe^- \rightarrow Me(s) \tag{8.14}$$

$$xH_2O \rightarrow (x/2)O_2 + 2xH^+ + 2xe^- \tag{8.15}$$

The acid is, therefore, regenerated in the anode process and can be recycled: the nett process is

$$MeO_x \rightarrow Me + (x/2)O_2 \tag{8.16}$$

Metals whose deposition potentials lie *negative* of the normal hydrogen potential can only be deposited from acid solutions if they show a large overpotential for hydrogen evolution. This allows such metals as Pb, Zn, Ni, Co, Cd, Cr, Sn and Mn (the last from neutral solution) to be extracted electrochemically, as well as more noble metals such as Au, Ag and Cu. In addition, the preparation of alloys by simultaneous deposition of two or more metals is possible.

The principles of electrowinning can be illustrated by the production of zinc. The cells used here are open lead- or plastic-lined concrete troughs 3m × 1m × 1.5m deep, in which the pure aluminium sheet cathodes and lead anodes are suspended from current collectors. The cells are charged with a solution of 95 g dm^{-3} zinc, in the form of zinc sulphate and 40 g dm^{-3} concentrated sulphuric acid. Application of a cell voltage of 3.5–4 V leads to zinc deposition with a current density of 0.5–1 kAm^{-2} and simultaneously evolution of oxygen at the anode, this latter being vented to the atmosphere. The deposition is continued until the zinc is reduced in concentration to 35 g dm^{-3} and the acid has increased in concentration to 135 g dm^{-3}, this solution then being removed and used to dissolve more of the roast oxide.

Although zinc possesses a high overpotential for hydrogen evolution, it is only just large enough for deposition to be the preferred process under the current densities used, and some hydrogen evolution is always observed. This actually serves a useful purpose in causing stirring of the solution and consequent improvement in mass transport. However, the process is sensitive to the deposition of impurity species that can lower the overpotential. In particular, solutions must be essentially free of Fe, Ni and Co ions.

The cell voltage can, in principle, be substantially reduced if, instead of oxygen evolution from lead, the oxidation of H_2 is used as the anode process. This process also generates H^+ ions, so the overall process will produce an acid-rich electrolyte as above, but only makes sense energetically if there is a supply of H_2 available, for example as a by-product of the chlor-alkali process.

8.3.2
Metal Purification in Aqueous Solution

If a copper anode and copper cathode are immersed in an acidified copper sulphate solution and a voltage applied, copper will dissolve at the anode and be deposited at the cathode, as shown schematically in Fig. 8.16. The process only requires a voltage of 0.1–0.2 V for current densities of 0.1–0.3 kAm^{-2}.

This process is the basis of the purification of pyrometallurgical copper, which is used as the anode. Only those impurities in the copper with normal potentials *negative* of the working potential of the anode (ca. +0.5 V) will dissolve as ions (including: Fe,

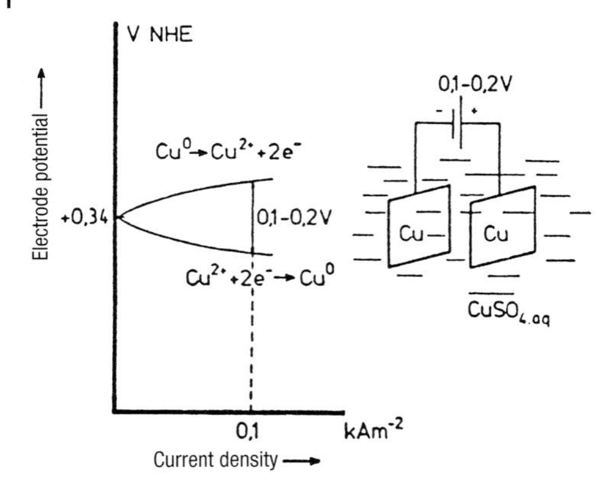

Fig. 8.16 Dissolution and deposition of copper.

Ni, Co, Zn and As). More noble metals, such as Ag, Au and Pt, remain behind in finely divided metallic form and collect under the anode as "anode sludge". However, at the working potential of the cathode (ca. +0.3 V), *only copper will deposit*, the remaining metals having normal potentials more cathodic than this value. This is shown schematically in Fig. 8.17. Given the small cell voltage, the specific energy consumption, ω, is very small (250 kWh/ton Cu), and the process highly attractive. From the technical standpoint, the cells are very similar to those used for zinc extraction, with the impure copper anodes being sheets several cm thick and the cathodes being initially thin sheets of pure copper. The anode sludge is worked up to extract the noble metals, and nickel and cobalt are extracted from the enriched electrolyte. Similar processes have also been developed for the purification of Ni, Co, Ag, Au, Pb and Zn.

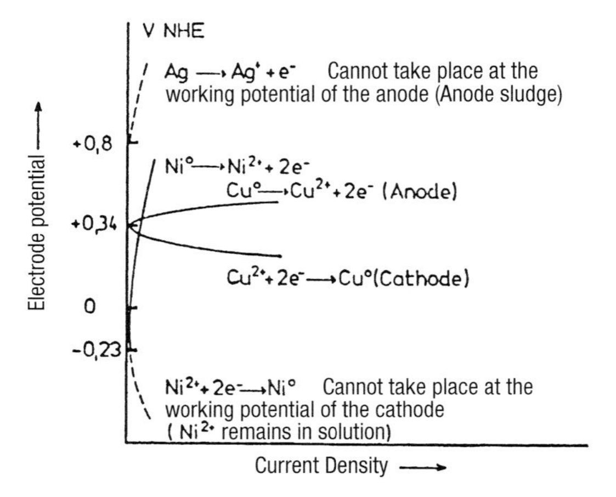

Fig. 8.17 Principles of the electrolytic purification of copper.

8.3.3
Molten Salt Electrolysis

Many metals, particularly those from the early main groups of the periodic table, cannot be deposited from aqueous solution (save as amalgams) since their normal potentials are too far negative of NHE. For the electrolytic preparation of such metals, we must then employ ionic systems with no free protons or water molecules, and no other components with relatively positive deposition or decomposition potentials. Such systems fall into two categories:

- Organic aprotic solvents containing suitable metal salts or organometallic compounds;
- Molten salts which dissociate into freely moving ions.

The first of these types has a number of disadvantages, including rather poor conductivity and relatively high cost, and has only found application in small scale niche areas. The second has found, by contrast, major application in the extraction of Al, Mg and Na, and, on the smaller scale, in the extraction of specialist metals, such as Li, Be, B, Ti, Nb, Ta and the rare earths.

The largest scale production is of Al. Unlike Mg or Na, however, which can be obtained from relatively low temperature molten salts or hydroxides, the main source of Al is the ore bauxite, Al_2O_3, whose melting point is very high (2050°C). This is far too high for electrolysis; there would be serious materials problems in fabricating appropriate cells, and the energy losses through thermal radiation would be extremely wasteful. As a result, a solvent must be found, and the traditional process is based on cryolite, Na_3AlF_6, which forms a eutectic mixture with 12–22.5 % Al_2O_3 melting at 935°C. This material has a higher decomposition voltage than Al_2O_3 itself, and can be recycled, though there is some loss through evaporation and decomposition in the cell.

The electrolysis is carried out in carbon-lined iron tanks between carbon anodes that project into the electrolyte melt, which in turn floats on the molten aluminium cathodes. The cell voltage is ca. 4.2 V (theoretical value 1.7 V), and current densities at the anode are 6.5–8 kAm^{-2} and at the cathode 3–3.5 kA m^{-2}. The main anode reaction is oxygen evolution, with the oxygen reacting immediately with the carbon to form CO and CO_2. However, the anode gas also contains fluorine and particulate oxide, both of which must be removed before venting, and the expense of this process has led to the development of other molten-salt electrolytes, of which the best known is the "Alcoa" process. This uses a melt of 10–15 % $AlCl_3$ in 1:1 LiCl/NaCl at 700°C and a bipolar configuration with carbon electrodes. The chlorine evolved at the anode is re-reacted with the Bauxite ore to regenerate $AlCl_3$.

Aluminium from the cryolite process still contains ca. 0.1 % Fe and 0.1 % Si, and further purification, if required, can be carried out electrochemically as for copper. In this case, however, the aluminium is molten, and a cell of the type shown in Fig. 8.18 must be set up. The molten anode is, in fact, an alloy of Al with ca. 60–70 % of other, more noble, metals, such as Cu, Si, Fe and Ti. This alloy is *denser* than the AlF_3/NaF/BaF_2 melt used as the electrolyte, and this in turn is denser than *pure* aluminium. Cells of the type shown in Fig. 8.18 are capable of producing at least 99.999 % Al.

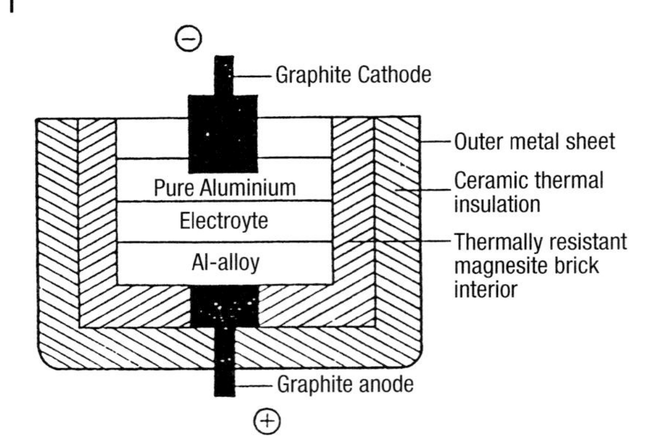

Fig. 8.18 Schematic representation of a cell for the refining of aluminium (the three-layer cell).

8.4 Special Preparation Methods for Inorganic Chemicals

In the previous sections, the most important industrial electrochemical processes have been discussed. We now turn to three further processes that either show interesting features (preparation of oxychloro-compounds, hydrogen peroxide and peroxysulphuric acid) or may become more important in the future (such as water electrolysis within the compass of the hydrogen economy). In addition to these, there are other compounds that can be prepared by electrochemical techniques on the industrial scale, including perborate ($NaBO_3 \cdot 4H_2O$), permanganate, braunstein (MnO_2, for use as a cathode material in batteries), Cu_2O (used as a fungicide and catalyst), chromic acid from Cr(III), and fluorine. The latter is made by electrolysis of a KF-HF mixture between steel and carbon electrodes between 70 and 110°C.

8.4.1 Hypochlorite, Chlorate and Perchlorate

If a NaCl solution is electrolysed in an undivided cell, the Cl_2 formed at the anode disproportionates in the presence of OH^- ions generated at the cathode:

$$Cl_2 + 2OH^- \rightarrow ClO^- + Cl^- + H_2O \tag{8.17}$$

The resultant hypochlorite can be further oxidised to chlorate:

$$2ClO^- + H_2O \rightarrow \frac{2}{3}ClO_3^- + \frac{4}{3}Cl^- + 2H^+ + \frac{1}{2}O_2 + 2e^- \tag{8.18}$$

However, this leads to the requirement of nine Faradays per mole of chlorate, whereas only six are needed if a chemical disproportionation route is followed:

$$2ClO^- + H_2O \rightarrow 2HClO + 2OH^- \tag{8.19}$$

$$2HClO + ClO^- \rightarrow ClO_3^- + 2Cl^- + 2H^+ \quad (8.20)$$

On the basis of the above reactions, we can see from (8.18) that *hypochlorite* is favoured by keeping the chloride concentration as high as possible. In addition, high current densities are employed owing to the fact that current-voltage curve for reaction (8.18) is flatter than for chlorine evolution. Reactions (8.19) and (8.20) can be suppressed by working at low temperature, and the resultant liquor can be used directly for bleach and disinfection purposes.

To enhance *chlorate* formation, the chemical processes (8.19) and (8.20) can be accelerated by raising the temperature to 80°C and by using cells with a high volume, since (8.18) is heterogeneous whereas (8.19) and (8.20) are homogeneous. The cathode in the chlorate cell is steel and the anode material graphite or Pt/TiO_2. Great care is needed on the choice of the steel: nickel/chrome steels, for example, do not work since leaching of Ni^{2+} into solution leads to catalytic decomposition of the chlorate to hypochlorite and O_2.

Perchlorate can be made by anodic oxidation of chlorate at Pt or PbO_2 electrodes, but a continuous process from NaCl in the same cell leads to poor yields, and it is essential to start with a solution of $NaClO_3$.

8.4.2 Hydrogen Peroxide and Peroxodisulphate

The chlorine-oxygen compounds of section 8.4.1 are used, like elemental chlorine, as bleach and disinfection agents. However, it has become clear that Cl_2 itself can leave behind highly poisonous organo-chlorine compounds in water, and the preferred chlorine-based bleaching agent today is ClO_2 in the so-called 'elemental-chlorine-free' (ECF) process. However, it would be environmentally preferable to do without chlorine altogether where possible, and attention has turned to 'total-chlorine-free' (TCF) processes using peroxides. H_2O_2 can be prepared electrochemically by the oxidation of sulphuric acid or the partial reduction of O_2. In the first case, oxidation of the singly ionised HSO_4^- ion takes place as:

$$2HSO_4^- \rightarrow H_2S_2O_8 + 2e^- \quad (8.21)$$

possibly by intermediate formation of the radical $SO_4^{\dot{-}}$ which then dimerises. High concentrations of HSO_4^- are needed in practice, and high anodic potentials are also necessary. A cell voltage of 5 V per cell is used and hydrogen is generated at the cathode. The anode must be chemically stable and have a high overpotential for oxygen evolution, and in practice only Pt and Pt-Ta-Ag alloys have the required properties.

H_2O_2 is obtained from the peroxodisulphate by hydrolysis

$$H_2S_2O_8 + 2H_2O \rightarrow 2H_2SO_4 + H_2O_2 \quad (8.22)$$

and separated by distillation. Owing to the high cost of electricity and the cost of the anodes, the proportion of H_2O_2 actually prepared by the above route has decreased steadily in recent years, with catalytic processes, such as the anthraquinone process, becoming dominant. The preparation of H_2O_2 by direct reduction of O_2 electrochemically is currently being technically developed. Its success depends on the electrocatalytic activity of the electrode material used as the cathode, and considerable accumulation of H_2O_2 can take place in the solution phase (see Section 6.4).

8.4.3
Classical Water Electrolysis

The electrochemical preparation of hydrogen and oxygen from water is technically an old process, but in recent decades the increasing need for hydrogen has been met by petroleum cracking or water-gas shift reactions rather than electrolysis. The electrolytic process is only competitive when the cost of electricity is low, and only in the neighbourhood of hydroelectric power stations can hydrogen, ammonia and other nitrogen compounds be prepared economically. However, the electrolysis of water has received considerable recent attention in the context of the so-called "hydrogen economy", in which hydrogen gas is employed as an energy vector.

Thermodynamically, a voltage of 1.23 V is necessary for the decomposition of water to H_2 and O_2, and this voltage is independent of pH. However, even in the case of "good" electrocatalysts, an overpotential of a few hundred mV is necessary, and for this reason, chloride ions cannot be tolerated, owing to interference at the anode by chlorine evolution. Moreover, sulphate-ion solutions show rather low conductivity, and acid electrolytes lead to corrosion problems, so the normal electrolyte in classical electrolysers is 6–8 M KOH. The cell construction is typically bipolar, with diaphragms separating anode and cathode compartments. The electrodes are formed from sheets of nickel or steel wire mesh pressed against the membrane to minimise the resistance, and the cell is designed to ensure that gas bubbles that form are removed rapidly from the electrolyte to avoid further increases in cell resistance. Stacks have been constructed capable of generating up to a few hundred cubic metres of H_2 per hour at temperatures of ca. 800 °C with current densities of 1–2.5 kAm^{-2} and cell voltages of 1.8–2.0 V. The energy requirement per cubic metre is ca. 4 kWh.

It should be pointed out that the evolution of D_2 and HD is kinetically slower than that of H_2, leading to an enrichment of the electrolyte in deuterium. This has been an important preparative route for D_2 and D_2O.

8.4.4
Modern Water Electrolysis and Hydrogen Technology

In the operation of coal and nuclear power stations, it is desirable to have some form of load levelling to compensate for strong peaks and troughs in energy demand, and a cost-effective possibility that has received considerable recent attention is the use of hydrogen gas as a *secondary energy vector*. Hydrogen is easily transported in pipes, and in energy terms, its transport is a factor of five cheaper than the costs of transporting electricity through high-voltage cables. Looking further ahead, the exhaustion of fossil fuels will offer the opportunity of a complete transition to the "hydrogen economy", in

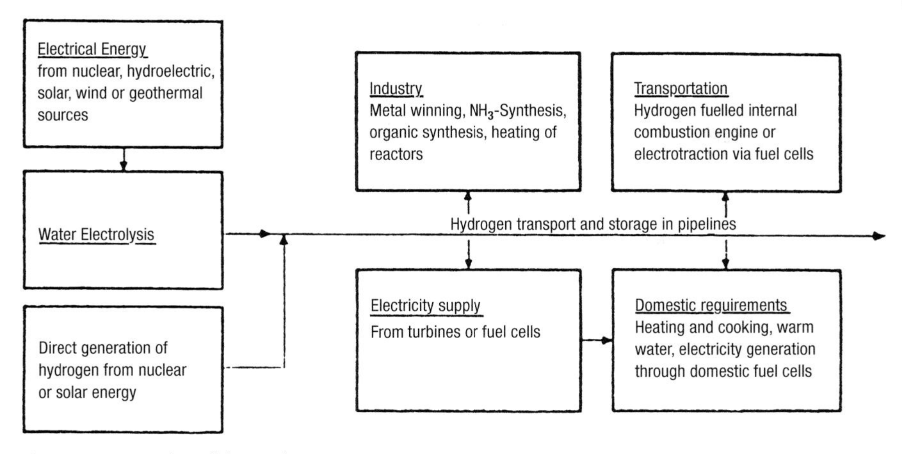

Fig. 8.19 Principles of the Hydrogen Economy.

which hydrogen gas, generated by nuclear power (including fusion reactions) or by solar or wind power operated water electrolysers is transported by pipes to consumers, who will use it to generate power in fuel cells, as an industrial raw material, as a source of process heat through combustion, as an energy vector for transport systems and to provide energy for the home, as shown in Fig. 8.19.

An additional driver for the adoption of the hydrogen economy is the urgent need to reduce CO_2 emissions into the atmosphere. These emissions are now known to be responsible for an appreciable fraction of the atmospheric warming currently taking place, and given that the combustion of hydrogen gives only H_2O, there are clear attractions in replacing fossil fuels with H_2. However, the eventual introduction of the hydrogen economy depends critically on improvements in the efficiency of current electrolysers, and in particular the urgent need to lower the cell voltage at which electrolysis is carried out. Currently, a voltage of 1.8 V is needed for electrolysis, but the recovery of electrical energy through a typical fuel cell would generate 0.8 V at best, giving an efficiency quotient of ca. 0.45, which is far too low.

Reduction of the electrolysis voltage can, in principle, be achieved in one or more of the following ways:

- Reduction of the overvoltage at the electrodes by the use of improved electrode structure or electrocatalysts;
- Reduction of the ohmic losses in the cell;
- Increasing the temperature, since $\partial E_{\text{cell}}/\partial T \approx -0.85$ mVK^{-1} for water electrolysis

Three main directions have been adopted:

- The further improvement of classical electrolysis cells by better materials (specially activated iron and nickel surfaces), improved gas removal from the cell and increase of the temperature to 80–100°C, giving a cell voltage of ca. 1.7 V at a current density of 2.5 kAm^{-2}.

- The use of highly active porous electrodes developed for fuel cell work (platinum or nickel-zinc alloys on the hydrogen side and Pt/IrO_2 on the anode) with a *thin polymer membrane* of high proton conductivity as electrolyte, as described above (see Fig. 8.5). The anode is supplied with demineralised water, and voltages as low as 1.55 V at 2.5 kAm^{-2} have been reported from laboratory prototypes working at 80°C.
- The electrolysis of steam at 700–1000°C can be carried out on porous catalyst layers deposited on oxide-ion conducting solid electrolytes. In laboratory experiments at 900°C, voltages of 1.2 V for current densities of 1 kAm^{-2} have been attained ($\partial E_{cell}/\partial T \approx -0.23$ mVK^{-1} for steam electrolysis).

As a final note, it is interesting that if PbO_2 is used instead of platinum as the anode in water electrolysis, the cell voltage rises above 2.4 V, at which point the reaction $H_2O \rightarrow 2H^+ + \frac{1}{3}O_3 + 2e^-$ becomes feasible. Production of ozone by this method is not cheap but does offer some advantages over the normal electrical discharge route.

8.5 Electro-organic Synthesis

8.5.1 An Overview of Processes and Specific Features

Electro-organic reactions were reviewed in Section 6.4, and from the characteristic features discussed there, certain special features can be identified as common to most electro-organic syntheses:

- Frequently the conductivities of the electrolytes used are low, particularly if based on non-aqueous solvents, leading to large ohmic potential drops in the cell, and hence to high energy costs unless the electrode separation is minimised;
- For conversion rates acceptable at a technical level, substrate concentrations must be high, a requirement often at variance with the need for high conductivity;
- The high reactivity of the radical product of the initial electron transfer process implies low selectivity in subsequent steps, and selectivity depends not only on a suitable choice of electrocatalyst and working potential but on concentration ratios of species at the phase boundaries.

A simple example of the influence of reaction path on the situation at the phase boundary has already been encountered in Section 8.1.3. Another possibility is that the initial electron transfer process may be followed either by electrophilic attack by solvent or by dimerisation to give the required product. In the first case, the reaction rate will be proportional to the concentration of the electrochemically generated primary product, whereas in the second case, it will be proportional to the *square* of the concentration. It follows that increasing the current density will favour the dimerisation process.

Table 8.1 Electro-organic synthesis taken to pilot scale or technical scale.

Electrolysis Product	Reactant	Electrolyte	Anode Material	Cathode Material	Current Density*	Diaphphragm
Oxidation						
Gluconic Acid	Glucose	$NaBr/H_2O/$ $CaCO_3$	Carbon	Carbon	0.3	No
Dialdehyde Starch	Starch	H_2SO_4/H_2O 10 %	Pb/PbO_2	Fe	0.5	Ion exchange
Hydrogenation						
m-aniline sulphonic acid	m-nitro-benzene sulphonic acid	H_2SO_4/H_2O 10 %	Pb	Cu	1.0	Yes
piperidine	pyridine	H_2SO_4/H_2O 5 %	Pb/PbO_2	Pb	1.5	Yes
1,2 Dihydro-phthalic acid	phthalic acid	H_2SO_4/H_2O 5 % Dioxane		Pb	1.0	Ion exchange
Dimerisation						
adipic acid dinitrile	acrylonitrile	$NaH_2PO_4/H_2O + NR_4^+$	Fe	Cd	2.0	No
Halogenation						
perfluoro-butyric acid	n-butyric acid	HF	Ni		0.2–0.4	No
perfluoro-octanoic acid	octanoic acid	HF	Ni		0.2–0.4	No
Organometallic						
tetra-ethyl lead	Pb/EtMgCl	THF/diglyme	Pb	Fe	0.03–0.07	No

kA m^{-2}

In spite of these complexities, a great many electro-organic syntheses have been optimised by appropriate choice of parameters and taken through to pilot-scale or technical stage, as shown in Table 8.1. As can be seen from the Table, conventional electrode materials, electrolyte systems (i.e. aqueous solutions with inorganic salts) and cell constructions have played the major role up to now. The low current densities in comparison with inorganic systems serve especially to lower the large cell voltages.

The future of electro-organic synthesis is difficult to predict. The previously expected lowering of energy prices has failed to materialise. However, the stimulus derived from this expectation has led to a vast array of research data and further development of attractive processes does seem likely. The impact of new cell designs (Section 8.6) and cell components (such as non-aqueous electrolytes, chemically modified electrodes, and SPE membranes) may also lead to new processes in the longer term.

8.5.2
Adiponitrile – The Monsanto Process

Acrylonitrile, $CH_2{=}CHCN$, can be reductively dimerised in aqueous solution to dinitrile-adipic acid. The basic reactions have already been discussed in Section 6.4.4, with the most important side reaction being the protonation of the initial anion:

$$CH_2{=}CHCN + e^- \rightarrow CH_2CHCN^{\bullet-}$$
$$CH_2CHCN^{\bullet-} + H^+ \rightarrow {}^{\bullet}CH_2CH_2CN$$

and

$$^{\bullet}CH_2CHCN + e^- + H^+ \rightarrow CH_3CH_2CN$$

Under normal circumstances, this sequence actually dominates, reducing the yield to a very low value. However, if as electrolyte a salt with a large unsolvated surfactant cation is used, the water is expelled from the region close to the electrode surface and dimerisation can take place as:

$$CNCHCH_2^{\bullet-} + CH_2{=}CHCN \rightarrow CN^{\bullet}CHCH_2CH_2CHCN^-$$
$$CN^{\bullet}CHCH_2CH_2CHCN^- + e^- \rightarrow {}^-CNCHCH_2CH_2CHCN^-$$
$$^-CNCHCH_2CH_2CHCN^- + 2H^+ \rightarrow CN{-}(CH_2)_4{-}CN$$

where protonation takes place further away from the electrode surface.

Technically, the electrolysis process takes place in an undivided bipolar cell with a current density of 2 kA m^{-2} and a cell voltage of 4 volts. The bipolar stainless steel plate electrodes are separated by 2 mm, and the cathode is coated with a layer of cadmium, which has a high overpotential for hydrogen evolution, thereby avoiding parasitic losses. The surfactant electrolyte is a 1 wt% solution of $[EtBu_2N{-}(CH_2)_6{-}NBu_2;Et]^{2+}$ phosphate with the necessary conductivity mainly attained with 10% Na_2HPO_4/boric acid. These latter materials have the advantage of inhibiting the corrosion of the stainless steel, a matter of some importance since otherwise, in the undivided cell, iron would dissolve from the anode and re-deposit on the cathode, lowering the overpotential for hydrogen evolution. To prevent any traces of iron reaching the cathode, the electrolyte also contains 0.5 wt% EDTA as complexing agent.

8.6 Modern Cell Designs

For competitive electrochemical production processes, a good space-time-yield is essential, particularly when a low current density has to be used. However, for conventional cell construction, with the electrodes present as flat plates, high A_v-values (A_v = electrode area per cell volume) can only be attained by small inter-electrode separation. Typical filter-press cells have A_v-values up to 2 cm^{-1}, and even for undivided capillary cells, the practical lower limit for inter- electrode separation is ca. 0.1 mm, giving the best A_v-values attainable as $\sim$ 5 cm^{-1}.

For divided cells, the smallest cell volume is associated with "zero-gap" cells, in which an SPE-membrane is used both as electrolyte and membrane. In such cells, the electrodes can take the form of porous catalysed carbon layers, with typical construction shown in Fig. 8.20. This type of cell is used in modern water electrolysers and in the generation of chlorine and ozone.

For higher A_v-values it is essential to move to three-dimensional electrode structures. Such structures can be formed from porous materials such as finely particular carbon, though unless the solution can flow freely through the pores, such electrodes show rapid decay in performance as the electroactive substrates becomes denuded from the interior of the electrode. Recently, thin very high surface-area porous layers

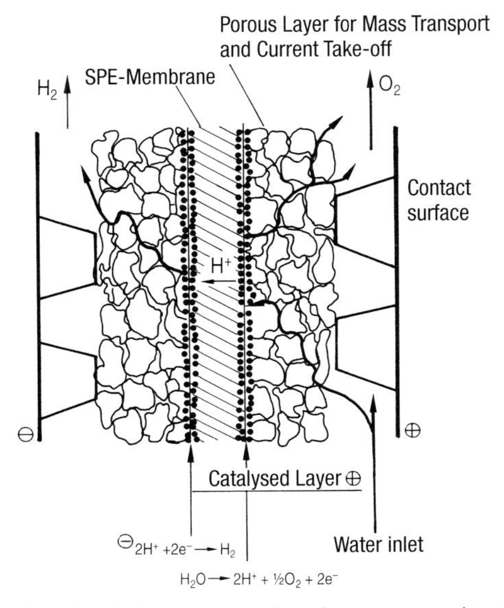

Fig. 8.20 Scheme for SPE-electrolysis (not to scale). The contact elements are also the current collectors.

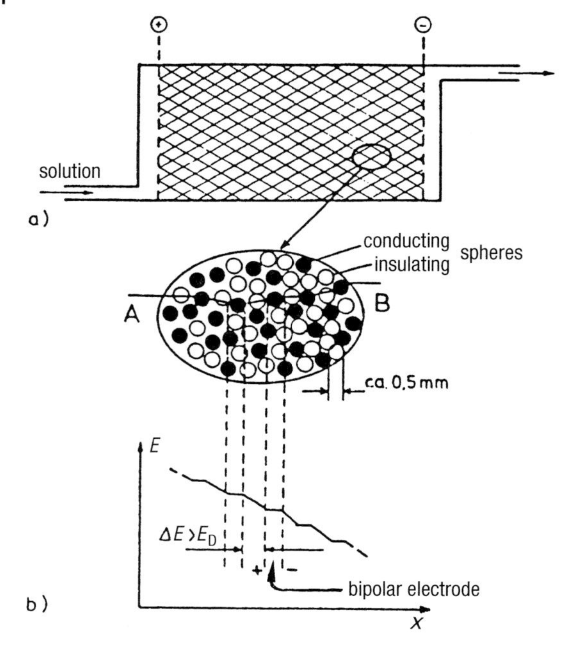

Fig. 8.21 Principle of an electrochemical fixed-bed reactor.

have been developed for use in fuel cells which counter this problem, with A_v-values of 30 upwards.

A second way around the problem of transport is the use of *particulate electrodes*, of which there are two different types: fixed-bed reactors and fluidised-bed reactors. Fixed-bed reactors are formed from a mixture of conducting and non-conducting spheres (e.g. carbon and plastic spheres of some 0.5 mm diameter) which is held immobilised between two porous plate electrodes through which the solution to be electrolysed flows. Provided the conductivity of this solution is not too high, and a sufficient voltage between the porous plates can be sustained, each conducting particle acts as a miniature bipolar electrode, as shown in Fig. 8.21. For the type of cell shown in 8.21, the overall potential drop is very high, ca. 60 V cm^{-1}, and such cells are normally operated only in undivided form. Fixed-bed reactors have found commercial use in the removal of heavy metals from waste water, and the best show A_v-values of 5 – 100 cm^{-1}.

The principle of the fluidised-bed reactor is shown in Fig. 8.22. In this system, conducting catalysed particles of diameter 0.1–1 mm are kept in activated suspension over a distributor plate through which a rapid flow of electrolyte is maintained. The substrate in solution is adsorbed onto the particles whilst they are in suspension, but periodically, the particles come into contact with the working electrode, at which point electron transfer to the adsorbed species takes place. Once the particles have lost contact with the electrode, the reaction product desorbs and the cycle begins again. The advantages of this reactor are that material transport and adsorption/desorption are

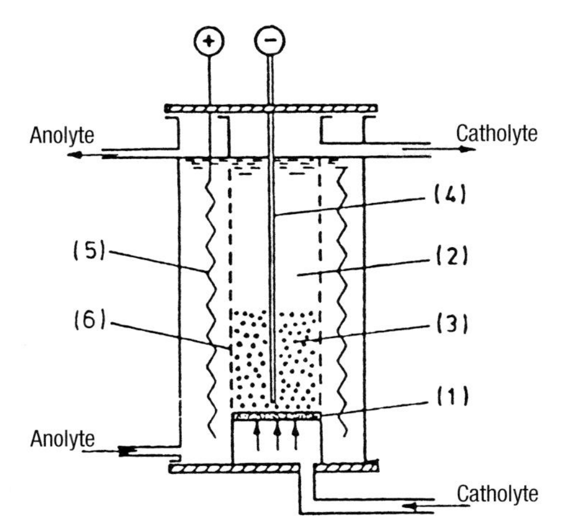

Fig. 8.22 Diagrammatic representation of an electrochemical fluidised-bed reactor: (1) distributor plate; (2) catholyte; (3) fluidised layer of catalyst-coated particles; (4) working electrode; (5) counter electrode; (6) diaphragm.

spatially separated from the electron transfer process (which is usually rapid), giving very high A_v-values of 10 – 200 cm^{-1} for low real current densities. This latter advantage allows very dilute solutions to be electrolysed.

The importance of increasing A_v-values is that the space-time-yield for electrochemical reactors in general compares unfavourably with chemical reactors. If the Monsanto process is taken as a model, then for $M = 108$, $n = 2$, $a = 0.9$ and $j = 0.2$ A cm^{-2}, values of the space-time-yield between 0.4 ($A_v = 1.5$ cm^{-1}) to 20 kg dm^{-3} hr^{-1} ($A_v = 50$ cm^{-1}) are obtained, whereas chemically catalysed reactors can show space-time yields of up to 1000 kg dm^{-3} hr^{-1}.

Finally, one important recent development has been in the use of micro-reactors in electrochemistry. Such reactors have linear dimensions in the order of mm, and the total reaction volume is of the order μL. The fabrication of such miniature systems has become possible by developments in micro-etching and casting technologies derived from the semiconductor industry, and the central idea is that the very short diffusion distances enhance the rates of diffusion-limited processes such as electrolyte transport. Scale-up is usually straightforward; in the simplest cases large numbers of micro-reactors can be put together in a modular fashion. Experimental micro-electrolysers based on thin-layer electrode stacks can give A_v values of up to 400 cm^{-1}, and have other advantages, such as very low ohmic losses, allowing the development of processes involving very low concentrations of electrolyte. In turn, this simplifies separation procedures.

8.7 Future Possibilities for Electrocatalysis

If a film containing a selective catalyst is attached to an electrode surface, the electrode is said to be 'chemically modified'. Furthermore, the structure, if polymeric, represents a transition from heterogeneous to homogeneous catalysis, with the catalytic centres now immobilised on the electrode. If such centres can show redox behaviour, electron transport can take place by a hopping mechanism throughout the film, as shown in Fig. 8.23. Of more interest from the catalysis perspective is the incorporation of highly selective catalysts such as enzymes in the electrode film; such enzyme films have already found commercial usage in highly specific sensors for such important biomolecules as glucose. By incorporating form-selective substrates such as zeolites, these catalysts can be made even more specific.

Another novel type of electrocatalyst is formed when a submonolayer of upd metal is deposited on a substrate. As might be expected, if several monolayers of metal are deposited, the catalytic effects are indistinguishable from the bulk deposit, but a submonolayer shows catalytic properties more typical of a surface alloy. One example is the combination of a platinum substrate with a sub-monolayer of OH- or O-adsorbing metals such as Ru, Sn, Sb, Bi, or As for the oxidation of CO to CO_2. Even a coverage of 10 % of co-catalyst can lead to a shift in oxidation potential of 300 mV or more. The upd metal may also alter the electronic structure of the surface, and this has been proposed as the mechanism whereby a submonolayer of Pb on Pt is so effective in promoting the oxidation of formic acid to CO_2.

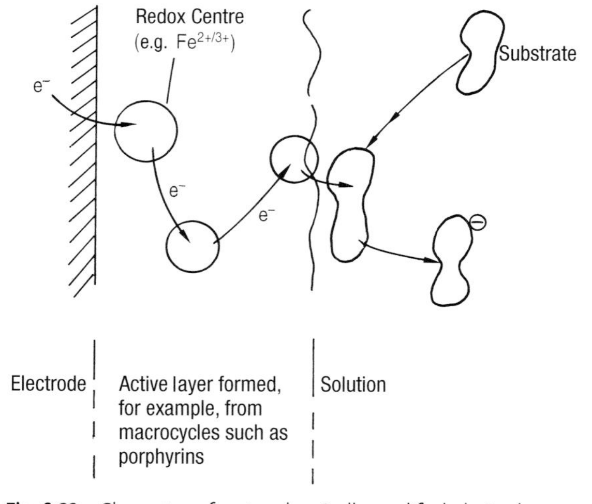

Fig. 8.23 Charge-transfer at a chemically modified electrode.

8.7.1 Electrochemical Modification of Catalytic Activity in Heterogeneous Chemical Reactions – The NEMCA Effect

Recently, it has been shown by Vayenas and co-workers that the rate of many heterogeneous chemical reactions can be altered by altering the potential between the (metallic) catalyst and a reference electrode. In many cases, this effect has been traced to the alteration of the concentration of surface adsorbed species on the catalyst, and this appears also to be bound up with changes to both the Fermi level of the catalyst and the electronic work-function, both being affected by a change in the electric field at the catalyst surface. The observation of an acceleration of a heterogeneous chemical reaction at an electrode surface by alteration of the electrode potential was, in fact, first reported in 1970 by Vielstich for the decomposition of HCHO at silver in alkaline solution. However, it was only with the investigations of gas reactions at metal catalysts on oxide-ion-conducting membranes between 500°C and 1000°C, where the potential is maintained between the catalyst on one side of the membrane and a reference Pt/H_2 electrode on the other, that the generality of the effect became clear. Subsequently studies by Lambert showed that a very similar effect could also be observed for cation-conducting superionic solids, such as Na β-alumina, where the mechanism has proved particularly suitable for investigation by surface-sensitive probes such as Auger and photoelectron spectroscopy. Given that the changes in rate and yield of the catalysed reaction far exceed the change in electrical current across the membrane, this effect has become known as the "Non-Faradaic Electrochemical Modification of Catalytic Activity" or NEMCA effect.

An example of this effect is shown in Fig. 8.24, in which the rate of oxidation of CO by oxygen on a Pt-catalyst on Y_2O_3/ZrO_2 is shown as a function of the potential difference between the catalyst and a Pt/H_2 reference. It can be seen that the basic behaviour is similar to the volcano plots shown in Section 4.5.4: r_0 is the rate in the absence of any potential on the electrode, and the ratio r/r_0 is plotted on a logarithmic scale. A second example is shown in Fig. 8.25: here the rate of chemical decomposition of HCHO to H_2 and CO on gold is measured as a function of the concentration of alkali and the electrode potential on gold by measuring the hydrogen yield through the DEMS technique (Section 5.4.4).

In a series of careful studies, Lambert has shown that metal ions derived from underlying superionic materials can diffuse onto the metallic catalyst and modify its behaviour equally profoundly. The experimental set-up used by Lambert is shown in Fig. 8.25a; a thin porous film ($\sim 1\mu m$) of the catalytically active metal is deposited onto a wafer of solid electrolyte (in this case a Na^+ ion conductor). This film forms the catalytically active working electrode of an electrochemical cell. The opposite side of the wafer carries a gold counter electrode and a gold reference electrode constituting a standard three-electrode cell as described earlier, and altering the potential between working and reference electrode allows control of the coverage of the former by sodium species, modifying its behaviour as a catalyst for the *gas-phase* reaction $A + B \rightarrow C + D$.

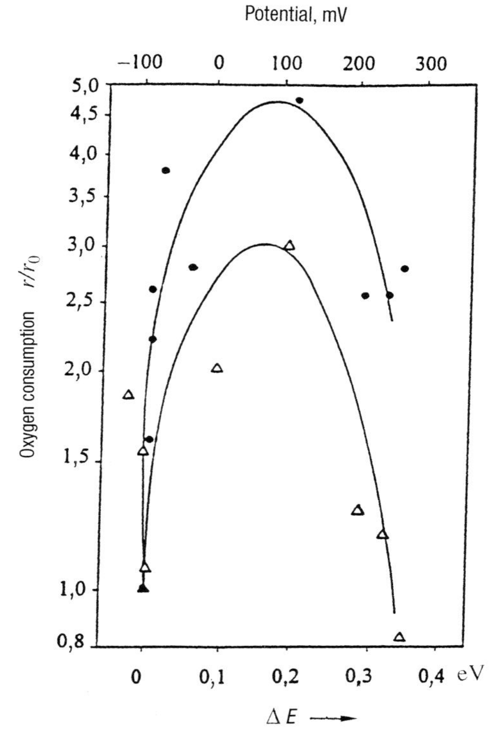

Fig. 8.24 Volcano plot for the purely chemical oxidation of CO on platinum dispersed on ZrO_2 as a function of the potential on the catalyst material; (•) T = 560°C; $r_O = 1.5 \times 10^{-9}$ g atom O s^{-1}; (Δ) T = 538°C; $r_O = 0.9 \times 10^{-9}$ g atom O s^{-1}; 0.21 % O_2 (from C.G. Vayenas, S. Bebelis, I.V. Yentekakis, P. Tsiakaras and H. Karasali, Plat. Met. Rev. **34** (1990) 122).

A simple example is the controlled electrochemical diffusion of K or Pb onto a platinum catalyst which alters the selectivity of the hydrogenation of ethyne to ethene. In the former case, photoelectron spectroscopy shows that the K wells up through the grain boundaries on the thin metal catalyst film when this is poised at a moderately negative potential with respect to a reference electrode on the other side of the superionic conductor. This process can be reversed with a timescale of minutes, corresponding exactly to the timescale over which the selectivity changes. The effect is substantial, selectivity changing from 10 % when the potential difference is +400 mV to 90 % at −400 mV. In the case of lead, the effect is even more marked, and mechanistic studies have shown that the electrochemical pumping of lead to the platinum surface leads to a substantial decrease in the adsorption energy of the ethene, allowing it to escape before further hydrogenation takes place.

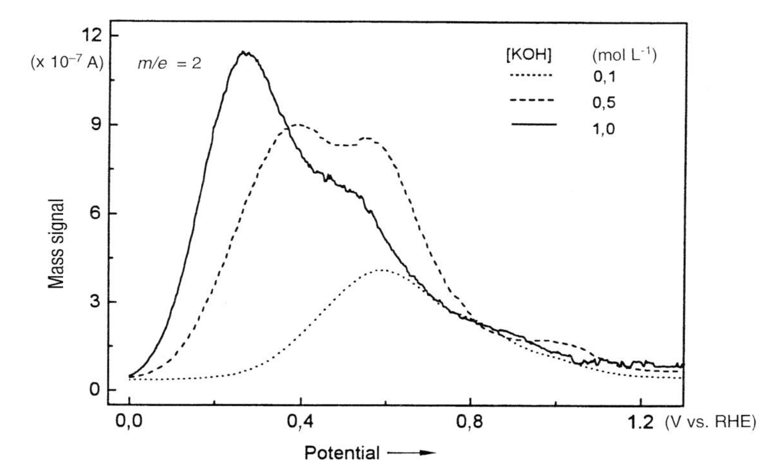

Fig. 8.25 Volcano plot of the NEMCA effect for hydrogen formation from 0.2 M formaldehyde on gold in KOH. The hydrogen formation is monitored using the DEMS technique in situ as the potential is scanned at 10 mV s^{-1} for three different KOH concentrations (from H. Baltruschat, N. Anastasijevic, M. Beltowksi-Brzezinska, G. Hambitzer and J. Heitbaum, Ber. Buns. Phys. Chem. **94** (1990) 996).

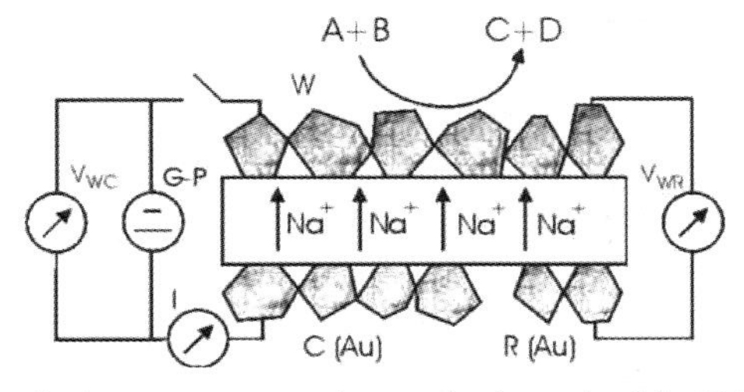

Fig. 8.25a Experimental set-up for the study of the NEMCA effect for metal ionic conductors (after Lambert).

8.8 Component Separation Methods

8.8.1 Treatment of Waste Water

The removal of undesirable impurities from liquid (usually aqueous) phases through electrochemical processes is based on the possibility of choosing suitable electrodes and potential/current conditions selectively to deposit or destroy these impurities. There are three distinct types of process:

1) The indirect electrochemical oxidation of inorganic or organic impurities;
2) Their direct electrochemical oxidation;
3) The cathodic removal of metal cations (usually those of heavy metals).

Waste-water treatment by *indirect electrochemical oxidation* has already been discussed in principle (Section 8.4.1). Into the contaminated water is inserted a graphite or dimensionally stable titanium anode at which chlorine or ClO^- is generated (if necessary with prior addition of Cl^- ions); these species are powerful oxidising agents, capable of destroying a wide range of biological micro-organisms as well as many organic chemicals in the homogeneous phase. The indirect homogeneous mechanism means that transport limitations are not so serious a problem, and even very dilute impurities can be removed. However, in recent years the risk of chlorination of certain types of organic impurity has been recognised: such chlorinated materials are likely to be toxic.

Indirect oxidation of species such as CN^- can also be achieved:

$$2CN^- + 5ClO^- + H_2O \rightarrow N_2 + 2CO_2 + 5Cl^- + 2OH^- \tag{8.23}$$

and amongst common organic impurities that can be oxidised in this way are phenols, mercaptans, napthenates, traces of heavy oils, aldehydes, carboxylic acids, nitriles, amines, and dyes.

If the decomposition potential of the solvent lies higher than that of the targeted impurity, or if a suitable electrocatalyst can be found, then *direct electrochemical oxidation* of impurities can be used in waste treatment. One application that has been investigated is the removal of CN^- by direct oxidation at graphite anodes:

$$2CN^- + 8OH^- \rightarrow 2CO_2 + N_2 + 4H_2O + 10e^- \tag{8.24}$$

Little industrial use seems to have been made of this type of process, save for the removal of traces of water from aprotic organic solvents such as acetonitrile and dimethylsulphoxide (DMSO); the decomposition potentials of these solvents lies well above that of water, which can, as a result, be removed by electrolysis. This type of process can be extended, for example, to the removal of thiophene from benzene, CCl_4 from acetic acid or chlorination products from caprolactam.

The cathodic reduction of metal cation concentrations in waste waters requires the potential to be lower than the standard electrode potential by an amount determined by the Nernst equation: $E = E^0 + (RT/nF) \ln a_{M^{2+}}$, where $a_{m^{2+}}$ is the target activity. There is, therefore, a substantial possibility of co-evolution of hydrogen, which can only be reduced if metals of high hydrogen-evolution overpotential are used and/ or purification is carried out at as high pH as possible. Hydrogen evolution will also be favoured by high current densities, since these lead, in turn, to high concentration overpotentials for the metal cations, especially if the latter are present only in small concentrations. Unfortunately, the low current densities dictated by this difficulty inevitably lead to low space-time-yields and hence high investment costs. From the engineering perspective, the use of modern cell designs with high A_v-values, such as fixed-bed or electrochemical fluidised bed reactors then becomes desirable. Thus, the concentration of Cu^{2+} in acidified sulphate solutions can be reduced from 2000 ppm to below 1 ppm by the use of a fluidised- bed reactor.

One other use of electrochemical purification is *electroflotation*: in this technique, waste water containing colloidal impurities or suspensions of dirt or mud is introduced into the electrolyte space between two electrodes at which hydrogen and oxygen are being rapidly evolved. The gas bubbles attach themselves to any solid impurity, causing it to float to the surface of the liquid in the cell from which it can be mechanically removed.

8.8.2
Electrodialysis

Dialysis is the separation of low-molecular weight material from solution through a selective membrane. If the material to be separated is ionic, then the process can usually be accelerated by an electric field, and can even take place against a concentration gradient; the main application for electrodialysis is the purification of salt water to give drinking water.

The principle of electrodialysis is shown in Fig. 8.26; during electrolysis the sodium-chloride solution in the middle compartment of the three-compartment cell becomes progressively less concentrated, whilst that in the outer compartments becomes more concentrated. In practice, about 100 to 200 three-compartment cells, each about 1 cm wide, are connected in series, and due to the high resistance of the cell membrane and middle compartment solution, a large voltage (> 100 V) is needed to drive the process.

Clearly, the current requirement for electrodialysis will scale linearly with initial salt concentration, making it less economic than distillation at high initial salt concentrations. The break-even point at the moment is ca. 1 % salt concentration (with a purification cost of ca. £ 0.5 per cubic metre), whereas sea-water has a salt concentration of ca. 3 %. In practice, therefore, electrodialysis is restricted to the purification of coastal groundwaters, though it has also been used for the *enrichment* of the salt concentration of sea water (up to 180–200 g dm^{-3}), prior to evaporation or direct use in the chlor-alkali process.

Electrodialysis is not restricted to desalination, but can be used in a variety of other demineralisation processes, including the removal of minerals from glucose solutions, glycerine, whey etc. A further use is the conversion of the salt into free acid and/or free base: an example industrially is the conversion of sodium sebacate into free sebacic acid, in this case using a two-compartment cell with cation-exchange membrane, so-

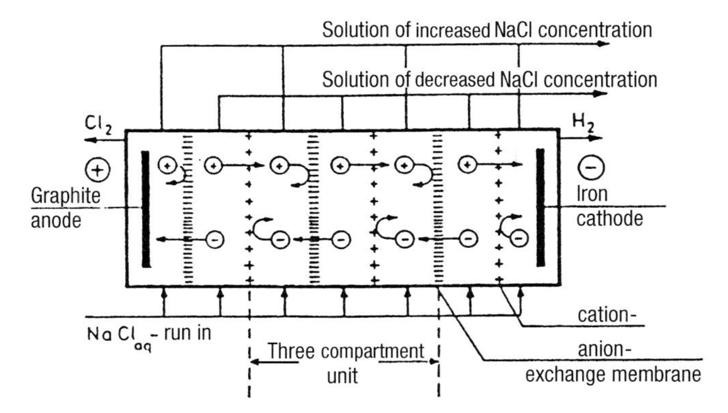

Fig. 8.26 Schematic diagram of electrodialysis equipment.

dium ions migrate through the membrane into the cathode compartment, forming NaOH, and leaving the sebacate ions behind in the anode compartment. These combine with protons generated at the oxygenevolving electrode to form the insoluble sebacic acid.

8.8.3
Electrophoresis

The principle of electrophoresis was described in Section 3.4.6: as a consequence of the specific adsorption of ions on colloidal particles present in solution, these particles acquire a charge and can migrate in the presence of an electric field. The simplest application of this process lies in the separation of the colloid from solution; an example is the removal of finely-dispersed kaolin from aqueous solution: the kaolin becomes negatively charged and migrates to a rotating cylindrical lead drum anode which half projects into the solution. As the cylinder rotates it carries the adherent kaolin powder out of solution and allows its removal through being scraped off the upper part of the cylinder.

If there are several different types of colloid in solution, then each will have a different charge due to specific adsorption effects and hence a different migration speed. This allows the development of separation procedures, which have proved especially valuable in medical and biological studies, where the separation of chemically very similar species such as different serums, hormones, antibodies, viruses can easily

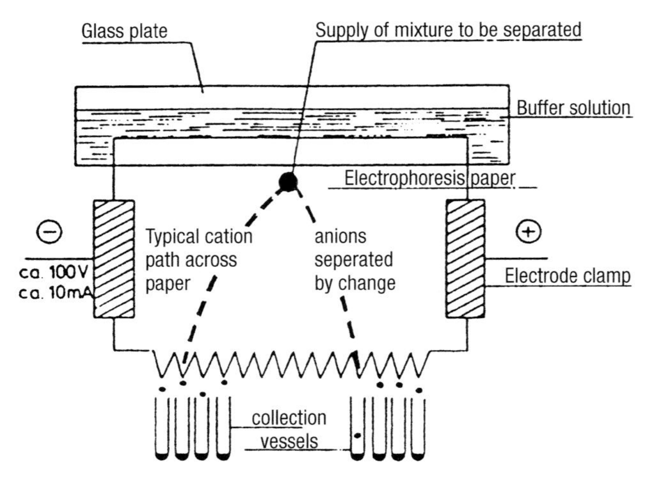

Fig. 8.27 Schematic diagram of equipment used for continuous separation by paper electrophoresis. The electrophoresis paper is held between two glass plates (upper part of the diagram) which also contains the buffer solution. This latter penetrates down the paper by capillary action, carrying the mixture to be separated with it. The lower edge of the paper is cut to form a series of triangles from the tips of which droplets of the separated components fall into appropriately placed glass tubes.

be carried out. The most common separation method uses a combination of chromatography and electrophoresis, and the principle is shown in Fig. 8.27. A droplet of the material to be separated is introduced in the middle of the cell onto an alumina or paper chromatography plate and eluted with a buffer solution in the presence of an electric field. As the eluant flows to the bottom of the cell, separation takes place and the separate components can be collected as shown.

8.8.4
Electrochemical Separation Procedures in the Nuclear Industry

The fuel rods normally used in today's nuclear reactors (light-water reactors) contain about 3 % ^{235}U-enriched uranium-238. During nuclear operation, some of the neutrons emitted from the ^{235}U during its fission are captured by ^{238}U nuclei, converting then to ^{239}Pu. Once the content of ^{235}U has fallen to ca. 1 % (with ca. 1 % Pu formed), the fission rate falls to the point at which the fuel rods must be exchanged. The uranium and plutonium (which is also fissile) from the spent rods can then be recovered after separation from the fission-products of ^{235}U and used in new rods, a route termed nuclear reprocessing.

Reprocessing is a chemical process. In the first stage, the spent rods are ground up and dissolved in hot concentrated nitric acid, with the U present as U(VI) and the Pu predominantly as Pu(IV). These can be separated from the non-volatile fission products of ^{235}U by exploiting the fact that both U(VI) and Pu(IV) can be ex-

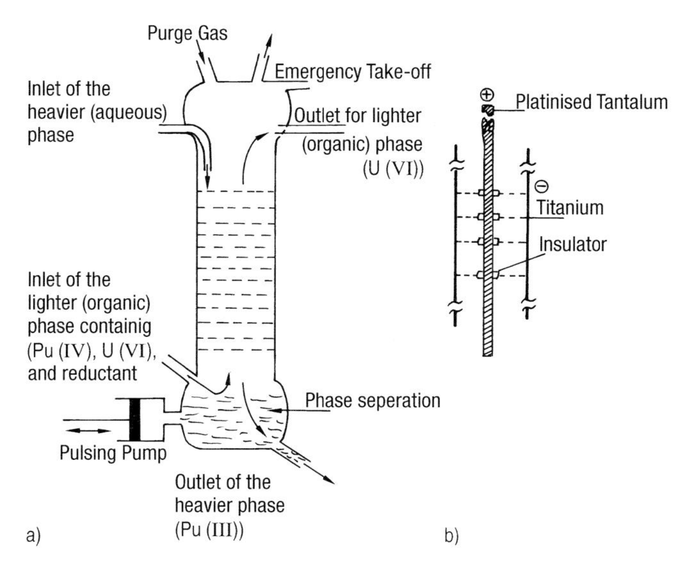

Fig. 8.28 (a) Schematic diagram showing the pulsing of the Purex liquid through a set of sieves placed in a vertical container; (b) the mid-part of the column in detail showing the electrochemical process (see text for details).

tracted from nitrate solutions into tri-butylphosphate (TBP) as stable complexes: $UO_2(NO_3)_2 \cdot (TBP)_2$ and $Pu(NO_3)_4 \cdot (TBP)_2$. Thus, by mixing the aqueous nitrate phase with a solution of TBP in a high boiling-point solvent such as kerosene, separation of these complexes from the remaining fission products can be achieved.

In order to separate the uranium and plutonium complexes in the organic phase, a reduction of Pu(IV) to Pu(III) is carried out. The Pu(III) is then extracted back into aqueous solution, this being referred to as the **P**lutonium-**U**ranium-**R**eduction-**Ex**trac-(Purex) process. The extraction equipment used is a column containing a series of sieves through which the organic and aqueous phases are pulsed by means of a pump to ensure adequate mixing, as shown in Fig. 8.28(a). The reduction agent can be chemical, introduced with the organic phase as shown at the bottom of the column, but reduction can also be carried out electrochemically, with the advantage that separation of unused reductant, which is, of course, contaminated with radioactive impurities, is not necessary. The electrochemical arrangement is shown in Fig. 8.28(b); the cathode is titanium and the anode platinised tantalum. The half-wave potential for Pu(IV) reduction at Ti is ca. 0.15 V vs. NHE, and limiting currents of 50 mA cm^{-2} are attainable, with an overall cell voltage of 2–3 V. The process was scaled up in the nineteen-eighties, allowing separation rates of up to 10 kg/day of Pu with 8 m high columns.

References for Chapter 8

General Texts on Industrial and Technical Electrochemistry and Cell Design

V.M. Schmidt: "Electrochemical Reactor Technology", Wiley-VCH, Weinheim, 2003.

D. Pletcher and F.C. Walsh: "Industrial Electrochemistry", 2nd. Ed., Chapman and Hall, London, 1990.

E. Heitz and G. Kreysa: "Principles of Electrochemical Engineering", Verlag Chemie, Weinheim, 1986.

M.I. Ismail (ed.): "Electrochemical Reactors; their Science and Technology", Elsevier, Amsterdam, 1989.

A. Schmidt: "Applied Electrochemistry", Verlag Chemie, Weinheim, 1976.

F. Hine: "Electrode Processes and Electrochemical Engineering", Plenum Press, New York, 1985.

F. Goodrich and K. Scott: "Electrochemical Process Engineering", Plenum Press, New York, 1995.

F. Beck, "Electro-organic Chemistry", VCH, Weinheim, 1974.

D. Hoormann, J. Jörrissen, H. Pütter: "Electrochemical Processes – New Developments and Tendencies", Chemie-Ingenieur Technick **77** (2005) 1363.

Ullmanns' Encyclopaedia of Technical Chemistry, Verlag Chemie, Weinheim. This is a comprehensive work in 24 volumes with various editions published to 1984 in German, and with the fifth edition completed in 1994 in English.

Dechema Monographs are a series of Monographs published by VCH, Weinheim and have been based, since 1981, on lectures given at the annual meetings of the Applied Electrochemistry Group in Germany.

The Chlor-alkali industry is covered in:

P. Schmittinger: "Chlorine" in "Ullmanns' Encyclopaedia", vol. **6A** (1986) pp. 399 – 481; vol. **8** (1999) pp 186 – 280.

P. Schmittinger (ed.): "Chlorine: Principles and Industrial Practice", Wiley-VCH, Weinheim, 2000.

"Modern Chlor-Alkali Technology", vol. **1**, 1980 (ed. M.O. Coulter) and vol. **2**, 1983 (ed. G. Jackson), Ellis Horwood, Chichester, U.K..

J. Newman and W. Tiedemann, in "Advances in Electrochemistry and Electrochemical Engineering", ed. H. Gerischer and C.W. Tobias, Wiley, New York, vol. **11** (1978).

D.L. Callwell in "Comprehensive Treatise of Electrochemistry" eds. J.O'M. Bockris, B.E. Conway, E. Yeager and R.E. White, Plenum Press, New York, vol. **2**, 1981, p. 105.

D.M. Novak, B.V. Tilak and B.E. Conway in "Modern Aspects of Electrochemistry", eds. B.E. Conway and J.O'M. Bockris, Plenum Press, New York, vol. **18**, 1982, p. 195.

W.N. Brooks: Chem. Brit. 1986 **22** 1095.

F. Gestermann: "Energy-saving industrial chlorine preparation with oxygen diffusion cathodes"; Gesellschaft Deutscher Chemiker, Monographie **23** (2002); German Patent DE 10152793 A1.

Metal Extraction

W.E. Haupin and W.B. Frank in "Comprehensive Treatise of Electrochemistry" eds. J.O'M. Bockris, B.E. Conway, E. Yeager and R.E. White, Plenum Press, New York, vol. **2** 1981, p.301.

Dechema Monographs **93**: "Electrochemistry of Metals – Electrowinning, Processing and Corrosion", **1983**, and **125**: "Electrochemical Material Extraction – Foundations and Technical Processes", **1992**.

Inorganic Industrial Electrochemical Processes

N. Ibl and H. Vogt in "Comprehensive Treatise of Electrochemistry" eds. J.O'M. Bockris, B.E. Conway, E. Yeager and R.E. White, Plenum Press, New York, vol. **2**, 1981, p.167.

C.H. Hamann, T. Röpke and P. Schmittinger: "Electrowinning of other Inorganic Compounds" *in* "Encyclopaedia of Electrochemistry", vol. 5, Wiley-VCH, Weinheim, 2005.

B.V. Tilak, P.W.T. Lu, J.E. Colman and S. Srinivasan in "Comprehensive Treatise of Electrochemistry" eds. J.O'M. Bockris, B.E. Conway, E. Yeager and R.E. White, Plenum Press, New York, vol. **2**, 1981, p. 1.

F. Gutmann and O.J. Murphy in "Modern Aspects of Electrochemistry", eds. R.E. White, J.O'M. Bockris and B.E. Conway, Plenum Press, New York, vol. **15**, 1983, p. 1.

J.A. McIntyre: Interface **4** (1995) 29.

Organic Industrial Electrochemical Processes

H. Lund and O. Hammerich (eds.): "Organic Electrochemistry", M. Dekker, New York, 2001.

A.J. Fry and W.E. Britton (eds.): "Topics in Organic Electrochemistry", Plenum Press, New York, 1986.

L. Eberson and H. Schäfer: "Organic Electrochemistry" in "Fortschritte der chemischen Forschung", **21** (1971), Springer Verlag, Berlin.

L. Eberson: "Electron Transfer Reactions in Organic Chemistry" in "Advances in Physical Organic Chemistry", vol. **18** (1982), eds. V. Gold and D. Bethell, Academic Press, London.

S. Torii: "Electro-organic Syntheses; Part I (Oxidations); Part II (Reductions)", VCH, Weinheim, 1985.

A.J. Bard and H. Lund (eds.): "Encyclopaedia of Electrochemistry of the Elements, Organic Section", vols. XI – XV, Marcel Dekker, New York, 1978 – 1984.

"Technique of Electro-organic Synthesis – Scale up and Engineering Aspects", eds. N.L. Weinberg and B.V. Tilak, John Wiley, New York, 1982 (Vol. **V**, Part III of the Series "Techniques of Chemistry").

H. Wendt: Electrochim. Acta **29** (1984) 1513.

D.E. Kyriacou and D.A. Jannakoudis: "Electrocatalysis for Organic Synthesis", John Wiley, New York, 1986.

C.H. Hamann and T. Röpke: "Preparation of Sugar Derivatives using the Amalgam Process", J. Carbohydrate Chem. **24** (2005) 13.

Novel Cell Designs and Future Progress

H. Wendt: Chem. Ind. Tech. **58** (1986) 644.

Dechema Monograph **94**: "Reaction Technology for Chemical and Electrochemical Processes"; see also monographs **97**, **98**, and **125**.

D. Pletcher: J. Applied Electrochem. **15** (1984) 403.

A.M. Couper, D. PLetcher and F.C. Walsh: Chem. Rev. **90** (1990) 837.

D.R. Rolison: Chem. Rev. **90** (1990) 867.

A. Adzit in "Advances in Electrochemistry and Electrochemical Engineering", ed. H. Gerischer and C.W. Tobias, Wiley, New York, vol. **17** (1984).

M. Fujihira in "Topics in Organic Electrochemistry", eds. A.J. Fry and W.E. Britton, Plenum Press, New York, 1986.

The NEMCA effect

C.G. Vayenas, S. Bebelis, I.V. Yentekakis and H.G. Lintz: Catal. Today **11** (1992) 303.

W. Vielstich: "Fuel Cells", Wiley, New York, **1970**, pp 93ff.

S.G. Neophytides, D. Tsiaplakides, P. Stonehart, M.M. Jaksic and C.G. Vayenas: Nature **340** (1994) 45.

Lambert RM, Williams F, Palermo A, Tikhov MS, Topics in Catalysis, **13** (1–2): 91–98 (2000).

Component Separations

G. Kreysa: Chem. Ing. Tech. **62** (1990) 357.

K.J. Müller and G. Kreysa: "Fixed-bed Reactors – Design Concepts and Industrial Uses", in DECHEMA Monograph **98** (1985) 367.

A. Schmidt: "Applied Electrochemistry", Verlag Chemie, Weinheim, 1965.

H.-J. Helbig: "Electrophoresis", in "Ullmann's Encyclopaedia of Technical Chemistry", Verlag Chemie, Weinheim, 1980.

F. Baumgärtner and H. Schmieder: Radiochimica Acta **25** (1978) 191.

H. Schmieder, M. Heilgeist, H. Goldacker and M. Kluth: DECHEMA Monograph **97** (1984) 217; see also ibid. **94** (1983) 253.

9
Galvanic Cells

If a chemical reaction takes place at two spatially separated electrodes, electrical power can be generated in an external conducting circuit connecting the two electrodes. In general, if the free energy associated with the nett chemical reaction is ΔG, then $\Delta G + nFE_c < 0$; only for a reversible reaction do we find that $\Delta G = -nFE_{c,0}$, where $E_{c,0}$ is the cell voltage at open circuit. Galvanic cells are, therefore, a means of directly converting chemical into electrical energy.

Electrochemical current sources were first described as such in the work of Galvani (1789), and shortly afterwards Volta constructed the first practical galvanic cell. Cells with larger current capacity were developed by Bunsen and Grove in the first half of the 19th century, and played an important role in early technical and scientific investigations of electricity.

Today, galvanic cells are used in three broad areas:

1. In the supply of electrical power to mobile systems such as cars, aeroplanes, space satellites, portable appliances etc
2. In the supply of power to stationary appliances beyond the reach of the electricity grid, such as sea-buoys, remote transmitters, automatic weather stations etc
3. In backup power sources to equipment that must be kept operational even in the event of a failure of the grid, such as lighting and equipment in hospital operating theatres, flight control equipment etc.

Conventional batteries, such as the Leclanché battery, the lead accumulator, the alkaline-manganese, NiCd, zinc-mercury and silver-zinc systems, remain of immense technical importance. However, new systems have been developed in response to particular needs: manned space-flight to the Moon required a much lighter system than that available at the time, and fuel cells based on cryogenic H_2/O_2 had to be provided. In addition, the lack of conventional batteries of sufficiently high power and energy density has inhibited the development of electrical vehicles capable of acceptable performance by modern standards. There is also continual pressure from manufacturers of modern portable electrical appliances for improved battery performance, pressure that has led to the introduction of novel battery types such as lithium-ion and Ni-metal-hydride (Ni-MH) batteries.

Electrochemistry. Carl H. Hamann, Andrew Hamnett, Wolf Vielstich

ISBN: 978-3-527-31069-2

9.1 Basics

A galvanic cell consists of two electrodes, together with the fuel or "electro-active components", the electrolyte, battery housing and, if applicable, the electrode separators. As an example, for the chlorine/hydrogen cell of Section 4.1, chlorine and hydrogen constitute the electro-active components; for conventional batteries, however, the electrochemically active components are incorporated into the electrode structure, and there are two broad categories of battery:

1. If, on application of an external polarity of opposite sense to that provided by the battery the electrochemical reactions can be reversed and the battery thereby re-charged, it is termed a *secondary* battery;
2. If one or both of the electrode reactions is *irreversible*, the battery is termed a *primary* cell. The two types are shown schematically in Fig. 9.1.

The chlorine/hydrogen cell shown in Fig. 4.1 is rather different. A galvanic cell such as this, which produces power as long as the electrochemical reactants are provided to the cell and the products removed, is termed a *fuel cell*. The advantage of the fuel cell over the battery is clear: provided there is a sufficient supply of fuel, electrical power can be produced continually and indefinitely.

Theoretically, there is almost an infinite number of possible electrode combinations that could give rise to galvanic devices, but technically usable systems require the fulfilment of a series of important additional caveats:

1. The electrode reactions must be rapid to avoid severe loss of cell voltage as current is drawn. For secondary batteries, the charging reaction must also be fast;
2. The two electrode processes must have equilibrium half-potentials sufficiently different from each other that a useful open-circuit cell voltage can be obtained. A good rule of thumb here is that an open-circuit voltage of

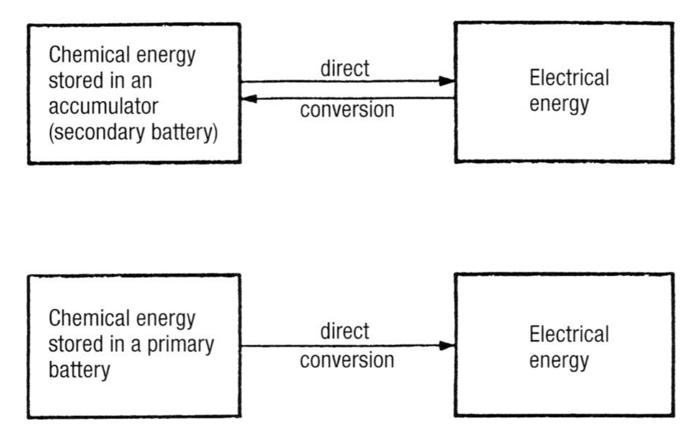

Fig. 9.1 Schematic representation of energy flows in primary and secondary battery systems.

a least 1 V is needed, corresponding to a ΔG value of -100 kJmol^{-1}. A minimum *working* voltage of ca. 0.5 V is desirable;

3. The active components of the cell should only react when the external circuit is *closed*; there should be *no* self-discharge;
4. The battery should have as large a power and energy density as possible;
5. The cell components should be cheap and easily available. They should also not be poisonous if possible and the battery should be disposable without environmental ill-effects.

9.2 Properties, Components and Characteristics of Batteries

9.2.1 Function and Construction of Lead-Acid Batteries

For the lead-acid battery, the active component of the positive electrode is lead dioxide, PbO_2, that of the negative electrode is elemental lead, and the electrolyte is aqueous sulphuric acid. The electrode reactions are:

$$PbO_2 + (2H^+ + SO_4^{2-}) + 2H^+ + 2e^- \underset{\text{charge}}{\overset{\text{discharge}}{\rightleftharpoons}} PbSO_4 + 2H_2O \tag{9.1}$$

$$Pb + (2H^+ + SO_4^{2-}) \underset{\text{charge}}{\overset{\text{discharge}}{\rightleftharpoons}} PbSO_4 + 2H^+ + 2e^- \tag{9.2}$$

The overall cell reaction is thus:

$$PbO_2 + Pb + 2H_2SO_4 \rightarrow 2PbSO_4 + 2H_2O \tag{9.3}$$

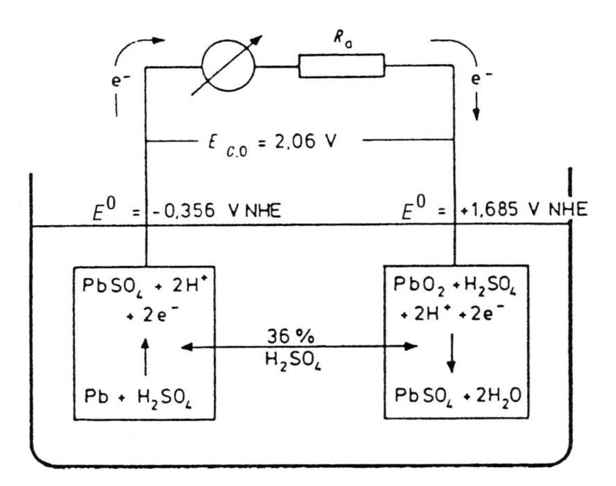

Fig. 9.2 Reaction scheme for a lead-acid accumulator, shown for discharge.

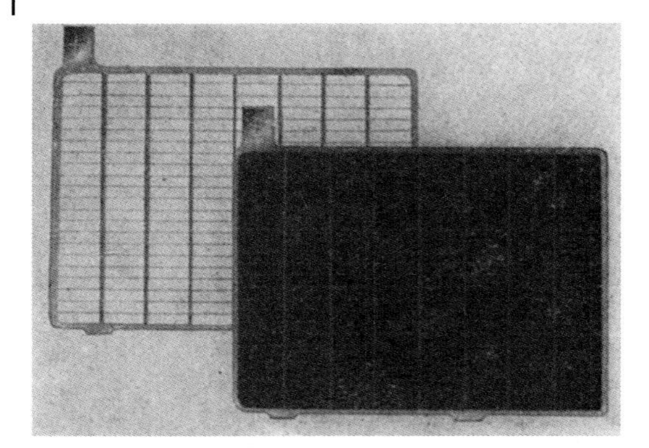

Fig. 9.3 Support plates for a lead-acid accumulator.

and this process is illustrated in Fig. 9.2. However, if a lead plate and a plate formed from compressed PbO_2 are placed in sulphuric acid, only a very low current is observed, since both plates are, for the most part, not "electrochemically active". Furthermore, during the first few charge-discharge cycles, the $PbSO_4$ formed would flake off the electrodes and collect at the bottom of the cell. In order to obtain active and stable electrodes, the electroactive components must be prepared in porous absorbent forms permitting the electrode reaction to take place over the highest possible surface area. Experience has shown that the best starting material is a paste of lead powder and finely divided PbO, the latter being partially converted to $PbSO_4$ by soaking in dilute sulphuric acid. This paste is spread onto a lead-alloy frame, as shown in fig. 9.3, and oxidation to PbO_2 or reduction to porous lead sponge then effected to form active electrodes.

Discharge cannot take place completely, since $PbSO_4$ is non-conducting, and in fact the mass-usage of lead-acid batteries is at best some 50 %, implying that at least 50 % of the material of such a cell remains unconvertible. Furthermore, during deep charge-discharge cycles, the maintenance of porosity is a major problem, particularly for the negative electrode, since the lead sponge that forms shows a preference for larger particle size. This process can be prevented by addition of large organic molecules, such as lignin sulphate, to the initial preparation process: such molecules, termed *expanders*, adsorb on the lead and prevent particle aggregation. Similarly, there is a tendency for the particles of $PbSO_4$ to aggregate; this can be alleviated by the addition of 0.5 % of $BaSO_4$ crystals, which act as crystallisation nuclei.

9.2.2
Function and Construction of Leclanché Cells

In a Leclanché cell, the electroactive component of the positive electrode is MnO_2, and that of the negative electrode is elemental zinc, with pH-neutral aqueous NH_4Cl as the electrolyte. The electrode processes are:

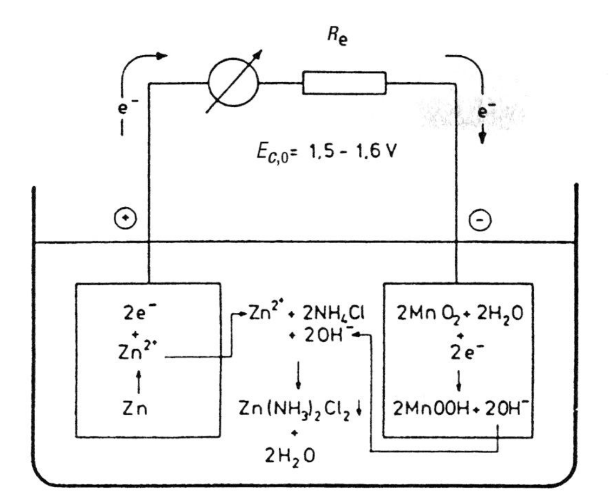

Fig. 9.4 Reaction scheme for a Leclanché cell.

$$\text{Anode}: \text{Zn} \rightarrow \text{Zn}^{2+} + 2e^- \tag{9.4a}$$

$$\text{Cathode}: 2\text{MnO}_2 + 2\text{H}_2\text{O} + 2e^- \rightarrow 2\text{MnOOH} + 2\text{OH}^- \tag{9.4b}$$

$$\text{Electrolyte}: \text{Zn}^{2+} + 2\text{NH}_4\text{Cl} + 2\text{OH}^- \rightarrow \text{Zn(NH}_3)_2\text{Cl}_2 + 2\text{H}_2\text{O} \tag{9.4c}$$

where the final reaction prevents the battery from being easily re-charged. The nett cell reaction is then

$$2\text{MnO}_2 + \text{Zn} + 2\text{NH}_4\text{Cl} \rightarrow 2\text{MnOOH} + \text{Zn(NH}_3)_2\text{Cl}_2 \tag{9.5}$$

and the process is shown schematically in Fig. 9.4.

Although the zinc electrode reaction is fast, and the product soluble, so that sufficient current can be obtained even with the use of compact rather than porous metal, this is not the case for the MnO_2 electrode. MnO_2 is not a good conductor, and it must be mixed as a porous microcrystalline powder with graphite and electrolyte to form a paste. In the familiar "dry-cell", this paste is formed into a cylinder, with a central carbon rod acting as current collector, as shown in Fig. 9.5. The outer casing of the cell is formed of zinc, and the NH_4Cl electrolyte, usually thickened with starch or methyl cellulose, separates anode and cathode. To prevent loss of electrolyte, which

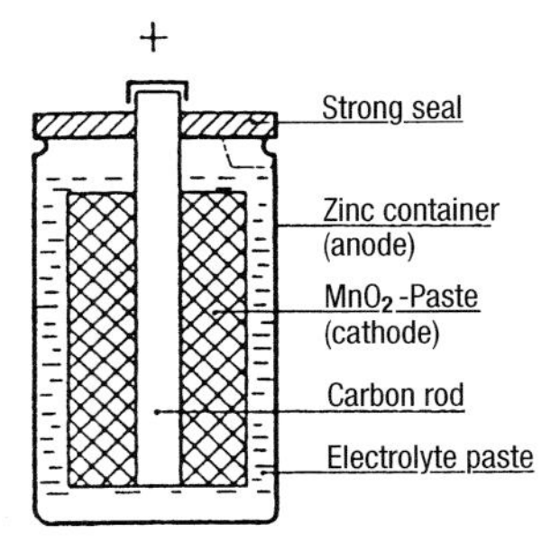

Fig. 9.5 Construction of a Leclanché cell.

can be forced out of the cell by side reactions as discussed below, the outer casing is nowadays normally surrounded by a steel container.

9.2.3
Electrolyte and Self-discharge

The electrochemically active components of the battery can react to form inactive products without passage of current under certain circumstances, a process termed generally "self-discharge". This can take place either through parasitic chemical reactions or through electrochemical corrosion. A simple example is the zinc electrode in Leclanché batteries, whose rest potential lies near to −0.8 V vs. NHE, well below the thermodynamic potential for hydrogen evolution in neutral solution (−0.4 V). The hydrogen overpotential on zinc is not, in fact, quite large enough completely to prevent hydrogen evolution at this potential, the nett discharge reactions being:

$$Zn \rightarrow Zn^{2+} + 2e^-; \quad 2e^- + 2H_2O \rightarrow H_2 + 2OH^-$$

Addition of inhibitors is, therefore, necessary to increase the overpotential and hence the shelf life of the battery. In principle, a similar situation obtains for the lead-acid battery, since the lead electrode lies below the hydrogen evolution potential and the PbO_2 electrode *above* the oxygen evolution potential. In fact, for both electrodes, overpotentials are high and the self-discharge rate for the lead-acid battery is lass than 0.5 % per day.

9.2.4
Open-circuit Voltage, Specific Capacity and Energy Density

In describing batteries, the most important characteristic is the open-circuit voltage, which can be calculated from the thermodynamic data for the electrode reactions. In practice, values less than those expected from the thermodynamic values are always encountered, due, for example, to the restricted ability of the electrodes to attain equilibrium or to the presence of parasitic reactions in the cell. The actual experimental open-circuit voltage will be represented by the symbol $E_{c,0}$ below.

A second important parameter is the specific capacity. The *theoretical* value of this, C_s^{th}, can be directly calculated for each electrode and for the overall cell reaction from the formula:

$$C_S^{th} = nF/M \tag{9.6}$$

where n is the number of electrons involved in each electrode reaction, and M is the molecular mass of the electroactive components. The units are usually expressed as ampère-hours/kg (Ah kg^{-1}) in which F takes the value 26.8 Ah. For the lead-acid battery, M is the total molecular mass of Pb (207), PbO_2 (239) and $2H_2SO_4$ (196) and $n = 2$, from which C_s^{th} has the value 83.5 Ah kg^{-1}.

The theoretical energy density is defined as the product of C_s^{th} for the cell and the *experimental* open-circuit voltage, $E_{c,0}$, and is expressed in units of Wh kg^{-1}; for the lead-acid battery, the theoretical energy density is 167 Wh kg^{-1} given a mean open-circuit voltage of 2 V. This is a rather low value, clearly arising from the high molecular weight of the components of the cell, and values in excess of 1000 Wh kg^{-1} are attainable in modern non-aqueous cells (see below). For the Leclanché cell, the value of $C_s^{th} = 155$ Ah kg^{-1}, and for $E_{c,0} = 1.58$ V, the theoretical energy density is 245 Wh kg^{-1}.

Of course, in practice, only a fraction of the theoretical capacity of a cell is obtained, since in addition to the active components of the cell, there are electrochemically *inactive* components of the two electrodes, the electrode current collectors, the separators, the electrolyte solvents, the battery housing etc. In addition, since the usage of electrode active mass and the cell voltage decrease as the discharge current increases, the experimental specific capacity and energy density depend sensitively on the discharge conditions. Various conventions have been suggested, which are dependent on the actual use to which the battery is to be put, but however defined, the actual energy densities achievable in practice are normally some 10–25 % of the theoretical values. For the lead-acid battery, for example, under two-hour discharge conditions, a value of 35 Wh kg^{-1} can be attained, and for the Leclanché battery under slow discharge, values of up to 80 Wh kg^{-1} are found.

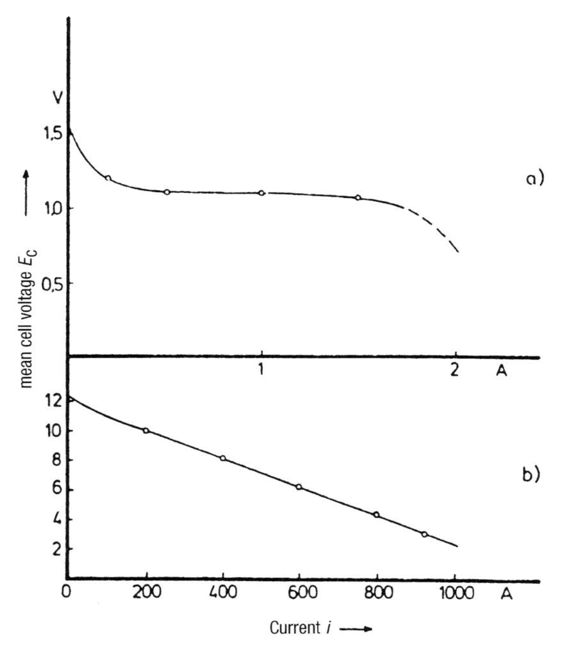

Fig. 9.6 (a) Current-voltage characteristic of a Leclanché cell; (b) the same for a six-cell lead-acid accumulator stack (12-V-45-Ah). Note the very different scales for the two current axes.

9.2.5
Current-Voltage Characteristics, Power Density and Power-density/Energy-density Diagrams

The current-voltage characteristic of a battery is a key datum, giving information on the current available at any given cell voltage: two examples, for the lead-acid and Leclanché cells, are shown in Fig. 9.6. Multiplication of the current by the corresponding voltage gives the electrical power, $P = i.E_c$, and the ratio of this power to the battery weight gives the power density in units of W kg^{-1}. There will normally be a maximum value for P; in fact, if there is a linear relationship between E_c and i, as for the lead-acid battery in Fig. 9.6, then the maximum power, P_{max} will be delivered for $E_c = E_{c,0}/2$. For the 12 V battery of Fig. 9.6, $P_{max} = 3.6$ kW, with a maximum power density of ca. 250 W kg^{-1}. In spite of the weight of the components, this is a very high value indeed, far higher than the corresponding value for the Leclanché cell of ca. 10 W kg^{-1}, but still significantly below that found for the internal combustion engine (ca. 1 kW kg^{-1}).

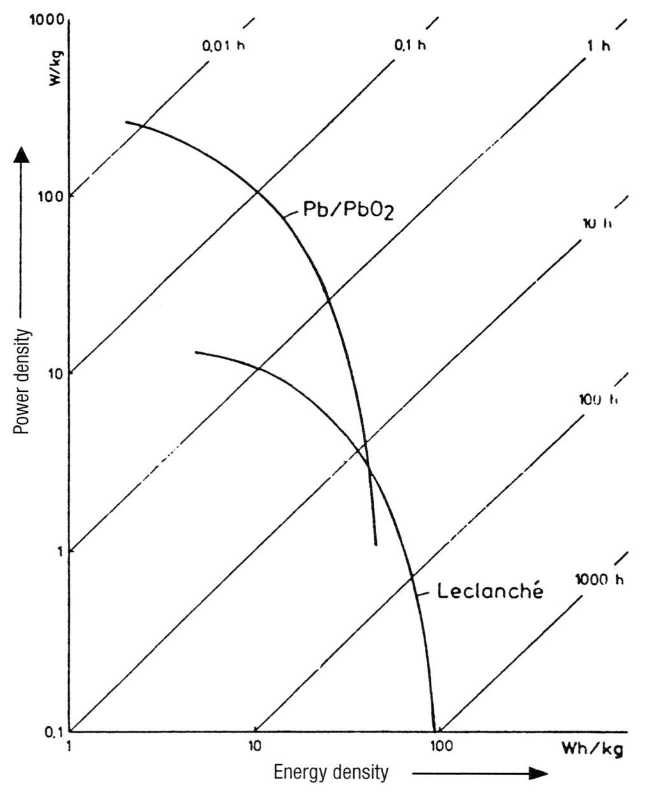

Fig. 9.7 Connection between the power density of a battery and the corresponding energy density for the lead accumulator and the Leclanché cell plotted on a double logarithmic graph.

The maximum power is found for the lead-acid battery at very high currents, and hence rapid discharge times, but at such high currents, the energy density of the battery is low, and in general there is an inverse relationship between energy density and power density for any given battery, as shown in Fig. 9.7. The type of graph shown in Fig. 9.7 is also of great importance in terms of the performance of the cell, but it should be remembered that all such data will be temperature dependent, showing, usually, a marked decrease as the temperature is lowered.

9.2.6. Battery Discharge Characteristics

The variation in battery voltage as a function of time for a constant discharge current or constant resistive load is termed the *discharge characteristic*. In an ideal case, the voltage would remain constant until the usable part of the electrode material became exhausted, and would then decay suddenly to zero. In practice, however, behaviour similar to that shown in Fig. 9.8 is seen, in which the cell voltage decays over time. This decay is due to two main effects:

- As the active component of the electrode is consumed, there is usually a reduction in the effective available surface area, and at constant external current, this corresponds to an increase in the real current density, giving rise to both poorer electron transfer kinetics and to concentration polarisation.

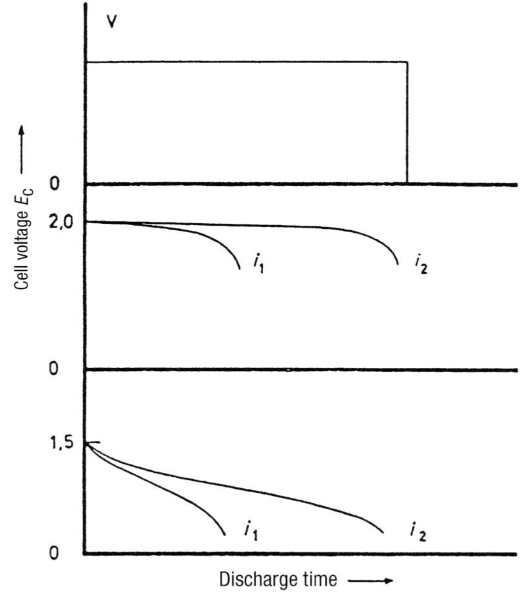

Fig. 9.8 (Top): Ideal discharge characteristic for a battery with constant current loading; (Middle): discharge characteristic of a lead-acid accumulator for two different discharge currents, with $i_1 = 2i_2$; (Below): discharge characteristic of a Leclanché cell.

– Initially, the discharge reaction takes place predominantly in the outer zones of the electrode, where mass transport is fastest. However, with increasing discharge, the electrode reactions take place increasingly within the electrode structure, leading again to a diffusion overpotential. This process is particularly marked for the lead-acid battery, where the product, $PbSO_4$, occupies three times the volume of elemental lead and 1.5 times the volume of PbO_2, leading to a narrowing of the pore diameters in the porous electrode structure.

9.2.7
Charge Characteristics, Current and Energy Yield and Cycle Number

The parameters described in the previous sections are defined for both primary and secondary batteries and for fuel cells, but secondary batteries have an additional set of characteristics.

The evolution of the charging voltage at constant current is shown for a lead-acid battery in Fig. 9.9; the charging voltage must, of course, be greater than $E_{c,0}$, and it is found that close to the end of the charging process, the charging voltage at constant current begins to increase rapidly, finally being sufficient to decompose the electrolyte.

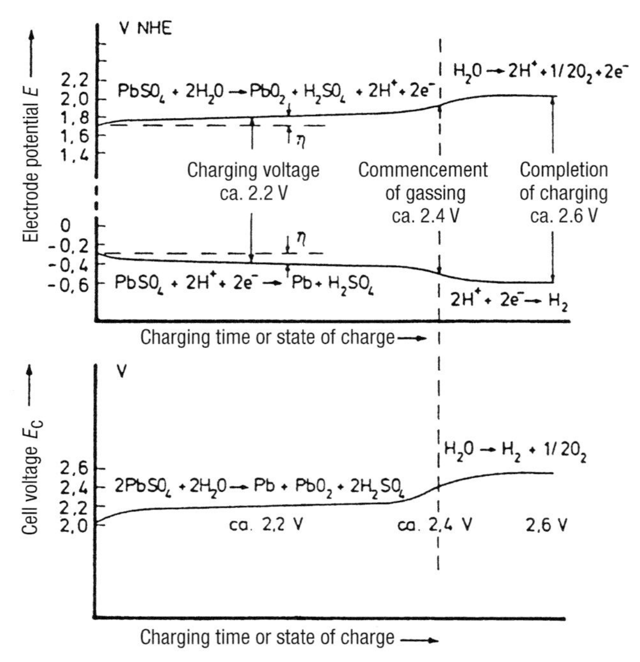

Fig. 9.9 Charging characteristics of a lead-acid accumulator, with electrode potentials and cell voltage corresponding to a constant five-hour charging.

If the voltage required for the latter is only slightly above that required for complete charging, then it will be very difficult to prevent some electrolysis taking place, a process termed "gassing" in the case of the lead-acid battery.

The current yield, A_i, is the ratio of the total charge passed during the discharge process to that passed during the charging process; since part of the charge passed towards the end of the charging process is likely to be used, for example, to electrolyse the solvent, $A_i < 1$: values of 0.9 are typically found for the lead-acid battery.

The energy yield, A_e, is the ratio of the *energy* generated during discharge to that required completely to charge the battery. This energy will have the form $\int E_c(t) \cdot i(t) \cdot dt$, and it is clear that A_e will be smaller than A_i (since the charging voltage is always larger than the discharge voltage for any current). Values of 0.8 are found for the lead-acid battery typically.

Finally, secondary batteries cannot, in general, be charged and discharged an indefinite number of times. In the lead-acid battery, for example, long term changes take place at the electrodes due to the different molar volumes of reactant and product. These lead to continual mechanical stresses on the electrodes, and finally to spalling or loss of electroactive material. Long term corrosion effects on the support framework also limit the performance of the cell, as does the slow formation, in a discharged or partially discharged battery, of an inactive form of $PbSO_4$, a process termed *sulphation*. These processes together ensure that exceeding 1500 charge/discharge cycles for lead-acid batteries is extremely difficult.

9.2.8 Cost of Electrical Energy and of Installed Battery Power

The cost per kWh of the electrical energy from a battery can be calculated for a primary battery directly from the costs of the battery and its energy content. For a secondary battery, the calculation is more complex, since to the costs of the charge must be added an appropriate proportion of the overall construction costs of the battery. A similar calculation is needed for a fuel cell, where the costs of the fuel and the investment costs of the cell must be included.

For a Leclanché cell, the costs of energy are in the region of ca. £200–£300 per kWh, which is on the cheap side for a primary battery. This value can be compared to a few pence per kWh for power from the electricity companies, and indicates how large a premium we are prepared to spend for the convenience of portable power. As might be expected, the lead-acid battery, as a secondary battery, is far cheaper, with costs as low as £0.20 per kWh possible, again in a favourable case. The main reason for the cheapness is the fact that over a sufficient number of cycles, the construction cost of the battery becomes a relatively small proportion of the total costs.

The cost per kW of installed power is a rather different calculation, emphasising the costs of the battery itself, and refers to the costs of that battery unit required to produce 1 kW of power for one hour. This calculation is particularly important for secondary batteries, with costs of £100–£150 per kW now found for the lead-acid battery.

9.3 Secondary Systems

9.3.1 Conventional Secondary Batteries

9.3.1.1 The Nickel-Cadmium and Nickel-Iron Batteries

The nickel-cadmium battery is based on the following reactions:

$$Cd + 2OH \underset{\text{charge}}{\overset{\text{discharge}}{\rightleftharpoons}} Cd(OH)_2 + 2e^- \tag{9.7}$$

$$2NiOOH + 2H_2O + 2e^- \underset{\text{charge}}{\overset{\text{discharge}}{\rightleftharpoons}} 2Ni(OH)_2 + 2OH^- \tag{9.8}$$

with the nett cell reaction during *discharge* being

$$Cd + 2NiOOH + 2H_2O \rightarrow 2Ni(OH)_2 + Cd(OH)_2 \tag{9.9}$$

and the schematic process is shown in Fig. 9.10. The nickel electrode process is complex: the Ni(III) oxyhydroxide is converted to $Ni(OH)_2$ by proton migration in a process that maintains the basic hydroxide structure. However, Ni(III) is a low spin $t_{2g}^6 e_g$ ion with a substantial Jahn-Teller distortion, leading to considerable flexing of the Ni-O framework during oxidation and reduction. Furthermore, K^+ ions can migrate into the hydroxide framework during charging, allowing some oxidation of Ni(III) to Ni(IV) in such species as $[Ni_4O_4(OH)_4](OH)_2K$. The electrodes themselves are formed initially from finely crystalline cadmium oxide powder or nickel hydroxide powder (the latter with graphite or nickel flake added to enhance the conductivity) pressed onto perforated steel sheet or porous nickel grids. One important feature is that partial discharge-charge cycles can lead to a rapid *decrease* in the usable capacity of the battery, which can only be reversed by a deep discharge-charge cycle.

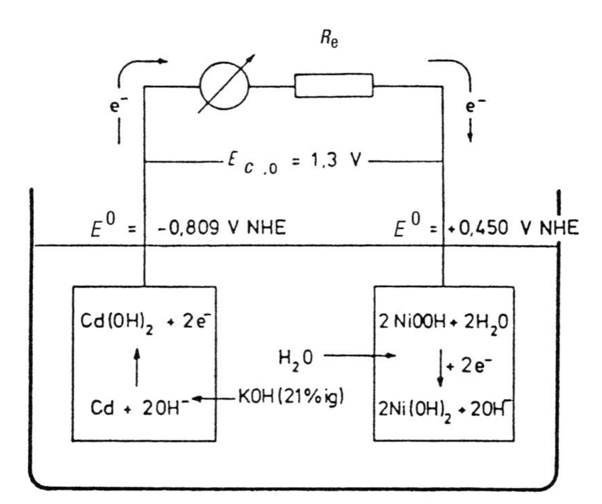

Fig. 9.10 Reaction scheme for the Nickel-Cadmium battery, shown for discharge process.

A related secondary battery is the Nickel-Iron system, which differs from the cadmium system only through the replacement of cadmium by iron. The corresponding anode reaction has the form:

$$Fe + 2OH^- \underset{\text{charge}}{\overset{\text{discharge}}{\rightleftharpoons}} Fe(OH)_2 + 2e^- \tag{9.10}$$

For both NiCad and FeCad batteries a particular problem is the parasitic electrolysis of water during the charging process. This electrolysis leads, in principle, not only to the evolution of gases that must be vented safely, but also to the loss of electrolyte, which must be replaced. The key to the development of a gas-tight and maintenance-free NiCad battery was the realisation that if *hydrogen* evolution could be inhibited in the first stage of over-charging, the *oxygen* evolved at the nickel electrode could be re-cycled to the cadmium electrode for re-reduction. The easiest way of inhibiting hydrogen evolution is through the incorporation of a higher electrochemical capacity at the cadmium electrode. This so-called "negative charge reserve" gives rise to characteristic charging curves for the NiCad battery that are shown in Fig. 9.11, and it is clear from these curves that evolution of oxygen takes place whilst the cadmium electrode is still charging. By suitable cell design, the oxygen generated at the anode as a consequence of over-charging can be transported efficiently to the cadmium electrode, where it can be reduced back to water:

$$1/2O_2 + H_2O + 2e^- \rightarrow 2OH^- \tag{9.11}$$

This process is reasonably rapid on cadmium at -0.8 V vs. NHE, and allows gas-tight operation of the NiCad cell.

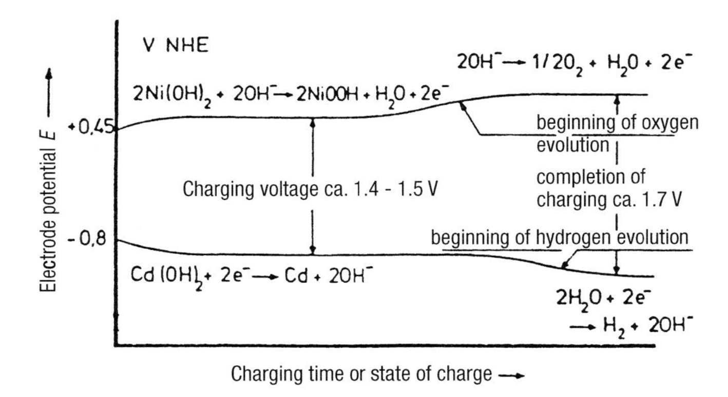

Fig. 9.11 Charging characteristic for the NiCad Battery, showing the higher capacity of the cadmium electrode (negative charge reserve).

9.3.1.2 The Silver-Zinc Battery

The theoretical capacity of the zinc electrode is far higher than lead (820 Ah kg^{-1} as against 259 Ah kg^{-1}) and in alkaline solution electrodeposition of zinc metal is facile, allowing us to develop a secondary battery. Another advantage is that the $Zn(OH)_2$ formed in the cell reaction:

$$Zn + 2OH^- \underset{\text{charge}}{\overset{\text{discharge}}{\rightleftharpoons}} Zn(OH)_2 + 2e^- \tag{9.12}$$

is soluble in excess alkali:

$$Zn(OH)_2 + OH^- \rightarrow [Zn(OH)_3]^- \tag{9.13}$$

together with $[Zn(OH)_4]^{2-}$ and other species. The positive electrode reaction of the silver-zinc battery has the (simplified) form:

$$AgO + H_2O + 2e^- \underset{\text{charge}}{\overset{\text{discharge}}{\rightleftharpoons}} Ag + 2OH^- \tag{9.14}$$

giving a theoretical energy density of 478 Wh kg^{-1}. In practice energy densities of 100–120 Wh kg^{-1} and maximum power densities of ca. 800 W kg^{-1} are obtained. Drawbacks to the battery include the cost, and the fact that the poise potential of the zinc electrode (−1.25 V vs. NHE) lies well below the H_2 evolution potential, leading to relatively rapid self-discharge. The number of re-charges is also limited, and the battery has only found real use in very specialised circumstances such as the Russian space programme and the powering of lunar modules during the Apollo mission.

9.3.2 New Developments

The main commercial goals of battery research are higher lifetimes, lower maintenance and, above all for vehicular traction, *high energy densities*. In addition, low cost and good availability of battery components are desirable.

9.3.2.1 Developments in Conventional Systems

The energy density of lead-acid batteries remains a major problem: the low mass usage of the electrode components and the lead framework used for electrode support both contribute to this low density, and attention has turned to identifying new corrosion-resistant materials that could be used for the latter.

Maintenance needs for the lead-acid battery usually arise primarily from the need to replace water lost through electrolysis during the charging process. Addition of components to the lead that increase the O_2-evolution overpotential (such as the replacement of part of the Sb by Ca) has been a particularly important development in starter motor batteries, but an alternative is to use a Pt- or Pd-impregnated catalyst in the gas vents to catalyse recombination of the evolved H_2 and O_2. Another possibility is to use the same basic idea as that underlying the gas-tight NiCd battery: this has been realised for the lead-acid battery by the use of gelling agents in the sulphuric acid, such as colloidal SiO_2. The gel physically retains O_2, allowing the concentration to build up over 50–100 cycles to the point at which loss of water is almost completely prevented.

The need for improvement in lifetime in lead-acid batteries used in traction has become apparent as experience has increased in this field. Conventional lead-acid batteries are normally unusable after several hundred cycles. We have already seen that the lifetime of lead-acid batteries is limited by such factors as spalling of electro-active material from the electrodes, corrosion of the support framework and sulphation. The second of these is particularly troublesome: the need for large currents for long periods leads to a marked increase in the temperature of the electrolyte, particularly in the central part of each cell, and some form of electrolyte stirring has proved necessary. The high temperatures also lead to degradation of the expanders (section 9.2.1), and periodic addition of such expanders, possibly automatically from a supply within the battery, is required. Finally, if gassing is not allowed for long periods, the acid concentration in the lower part of the lead-acid cell can rise in comparison with the upper part, leading both to enhanced local corrosion and sulphation. Again, this can be helped by forced circulation of the electrolyte.

These improvements have led to a substantial increase in battery life in electric buses on city routes, where lifetime ranges of up to 140,000 km have been reported, corresponding to 2000 charge-discharge cycles, with an energy density of 35 Wh kg^{-1}. This latter can be further improved by attention to the weight of non-electrochemically active components of the battery.

The fact that Leclanché cells are *primary* means that once used they must be disposed of, a fact that has led to increasing environmental concerns. Indeed, it is environmental concerns that may well spell the end of the primary Leclanché cell in the near future, as there is a concerted move towards re-cycling in most advanced industrial nations, and a strong research interest in supplying improved re-chargeable batteries as a result. That said, however, re-chargeable cells based on the Leclanché principle have proved extremely difficult to develop: a concerted research effort by Kordesch at the University of Graz has led to the development of so-called RAM cells, now marketed by Raynovac and based on the structure:

$$Zn(s)|KOH(aq.)|MnO_2(s),\ C(s)$$

Provided that the discharge is not permitted to go beyond MnO(OH) at the cathode, then this cell can be re-charged, though it is found that the capacity of the cell falls steadily with cycle number, particularly if relatively deep discharge is permitted. However, for applications such as mobile phones, personal audio, electronic organisers, cameras, toys and games they are reasonably suited, and have had some market penetration, helped by their excellent shelf-life as compared to other secondary systems.

The environmentally undesirable effects of cadmium have led to experimental systems designed to replace the NiCd systems with novel systems based on nickel/metal hydrides, the **NiMH batteries** The possibility of storing hydrogen in metals or alloys such as Pd, Ni or $NiTi_2$ has been known for many years. If the Cd electrode of a Ni/Cd battery is replaced by a hydrogen storage electrode with a similar potential, a nickel-metal-hydride battery is obtained. Not only do these have environmental advantages, but they also show increased capacity and longer lifetime compared to the Ni/Cd system.

During charging, hydrogen atoms generated at the cathode diffuse into the metal lattice to form a metal hydride, rather than combine to form gaseous H_2. During discharge, this reaction is reversed, with the stored hydrogen being oxidised at the metal surface. The charging reactions are:

$$Ni(OH)_2 + OH^- \rightarrow NiOOH + H_2O + e^-$$
$$H_2O + M + e^- \rightarrow MH + OH^-$$
$$\text{nett reaction}: Ni(OH)_2 + M \rightarrow NiOOH + MH$$

where M is the storage metal or alloy. The most effective alloys for hydrogen storage are the so-called AB_5 alloys, of which $LaNi_5$ is the most familiar. However, $LaNi_5$ itself is not stable against continued charge/discharge cycles, and corrodes in the strongly alkaline medium. By alloying small quantities of cobalt and other transition elements, and using commercially much less expensive "mischmetall" rather than pure lanthanum (mischmetall being obtained directly from the ores without separation into individual lanthanide metal components), reasonably inexpensive metal hydride systems can be manufactured that are stable to 1500 charge/discharge cycles. The performance of NiMH batteries is now comparable to high-quality NiCd batteries (see Table 9.1), but like the latter they have the disadvantage of rather rapid loss of capacity during storage in the charged state (ca. 20 % per month).

This loss is caused by (a) the self-discharge reaction of the charged NiOOH electrode: $6NiOOH \rightarrow 2Ni_3O_4 + 3H_2O + {}^1/_2O_2$; (b) an ionic short-circuit of the cell arising from the ammonia-nitrite redox cycle: $6NiOOH + NH_3 + H_2O + OH^- \rightarrow 6Ni(OH)_2 + NO_2^-$; $NO_2^- + 6MH \rightarrow NH_3 + H_2O + OH^- + 6M$, where the ammonia originates as an impurity in the KOH solution; (c) the chemical reduction of NiOOH by hydrogen gas slowly evolved from the hydride electrode: $2NiOOH + H_2 \rightarrow 2Ni(OH)_2$.

Given the other advantages, however, of the nickel-metal-hydride systems, it is understandable that both they and another new system of importance, the lithium-ion battery (see 9.3.2.4), are replacing NiCd cells in many applications.

9.3.2.2 **Experimental Secondary Batteries with Zinc Anodes**

If the rather expensive silver in the silver-zinc battery is replaced by the less costly NiOOH, we obtain a battery with an even higher energy density. This nickel-zinc battery has the reaction

$$2NiOOH + Zn + 2H_2O \underset{\text{charge}}{\overset{\text{discharge}}{\rightleftharpoons}} Zn(OH)_2 + 2Ni(OH)_2 \qquad (9.15)$$

and an open-circuit voltage of 1.73 V. This gives a theoretical energy density of 362 Wh kg^{-1}; in practice values of up to 100 Wh kg^{-1} are attainable. The battery itself is constructed in a similar manner to the silver-zinc and lead-acid batteries, with the electrodes in the form of parallel plates separated by a diaphragm to prevent internal shorting. Its main problem is low cyclability, primarily associated with the zinc electrode, which shows progressive morphological changes during successive charge-discharge cycles. The most serious difficulty of this sort is the formation of "dendrites", which are zinc needles growing outwards from the electrode surface, leading to destruction of the separator and internal short circuits in the cell. This can be ameliorated by the incorporation of the zinc into a matrix, which traps the zinc oxide/hydroxide preventing its dissolution, and hence inhibiting dendrite formation. Other methods that have been investigated include forced convection in the anode compartment and pulsed recharging protocols.

Another experimental system is the zinc-bromine battery, with a cell reaction:

$$Zn + Br_2 \underset{\text{charge}}{\overset{\text{discharge}}{\rightleftharpoons}} ZnBr_2 \tag{9.16}$$

The electrolyte is aqueous zinc bromide, and the open-circuit voltage is ca. 1.8 V. The theoretical energy density is 440 Wh kg^{-1}, but realisation of the battery has proved difficult, and the current development goal is only 80 Wh kg^{-1}, with prototype batteries of 0.5 – 20 kWh being constructed. The battery itself differs considerably in design from traditional accumulators of the type described above, since the bromine is stored *outside* the cell, as shown in Fig. 9.12, and the electrolyte is also circulated at the zinc electrode to inhibit dendrite formation. Storage of the bromine is achieved by addition of a large organic cation such as *N*-methyl or *N*-ethyl-morpholine, which converts the bromine into a Br_3^- complex with an oily consistency and higher density than the aqueous medium. This oil can be stored as the lower layer in a reservoir. During discharge, this layer is remixed with the aqueous electrolyte and fed through a highly porous carbon layer to one side of the electrodes, which are arranged in bipolar form.

The bromine can, in principle, be replaced by chlorine, with the chlorine stored as the bright yellow chlorine hydrate, $Cl_2 \cdot 5.75\ H_2O$. This latter is stable below 9.6°C at atmospheric pressure, and can be formed simply by passing chlorine into cold water. The chlorine is released by simply heating the water. Both the Zn/Br_2 and Zn/Cl_2 batteries have not been commercially developed as yet.

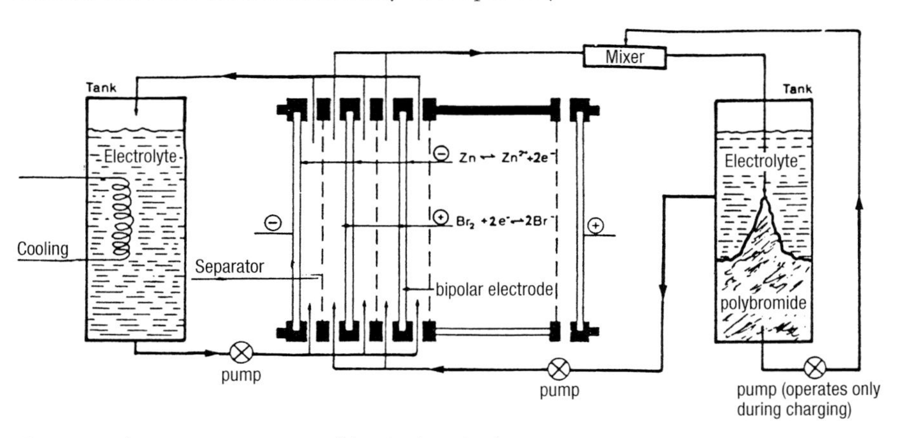

Fig. 9.12 Schematic representation of the zinc-bromine battery.

9.3.2.3 **Sodium-Sulfur and Sodium Nickel Chloride (ZEBRA) Batteries**

The central idea of the Na-S battery is the use of *molten* sodium at 300 – 350°C as the negative electrode:

$$2\text{Na} \underset{\text{charge}}{\overset{\text{discharge}}{\rightleftharpoons}} 2\text{Na}^+ + 2e^- \tag{9.17}$$

The sodium metal is in contact with a solid Na^+-conducting membrane which serves both as electrolyte and separator from the cathode. During discharge, Na^+ ions diffuse through the membrane and react with liquid sulfur impregnated into carbon felt, with the latter providing electronic conductivity:

$$3\text{S} + 2\text{Na}^+ + 2e^- \underset{\text{charge}}{\overset{\text{discharge}}{\rightleftharpoons}} \text{Na}_2\text{S}_3 \tag{9.18}$$

together with other polysulfides. The open-circuit voltage of this battery is 2.1 V, and the theoretical energy density an extremely high 790 Wh kg^{-1}. A schematic of an experimental sodium-sulfur battery is shown in Fig. 9.13: the usual solid ionic conductor is sodium-β-alumina (see section 7.1), and very carefully controlled sintering conditions are required to ensure long lifetimes and high cyclability for the battery. These sintering conditions unfortunately preclude the fabrication of large devices, and the energy content per cell is currently limited to ca. 100 Wh. Unfortunately, in spite of the good results obtained with single cells and complete batteries, the technical development of this interesting system is now in abeyance.

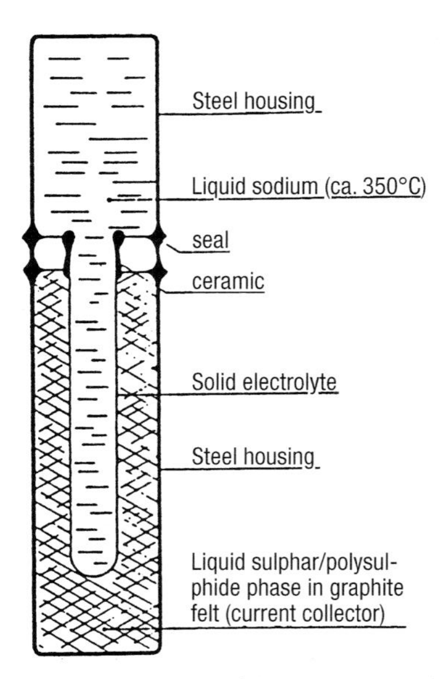

Fig. 9.13 Principles of the construction of a sodium-sulphur battery. More recent versions have dimensionally larger electrolyte tubes which can serve, at the same time, as reservoirs for the sodium.

The problems associated with the sulphur electrode can be improved by the use of a closely related battery system, the sodium-nickel-chloride system or ZEBRA ("Zero Emission Battery Research Activities") battery. The cell reaction takes the form:

$$2Na + NiCl_2 \underset{\text{charge}}{\overset{\text{discharge}}{\rightleftharpoons}} 2NaCl + Ni \tag{9.19}$$

and the cell uses a double electrolyte of β-alumina and $NaAlCl_4$, the latter having a melting point of 200°C. The open-circuit voltage of such cells is 2.58 V, and experimental batteries of 120 Wh kg^{-1} and power densities of 200 W kg^{-1} have been constructed capable of 500 charge-discharge cycles. They are also more stable against both over-charging and over-discharging and have shown promising performance in traction applications.

9.3.2.4 Lithium Secondary Batteries

In the last few decades non-aqueous systems of very high potential power densities have been developed based on lithium. This metal combines an extremely negative electrode potential ($E^0 = -3.045$ V vs. NHE) with a very low atomic weight ($M = 6.939$) to give an exceptionally high theoretical specific capacity C_s^{th} of 3862 Ah kg^{-1}. The electrode potential of Li implies that in combination with a suitable cathode material, battery voltages in excess of 4 V should be possible provided a stable organic electrolyte can also be identified. (Clearly, water is not possible, since such a cell voltage would lead to rapid internal electrolysis). These systems have a wide variety of applications: primary or secondary cells based on lithium are employed for portable power systems, for watches, calculators, cameras, memory backup systems, pacemakers and the like, where lightness, long shelf life and mechanical reliability are essential. In such cells, the positive electrode normally acts to *intercalate* lithium, and the negative electrode may be either lithium metal or a second intercalation agent such as carbon, the latter being preferred for *secondary systems* (see below). A simple (primary) example of a lithium battery might have the form:

$$Li(s)|LiPF_6 \text{ (inert organic solvent)}|MO_2(s)$$

$$\text{negative electrode : } xLi(s) \rightarrow xLi^+(\text{solvent}) + xe^- \tag{9.20}$$

$$\text{positive electrode : } xLi^+(\text{solvent}) + xe^- + MO_2(s) \rightarrow Li_xMO_2 \tag{9.21}$$

where M is a transition metal of variable oxidation state. Clearly, to operate effectively, Li_xMO_2 should ideally be electronically conducting, or at least miscible with an appropriate inert conducting adduct, and its formation, through Li-ion diffusion, must be reasonably facile, and ideally highly reversible if the battery is to be re-chargeable. This latter suggests that layered or channel structures are likely to be the most appropriate. Suitable metal oxides include MnO_2 and V_6O_{13}; higher voltages may be obtained by using $Li_{1-y}CoO_2$ and $Li_{1-y}NiO_2$, since these oxides possess the transition-metal in a high oxidation state, which allows a high cell voltage to develop (given that the intercalation process is, in essence, reduction of the metal). The oxides are also capable of accommodating large quantities of lithium per formula unit, and have low formula weights, giving rise to high power and energy densities. Importantly, while they can intercalate lithium relatively easily, they do not co-intercalate solvent, since the interlayer separation in these structures is too small. They are also stable in contact with the solvent, are of low cost, are easily fabricated into electrodes and are neither of them environmentally problematic. Some, such as MnO_2 show spin-state changes that lead to Jahn-Teller type distortions: the spinel $LiMn_2O_4$ is cubic, for example, but $Li_2Mn_2O_4$ is tetragonal. Such changes do lead to excellent voltage-discharge characteristics, but militate against reversibility, and considerable effort is now being directed to optimising the two features together.

The solvent used may be an organic ether or a cyclic or acyclic organic carbonate (such as propylene carbonate or diethyl carbonate respectively), which must, of course, be thoroughly dried. Conducting salts include $LiPF_6$, $LiBF_4$, $LiAsF_6$ or $LiClO_4$, though the last two have proved less suitable owing respectively to toxicity and explosion risk. Polymeric ethers, such as Li-salt impregnated polyethylene oxide, have particularly attractive properties for certain applications, and particularly when combined with electronically conducting intercalation polymers on the cathode side, have the immense advantage of being capable of continuous fabrication using fast polymer-film processing techniques. Such thin-film batteries can be fabricated into essentially arbitrary shapes, giving considerable design flexibility, and they have large surface areas, which makes thermal management more straightforward. However, there are disadvantages: the initial manufacturing set-up costs are high, and the need for critical gas-tightness means that manufacturing tolerances are very tight.

A second type of system, developed from the analogous Na/S battery described above, was initially based on the cell reaction:

$$2\mathrm{Li} + \mathrm{S} \rightarrow \mathrm{Li_2S} \tag{9.22}$$

with a LiCl/KCl eutectic electrolyte and a cell voltage of 2.25 V. This cell had an extremely high theoretical energy density (2624 Wh kg^{-1}) and a high power density, but suffered from the high vapour pressure of sulphur and, as above, the solubility of Li in the electrolyte. Both of these problems were addressed in the LiAl/FeS system, based on Li alloy anodes and FeS_2 cathodes with a lower temperature molten-salt eutectic. This battery has a high specific power, high rate capability and long shelf life, and has exciting possible applications to traction. The cell reactions are:

$$\text{anode:} \quad \mathrm{LiAl} \rightarrow \mathrm{Li^+} + \mathrm{Al} + e^- \tag{9.23}$$

$$\text{cathode:} \quad 2\mathrm{Li^+} + 2e^- + \mathrm{FeS} \rightarrow \mathrm{Li_2S} + \mathrm{Fe} \tag{9.24}$$

$$\text{or} \quad 4\mathrm{Li^+} + 4e^- + \mathrm{FeS_2} \rightarrow 2\mathrm{Li_2S} + \mathrm{Fe} \tag{9.25}$$

with a (relatively) low m.p. electrolyte of LiCl-LiBr-KBr (400–450°C). However, the cell voltage is only 1.3 V for cathode reaction (9.24) and 1.6 V for reaction (9.25), and the energy density therefore rather low, particularly for reaction (9.24). In addition, the cells are not tolerant to overcharge, and there is a number of rather difficult materials problems to overcome. Nevertheless, developments of this battery have proceeded apace, and demonstrator units are now being contemplated, specifically with traction in mind. However, although the $LiAl/FeS_2$ system has considerable potential as a traction-power battery, there are inherent difficulties of cost that need to be addressed. Lithium is not a rare metal, but it is not inexpensive, and there are considerable engineering and scientific problems that need to be overcome before a reliable re-chargeable traction system can be developed.

The basic difficulty of the above cells is that the replating of lithium leads to loss of electrical contact between particles due to the formation of insulating layers that presumably arise from chemical reaction with the solvent or adventitious water. Multiple cycling also leads to surface expansion of the lithium anode with associated non uniformities (such as dendrite formation). Such non-uniformities can lead to hot spots, and in turn to disastrous failure of the battery by explosion or fire.

The above type of problem has been tackled by replacing the lithium anode by a second intercalate, such as Li_xC_6, giving a battery with an anode reaction:

$$Li_xC_6 \rightarrow Li_{x-y}C_6 + yLi^+ + ye^- \tag{9.26}$$

This type of cell, the so called **Li-ion battery**, is shown schematically in Fig. 9.14a, and the charge- discharge cycle for one example: Li/C:$LiCoO_2$ is shown in Fig. 9.14b. In this figure, the potentials of both half cells are given (referred to the Li/Li^+ standard potential) as well as the resultant cell voltage.

Li-ion batteries do allow high discharge rates. A 1.26 kWh launch-battery (36 V, 35 Ah,18 kg) can deliver 350 A for 5 min (29 Ah). In Table 9.1 data are compared for small NiCd-, NiMH and Li-ion batteries.

Table 9.1 Comparison of three commercially available rechargeable battery types (4/3 A – size in roll-form, diameter 17 mm, height 67 mm, electrode length in total about 20 cm, gas-tight).

	Lithium-Ion	**Nickel-Cadmium**	**Nickel-Metal hydride**
Cell Voltage:			
Charged; open circuit	4.2	1.4	1.4
After 5 hours discharge (C/5)	3.6	1.2	1.2
Working temperature [°C]	– 20 to 60	– 40 to 60	– 20 to 60
Capacity [Ah]	1.2	1.7	3.5
Weight [g]	36	42	55
Energy Density			
[Wh/kg]	120	49	76
[$Wh \cdot dm^{-3}$]	285	134	275
Power density [W/kg]	230 (2C)	390 (8C)	210 (3C)
Self-discharge at 20°C as percent diminution per month	5–10	15–20	20
Cyclability as number of charge-discharge cycles to 80 % residual capacity	> 700	500–700	> 700

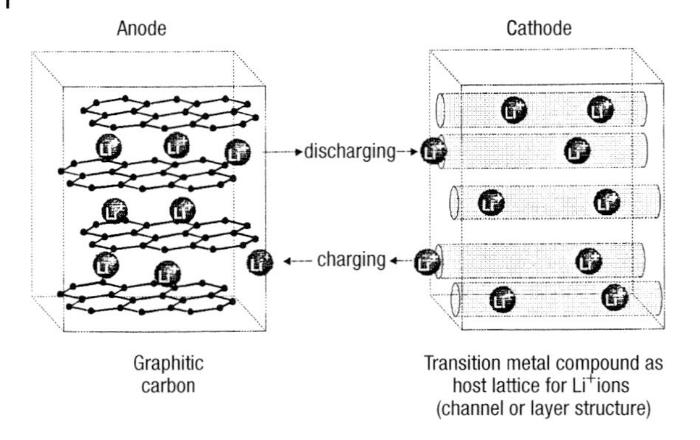

Fig. 9.14a Schematic representation of a lithium-ion battery.

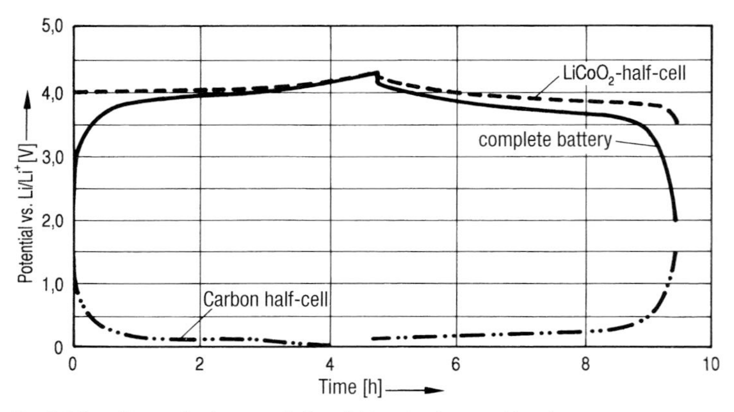

Fig. 9.14b Charge-discharge cycle for a lithium-ion battery. Also shown are the potentials of the $LiCoO_2$ and carbon electrodes with respect to the Li/Li^+ standard potential.

9.3.2.5 **Redox Storage Systems**

In redox storage systems, the charge-discharge cycles involve alteration in the valence states of two species in solutions separated by a membrane. The most familiar example is the chromium-iron system in which the charging process oxidises Fe(II) to Fe(III) and reduces Cr(III) to Cr(II) in solutions of the chlorides in aqueous HCl. The two solutions are pumped through the anode and cathode compartments respectively during charge and stored in separate containers. During discharge, they are again pumped through the respective electrode compartments. Such systems have some advantages: they avoid the problems associated with long-term alteration in electrode morphology found in conventional secondary batteries, and the electrodes themselves can be simply constructed from polymer-bound graphite. The membrane itself should be permeable to chloride ions but not to the two metal ions, which is a significant materials problem, and there are also problems maintaining the stoicheiometry of the cell reaction arising from the competitive evolution of hydrogen during the reduction of Cr(III) to Cr(II) ($E_0 = -0.41$ V). This can be counter-balanced by controlled evolution of Cl_2 and the recombination of the two gases restores the overall material balance. The energy density is poor – about 30 Wh for 2 kg of solution –, and the main thrust in development has been as a storage system for photovoltaic devices.

9.3.3
Summary of Data for Secondary Battery Systems

A plot of energy density against power density for some secondary systems is shown in Fig. 9.15, and the data for individual systems is tabulated below.

The data for each battery is given under ten headings:

1. The electrolyte used and the working temperature of the cell;
2. The cathode reaction, standard potential and theoretical specific capacity
3. The anode reaction and analogous values

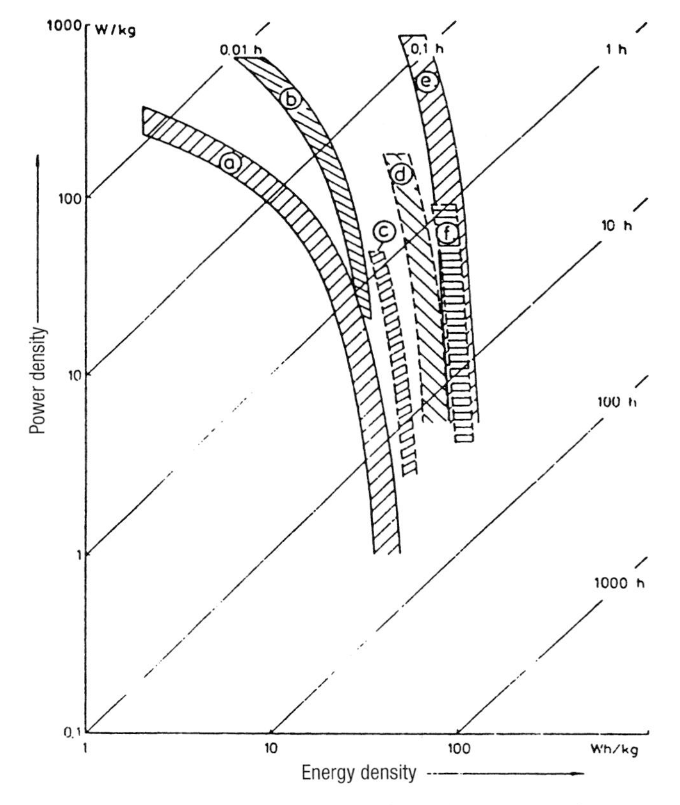

Fig. 9.15 Energy-density vs. Power-density for the following secondary batteries: (a) lead-acid; (b) NiCad; (c) Ni/Fe; (d) Ni/Zn; (e) Ag/Zn; (f) Na/S. The data for the Ni/Fe system is derived from a recent improved performance as compared to that described in Table 9.1.

4. Cell reaction, the experimental open-circuit voltage and the theoretical energy density
5. Experimental energy density for one and five-hour discharge times
6. Power density for short period discharge
7. Lifetime (cycle number) and energy efficiency
8. Cost per kWh of energy taken form the cell and cost per kW of installed power (both in £ Sterling)
9. Status of the battery and examples of usage or potential use.
10. Remarks

$Pb/H_2SO_4/PbO_2$

1. Aqueous sulphuric Acid ($\rho = 1.28$ g cm^{-3}); room temperature
2. $PbO_2 + H_2SO_4 + 2H^+ + 2e^- \rightarrow PbSO_4 + 2H_2O$; +1.685 V vs. NHE; 224 Ah kg^{-1}
3. $Pb + H_2SO_4 \rightarrow PbSO_4 + 2H^+ + 2e^-$; −0.356V vs. NHE; 259 Ah kg^{-1}
4. $Pb + PbO_2 + 2H_2SO_4 \rightarrow 2PbSO_4 + 2H_2O$; 2.06 V; 167 Wh kg^{-1}
5. 40 Wh kg^{-1}; 35 Wh kg^{-1}
6. 250 Wh kg^{-1} (corresponding to 300 mA cm^{-2})
7. 300 – 1500 cycles; 70 – 80 %
8. £0.20 – £0.40 per kWh; £100–£150 per kW
9. Very broad range of applications including starter batteries, traction units in milk floats and pallet trucks, emergency back-up units etc. The most frequently encountered secondary system
10. Must be developed to higher energy densities if it is to find application in traction. At the present time NiMH systems are prefered.

Cd/KOH/NiOOH

1. Aqueous KOH ($\rho = 1.17$ g cm^{-3}); – 40 to + 45°C
2. $2NiOOH + 2H_2O + 2e^- \rightarrow 2Ni(OH)_2 + 2OH^-$; +0.45 V vs. NHE; 294 Ah kg^{-1}.
3. $Cd + 2OH^- \rightarrow Cd(OH)_2 + 2e^-$; −0.809 V vs. NHE; 477 Ah kg^{-1}
4. $Cd + 2NiOOH + 2H_2O \rightarrow 2Ni(OH)_2 + Cd(OH)_2$; 1.3 V; 244 Wh kg^{-1}
5. 35 Wh kg^{-1}; 32 Wh kg^{-1}
6. normally ca. 260 W kg^{-1}; can be up to 700 W kg^{-1}
7. up to 3000 cycles; 65 %
8. £0.25 – £0.50 per kWh; £400–£500 per kW
9. Broad range of uses including starter and storage batteries in aircraft and space craft; electronic components etc.
10. Robust and long-lived system, but in many applications now being replaced by NiMH and Li-ion.

Metal-hydride/KOH/NiOOH

1. Aqueous KOH ($\rho = 1.17$ g cm^{-3}); – 20 to + 50°C
2. $NiOOH + H_2O + e^- \rightarrow Ni(OH)_2 + OH^-$; +0.45 V vs. NHE; 294 Ah kg^{-1}.

3. $MH + OH^- \rightarrow M + H_2O + e^-$; ca. − 0.8 V vs. NHE
4. $MH + NiOOH \rightarrow M + Ni(OH)_2$; ca. 1.3 V; 278 Wh kg^{-1}
5. ca. 65 Wh kg^{-1}
6. normally ca. 200 W kg^{-1}
7. > 700 cycles
8. ??
9. Likely to replace Ni/Cd batteries owing to better environmental impact and properties
10. Self-discharge quite high (c. 20 % per month)

Fe/KOH/NiOOH

1. Aqueous KOH ($\rho = 1.2$ g cm^{-3}); room temperature
2. $2NiOOH + 2H_2O + 2e^- \rightarrow 2Ni(OH)_2 + 2OH^-$; +0.45 V vs. NHE; 294 Ah kg^{-1}.
3. $Fe + 2OH^- \rightarrow Fe(OH)_2 + 2e^-$; −0.877 V vs. NHE; 960 Ah kg^{-1}
4. $Fe + 2NiOOH + 2H_2O \rightarrow 2Ni(OH)_2 + Fe(OH)_2$; 1.36 V; 265 Wh kg^{-1}
5. ca. 30 Wh kg^{-1}; 23 Wh kg^{-1}
6. 100 W kg^{-1}
7. More than 2000 cycles; 50 %
8. Comparable or slightly cheaper than Ni/Cd

Zn/KOH/AgO

1. Aqueous KOH ($\rho = 1.45$ g cm^{-3}); room temperature
2. $AgO + H_2O + 2e^- \rightarrow Ag + 2OH^-$; 0.608 V vs. NHE; 432 Ah kg^{-1}
3. $Zn + 2OH^- \rightarrow Zn(OH)_2 + 2e^-$; −1.25 V vs. NHE; 820 Ah kg^{-1}
4. $AgO + Zn + H_2O \rightarrow Ag + Zn(OH)_2$; 1.86 V; 478 Wh kg^{-1}
5. 100 – 120 Wh kg^{-1}; 80 – 100 Wh kg^{-1}
6. 500 – 800 W kg^{-1}
7. up to 100 recharge cycles
8. £25 per kWh; £1500 per kW
9. Specialist uses in air and space travel, weapons technology, high-value electronic equipment etc.
10. The most expensive of the commonly encountered secondary systems, but with a very high power density. The cathode reaction is highly simplified above.

Zn/KOH/NiOOH (The nickel-zinc accumulator)

1. Aqueous KOH ($\rho = 1.2$ g cm^{-3}); room temperature
2. $2NiOOH + 2H_2O + 2e^- \rightarrow 2Ni(OH)_2 + 2OH^-$; 0.45 V vs. NHE; 294 Ah kg^{-1}.
3. $Zn + 2OH^- \rightarrow Zn(OH)_2 + 2e^-$; −1.25 V vs. NHE; 820 Ah kg^{-1}
4. $Zn + 2NiOOH + 2H_2O \rightarrow 2Ni(OH)_2 + Zn(OH)_2$; 1.73 V; 326 Wh kg^{-1}
5. 80 Wh kg^{-1}; 60 Wh kg^{-1}
6. 200 W kg^{-1}
7. up to 200 cycles; c. 55 %

8. £250 – £400 per kW installed power
9. Experimental battery for traction

Li/organic electrolyte/oxide

1. Cyclic or polymeric ethers such as derivatives of tetrahydrofuran or polyethylene oxides or cyclic/acyclic carbonates; – 20 to 55°C;
2. $xLi^+ + MO_y + xe^- \rightarrow Li_xMO_y$; usually ca. 1 V vs NHE; ca. 180 Ah kg^{-1}
3. $Li \rightarrow Li^+ + e^-$; −3.045 V vs. NHE; 3862 Ah kg^{-1}
4. $xLi + MO_y \rightarrow Li_xMO_y$; c. 3.5 V (o.c.v. 4.1 V); 750 Wh kg^{-1}
5. 80–90 Wh kg^{-1}
6. ca. 100 W kg^{-1}
7. Single cells from 400–1200 cycles
8. Currently up to £3000/kWh for button cells
9. Very wide range of applications both as primary batteries in electrical equipment and proposed secondary batteries in traction.
10. Low self-discharge rate of ca. 5–10 % per month.

9.4 Primary Systems other than Leclanché Batteries

9.4.1 Alkaline-Manganese Cells

The alkaline-manganese primary battery is a further development of the Leclanché cell that still employs zinc and MnO_2 as the electrode materials. The main differences are:

(a) the use of KOH as the electrolyte, which allows further reduction of the MnO_2 at the cathode to Mn(II) as

$$MnO_2 + H_2O + e^- \rightarrow MnOOH + OH^- \quad (9.27)$$

$$MnOOH + H_2O + e^- \rightarrow Mn(OH)_2 + OH^- \quad (9.28)$$

(b) the use of metallic zinc powder suspended in alkali as the anode rather than a zinc sheet

(c) the cell arrangement is the opposite of the Leclanché cell, as shown in Fig. 9.16.

These differences lead to an improvement of some 50 % in properties as compared to the original Leclanché cell with the same open-circuit voltage. Interestingly, the manganese-alkali cell can, in principle, also be used as a secondary battery as described above, provided a membrane is used that is insensitive to dendrite growth and provided discharge only to Mn(III) is permitted.

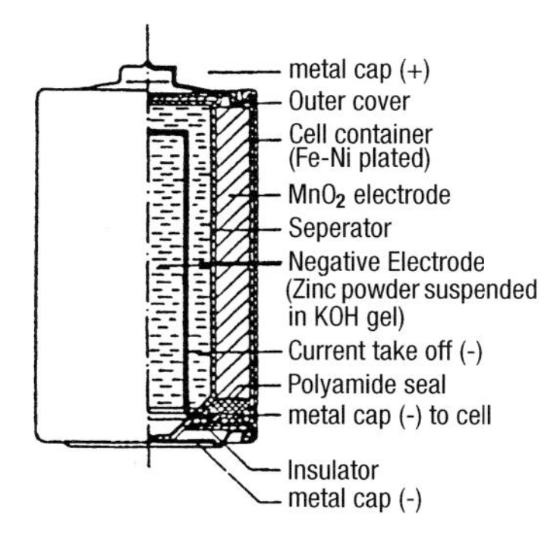

Fig. 9.16 Cross-section of one form of the alkaline-manganese battery.

9.4.2
The Zinc-Mercury Oxide Battery

The electrolyte of this battery consists of KOH in which ZnO is dissolved impregnated in a membrane. The anode is zinc powder suspended in alkali as above and the cathode consists of 90 % red mercury oxide/10 % powdered graphite, the latter acting as to facilitate electronic conduction. The cathode reaction is:

$$HgO + H_2O + 2e^- \rightarrow Hg + 2OH^- \qquad (9.29)$$

and the battery is constructed as in Fig. 9.17.

The battery is usually referred to as the Mallory battery from its main manufacturer, and has a high realisable energy density of 110 Whkg^{-1} ($W_s^{th} = 241$ Whkg^{-1}). It also has a high shelf-life but its high price has restricted its applications to specialised uses, particularly in button cells for hearing aids, pocket calculators, cameras and heart pacemakers. In recent years, acute concerns about the environmental impact of mercury have substantially reduced the market penetration of this battery, and it has been superseded by lithium batteries in almost all areas.

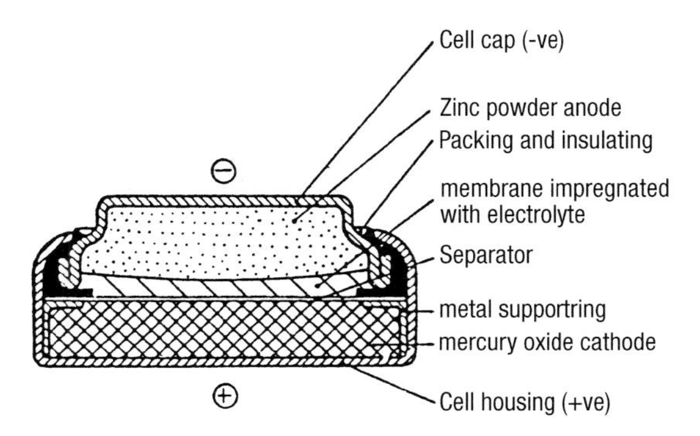

Fig. 9.17 Cross-section through a zinc-mercury-oxide button cell.

9.4.3
Lithium Primary Batteries

In addition to the primary/secondary Li batteries described above (Section 9.3.2.4), there are several high-performance primary lithium batteries that have been developed for specialised applications. Medium energy/high power systems are of particular military importance, since they allow very large current pulses to be drawn and, at the same time, are very light. Such batteries are based on Li anodes, and inert cathodes capable of reducing such species as $SOCl_2$ or SO_2; experimental Li/BrF_3 batteries capable of up to 1000 Wh kg^{-1} are now being developed. These batteries are usually supplied in "reserve" form: that is, the battery is supplied without electrolyte, and will operate once the electrolyte is added.

In the case of the Li-SO_2 systems, the anode is a lithium foil separated by a polypropylene sheet from the porous carbon cathode and the entire assembly is held in a gas-tight housing under 1 bar pressure. The electrolyte is a mixture of propylene carbonate, acetonitrile and SO_2 with 1.8 M LiBr to provide adequate conductivity, and the overall reaction is $2Li + 2SO_2 \rightarrow Li_2S_2O_4$; $E_{c,0} = 2.9$ V.

For the Li/$SOCl_2$ battery, an electrolyte of the form $SOCl_2/LiAlCl_4$ must be added, whereupon the battery generates electricity from the cell reaction:

$$4Li + 2SOCl_2 \rightarrow 4LiCl + SO_2 + S \tag{9.30}$$

High energy/high power batteries based on lithium have also been developed. The simplest of these was the Li/Cl_2 battery, which used a LiCl/KCl/LiF eutectic electrolyte melting at 450°C. The overall cell reaction is similar to the Zn/Cl_2 cell described above, but the cell voltage is very high (3.46 V), and the theoretical energy density enormous (2200 Wh kg^{-1}). However, the actual operating temperature has to be rather high (600°C), and there are severe corrosion problems caused in part by the solubility of the lithium in the electrolyte. This type of cell has been developed into the Sohio battery, which uses a rather lower melting-point electrolyte, LiCl/KCl at an operating temperature of 400°C, and a LiAl alloy as the anode, to reduce solubility problems. The overall cell reaction for this system is:

$$4LiAl + 2Cl_2 \rightarrow 4LiCl + 4Al \tag{9.31}$$

with a voltage of 3.2 V and a high peak current density of 2 Acm^{-2}. The main disadvantages of this cell are that the discharge voltage is dependent on the state of charge and the energy density is considerably lower (62 Wh kg^{-1}).

9.4.4
Electrode and Battery Characteristics for Primary Systems

As for the secondary cells above, we conclude this section by listing the characteristics for primary batteries.

The data for each type of battery are listed in the following order:

1. Electrolyte and working temperature used
2. Cathode reaction, corresponding *thermodynamic standard potential* and theoretical specific capacity
3. Anode reaction with analogous values
4. Cell reaction, *experimental* open-circuit voltage and theoretical energy density
5. Limiting energy density corresponding to slow discharge
6. Short-time power density ($W\ kg^{-1}$)
7. Shelf-life (yr) corresponding to storage at normal temperatures
8. Cost per kWh of energy drawn from the system (£/kWh)
9. Status of battery and examples of use
10. Remarks

$Zn/NH_4Cl/MnO_2$

1. 25 % NH_4Cl solution with 9 % $ZnCl_2$ addition; – 15 to + 40°C
2. $2MnO_2 + 2H_2O + 2e^- \rightarrow 2MnOOH + 2OH^-$; ca. +1.1 V vs NHE; 308 $Ah\ kg^{-1}$
3. $Zn \rightarrow Zn^{2+} + 2e^-$; −0.76 V vs. HE; 820 $Ah\ kg^{-1}$
4. $2MnO_2 + Zn + 2NH_4Cl \rightarrow 2MnOOH + Zn(NH_3)_2Cl_2$; 1.5 – 1.6 V; 245 $Wh\ kg^{-1}$
5. 80 $Wh\ kg^{-1}$
6. 10 $W\ kg^{-1}$
7. 3 yr
8. £200–700/kWh
9. Wide range of uses, mainly in portable electrical equipment.
10. The data above refer to the high-current single cells.

$Zn/KOH/MnO_2$

1. KOH (density: 1.2 gcm^{-3}); – 15 to + 40°C
2. $MnO_2 + 2H_2O + 2e^- \rightarrow Mn(OH)_2 + 2OH^-$; ca. 0.7 V vs. NHE; 617 $Ah\ kg^{-1}$
3. $Zn + 2OH^- \rightarrow Zn(OH)_2 + 2e^-$; −1.25 V vs. NHE; 820 $Ah\ kg^{-1}$
4. $MnO_2 + Zn + 2H_2O \rightarrow Mn(OH)_2 + Zn(OH)_2$; 1.58 V; 450 $Wh\ kg^{-1}$
5. 100 $Wh\ kg^{-1}$
6. up to ca. 30 $W\ kg^{-1}$
7. 2–5 yr
8. £200–800/kWh
9. Wide range of uses, particularly in electrical equipment with high energy/current requirements
10. The data above refer to single cells.

Zn/KOH/HgO

1. KOH (density: 1.2 gcm^{-3}); 0 to 40°C
2. $HgO + H_2O + 2e^- \rightarrow Hg + 2OH^-$; +0.098 V vs. NHE; 247 $Ah\ kg^{-1}$
3. $Zn + 2OH^- \rightarrow Zn(OH)_2 + 2e^-$; −1.25 V vs. NHE; 820 $Ah\ kg^{-1}$
4. $HgO + Zn + H_2O \rightarrow Hg + Zn(OH)_2$; 1.35 V; 241 $Wh\ kg^{-1}$

5. 110 Wh kg^{-1}
6. $<$ 10 W kg^{-1} (button cells)
7. 3–7 yr
8. Up to £10,000/kWh for the smallest cells
9. Range of uses, but now restricted to high-value electronic equipment with very demanding space restrictions, such as hearing aids.

Li/inorganic electrolyte/$SOCl_2$

1. $LiAlCl_4$ dissolved in $SOCl_2$; -40 to $+50°C$
2. $2SOCl_2 + 4e^- \rightarrow SO_2 + S + 4Cl^-$; Not accurately known; 450 Ah kg^{-1} based on $SOCl_2$
3. $Li \rightarrow Li^+ + e^-$; not accurately known in the solvent; 3862 Ah kg^{-1} based on lithium
4. $4Li + 2SOCl_2 \rightarrow 4LiCl + SO_2 + S$; 3.65 V; 1470 Wh kg^{-1}
5. ca. 500 Wh kg^{-1}
6. $>$ 100 W kg^{-1}
7. 5–10 yr
8. £450–1000/kWh
9. Predominantly larger units ($>$ 2000 Ah) for military use
10. Significantly larger energy and current densities compared to more familiar primary systems.

9.5 Fuel Cells

The possiblity of generating electrical energy at will by continually feeding electrochemically active material to a suitable cell has interested the scientific world from an early date. The first experimental realisation of this fuel-cell idea was achieved by Grove, who constructed a series of hydrogen-oxygen fuel cells between 1839 and 1842, based on platinum electrodes immersed in aqueous acid electrolyte. Even after the discovery of electricity generation from the dynamo, research in fuel cells continued, driven by the desire directly to convert the then primary fuel, coal, more efficiently to electricity. However, even in molten alkali electrolytes, the electrochemical oxidation of coal proved impossible to effect at sufficiently low potentials, and electrode fouling by sulphur impurities remained a major problem.

The modern development of fuel cells dates from the early 1950s, receiving a strong boost from the requirements of the space programme for a system with energy density in the kWh/kg region. The hydrogen-oxygen fuel cell, with a rest potential near 1 V and a theoretical energy density of 3000 Wh kg^{-1} proved highly attractive for this purpose, and its development has led to the strong modern interest in fuel-cell technology.

9.5.1
Fuel Cells with Gaseous Fuels

The main fuel used in such cells is hydrogen, since at noble metal electrodes such as platinum this has extremely facile electrode kinetics at low temperatures. Other fuels as methane, ethane, natural gas and CO could be used at the anode, but at temperatures below ca. 200°C, which represents the upper limit of concentrated aqueous electrolytes, the electrochemistry of these species is sluggish, and only hydrogen, and more recently methanol, have been found to have electrode kinetics sufficiently facile to give good cell voltages. Similarly, at the cathode, oxygen, usually supplied as air, is the primary combustant. In principle, chlorine might be used, but corrosion and environmental problems have severely limited its applicability. However, for oxygen reduction at lower temperatures, the electrode kinetics at the cathode are poor, and if flat electrode surfaces are used, currents of only mA cm^{-2} are attained, which must be improved by better electrode design.

9.5.1.1 Gas Diffusion Electrodes

Oxidation of hydrogen or reduction of oxygen must take place at a three-phase boundary consisting of the electrode itself, acting as source or sink for electrons, the electrolyte solution, containing OH^- or H^+ ions and water, and the substrate gas, H_2 or O_2. The reason for the three-phase system is that both H_2 and O_2 are poorly soluble, with concentrations of ca. 10^{-3} M at one atmosphere pressure. This low solubility severely inhibits mass-transport, and only where liquid, gas and electrode are present together can the effective concentration of the gas become sufficient to sustain a high current density.

Clearly, as indicated in Chapter 8, the spatial extent of the three-phase zone can be substantially extended by the use of porous electrodes placed between electrolyte and gas chambers as shown in Fig. 9.18: the faradaic reactions take place in an annulus

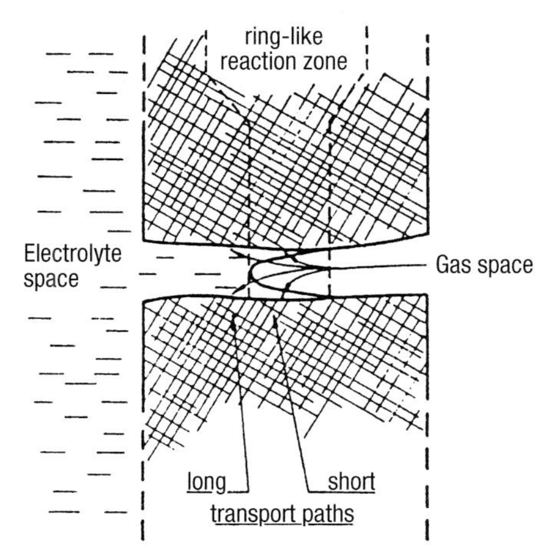

Fig. 9.18 Single pore of a gas diffusion electrode.

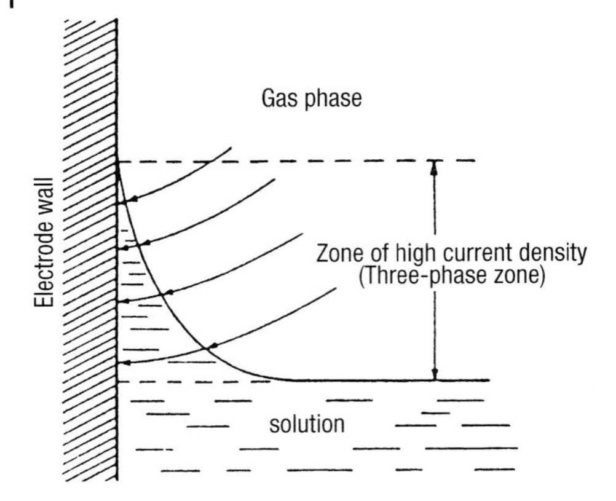

Fig. 9.19 Three-phase zone in a gas-diffusion electrode.

where the electolyte layer is thin enough to sustain rapid mass transport as shown in Fig. 9.19. In order to attain the highest possible current densities, these porous electrodes must satisfy two basic requirements:

(a) the catalyst used must be as active as possible;
(b) during operation of the electrode, the pores must neither fill up with electrolyte due to capillary action ('flooding'), nor must the gas pressure become so large that electrolyte is completely driven out of the pores.

For the oxidation of hydrogen, the platinum metals and their alloys, as well as tungsten carbide and specially prepared nickel are suitable catalysts, whereas for the reduction of oxygen, platinum metals, nickel and silver are all fairly effective. These catalysts are used in the form of small particles embedded in porous supporting materials such as sintered nickel or activated carbon, the latter usually being bound to form a self-standing layer with a polymer such as PTFE. The amount of the latter can be used to vary systematically the hydrophobicity of the electrode, and hence to allow the variation of gas pressure in the pores. Control of the latter can also be effected by using two-layer porous electrodes, with the layer next to the gas chamber having rather wider pores than that next to the liquid electrolyte. The former layer can also contain additional PTFE to ensure a high degree of hydrophobicity and hence low leakage.

A schematic diagram and photograph of a porous electrode for use in an alkaline fuel cell are shown in Fig. 9.20, and a typical cell arrangement is shown in Fig. 9.21. For this electrode, a silver net is used both as support and current collector, and the electrode shown is only 1 mm thick, weighing ca. 0.3 g cm^{-2}. Such electrodes are capable of sustaining current densities of 1–2 A cm^{-2}, though in practice they are operated at lower currents to avoid high overpotentials.

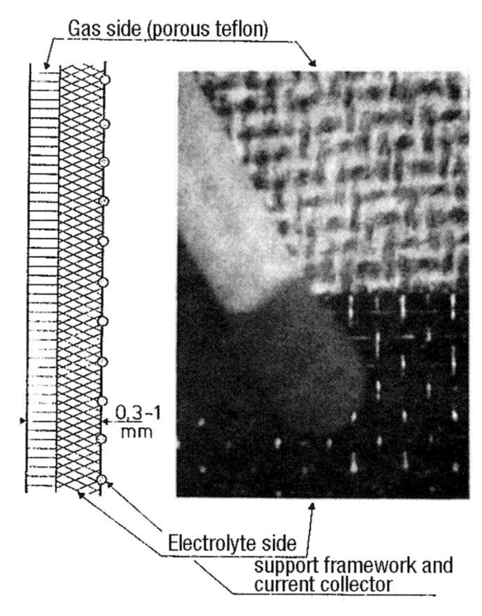

Fig. 9.20 Schematic diagram and photograph of a gas-diffusion electrode for oxygen reduction showing the hydrophobic outer uncatalysed layer and the inner catalysed layer next to the electrolyte.

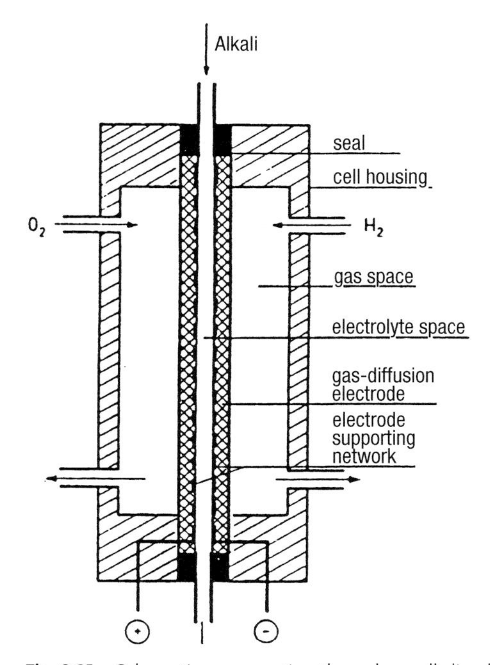

Fig. 9.21 Schematic cross-section through an alkaline hydrogen-oxygen fuel cell.

9.5.1.2 The Apollo Cell

The most influential use of hydrogen-oxygen fuel cells has been in the American Space Programme, particularly the Apollo missions to the Moon. The construction of the cell showed a strong adherence to the principles outlined above: electrodes were formed of porous sintered nickel and the operating temperature was well above 100°C, necessitating a concentrated (75 %) KOH solution. A total of 31 single cells were connected in series to form a stack. Both gases (H_2 and O_2) were stored cryogenically and supplied to the fuel cell under pressure, with the hydrogen being recycled from the anode exhaust. Temperature regulation was through a cooling system using nitrogen gas, which passed at pressure through an internal circulation system and thence to the outside of the space ship. The unit had a nominal power of 1.12 kW at 28 V, with each cell delivering 0.9 V at 100 mA cm^{-2}; it weighed 110 kg.

The reactions in the alkaline fuel cell are:

$$H_2 + 2OH^- \rightarrow 2H_2O + 2e^- \tag{9.32}$$

$$\begin{array}{c} \frac{1}{2}O_2 + H_2O + 2e^- \rightarrow 2OH^- \\ \hline H_2 + \frac{1}{2}O_2 \rightarrow H_2O \end{array} \tag{9.33}$$

At the hydrogen anode, water is formed. A part of it will react with the oxygen at the cathode but the rest will leave the cell as reaction product in the circulating electrolyte (Fig. 9.21).

9.5.2 Modern Developments

These have been driven by the necessity of lowering both construction and operating costs, essential steps if fuel cells are to be more widely commercialised. Construction costs can be lowered by either using much lower loadings of noble metal catalysts (values of 0.1 – 0.5 mg cm^{-2} must be attained without degradation of performance) or by using catalysts based on non-noble metals. The operating costs can be lowered by using hydrogen prepared by reforming methane or naphtha rather than electrolytic hydrogen, and by using air rather than pure oxygen as the combustant. For traction, methanol could serve as the primary fuel, also reformed in the vehicle. Reformation of hydrocarbons requires temperatures in the range 650°C – 800°C, whereas methanol can be reformed at much lower temperatures (300°C). For hydrocarbons, the primary reaction involves steam and an appropriate catalyst:

$$C_nH_{2n+2} + nH_2O \rightarrow nCO + (2n+1)H_2 \tag{9.34}$$

$$CO + H_2O \rightarrow CO_2 + H_2 \tag{9.35}$$

Reforming lowers the operating costs from ca. £0.50/kWh for electrolytic hydrogen to a tenth of this value, though the costs of the reformer and desulphuriser systems (this latter to avoid poisoning the Pt surface) add substantially to the construction costs.

There are five different types of fuel cell to be discussed:

(a) **A**lkaline **F**uel **C**ells (**AFC**) with 30 % aqueous KOH as electrolyte and a working temperature of up to 90°C;
(b) **S**olid **P**olymer **F**uel **C**ells (**SPFC**) using an SPE membrane as the electrolyte and operating at up to 80°C;
(c) **P**hosphoric **A**cid **F**uel **C**ells (**PAFC**) with phosphoric acid as the electrolyte and operating temperatures of up to 220°C;
(d) **M**olten **C**arbonate **F**uel **C**ells (**MCFC**) with molten Li/K carbonate as the electrolyte and operating temperatures of ca. 650°C;
(e) **S**olid **O**xide **F**uel **C**ells (**SOFC**) using oxide-ion conducting ceramics as electrolytes and operating at up to ca. 1000°C.

The AFC and SPFC cells are usually referred to as low-temperature systems; they give good energy and power densities and are particularly suitable for traction, especially in military applications where silence of operation and low thermal signature are important.

The PAFC is an intermediate temperature device and the MCFC and SOFC are referred to as high-temperature cells, which are likely to find primary applications as stationary power sources in, for example, load levelling systems. The waste heat from the PAFC can serve for space heating, and for the high-temperature systems, the waste heat can serve to operate a steam turbine, adding further to the energy conversion efficiency.

The *alkaline fuel cell* has already been illustrated in Fig. 9.21, Raney nickel or Pt-metals are used as catalysts; cells are connected in a bipolar configuration to give a complete stack. If the cell is operated with air rather than pure oxygen, however, a significant problem arises as the CO_2 in the air dissolves in the KOH, eventually leading to precipitation of K_2CO_3 in the electrode pores ("fouling"). It is, therefore, necessary to pass the air through a scrubber before it enters the cell.

The construction of the *Solid Polymer Fuel Cell* has already been described in Section 7.2 & Fig. 8.20: the electrolyte is a thin layer of proton-conducting polymer on both sides of which are attached thin porous layers of either particulate noble-metal catalyst (at a loading of 0.1 to 0.5 mg cm^{-2}) or such catalyst dispersed on a porous carbon/Nafion composite (with the metal loadings reduced by an order of magnitude in recent cells). In order to complete the membrane-electrode-assembly (MEA) diffusion layers of the same size (ca. 100 μ), out of Teflon-bonded carbon powder, are pressed against the two catalyst layers.

The current collectors are usually formed from graphite. The product water from the electrochemical reaction appears in acid cells at the *cathode*, but it has been found necessary to humidify the anode gas to prevent the membrane on the *anode* side becoming dehydrated.

The cell reactions for the acid hydrogen/oxygen SPE cell are:

$$H_2 + 2H_2O \rightarrow 2H_3O^+ + 2e^- \qquad (9.35)$$

$$\frac{\frac{1}{2}O_2 + 2H_3O^+ + 2e^- \rightarrow 3H_2O}{H_2 + \frac{1}{2}O_2 \rightarrow H_2O} \tag{9.36}$$

The extremely thin MEAs permit a compact construction of the cell elements whilst at the same time allowing very low ohmic cell resistance. At least for a short time, at temperatures of between 80 and 100°C, it is possible to draw currents of up to 1.5 A cm^{-2}. By optimisation of the catalyst structure, only 0.1 – 0.5 mg Pt cm^{-2} are necessary for such activity. Especially favourable systems data can be attained by the use of pure hydrogen and oxygen, which is feasible, for example, for underwater propulsion (Fig. 9.22).

For terrestrial uses, the hydrogen is normally obtained from the reformation of natural gas and biomass. However, reformers always produce a small amount of CO as well as CO_2 and CO is particularly strongly adsorbed on platinum up to potentials of 500 – 600 mV at room temperature. This blocks the electrode, reducing to a low level its ability catalytically to oxidise hydrogen, and it has proved essential to develop CO-tolerant anodes. The classic example is a mixture of platinum and ruthenium, as much as possible in alloyed form (see section 6.5). More favourable oxidation potentials for hydrogen are possible not only with ruthenium, with its relatively weak bonding to CO, but also for other non-noble metals alloyed with platinum as well.For pure platinum, it can be seen that already 100 ppm of CO in the fuel feed are sufficient at

Fig. 9.22 PEM H_2/O_2-fuel cell system for 35–50 kW nominal ouput, 72 cells, 1163 cm^2 electrodes using Nafion 117 at 80°C, 2.3 bar O_2, and 2 bar H_2 [K. Strasser, "PEM fuel-cell systems for submarines" *in* Handbook of Fuel Cells, vol. 4, eds. W. Vielstich, H. Gasteiger and A. Lamm, Wiley, Chichester, UK, 2003].

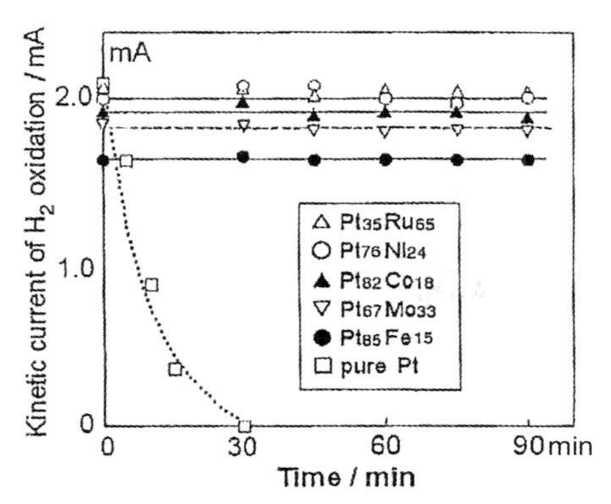

Fig. 9.23 CO-tolerant catalysts: time dependency for H_2-oxidation in 0.1 M $HClO_4$ saturated with 100 ppm CO/H_2 at rotating electrodes held at 20 mV RHE and at room temperature [H. Igarashi, T. Fujino, H. Ushida and M. Watanabe, PCCP **3** (2001) 306].

room temperature temporarily to poison the catalyst surface strongly. In addition to ruthenium alloys, at would appear from Fig. 9.23 that alloys of platinum with iron, cobalt and molybdenum are also effective in reducing this poisoning.

The phosphoric-acid electrolyte of *phosphoric acid fuel cells* is impregnated into a porous inert matrix, on both sides of which are placed thin gas diffusion electrodes having a total precious metal catalyst loading of no more than 0.75 mg cm^{-2}. On the cathode side, this catalyst contains a number of additions, including Cr, Mo and Ga, and for both electrodes, carbon current collectors are also used to channel the gases to the electrodes as shown in Fig. 9.24; the current-voltage curve for this type of fuel is shown in Fig. 9.25. As in the solid- polymer cell, CO_2 does not present a problem in the acidic electrolyte, and at the operating temperature of ca. 200°C, CO is also quite rapidly oxidised at the platinum anode, permitting a rather higher concentration of CO in the anode feed gas to be tolerated (up to ca. 1%).

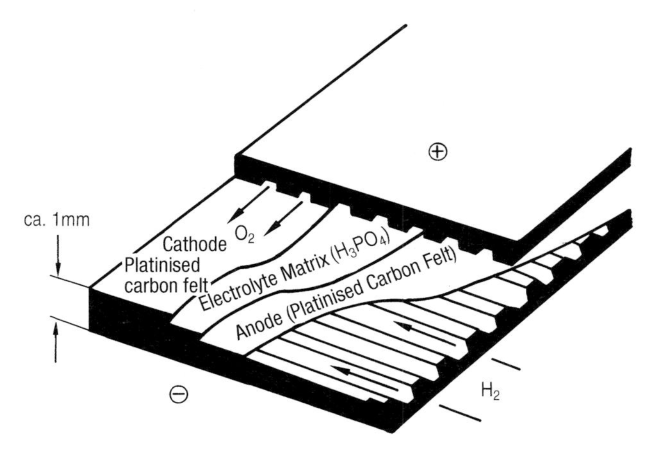

Fig. 9.24 Arrangement of the electrolyte and electrodes in a PAFC.

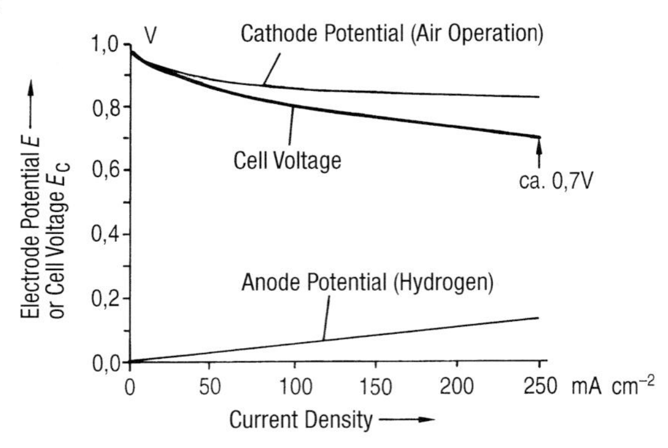

Fig. 9.25 Current-potential behaviour for the two electrodes of a PAFC and the nett current-voltage characteristic for the complete cell.

The advantages of PAFCs lie in the low loading of precious-metal catalyst needed, and the cheapness and relative simplicity of the cell construction. Small commercial power units, of up to 11 MW, have been installed in various places around the world in order to provide market testing, and combining the electrical energy with heat energy produced by the cell allows efficiencies of up to 80 % to be attained. However, the cost remains uncompetitive: current installation costs for a complete system, including gas reformer train, are ca. \$2,000 (~£1,500) per kW, whereas a corresponding gas-turbine system operating in Combined Heat and Power (CHP) mode can achieve 85 % efficiency at about half the cost.

The construction of the *Molten Carbonate Fuel Cell* consists of a ceramic matrix (usually $LiAlO_2$) impregnated with a Li_2CO_3/K_2CO_3 eutectic to which Na_2CO_3 can be added. The matrix lies flat on a porous nickel anode, allowing some of the molten carbonate to be drawn down by gravity into the nickel. The cathode is porous $Li_xNi_{1-x}O$, into which the molten carbonate is drawn by capillary action to form a three-phase boundary, and the cell reactions at the operating temperature of 600°C are:

Anode $$H_2 + CO_3^{2-} \rightarrow H_2O + CO_2 + 2e^- \quad (9.37)$$

$$CO + CO_3^{2-} \rightarrow 2CO_2 + 2e^- \quad (9.38)$$

Cathode $$1/2O_2 + CO_2 + 2e^- \rightarrow CO_3^{2-} \quad (9.39)$$

Note that this implies the necessity of circulating CO_2 from anode compartment to cathode compartment. A major advantage of the higher temperature of operation is the ability rapidly to oxidise CO, allowing a much simpler reformer train. In fact, at 600°C, the reformer process can in principle take place directly in the cell, with the heat generated by the cell being used to offset the endothermic reformer reaction, a process termed *internal reforming*.

The higher operating temperature of the MCFC leads to a reduction in the thermodynamic voltage to 1.04 V, and achievable voltages under practical operation are estimated to be in the range 0.8 – 0.9 V. In combination with internal reforming, this gives an efficiency of some 60 % for natural gas to electrical power, a value that can be enhanced by combined cycle operation using the waste heat from the cell. With coal as fuel, the direct efficiency is some 50 %, which again can be raised by combined cycle operation.

The *solid oxide fuel cell* represents a further development of fuel cell technology, in which the electrolyte is an oxide-ion conducting ceramic, such as Y_2O_3(15 %)/ZrO_2 (YSZ) in the form of a plate or tube, covered with porous catalyst layers on both sides. Early forms of the cell used cylindrical tubes of ceramic and operated at 1000°C to achieve adequate ceramic conductivity, as shown in Fig. 9.26a, but more recent cells have had a flat configuration, as shown in Fig. 9.26b. In this configuration, operation at a lower temperature of 800°C has proved necessary to solve problems associated with cell sealing and the different thermal expansion coefficients of the cell components. This lower temperature, in turn, demands the use of very thin electrolyte layers to ensure adequate internal cell conductivity. Research has been carried out into improved electrolytes, which must: (a) show high oxygen ion

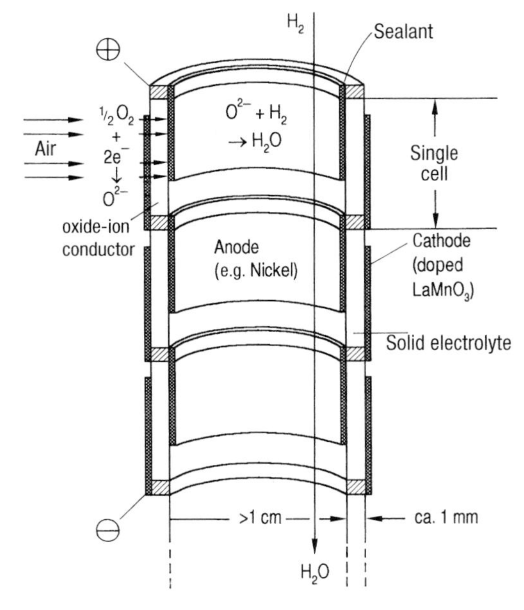

Fig. 9.26a Schematic arrangement of a SOFC with tubular construction.

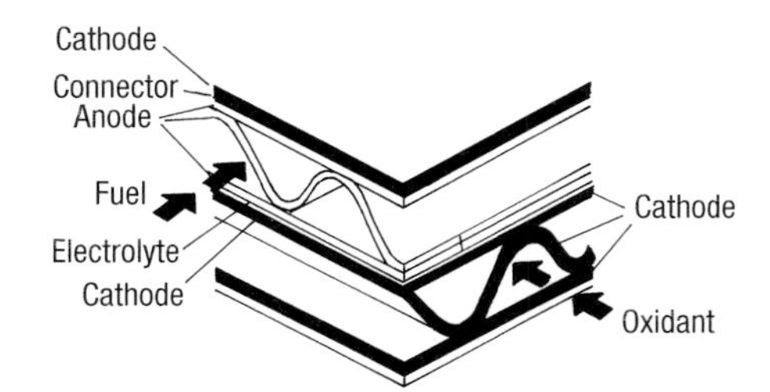

Fig. 9.26b More recent flat configuration for SOFC operation.

conductivity and minimal electronic conductivity; (b) good chemical stability with respect to the electrodes and inlet gases; (c) have a high density to inhibit fuel crossover; (d) have a thermal expansion compatible with other components. Materials such as Bi_2O_3 and doped-CeO_2 do show higher oxide-ion conductivity at lower temperatures than stabilised zirconia, and recent suggestions have included LaGa-based perovskites and Gd-doped ceria, both of which show better oxide-ion conductivity at low temperatures, but in the former case show reactivity to nickel and in the latter case rather undesirable electronic conduction and thermal expansion properties.

The anode catalyst is nickel, which is incorporated into a YSZ-matrix to ensure matching of the expansion coefficients between electrolyte and electrode. This electrode structure, which shows both ionic and electronic conductivity, is termed a "cermet", and a similar structure is used for the cathode. The presence of high-surface-area nickel leads to problems with carbon deposition particularly when internal reforming or direct oxidation of methane takes place; this can be ameliorated by incorporating ceria into the anode, which has a high activity for oxidising hydrocarbons, but its low inherent conductivity means that other metals such as copper must be added. The main difference in the cathode structure is the use of electronically conducting oxides such as $LaMnO_3$, $La_{1-x}Sr_xMnO_3$ or $LaCoO_3$ as the catalysts. These catalysts, however, do not show a satisfactory activity at 800°C, and increasing activity by modification to the materials (such as the recent introduction of LaSrCr manganates) and/or the addition of a small loading (ca. 0.5 mg cm^{-2}) of finely divided platinum are essential, especially with the drive to lower temperature operation to solve major materials problems associated with sealing and interconnect. Finally, both anode and cathode are connected to the electrolyte through so-called "interconnection materials", comprised of mixed oxides, such as $La_{1-x}Sr_xMnO_3$, $La_{1-x}Sr_xCrO_3$ and $CoCr_2O_4$. The final cells are currently capable of delivering very high current densities, ca. 1 A cm^{-2}, at a cell voltage of ca. 0.6 V at the higher temperature end.

9.5.2.1 **Summary of Fuel-Cell Characteristics using Gaseous Fuels**

The table below summarises current data on the fuel cells listed above; details of the reaction mechanisms are to be found in sections 6.2 and 6.3. The data are listed in the following order:

1. Working temperature
2. Electrolyte
3. (a) Anode; (b) Cathode materials
4. Additional construction materials
5. Main additional components
6. Oxidant, primary fuel and cost per kWh
7. Status
8. Current costs
9. Uses
10. Fuel conversion efficiency of: (a) cell alone; (b) cell and reformer
11. Problems
12. Countries where development is taking place
13. Remarks

Alkaline Fuel Cells (AFC)

1. Up to 90°C; Low temperature system
2. 30 % KOH
3. PTFE-bonded carbon with noble-metal loading
4. Polymer materials
5. Water evaporator
6. Pure O_2; Pure H_2: ca. £0.5 per kWh
7. Commercial units available, up to ca. 100 kW.
8. £1,500 – £2,500 per kW
9. Specialist markets such as military and space uses.
10. (a) up to 55 %
11. Price
12. USA, Canada, Germany
13. Mature technology; Feed gases should only contain small quantities of CO and CO_2.

Solid Polymer Fuel Cells (SPFC)

1. Up to 80°C; Low temperature systems
2. Proton-conducting polymer membrane
3. Thin noble metal layers that may be dispersed on carbon
4. Polymeric materials
5. Water separator
6. Oxygen or air; pure H_2: ca. £0.5 per kWh
7. Commercial units available, up to 500 kW
8. £2,500 – £5,000 per kW for civilian uses
9. Specialised markets: military uses, particularly in submarine operation; Space flight
10. (a) 55 %
11. Price
12. USA, Canada, Germany
13. For operation without CO-tolerant anodes the CO content of the fuel must lie below 100 ppm.

Phosphoric Acid Fuel Cells (PAFC)

1. Up to 220°C; Medium temperature system
2. Concentrated phosphoric acid in a support matrix
3. Graphite felt with low noble-metal loadings
4. Polymeric materials
5. Water separation; heat exchanger; reformer
6. Oxygen or air; H_2, methane, natural gas; with a reformer, costs are ca. £0.05 per kWh
7. Commercialisation just commencing for smaller units (200 kW); prototypes up to 11 MW constructed
8. £2,000 per kW
9. Local residential power supplies (200 kW); Small distributed power supplies (up to 11 MW)

10. (a) 55 %; (b) 40 %
11. Reliability; Lifetime; Maintenance costs; Price
12. USA, Japan
13. CO content of the fuel gas must lie below ca. 1 %

Molten Carbonate Fuel Cells (MCFC)

1. Up to 650°C; (high-temperature system)
2. Molten Li/K carbonate in a ceramic tile
3. (a) Nickel; (b) Lithiated nickel oxide
4. Ceramic; Steel
5. Water evaporator; Heat exchanger; reformer; combined cycle possibility for heat usage
6. Oxygen or air; Hydrogen, methane or natural gas With a reformer, costs are ca. £0.05 per kWh
7. Experimental system; prototypes of up to 2 MW
8. Overall costs not yet available
9. Power station; second generation design. Also for load-levelling
10. (a) 55 – 65 % (b) up to 55 %
11. Stability of electrodes and electrolyte matrix; costs must be brought down to £500–£1,000 per kW to match the costs of conventional power stations
12. USA, Japan, Holland, Italy, Germany
13. There are still considerable development problems to be overcome. If internal reforming can be introduced, this will increase the efficiency quoted in 10(b) by some 5 percentage points. By contrast, the use of coal as the primary fuel lowers this efficiency by the same amount

Solid Oxide Fuel Cells (SOFC)

1. Up to 1000°C; High-temperature system
2. Oxide-ion conducting ceramic, conventionally $ZrO_2.15\,\%Y_2O_3$
3. (a) Nickel; (b) Sr-doped $LaMnO_3$
4. Ceramics; steel
5. Water evaporator; heat exchanger; reformer; possibility of using combined cycle systems to exploit high-grade waste heat.
6. Oxygen or air; hydrogen, methane or natural gas Costs in the region of £0.05 per kWh with the use of reformer hydrogen
7. Experimental system, with prototype systems up to 200 kW
8. Overall costs not available
9. Stationary power systems (3rd Generation), Power supply to motor vehicles
10. (a) 55 – 65 %; (b) 55 %
11. Electrode activity at lower temperatures; sealants; ohmic resistance of electrolyte at lower temperatures
12. USA, Japan, Germany, UK
13. Many problems still remain to be overcome in the development stages; big push to get the operating temperature down to 650°C.

9.5.3
Fuel Cells with Liquid Fuels

Costs and engineering complexity are the main problems in the further exploitation of fuel cells, and to reduce both, attention has focussed on the possibility of replacing hydrogen as the fuel by a liquid that can either be introduced directly to the anode or is highly soluble in the electrolyte. Instead of gas storage, or liquid storage with a reformer, it would suffice to have a simple conventional fuel tank feeding directly to the stack. For simplicity, such a cell would use air as the combustant, and the design principles can then be similar to those shown in Fig. 9.21, with a porous anode being fed from the back by the fuel and with products conducted away from the back. An alternative is to place a two-phase electrode in the electrolyte space between two oxygen electrodes; in this case, the fuel must be soluble in the electrolyte. It is also necessary, since contact between the fuel and the *cathode* can take place, to prevent any back-reaction taking place at the cathode, giving rise to a mixed potential. This can in principle be secured by the use of oxygen electrocatalysts that are selectively *inactive* for the fuel. Such selective electrodes can also be used in the some of the cell designs discussed above, with the further advantage that single-chamber operation becomes possible, with concomitant simplification of cell design.

The first liquid-fuel stacks were constructed about thirty years ago for military purposes, and used hydrazine as the fuel and KOH as the electrolyte, with the anode reaction:

$$N_2H_4 + 4OH^- \rightarrow N_2 + 4H_2O + 4e^- \qquad (9.40)$$

This gave a theoretical energy density of 3850 $Whkg^{-1}$, and 10 kW prototypes achieved an energy density of 500 W kg^{-1}. Other fuels that were investigated at this time included simple liquid CHO compounds (methanol, CH_3OH; glycol, CH_2OH-CH_2OH; formaldehyde, HCHO; formic acid, HCOOH) and even alkali-metal amalgams, the driving force for the latter being the interest in replacing hydrogen as the by-product in the chlor-alkali process by electricity.

For civilian use, hydrazine is simply too toxic, and attention has focussed on the liquid CHO compounds. Using such compounds in alkaline electrolytes leads, however, to the problem that carbonates are formed, and a part of the electrolyte is consumed as well. For the case of methanol oxidation, for example, which is relatively facile in alkaline solution, the reaction at the anode, which in acid solution is:

$$CH_3OH + H_2O \rightarrow CO_2 + 6H^+ + 6e^- \qquad (9.41)$$

now becomes:

$$CH_3OH + 8OH^- \rightarrow CO_3^{2-} + 6H_2O + 6e^- \qquad (9.42)$$

which can also lead to deposition of the relatively insoluble K_2CO_3 unless the carbonate can be removed. Attention has therefore focussed on the use of acid electrolytes, and most recently on the use of SPE-membranes as electrolyte and methanol as the fuel: the *Direct Methanol Fuel Cell* (DMFC). Methanol can be easily and cheaply synthesised from primary sources, and has a theoretical energy density of 6 kWh kg^{-1}, and it has become the most carefully studied of the liquids listed above. The anode reaction is given in (9.41) and the corresponding cathode reaction is oxygen reduction, giving a nett cell reaction:

$$CH_3OH + 3/2O_2 \rightarrow CO_2 + 2H_2O$$

The anode reaction involves at least three pathways in parallel. CO, formaldehyde and formic acid are formed as the main intermediaries (see chapter 6.4), and conventional platinum electrodes have to be replaced by CO-tolerant catalysts: PtRu alloys and ternary alloys have been found to be sufficiently active for reasonable performance at temperatures above $\sim 60\,^{\circ}C$.

The necessary complete cell design for a DMFC with polymer membrane as electrolyte is shown in Fig. 9.27. The water demand at the anode, which arises because of equation 9.41 and the electro-osmotic migration of protons through the membrane from anode to cathode, is regulated by the corresponding admission of methanol/water mixtures in the fuel supply. Methanol and a considerable proportion of the CO_2 generated at the anode can also diffuse though the porous membrane to the cathode side. If conventional platinum-based catalysts are used for oxygen reduction, then such catalysts can also oxidise methanol at the same time as reducing oxygen allowing a nett current to flow even in the absence of any external loading on the cell. The cathode forms a *mixed potential* which lies more than 100 mV more negative than that encountered in the absence of this *crossover* effect.

Clever circulation design of the reactants permits the construction of reliable DMFC systems. One example is shown in Fig. 9.28. With the help of convection pumps and installed electronics, the design temperature of 60°C can be maintained automatically with a current density of 200 mA cm^{-2}. The complete system offers a nett output of 1 Wh cm^{-3} of methanol feed with the use of 1 mg cm^{-2} noble metal on the electrodes. A comparison of the efficiency of a single cell for the temperature range 90 – 120°C is carried out in Fig. 9.29.

With, for example, glycol as a fuel, clearly lower electrical performance is obtained, and the use of ethanol as a fuel has suffered up to now from its incomplete conversion to CO_2 with acetic acid as a major by-product.

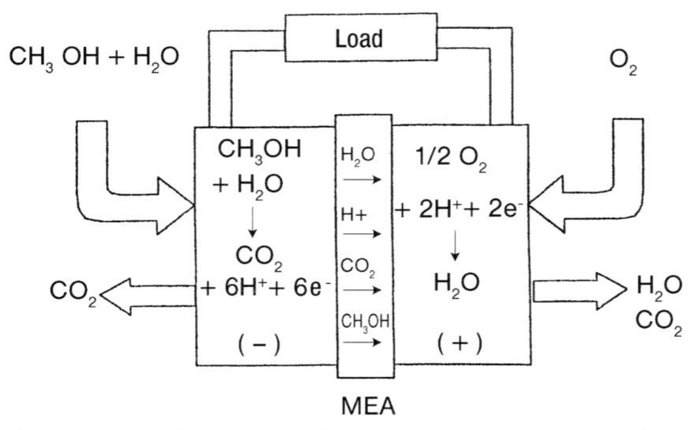

Fig. 9.27 Reaction scheme of a methanol/oxygen (air) cell (DMFC); the diffusion of methanol through the membrane is termed *crossover.*

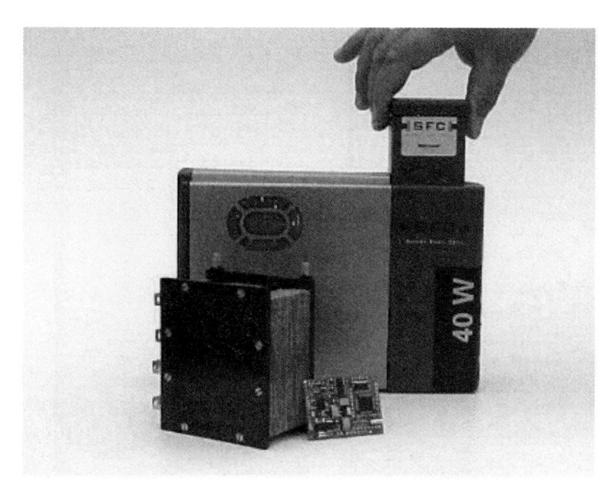

Fig. 9.28 Complete DMFC-system from Smart Fuel Cell AG (Brunnthal, Germany), incorporating a 2.5 L methanol tank and showing a 12 V stack and electronic circuit diagram in the foreground.

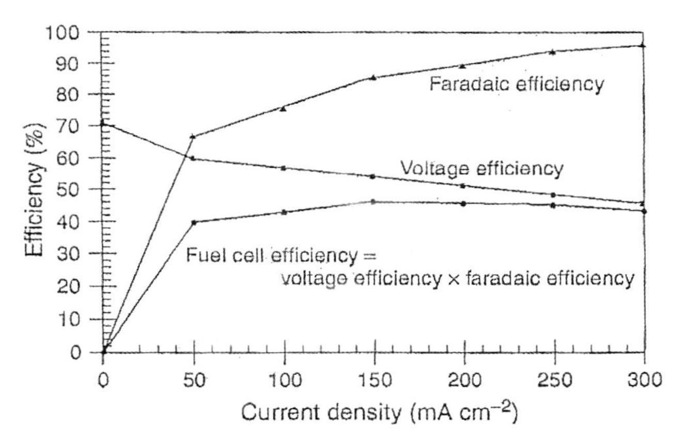

Fig. 9.29 Efficiency in percent of a direct methanol fuel cell (DMFC) with reference to faradaic current, cell voltage and electrical load at 90 – 120°C [J. Muller, G. Frank, K. Colbrow and D. Wilkinson, "Transport/kinetic limitations and efficiency losses", in Handbook of Fuel Cells, vol. 3, eds. W. Vielstich, H. Gasteiger and A. Lamm, Wiley-UK. Chichester, 2003].

9.6 Primary and Secondary Air Batteries

These systems normally consist of an air cathode and, as the anode, a primary or secondary electrode. The advantage of such batteries is that air, as the positive active mass, is cost-free, available in unlimited quantities, and does not contribute to the weight of the battery.

9.6.1. Metal-Air Primary Batteries

Most familiar are systems based on zinc/alkali, which were originally developed during the first-world war:

$$\text{Anode}: \quad Zn + 2OH^- \rightarrow Zn(OH)_2 + 2e^- \tag{9.43}$$

$$\text{Cathode}: \quad 1/2O_2 + H_2O + 2e^- \rightarrow 2OH^- \tag{9.44}$$

$$\text{Nett Reaction}: \quad Zn + 1/2O_2 + H_2O \rightarrow Zn(OH)_2 \tag{9.45}$$

This cell usually has an experimental open-circuit voltage in the range 1.45–1.5 V and a theoretical energy density of 960 Wh kg^{-1}. Its construction is similar to the zinc-manganese cell described earlier in this chapter, with the central MnO_2/graphite paste replaced by a gasdiffusion electrode in the form of a cylinder of porous hydrophobic activated carbon connected to an external air supply through its top surface. The diffusion path of air through the electrode is quite long, and the achievable current density is limited to ca. 10 mA cm^{-2}, making any further catalytic activation of the carbon unnecessary. This low current density restricts such cells to applications requiring long-term provision of low currents, including emergency lighting in buildings and electric fences for cattle control.

The development of much higher current-density air electrodes as part of the fuel-cell programme has led to the production of zinc-air cells of much better performance. In such modern cells, the zinc and air electrodes are interchanged, with the central zinc electrode being surrounded by a porous-film air electrode, the two being kept apart by a separator impregnated by aqueous KOH. The cell construction is such that both metal anode and electrolyte can be easily replaced after discharge, giving mechanically rechargeable cells. Larger cells of this type, up to 100 Ah capacity, have been developed mainly for military use; they can reach energy densities of 200 Wh and short-term power densities of 80 W kg^{-1}. Commercial uses include batteries for hearing aids and watches.

In addition to zinc, magnesium and aluminium have been developed in prototype systems. The low atomic weight and high valence of aluminium make it particularly attractive, with a theoretical specific capacity of 3000 Ah kg^{-1}. In alkaline solution, pure aluminium shows strong corrosion, with extensive hydrogen evolution, but this can be reduced to acceptable limits at high current loading. However, in spite of the good energy and power density of aluminium-air batteries, they have not been developed successfully for traction using KOH as electrolyte owing to the difficulty of separating and working up the colloidal $Al(OH)_3$ product. More promising appears to be the use of neutral solutions, such as sea water. Here, the $Al(OH)_3$ product forms a gel, and metal and electrolyte must be replaced mechanically after discharge of the cell. By using open cells in sea water, the electrolyte can be replaced continuously, allowing such cells to be recharged by replacement of the aluminium alone. Interestingly, in neutral solution, aluminium passivates, owing to the formation of a insulating oxide layer; it has been found, however, that small amounts of alloyed base metal, such as In, Ga, Tl, Zn or Cd can prevent this passivation (as indeed can amalgamation with mercury). Such Al-air batteries may find application in sea and coastal regions, particularly for emergency back-up systems.

9.6.2. Metal-Air Secondary Systems

The use of zinc, cadmium or iron in alkaline solution allows, in principle, the recharge of a metal-air system giving a secondary battery. The problem is the stability of the oxygen electrode: during the charging process, oxygen is evolved and escapes, and this process can lead to the destruction of the catalytic centres active for oxygen reduction, and to a drastic lowering in cell performance. One way forward is to have a composite air electrode, in which a second porous layer is added with a catalyst having a lower overpotential for oxygen evolution than the active layer for oxygen reduction. In order to prevent too high a potential developing on the oxygen-reduction layer, the oxygen evolution catalyst layer must be placed on the electrolyte side of the cell; this allows, for example, the use of catalysed activated carbon as the oxygen reduction layer and porous nickel as the oxygen evolution material. The properties of a zinc-air secondary battery are shown in Fig. 9.30; experimental systems of this type show energy densities near 100 Wh kg^{-1} and power densities of ca. 100 W kg^{-1}, suggesting that these systems may be suitable for electrical traction.

An important development for such cells would be true bifunctional catalysts, and the SOFC programme has shown that such materials as $La_{0.6}Ca_{0.4}CoO_3$ dispersed on Vulcan XC-72 carbon may have the desired properties. As the anodes, zinc-paste electrodes formed from ZnO/PTFE/PbO/cellulose have been developed: the cellulose helps suppress the formation of dendrites, and such cells have shown more than 100 charge/discharge cycles (charging voltage 2 V; discharge voltage 1.2 V) at constant capacity. A power density of 100 mW cm^{-2} appears achievable.

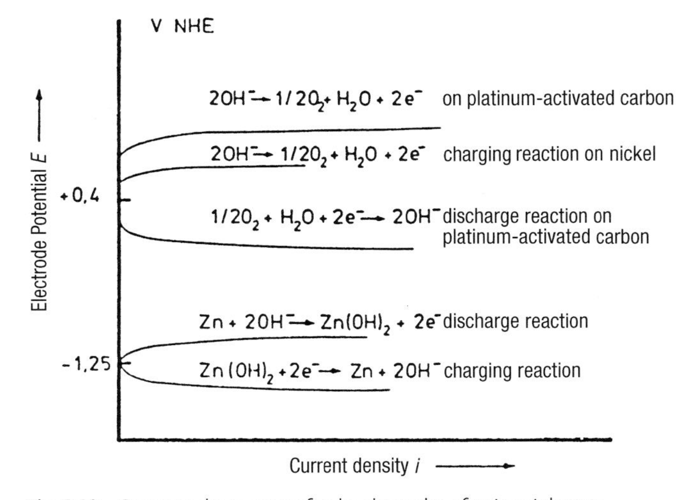

Fig. 9.30 Current-voltage curves for the electrodes of a zinc-air battery.

9.7 Efficiency of Batteries and Fuel Cells

The thermodynamic or ideal efficiency of a battery or fuel cell is the ratio of the maximum otainable electrical energy

$$-nFE_c^0 = \Delta G = \Delta H - T\Delta S \tag{9.46}$$

to the reaction enthalpy, ΔH, of the reaction:

$$\xi_{th} = \frac{\Delta G}{\Delta H} = 1 - \frac{T\Delta S}{\Delta H} \tag{9.47}$$

The thermodynamic efficiency can, therefore, depending on the sign of ΔS, be smaller than, equal to or greater than unity. For the common situation of $\Delta G, \Delta H < 0$, $\xi_{th} > 1$ for $\Delta S > 0$.

If the "thermal" cell voltage is defined as $E_H^0 = -\Delta H/nF$, then

$$\xi_{th} = \frac{E_c^0}{E_H^0} \tag{9.48}$$

and generally, for $\Delta S < 0$, ie $|\Delta G| < |\Delta H|$, then $E^0 < E_H^0$, with the reverse true for $\Delta S > 0$. As an example, for the fuel cell reaction $H_2 + 1/2O_2 \rightarrow H_2O$, $E_c^0 = 1.23$ V and $E_H^0 = 1.48$ V at room temperature, giving $\xi_{th} = 0.83$. For the methanol fuel cell described above, $\xi_{th} = 0.97$ for liquid methanol fuel. The relationships between the parameters discussed above is shown for $\Delta S < 0$ in Fig. 9.31 .

Once the cell is under load, the efficiency will fall: including the internal resistive loss in the electrolyte as an overpotential $\eta_{res} = iR_E$, and taking into account the activation and diffusion overpotentials at the two electrodes (η_P in Fig. 9.31) the *load* efficiency, ξ_{load}, is defined with respect to the actual cell voltage $E_{cell}(i)$ at current i as:

$$\xi_{load} = -\frac{nFE_c(i)}{-nFE_H^0} = -\frac{nF((E_c^0 - \sum|\eta(i)|)}{-nFE_H^0} = \frac{(\Delta G + nF\sum|\eta(i)|)}{\Delta H} \tag{9.49}$$

For the alkaline H_2/O_2 fuel cell (AFC), for example, technically interesting current densities can be obtained at cell voltages of ca. 0.8 V, giving a load efficiency of $0.8/1.48 = 0.54$.

The *actual or effective* efficiency, ξ_{eff}, corrects for the usage of active mass in the battery or fuel cell: in the case of the AFC, at least 95 % of the gases are used, giving $\xi_{eff} = 0.95 \times 0.54 = 0.513$. Further losses will include the provision of power to fuel supply systems, voltage regulation etc., accounting for a further 10 % loss, giving a final efficiency for the AFC at 0.8 V of 0.46.

The nett heat balance per unit time of a cell can be written:

$$W = \frac{iT\Delta S}{nF} + i\sum|\eta(i)| + i^2R_E = i(E_H^0 - E_c^0) + i\sum|\eta(i)| + i^2R_E \tag{9.50}$$

where we have included the internal cell resistance explicitly. The units of W are watts, and W is always positive. Its value is of great importance in engineering design.

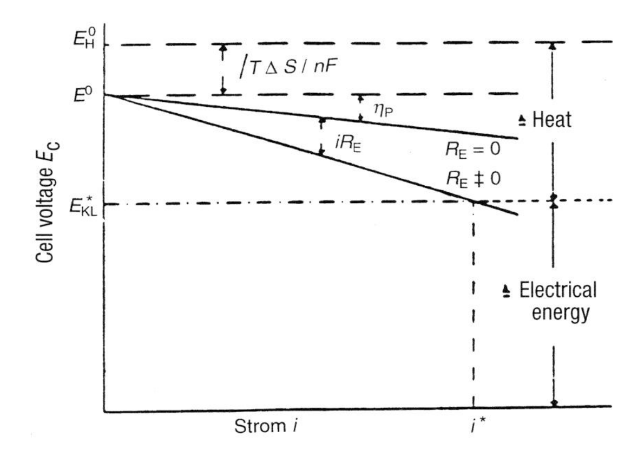

Fig. 9.31 Dependence of the cell voltage of a galvanic cell (with reaction entropy $\Delta S < 0$) on the current load, $i \cdot R_E$ is the electrolyte resistance, η_P the overpotentials at the two electrodes. The other symbols are defined in the text.

The efficiency of a battery or fuel cell may be compared to the thermodynamic efficiency of a heat engine, such as an internal combustion engine, which has the value:

$$\xi_{HE} = \frac{\text{Temperature difference between hot and cold sinks}}{\text{Temperature of hot sink}} = \frac{T_h - T_c}{T_h} \tag{9.51}$$

Typically, the theoretical efficiency for an internal combustion engine, the Carnot factor, is ca. 0.5, but in city driving, for example, this can drop to as low as 0.1. The advantage of a fuel cell is thus its ability to convert chemical to electrical energy with high efficiency, particularly under part load.

9.8 Super-capacitors

A super-capacitor is, in terms of its function, intermediate between a rechargeable battery and an electrostatic capacitor. For any capacitor

$$C = q/\Delta\varphi = Q/V \tag{9.52}$$

where the capacitance, C, is the ratio of the charge q on the electrode or, equivalently, the charge Q placed on the positive plate (which induces an equal and opposite charge on the negative plate) divided by the potential difference across the double layer, or equivalently the capacitor voltage between the plates. There is a linear relationship between charge Q and voltage V, and the energy E_C stored in the capacitance is

$$E_C = \int V(Q)dQ = \frac{1}{C}\int QdQ = \frac{1}{2}Q^2V/C = \frac{1}{2}CV^2 \tag{9.53}$$

For an *ideal* secondary battery, in comparison, the voltage V would remain constant during charge and discharge, and the stored energy $E_B = QV$. The electrostatic capacitor is characterised by the charging of two plates lying opposite to each other and separated by a dielectric. The electrochemical capacitor, consisting of two porous layers on either side of an ionically conducting electronically insulating layer, corresponds in its construction (as shown in Fig. 9.32) to the membrane-electrolyte-assembly of an SPE electrolyser (Fig. 8.20) or PEM fuel cell. Critical to the functioning and dimension of these capacitors is the reversibility and magnitude of charge stored as a function of potential. The performance that can already be obtained from a pure double-layer capacitor with porous carbon layers is shown in Fig. 9.33. The classic example of a reversible redox-pseudo-capacitance of a super-capacitor is hydrated RuO_2 in sulphuric acid.

Up to 4 F cm^{-2} can be stored, corresponding to an energy density of 5 Wh/litre. Interestingly, when used for short periods (in the order of milliseconds), a power storage capacity of 10^5 W/litre is attainable. Applications of super-capacitators as additional short-time power sources are suggested, e.g., for different kinds of automobils.

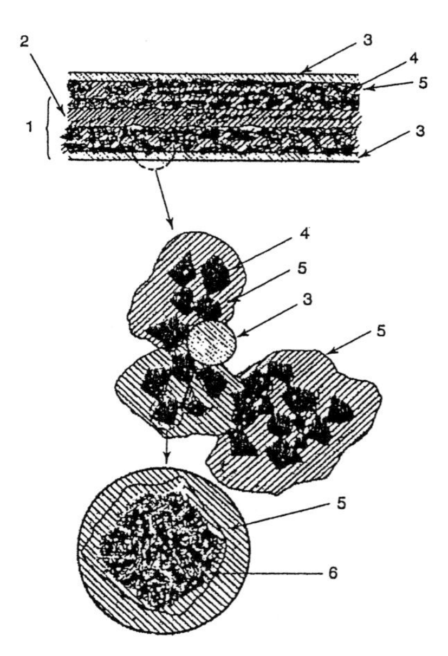

Fig. 9.32 Electrochemical capacitor with polymer membrane as electrolyte: (1) membrane-electrolyte-assembly (MEA); (2) membrane; (3) Current collector; (4) Metal-oxide particles; (5) polymer electrolyte; (6) porous support [J.A. Kosek, B.M. Dweits and A.B. LaConti, "Technical characteristics of PEM Electrochemical capcitors", in Handbook of Fuel Cells, vol. 2, eds. W. Vielstich, H. Gasteiger and A. Lamm, Wiley-UK. Chichester, 2003].

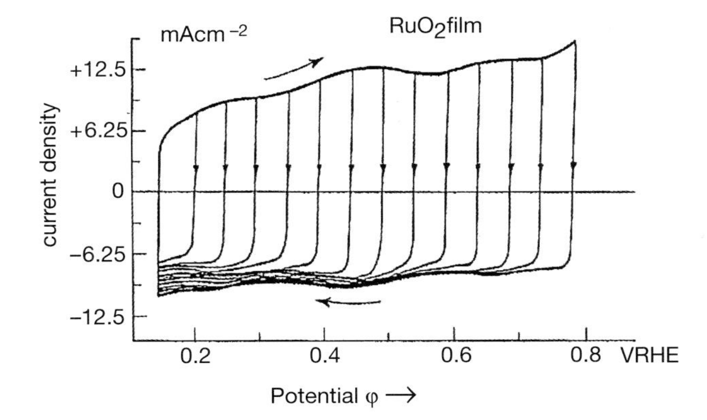

Fig. 9.33 Surface related capacitance of supported carbon fibres in a double-layer-capacitor with 1M H_2SO_4; true surface area 2,500 m^2g^{-1} [B.E. Conway, V. Birss and J. Wojtowicz, J. Power Sources **66** (1997) 1 and B.E. Conway, "Electrochemical Supercapacitors", Kluwer Academic-Plenum Press NY, 1999].

References for Chapter 9

General References to batteries and fuel cells

C.A. Vincent and B. Scrosati: "Modern Batteries", 2nd Edition, Arnold, London, 1997.

D. Linden and T. Reddy (eds): *Handbook of Batteries*, McGraw-Hill, New York, 2001.

I.O. Besenhard (ed.): *Handbook of Battery Materials*, Wiley-VCH, Weinheim, 1988.

D. Berndt: "Galvanic Cells: Primary and Secondary Batteries", in Ullmanns' Encyclopaedia, **A3** (1985) 343.

Philosophical Transactions of the Royal Society, **354** (1996) pp. 1513–1712; "Materials for Electrochemical Power Systems".

References to Specific Battery Systems

G. Nazri and G. Pistoia (eds.), *Lithium Batteries, Science and Technology*, Kluwer Academic Publishers, New York, 2004.

K.V. Kordesch: *Batteries, vol. 1: Manganese Dioxide*, Marcel Dekker Inc., New York, 1974.

K.V. Kordesch: *Batteries, vol. 2: Lead-acid Batteries and Electric Vehicles*, Marcel Dekker Inc., New York, 1977.

H. Tuphorn: "Sealed Maintenance-Free Lead-Acid Batteries", J. Power Sources **23** (1988) 143.

T. Allmeidinger: "Electrochemical Energy Storage" in Chem. Ing. Tech. **63** (1991) 428.

J.F. Cole: J. Power Sources **40** (1992) 1.

A. Fleischer and J.J. Lander (eds.): *Zinc-Silver oxide Batteries*, J. Wiley, New York, 1971.

F. von Sturm: "Silver-Zinc Scondary Batteries" in *Comprehensive Treatise of Electrochemistry*, J.O'M. Bockris, B.E. Conway, E. Yeager and R.E. White (eds.), Plenum Press, New York, vol. **3**, 1981, p. 407

S. Ruben: "Sealed Mercury Cathode Dry Cells" in *Comprehensive Treatise of Electrochemistry* eds. J.O'M. Bockris, B.E. Conway, E. Yeager and R.E. White (eds.), Plenum Press, New York, vol. **3**, 1981, p. 233.

C.Fabjan and K.V. Kordesch: "Conception, development and possibilities for the zinc-halogen storage system" in Dechema Monographs **109** 335 (1987) and **124** 455 (1991).

J.L. Sudworth and A.R. Tilley: *The Sodium Sulphur Battery*, Chapman and Hall, New York, 1985.

S. Mennicke: "Electrolyte and Electrodes for the Na/S-battery", Dechema Monograph **109** 409.

C.-H Dustmann, "Advances in ZEBRA Batteries", J.Power Sources **127** (2004) 85.

M. Wakahira and O. Yamamoto (eds.): *Lithium-Ion Batteries*, Wiley-VCH, Weinheim, 1998.

G. Mamantov and A.J. Popov (eds.): *Chemistry of Non-aqueous Solutions*, VCH, Weinheim, 1994.

G.W. Heise and N.C. Cahoon (eds.): *The Primary Battery*, J. Wiley and Sons, New York, 1970.

J.P. Gabano ed.: *Lithium Batteries*, Academic Press, New York, 1983.

J.L. Sudworth: "ZEBRA Batteries" in J. Power Sources **51** (1994) 105.

H. Cnobloch, H. Nischik, K. Pantel, K. Ledjeff. A. Heinzel and A. Reiner; "Iron-Chromium Redox Storage Systems" in Dechema Monographs **109** 427 (1987).

Specific References to Fuel Cell Systems

W. Vielstich, H. Gasteiger and A. Lamm (eds.), *Handbook of Fuel Cells, Fundamentals,Technology and Applications*, Wiley UK. Chichester, 2003.

K.V. Kordesch and G. Simader: "Fuel Cells and their Applications", Wiley-VCH, Weinheim, 1996.

W. Vielstich and T. Iwasita: "Fuel Cells" in G. Ertl, H. Knözinger and J. Weitkamp (eds.): *Handbook of Heterogeneous Catalysis*, Wiley-VCH, Weinheim 1997.

A. Hamnett: Phil. Trans. Roy. Soc. **A354** (1996) 1653.

Berichte der Bunsengesellschaft für Physikalische Chemie, **94** No. 9, 1990.

A.J. Appleby and F.R. Foulkes: "Fuel Cell Handbook", Van Nostrand, New York, 1989.

L.J.M.J. Blomen and M.N. Mugerwa eds.: "Fuel Cell Systems", Plenum Press, New York, 1993.

K. Prater: J. Power Sources **51** (1994) 129.

A. Hamnett and G.L. Troughton: "The Direct Methanol Fuel Cell", in Chem. Ind. **1992** 480.

W. Vielstich: "Fuel Cells", Wiley Interscience, New York, 1970.

10
Analytical Applications

Electrochemical technology can be used in a wide variety of ways to contribute to the solution of analytical problems. It can be used to determine the position of the end-point of a titration through conductimetric measurements (Section 2.1.2), potentiometric measurements (Chapter 3), or through amperometric measurements (see below). Alternatively, measurement of the signal arising from an electrochemical process can be used directly to determine the amount of material present, as in polarography (Section 10.2). A third possibility is the use of electrochemical devices as sensors to monitor the concentrations of key species required in the control of chemical processes (Section 10.3).

Analytical applications of electrochemistry are found in many areas, including environmental monitoring, metallurgy, geology, pharmacy, medicinal chemistry and biochemistry, and include the quantitative determination of anions, cations and a wide range of inorganic and organic compounds, often in trace amounts. Electrochemical methods have the great advantage of being usable in highly coloured or opaque solutions, and can often give simultaneous recordings of several different species. In addition, the apparatus is often quite inexpensive, allowing extensive field measurements to take place; even where equipment is less portable, the number of steps necessary to treat the sample before measurement is usually smaller than for spectroscopic techniques.

This chapter cannot hope to cover all the existing methods and their variants in electro-analytical chemistry. Rather, the basis of a number of core techniques will be given, in order to help create a framework of knowledge to which will be added additional specific examples where appropriate.

10.1
Titration Processes using Electrochemical Indicators

The techniques of conductimetric and potentiometric titration have been summarised earlier. Amperometric titrations, however, depend on the existence of a diffusion-limited current as a measure of the concentration of a species involved in a titration process. In the simplest case, as the titration proceeds from the initial point (where f, the fractional progress of the titration is zero) to the end-point ($f = 1$), the limiting

Electrochemistry. Carl H. Hamann, Andrew Hamnett, Wolf Vielstich

ISBN: 978-3-527-31069-2

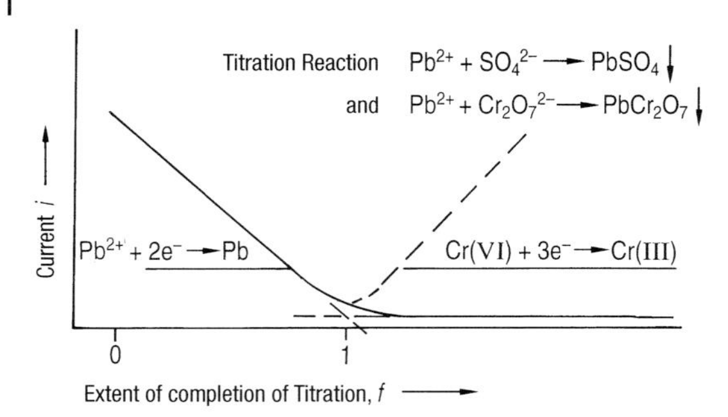

Fig. 10.1 Evolution of current in an amperometric titration with fractional progress of the titration, f, where $f = 1$ corresponds to the end-point of the titration.

current will decrease to close to zero. An example of this is shown in Fig. 10.1, where the limiting current for reduction of solution Pb^{2+} decreases towards zero with addition of sulphate ion as precipitant. Interestingly, if dichromate is used as precipitant, the limiting current rises again with addition of excess $Cr_2O_7^{2-}$ owing to reduction of the latter to Cr^{3+}. Where only the titrant is electrochemically active, the behaviour is rather different, with the limiting current very low until the end-point is reached, whereupon it rises strongly.

These measurements require constant hydrodynamic conditions (i.e. δ_N must be constant, see Chapter 4), and either rotating disc or micro-electrodes can be used. To avoid hydrogen evolution interfering with the measurement, a disc of glassy carbon covered with mercury can be used as the working electrode.

A general problem with this technique is that the end-point is frequently not well marked, and a variant technique, the biamperometric or "dead-stop-method", has been

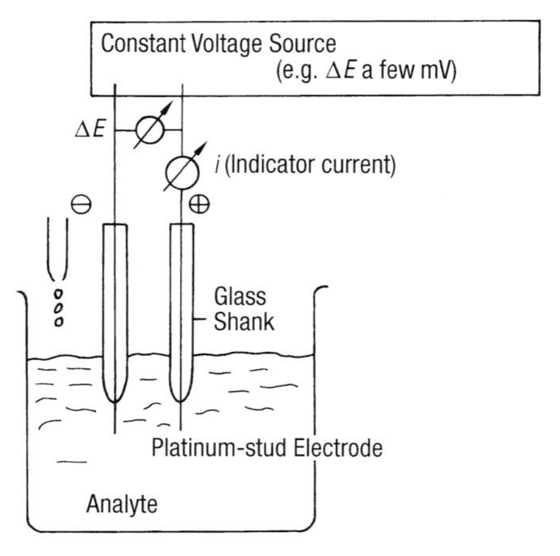

Fig. 10.2 Diagrammatic representation of the dead-stop process in amperometric titration. The electrodes are generally up to about 10 mm^2 in area.

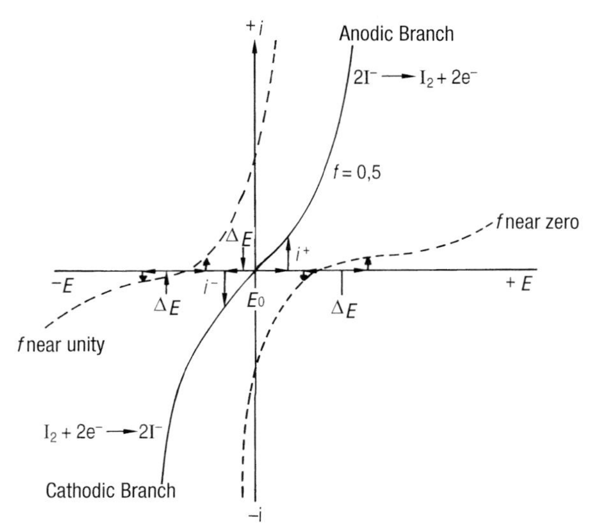

Fig. 10.3 Variation of current with potential for various values of f (see text) in dead-stop titrations.

developed, the principle of which is shown in Fig. 10.2. The method exploits the fact that for two electrodes connected in series as shown, the current flowing at both must be equal in magnitude and opposite in sign. The method is best exemplified with a concrete example: the behaviour of I^-/I_2 solutions. If the concentrations of I^- and I_2 are equal, then the current potential curve is shown as the continuous line in Fig. 10.3. In this case, connection of the two electrodes in series leads to the passage of current i^- at the cathode and i^+ at the anode, and the potential difference between the two electrodes, ΔE, is also shown in Fig. 10.3. For the titration reaction

$$I_2 + 2S_2O_3^{2-} \rightarrow 2I^- + S_4O_6^{2-} \tag{10.1}$$

close to $f = 0$, the same value of ΔE gives a much smaller current (see Fig. 10.3), and a similar situation exists for f close to one. At $f = 0$ and $f = 1$, no current will flow, since in the former case, there is no I^- to oxidise, and in the latter case, no I_2 remains to be reduced. The plot of current flowing for a fixed ΔE vs. f for the titration is shown in Fig. 10.4: clearly a maximum in the current is found for $f = 0.5$; note that although the potential difference between the two electrodes is fixed, the potential of the two electrodes with respect to a fixed reference will vary in the course of the titration.

An alternative is to fix the current flowing between the electrodes (usually in the region of a few $\mu A\ mm^{-2}$) and measure the changing value of ΔE. In the ideal case of Fig. 10.3, it is clear that ΔE will initially be large, since solvent decomposition will have to take place to supply an anodic reaction, and its value will decrease as f approaches 0.5. Above this value, ΔE will again increase, reaching a high constant value at the end point.

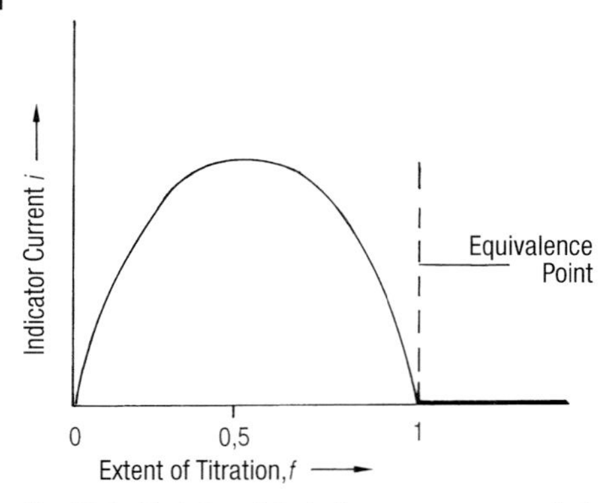

Fig. 10.4 Variation of the indicator current seen during a bi-amperometric or dead-stop titration with the fractional progress of the titration, *f*.

10.2
Electro-analytical Methods

The concentration of an electro-active substrate is mirrored directly in the limiting current, as indicated in the previous section, and this forms the basis of the important technique of polarography. In addition, we will describe analytical techniques (coulometry and electrogravimetry) based on a direct measure of the total charge required for electrochemical conversion, and one technique (chronopotentiometry) based on the transition time for such conversion.

10.2.1
Polarography and Voltammetry

Polarography and the more generic voltammetry in their different variants have remained the most familiar and frequently employed investigative tools in electrochemistry. The term voltammetry in electroanalytical chemistry refers to any method in which the dependence of current on electrode potential is measured. The term polarography, as normally used, is restricted to the use of a *mercury* working electrode, and, like cyclic voltammetry, is a special case of a potentiodynamic technique. Polarography has its origin in the fundamental investigations of the mercury-electrolyte interface carried out by J. Heyrovsky over a period of twenty years, and led to his winning the Nobel Prize for chemistry in 1960.

10.2.1.1 The Principles of dc Polarography

The principles of *dc* polarography are illustrated in the classical arrangement shown in Fig. 10.5; this shows a dropping mercury on which is imposed a linear negative-going potential ramp, usually of the order of 200 mV per minute. Mercury, as an electrode material, has two important advantages: the first is that the kinetics of the hydrogen

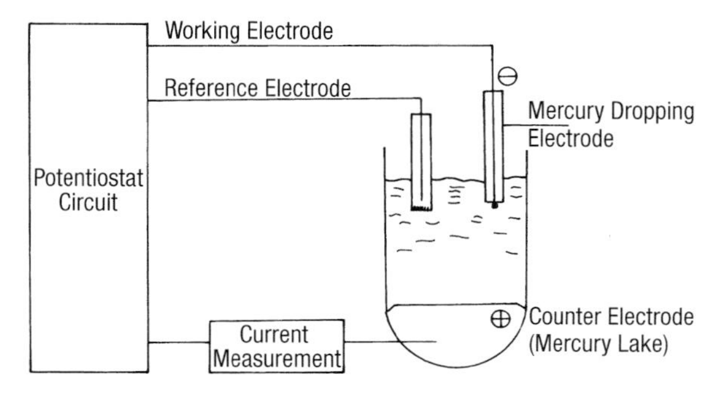

Fig. 10.5 Schematic representation of the set-up for dc polarography.

evolution reaction are extremely slow on mercury, allowing study of aqueous electrolytes in neutral or alkaline solutions over a wide, if rather negative, potential range, from ca. −2.0 to ca. 0 V vs. SCE, above which mercury dissolves. The second advantage is that by employing a simple capillary tube, as shown in Fig. 10.5, the liquid mercury forms drops with a lifetime of a few seconds. As each drop forms and falls from the electrode, the surface of the mercury is continually kept clean and free of adsorbed impurities, which would otherwise poison the mercury. Additional advantages include the facts that (a) currents through the drops (which are of the order of one mm diameter) are small (of order 100 μA), so the composition of the electrolyte does not change during a measurement, and (b) that mercury is a relatively inert metal, reacting chemically with few electrolytes.

In the original arrangement, the working dropping mercury electrode and the combined reference and counter electrode are immersed as shown in a chloride-containing electrolyte, and the potential difference between the two steadily increased. Since the area of the counter electrode is so much larger than the working electrode, its potential is essentially constant and determined by the Hg_2Cl_2/Hg couple. The effect of increasing the potential difference is then to drive the working electrode potential to steadily more negative values. Meanwhile, the current is constantly monitored. Since the drops are continually growing and then falling from the capillary, and the current is measured instantaneously without any smoothing, the measured current response actually consists of a series of equi-spaced maxima and minima giving a saw-tooth response.

Originally, the potential ramp was generated mechanically with a motor driven potentiometer, but in modern systems, the potential ramp is generated electronically, and a standard three-electrode system is used with a silver-silver chloride reference electrode and a mercury-sea counter electrode. The current is normally recorded with an x-t recorder. To avoid the sawtooth appearance of the earlier polarograms, the drops are usually forcibly dislodged from the capillary with a mechanical hammer, and the current may be measured for each drop immediately before it is removed.

An idealised polarogram is shown in Fig. 10.6, which shows the current response for an solution containing 1 M KCl as the supporting electrolyte and 0.001 M $ZnSO_4$. It is clear that immediately after the start of the negative-going potential sweep (for which the x-axis is conventionally reversed so that more negative potentials are shown to the *right-hand side*), little current passes until the potential reaches the region close to – 1.0 V where Zn^{2+} ions are reduced. The small current at more positive potentials (i.e. to the *left* of the *x*-axis) corresponds to the charging of the mercury double layer, but once the deposition potential of Zn is reached, the current rises rapidly until the diffusion-controlled limit is reached, giving rise to a diffusion current i_d. Only when the reduction potential of a further electrochemical process is reached (here reduction of K^+ ions to form a potassium amalgam) can the current rise further. There are thus two pieces of information to be derived from the polarogram: the *potential* at which the current-step occurs, which gives information about the *chemical origin* of the reduction process, and the *magnitude* of the diffusion-limited current, which yields information about the *concentration* of the species being reduced.

The potential at which the current has risen to half the diffusion-limiting value is termed the half-wave potential, $E_{1/2}$. Its value is found to be very close to the standard redox potential for the electrochemically active couple if both reduced and oxidised species remain in solution:

$$E_{1/2} = E^0 + (RT/nF)\ln(D_{\text{red}}/D_{\text{ox}})^{1/2} \tag{10.2}$$

However, as is the case with zinc, where the reduced species is an amalgam, the half-wave potentials are somewhat more positive in value, though they are still reproducible and have been tabulated.

For analytical purposes, the most important measurement is the magnitude of the limiting current. Ilković established from the diffusion equations that this limiting current has the value:

$$i_d = AnD^{\frac{1}{2}}m^{\frac{2}{3}}\tau^{\frac{1}{6}}c^0 \tag{10.3}$$

where A is a constant, m is the mass flow rate of mercury from the capillary, τ the drop time and c^0 the concentration of species being reduced. This equation can be derived straightforwardly from the expression in Section 4.3.7 for i_{lim} for diffusion to a *plane* surface (an assumption justified by the fact that δ_N is very much smaller than the drop radius r). The area of the plane surface is $4\pi r^2$, which clearly varies

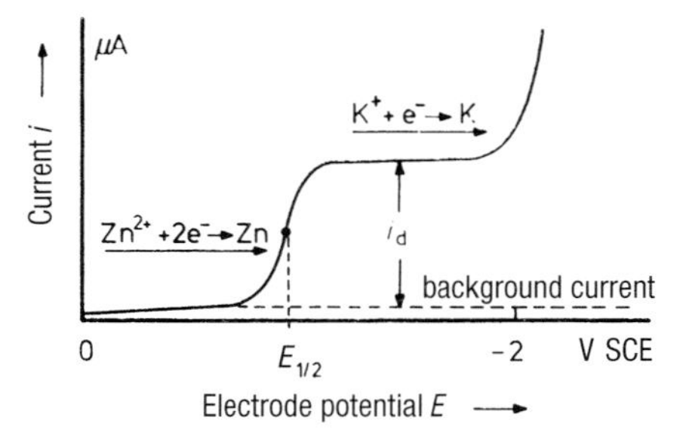

Fig. 10.6 Polarogram of a solution containing zinc ions and with KCl as supporting electrolyte; i_d is the mean diffusion-limited current for zinc reduction, and $E_{1/2}$ the half-wave potential. The background current is shown as a *dashed curve*, and corresponds to the charging current of the mercury droplet.

with time, and from the relationship between the mass mt of the drop at any time t, $r^2 = \left(\frac{3mt}{4\pi\rho_{Hg}}\right)^{\frac{2}{3}}$, where ρ_{Hg} is the density of liquid mercury. Placing this value for r^2 in the expression for i_{lim} and putting t equal to the drop time τ gives the Ilković equation. If the units of D are $cm^2 s^{-1}$, m mg s^{-1}, τ seconds and c^0 mmol dm^{-3}, the value of A, after making other small corrections, is given by Ilković as 708. Averaging over the lifetime of the drop reduces this factor to 607.

It would be unusual, however, to use the Ilković equation directly; normal practice is to use a series of calibration solutions. It is also normal to remove oxygen from the solution before measurement is carried out: oxygen is reduced at mercury first to HO_2^- at –0.2 V and then to OH^- at ca. −1.0 to −1.3 V vs. SCE in neutral solution. A further practical problem in analytical polarography is that as the reduction current rises towards the diffusion-controlled limit in certain solutions, convective vortices form close to the mercury drop; these lead to a considerable enhancement of mass transport and to the frequent observation of "polarographic maxima", in which the current rises above the diffusion-controlled limit a little above $E_{1/2}$, and then falls to i_d at more negative potentials. This effect can normally be eliminated by addition of a small quantity of surfactant to the solution.

Polarography is very sensitive; even the simple arrangement of Fig. 10.5 can be used for concentrations as low as 10^{-5} M, and modern electronically enhanced systems can push this limit down by orders of magnitude, as described below. A further advantage of the technique is the possibility of detecting several species simultaneously, provided the $E_{1/2}$ values are reasonably well-separated. This is illustrated in Fig. 10.7, in which Tl^+, Cd^{2+}, Zn^{2+} and Mn^{2+} are determined during the same potential sweep (lasting some 10 minutes). The rapidity of the technique allows its use in serial analysis, and analytes are not limited to metal cations but also include anions and neutral substances, as well as organic compounds, provided the latter have a suitable electro-active functional group.

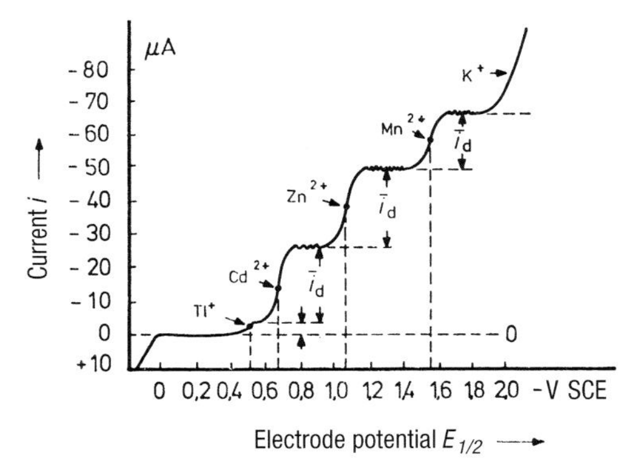

Fig. 10.7 Polarogram showing the mean current for a solution containing the several ions shown; the corresponding limiting currents and half-wave potentials for each species are also shown.

10.2.1.2 Modern Instrumental Improvements in Sensitivity and Resolution in Polarography

Since the main limitation in sensitivity is the ratio of the faradaic to the capacitive current, the sensitivity can be increased either by increasing the first of these or decreasing the second.

Reduction of the capacitive current has been a primary goal of modern instrumental developments in polarography. In the simplest case, given that the main part of the capacitive current falls as $t^{-1/3}$ (since $i_c \sim C_d \cdot dA/dt$, where A is the drop area) and the Faradaic current rises as $t^{1/6}$ during the growth of each drop, optimal sensitivity will be obtained if the current is measured as close as possible to the point at which the drop falls. This can be most easily controlled by electro-mechanically dislodging the drop, so the required measurement can be carried out reproducibly and accurately.

A further variant is pulse polarography, in which narrow rectangular potential pulses of steadily increasing magnitude are applied to each successive drop, as shown in Fig. 10.8. On application of a potential step, the current response can, in effect, be "factored" into the capacitive and faradaic parts, with the former decaying as $e^{-t/RC}$, and the latter decaying as $t^{-1/2}$. Provided R is kept small, by using high background electrolyte concentrations, the capacitive component of the current will decay essentially to zero during the course of the potential pulse, and measurement of the current remaining towards the end of the pulse will yield essentially the pure Faradaic part. This measurement can increase the sensitivity of polarography as compared to the *dc* technique by some two orders of magnitude.

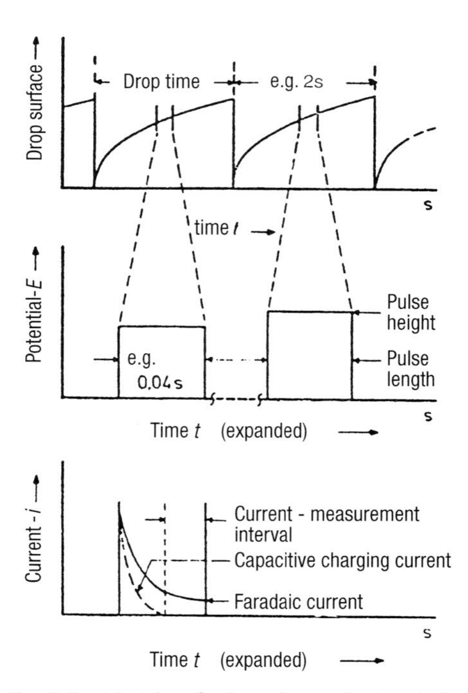

Fig. 10.8 Principles of pulse polarography (see text).

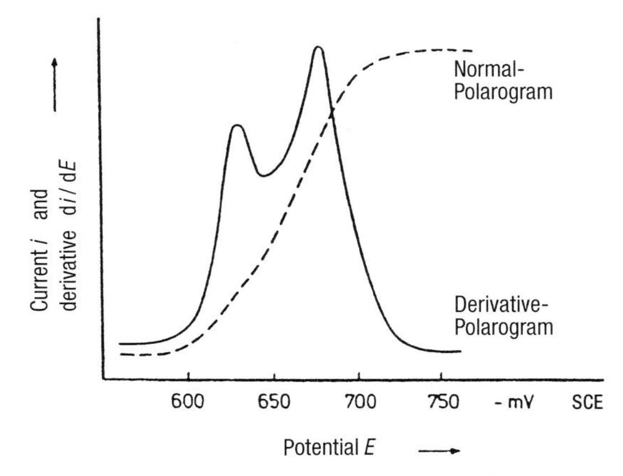

Fig. 10.9 Derivative polarogram of a solution of In^{3+} ($E_{1/2} = -630$ mV) and Cd^{2+} ($E_{1/2} = -680$ mV) in LiCl supporting electrolyte. In normal polarography, the two peaks would be practically inseparable.

In addition to increasing the sensitivity, it is often desirable to increase the resolution of the technique, particularly if two analytes have $E_{1/2}$ values differing only slightly. The simplest way in which this can be done is by differentiation of the current with respect to potential; a derivative polarogram is shown in Fig. 10.9, demonstrating the resolution of In^{3+} ($E_{1/2} = -630$ mV) from Cd^{2+} ($E_{1/2} = -680$ mV). Enhanced resolution can also be achieved with ac polarography, in which a small ac potential of frequency below ca. 100 Hz is superimposed on the linear potential sweep. The measured ac component of the resultant current is also, in essence, the *differential* of the current step, allowing better separation of signals from different species and improved precision in the measurement of $E_{1/2}$. The basis of the technique is shown in Fig. 10.10, which shows that a peak in the *ac* response is expected at $E_{1/2}$, since it is

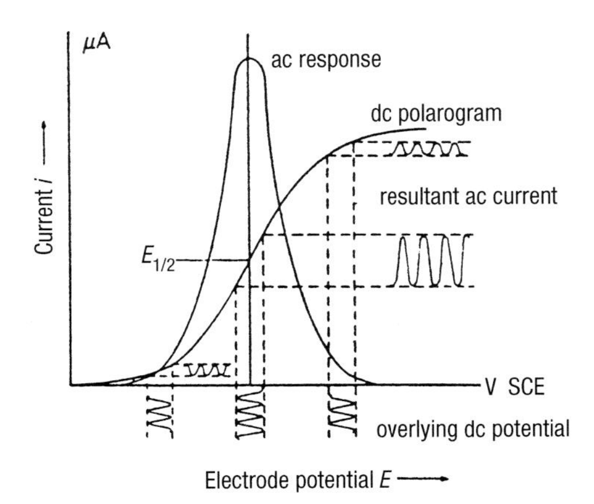

Fig. 10.10 Principles of ac polarography (see text).

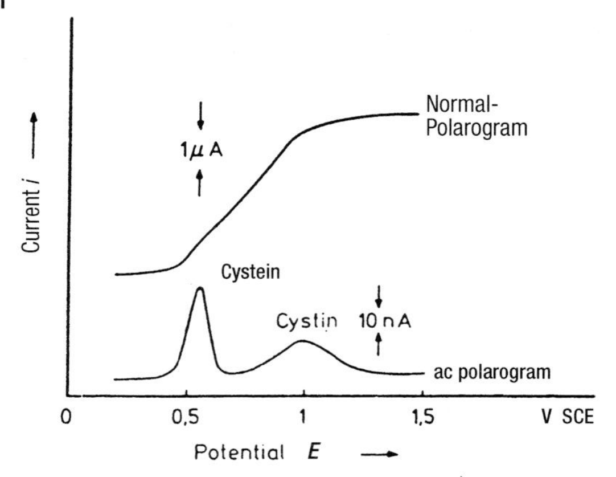

Fig. 10.11 AC polarogram of the system 10^{-4} M cystein and 6×10^{-4}M cystine.

here that changes in the potential lead to the largest changes in current. The value of ac polarography in the resolution of close-lying maxima is shown in Fig. 10.11, showing the resolution of cystein and cystine.

It should be emphasised that the *sensitivity* of *ac* polarography is no greater than the classical *dc* form since the capacitive component of the current also varies sinusoidally. To eliminate the latter, a technique termed *differential pulse polarography* has been developed in which the sinusoidal voltage superimposed on the *dc* ramp is replaced by a series of rectangular pulses, as shown in Fig. 10.12. Measurement is taken just before each step up or down in potential to minimise the capacitive current, and successive measurements subtracted to give a *differential* signal. The sensitivity of this technique is very high; in favourable cases, concentrations of 10^{-8} M can be measured.

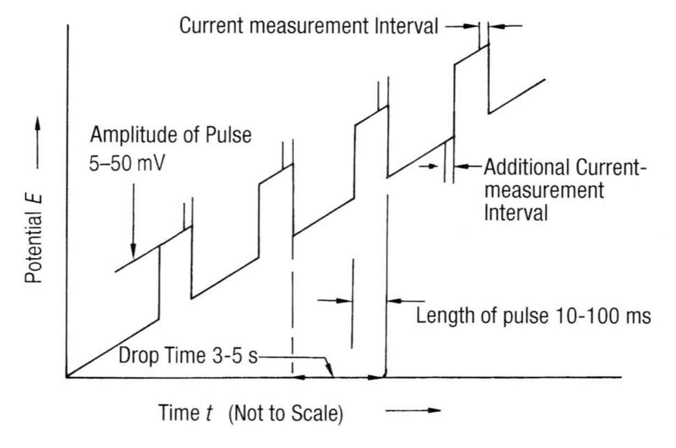

Fig. 10.12 Change of potential with time at the mercury droplet and the measurement interval in differential pulse polarography (see text).

10.2.1.3 Polarographic Methods using a Stationary Mercury Electrode

The sensitivity of polarography can also be increased by increasing the faradaic current i_f. In principle this can be achieved through increasing the convective current, for example by using a rotating mercury electrode, though this has not been extensively explored. The main method whereby the Faradaic component of the current can be increased is through a technique termed stripping voltammetry: the very dilute analyte is first electro-deposited onto a stationary hanging mercury drop at an appropriate potential for a pre-determined period of time under highly reproducible conditions.

In the case of *anodic stripping voltammetry*, a reducible substance, such as a metal cation, is deposited at potentials ca. 200 – 400 mV negative of the half-wave potential, enriching the drop in the reduced form, which may be held as an amalgam, or a metal-film or precipitate on the mercury surface. The potential is then swept in a *positive* direction and exhibits a *positive* current as the reduced analyte re-dissolves, as shown schematically in Fig. 10.13. The peak current, i_p, is proportional to the initial concentration of the analyte provided the deposition conditions are identical from sample to sample. This apparently simple modification increases the sensitivity by an astonishing three to four *orders of magnitude* cadmium, for example, can easily be detected at the parts per billion (ppb) level. Further gains in sensitivity can be obtained by using differential-pulse anodic stripping voltammetry (DPASV), for which sensitivities of 10^{-10} M can be achieved, corresponding to only ca. 5×10^{-12} g cm^{-3} of metal ion.

Cathodic stripping voltammetry is slightly different, in that the immobilisation reaction involves a chemical reaction with electro-oxidatively generated species at the drop surface. The simplest case is the direct reaction with mercurous ions: chloride ion can be determined by reactions of the form:

$$2Hg \rightarrow Hg_2^{2+} + 2e^- \qquad (10.4)$$

$$Hg_2^{2+} + 2Cl^- \rightarrow Hg_2Cl_2 \qquad (10.5)$$

Following deposition of the insoluble mercurous chloride, the potential is ramped *negative*, re-reducing the mercurous salt and releasing chloride.

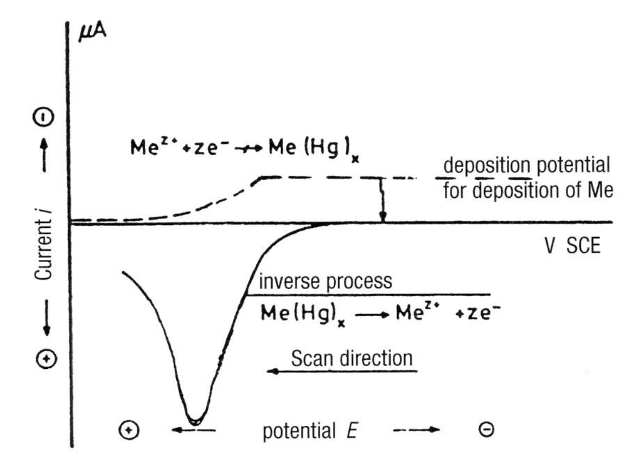

Fig. 10.13 Principle of anodic stripping voltammetry (see text).

Of course, the use of a stationary rather than dropping electrode does increase the likelihood of electrode poisoning, and frequent renewal of the surface remains highly desirable. In the case of mercury, simply releasing the drop and forming another at the end of each measurement is possible; for mercury *film* electrodes, re-deposition of mercury on an easily cleaned and polished surface such as glassy carbon has proved effective. Noble-metal electrodes, which have obvious advantages at positive potentials where mercury cannot be used, must also be continually cleaned if they are to be used for stripping voltammetry: either they must be mechanically polished at frequent intervals, or a potential programme devised that always leaves a reproducible clean surface.

10.2.1.4 **Final Remarks on Polarography**

The range of application of polarographic techniques is huge, including not only the determination of a very wide range of metals, allowing the use of polarography in metallurgy, mineralogy, water analysis and environmental chemistry, but also the determination of anions and neutral inorganic and organic molecules. This latter type of application has led to a marked increase in interest in the use of polarography in areas of analytical biochemistry and pharmacology, as well as in the petrochemical and ceramic industries, in clinical toxicological analysis and in forensic trace analysis. Furthermore, the possibility of using micro- and ultramicro-electrodes can also lead to very small capacitive currents, as described in section 4.3.8.2, where a mercury drop of 7 μm permits rapid analysis of 10^{-7} M solutions of Pb^{2+} and Cd^{2+}.

The fact that many complex organic molecules only possess a single electro-active functional group may allow the specific determination of such a molecule even in a mixture, without prior chemical separation. This combination of rapidity of measurement with selectivity is exceptionally attractive in the modern chemical industry.

Finally, although the analytical uses of polarography are now the main reason for its continued use, it should be emphasised that the early study of adsorption at mercury was a key element in increasing our understanding of the electrical double layer, and polarography has also been used to investigate a very wide range of diffusion processes and electron-transfer rates.

10.2.2
Further Methods - Coulometry, Electrogravimetry and Chronopotentiometry

The technique of coulometry is based on the fact that the amount of charge passed is directly related to the mass of substance, m, undergoing electrochemical conversion:

$$Q = \int i dt = \frac{m}{(M/nF)} \tag{10.6}$$

where M is the molar mass. For a charge of 10^{-3} C, and for $n = 1$, this corresponds to 1.0364×10^{-8} moles of substance, or ca. a microgram for $M = 100$.

Measurement of the charge is straightforward using electronic integration of the current. However, equation (10.6) is only applicable if the current yield for the desired reaction is 100 %. It is critical to the applicability of the technique that such common side reactions as hydrogen or oxygen evolution, reduction of dissolved oxygen, dissolution of the electrode metal in the presence of complexing agents, and the back reaction of the electro-generated species with adventitious impurities be minimised. Critical also is the requirement that working and counter-electrode compartments be kept separate with a diaphragm to minimise diffusion. In addition, the requirement that electrochemical conversion take place until complete exhaustion of the analyte is very difficult to satisfy; even with vigorous stirring, the process would take a very long time, and as a result, the *direct* use of equation (10.6) has not proved a very useful approach.

Several methods have, therefore, been developed to turn coulometry into a usable technique. In the simplest case, *potentiostatic coulometry*, the potential is maintained constant within the limiting current region, and convective control used to keep δ_N constant. The limiting current then takes the value:

$$i(t) = \frac{nFDAc(t)}{\delta_N} \tag{10.7}$$

where only the concentration of analyte is explicitly time dependent. If the volume of the solution containing the analyte is V, we have, from Faraday's law:

$$i = \frac{dQ}{dt} = -nF \cdot \frac{d(Vc)}{dt} \tag{10.8}$$

since Vc is the total amount of material present in solution at time t. Combining (10.7) and (10.8), we find:

$$\frac{dc}{dt} = -\frac{DAc}{V\delta_N} \tag{10.9}$$

and integrating from time $t = 0$, $c = c_0$

$$c = c_0 \exp\left(-\frac{Dat}{V\delta_N}\right) \equiv c_0 e^{-kt} \tag{10.10}$$

From (10.10), substituting in (10.7), we find

$$i = i_0 e^{-kt} \tag{10.11}$$

whence

$$Q = \int_0^\infty i_o e^{-kt} dt = \frac{i_0}{k} \tag{10.12}$$

Hence, instead of measuring the charge corresponding to complete discharge of analyte, only the initial current i_0 need be measured, together with the rate constant k, for which measurements of the current over a few tens of minutes will normally suffice. The accuracy with which equation (10.11) is obeyed is also a useful check that there are no significant side reactions. The main area of application of potentiometric coulometry is in the determination of small amounts of non-ferrous, noble or actinide metals, where the electrochemical process may exploit a redox change in the analyte or deposition onto a mercury, noble-metal or glassy-carbon electrode. Sensitivities in the region of 10^{-9}–10^{-10} M are achievable.

A variant method is *galvanostatic coulometry*, in which an active reagent, such as I_2, is generated in situ in the solution, electrochemically, at a constant rate. This is really an alternative version of a normal titration process, with addition from a burette being replaced by electrochemical generation. It follows that determination of the end-point must be carried out by an auxiliary experiment, in which the concentration of analyte of reagent is monitored spectrophotometrically, potentiometrically or in some other way. This method does have advantages over normal titrimetry where, for example, the titrant solution might be difficult to prepare or where only small amounts of titrant are necessary, and errors in measurement are therefore large.

Another variant is *electrogravimetry*, in which a solution species is completely deposited on an electrode surface, and the weight of the deposit determined by difference. As with potentiostatic coulometry, the complete deposition of material is often not practical and use can be made of equations similar in principle to (10.8) to (10.12). Recently, the technique of quartz-crystal microgravimetry has been introduced into electrochemistry: in this method, the resonance frequency of a quartz crystal, which depends on the mass of a metal electrode evaporated onto the crystal, can be monitored (see Section 5.4.5.2). The change in resonant frequency is given by the Sauerbrey equation:

$$\Delta f_0 = -A \cdot f_0^2 \cdot \Delta m \tag{10.13}$$

where f_0 is the frequency measured for the electrode at the beginning of the deposition process and Δm the change in mass of the electrode; A is a constant, and the negative sign implies that the resonant frequency falls as deposition takes place. Given that Δm is proportional to $q(t)$ (the charge passed at time t) measurement of Δm as a function of time over a few tens of minutes will permit c_0 to be calculated.

A range of techniques based on the determination of the transition time, τ, in galvanostatic experiments can be used for analysis, such techniques having the generic descriptor *chronocoulometry*. The concept of a transition time was introduced in Section 4.3.4: if the potential of the electrode is monitored during the galvanostatic conversion of an analyte, there is a marked change in the potential once the analyte is exhausted and, for example, solvent decomposition begins to take place to maintain the current. Measurement of can thus give a measure of initial analyte concentration, with a limit of 10^{-4} – 10^{-5} M. Sensitivity may be increased by pre-concentration techniques, as described in Section 10.2.1.3; again, as the deposited film or amalgam becomes exhausted during *galvanostatic* discharge, there is a marked change in potential. This latter method is termed *stripping chronocoulometry* or *potentiometric stripping analysis* (PSA): transition times from 10^{-3} s to 100 s are easily handled, and concentrations as low as 10^{-8}–10^{-9} M are measurable.

In principle, the galvanostatic dissolution process following electrochemical preconcentration can be replaced by a chemical dissolution induced by the addition of a reactant to the electrolyte, a process referred to as *chemical stripping chronocoulome-*

try, c-SCC or chemical potentiometric stripping analysis, c-PSA. The electrode is placed at open-circuit and its potential followed: once the analyte in the film is exhausted, the open-circuit potential will jump to that associated with the redox couple of the added reagent. A simple example is the oxidation of metal in amalgam by dissolved oxygen in the electrolyte; the potential of the electrode will remain close to that of the foregoing electro-deposition process until the metal is completely re-dissolved, whereupon it will move to the oxygen potential.

10.3 Electrochemical Sensors

Sensors now have a major role in medicine, environmental monitoring and industrial process control. Next to thermal, gravimetric and optical sensors, electrochemical sensors now play their own substantial part, particularly in the monitoring of analyte concentration.

10.3.1 Conductivity and pH Measurement

The basis of conductivity measurement has already been discussed in Sections 2.1 and 2.2. Provided the concentration of ions is not too high, the ionic conductivity measured is approximately proportional to the concentration of dissolved (strong) electrolyte, and conductivity measurements are, therefore, suitable for the monitoring of salts, acids or bases both in aqueous effluent and as intermediates in chemical processes. Examples of industrial use include: control of bath composition in electro-plating technology; monitoring of water quality in rivers and reservoirs; and a wide range of uses in the chemical, petrochemical and paper and textile industries. Naturally, the design of the conductivity cells must satisfy the particular requirements of the application, for example in rapidly flowing, viscous or dirty liquids; in practice this might mean the use under appropriate circumstances of mechanical wipers or even ultrasonic cleaning of the equipment.

The technology of pH meters has been described in detail in Section 3.6.7, and the use of such sensors is extremely widespread, so much so that only some selected applications can be given:

1. During the treatment of drinking water, purification and de-colouration frequently involve addition of either aluminium or iron salts. To be successful, this addition must be carried out in slightly acid conditions (pH 4.5); the hydroxides are then precipitated by the addition of chalk, again under careful pH control; these hydroxides carrying down with them particulate and colloidal impurities.
2. Acidic or alkaline effluents pose a serious environmental problem, and pH sensors can now be used that automatically ensure that effluent is only discharged within narrowly defined pH bands.

3. The pH of the gastro-intestinal tract has important diagnostic uses in medicine, and can now be measured by swallowing a pH sensor attached to a radio-transmitter. Normally an antimony pH electrode is used in this application, the potential of which is used to encode the frequency of the micro-transmitter.

10.3.2 Redox Electrodes

If a solution contains a species that exists in two different oxidation states and exhibits rapid electrode kinetics at an appropriate electrode, then the potential of that electrode will be determined by the Nernst equation for that species (assuming all other components are not electro-active):

$$E = E^0 + \frac{RT}{nF}\ln\left(\frac{a_{ox}}{a_{red}}\right) \tag{10.14}$$

An interesting example is the control of effluent from chrome-plating baths. Such effluent contains initially primarily Cr(VI), which must first be reduced to Cr(III) by addition of sulphurous acid or SO_2. This reduction is monitored by measuring the electrode potential of the Cr(VI)/Cr(III) couple until the potential $E = E^0 - 0.1$ V, which corresponds to $a_{Cr(VI)}/a_{Cr(III)} \approx 10^{-5}$. Finally, the solution is neutralised to precipitate $Cr(OH)_3$.

Clearly, if the redox potential falls below the hydrogen evolution potential at the pH of the solution, then evolution of hydrogen at the electrode may take place, provided the overpotential is sufficiently small. In practice, this possibility is characterised by the "rH" value, the latter being defined as the negative *decadic* logarithm of the pressure of hydrogen in equilibrium with water *at that pH* and at the potential determined by the redox reaction of interest. Evidently, the larger rH, the more serious will hydrogen evolution interference be, and the more likely is the reduction of the solution.

10.3.3 Ion-sensitive Electrodes

If ions can penetrate the boundary between two phases, then an electrochemical equilibrium will be set up, in which the two phases may acquire different potentials (see Chapter 3). If only *one* type of ion can be exchanged between the two phases, then the potential difference set up between the phases depends only on the activities of that ion in the two phases. This process was explored in some detail in Section 3.3, with particular regard to an ion-exchange membrane separating two solutions with different ionic activities a^{I} and a^{II}. Provided the membrane is only permeable to this single type of ion, the potential difference across the membrane has the value:

$$\Delta E = \frac{RT}{zF}\ln\left(\frac{a^{II}}{a^{I}}\right) \tag{10.15}$$

Provided the activity of the ion in phase I is kept *constant, then the activity in phase II, a_x, is related to ΔE by:*

$$\Delta E = \frac{RT}{zF}\ln\left(\frac{a_x}{a^{\mathrm{I}}}\right) = \text{const.} + S \cdot log_{10} a_x \tag{10.16}$$

where $S = 59.0/z$ mV at 298 K. ΔE can be measured as the potential difference between two identical reference electrodes placed in the two phases. By using a series of calibration solutions, the activity of the ion in any unknown solution can be measured directly, though it should be recognised that only at constant ionic strength will there be any clear cut relationship between activity and concentration.

If there are other ions present for which the membrane is non-permeable, then such ions will have no effect on the measured potential difference. However, in reality, no membrane is truly permeable only to a single type of ion, and completely impermeable to all others, and corrections must be made for *interference* effects. These corrections are applied through the Nikolski equation:

$$\Delta E = \text{const.} + S \cdot log_{10} a_x + S \cdot \sum_i log_{10}\left\{K_i a_i^{z/z_i}\right\} \tag{10.17}$$

where the summation is over all types of interfering ions, i, K_i is the selectivity coefficient (determined empirically), and z_i the charge on the ion of type i.

10.3.3.1 Glass-membrane Electrodes

These have similar properties to the glass pH electrode already described in Section 3.6.7. The simplest example of a cation-sensitive electrode is the sodium-sensitive system (the pNa electrode), whose construction is essentially identical to the pH electrode, with the exception of the glass used (silica with 10% Na_2O and Al_2O_3) and the fact that the inner reference solution has a fixed sodium-ion activity. Interference from H^+ ions is high, with $K_i \approx 10^3$; the activity of H^+ should, therefore, be maintained some four orders of magnitude less than the sodium-ion activity to be determined. Silver ions also interfere badly, with $K_i \approx 500$. However, for K^+, Li^+, Cs^+, and Tl^+, K_i values of ca. 10^{-3} are found, with even smaller values for Rb^+ and NH_4^+. Clearly, all results with this type of sensor should be assessed critically, though with proper precautions, the range of activities measurable extends from 1 down to 10^{-8} M.

The advantages of the pNa electrode as compared to other methods for the determination of sodium include the simplicity of the instrumentation, the robustness of the technique (in process control or effluent analysis, for example), the rapidity of the measurement (usually just seconds) and the large concentration range. A good example of the use of this type of electrode is in the monitoring of blood pNa values during post-operative care, for which a flow-through cell can be designed.

Although the pNa electrode is insensitive to ions other than H^+ and Ag^+, it has not proved possible to find glass membranes with the same selectivity for other ions, such as K^+. Glass membranes showing good sensitivity to a set of univalent ions have, however, been identified, and may find uses in the monitoring of the total content of all such ions in a sample. Alternatively, if there are only two such ions present in a sample, such as Na^+ and K^+ in blood, then the use of both a pNa electrode and a non-selective electrode can permit K^+-ion determination by difference.

10.3.3.2 **Solid- and Liquid-membrane Electrodes**

Such insoluble materials as Ag_2S, CuS, CdS, PbS, LaF_3, AgCl, AgBr, AgI and AgSCN have all been tested as cation-exchange membranes, and can be incorporated into sensors in the form of single-crystal or compressed powder discs as shown in Fig. 10.14. These materials are all ionic conductors, though the conductivity is extremely small at room temperature, and mainly takes place through the migration of point defects in the lattice. However, the ionic conductivity, and hence the speed of response, can be increased by incorporating aliovalent ions into the lattice; as an example, the fluoride-ion sensitive LaF_3 can be doped with Eu^{2+} or Ca^{2+} ions. Sensors for the detection of Ag^+, Cu^{2+}, Cd^{2+}, Pb^{2+}, S^{2-}, F^-, Br^-, I^-, SCN^- and CN^- ions can be constructed from such materials, with the sensitivity to anions arising from equilibria at the membrane surface such as:

$$Ag_2S2Ag^+ + S^{2-} \tag{10.18}$$

The measurement range of such sensors lies in the range $10^{-6} - 1$ M, but interference effects are frequently encountered.

The membranes themselves are usually a millimetre or so thick, and in addition to single-crystal or pressed powder form, it has been found possible to use insoluble metal salts embedded in an inert matrix such as silicone rubber or polyethylene. This has considerable advantages in terms of membrane fabrication.

In addition to using an inner electrode, as shown in Fig. 10.14, it has also proved possible to fabricate sensors by directly contacting the membrane with a wire to form an ohmic contact. Such systems are found to show complex temperature (and time) dependencies, necessitating frequent re-calibration, but they are extremely simple to fabricate. An example is a silver sensor that can be made by attaching a wire to the back of a compressed graphite/PTFE disc, on the front side of which the silver halide is rubbed in to the surface. This simplicity of construction lends itself well to miniaturisation (see also Section 10.3.3.3 below): simply coating a wire with a sensor film of PVC containing an ion-exchanger is sufficient to generate a robust, though rather irreproducible, miniature sensor, and such sensors have found application in, for example, toxicological analysis.

In addition to solid membranes, immiscible liquid (organic) phases with ion-exchange properties can also be used, with such phases stabilised against the external solution phase within a polymer or ceramic membrane. As an example, a calcium-sensitive electrode can be constructed using the calcium salt of didecylphosphoric acid as a specific chelating agent in di-octylphenylphosphonate, these chemicals being stabilised by incorporation into a porous polymeric matrix of PVC or silicone rubber. Non-ionic chelating agents can also be used; the best known example is the macrocyclic ionophore valinomycin, usually immobilised in phenylether, which is strongly selective towards potassium, allowing the construction of a K^+-ion selective electrode

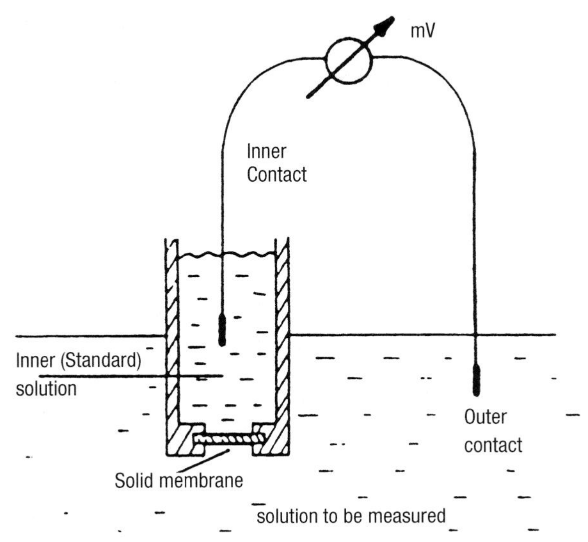

Fig. 10.14 Use of a solid-membrane as an example of an ion-selective electrode.

with low interference even from H^+. Other membranes have been devised with strong selectivity for ions such as NO_3^- (using ammonium tetradecylammonium nitrate), NH_4^+ (using nonactin as the active ionophore), Ca^{2+} using didecyl phosphoric acid in di-octyl phenylphosphonate, etc.

Additional selectivity can be attained by using composite membranes in which an enzyme present in the outer part of the membrane catalyses a specific chemical reaction to generate a product ion that can be detected using an internal ion-selective membrane. A well-known example is the detection of urea selectively using *urease* as the enzyme catalyst

$$CO(NH_2)_2 + 2H_2O \xrightarrow{\text{urease}} 2NH_4^+ + CO_3^{2-} \tag{10.19}$$

The ammonia generated can then be detected by an ammonia-selective membrane electrode as described above. Similarly, enzyme reactions generating protons can be followed with glass or other proton-selective membranes. There is a multiplicity of selective electrodes that can be made in this way, with substrates including aliphatic alcohols, acetylcholine, amygdalin, asparagine, glucose, glutamine, and penicillin. Biosensors for the detection of antibodies or hormones are also being developed, and this whole area is of huge modern importance in clinical analysis.

10.3.3.3 **ISFETs**

The miniaturisation of ion-selective electrodes to permit their use in clinical and other applications, where smallness of size is critical, has not proved straightforward, owing to the fact that simply reducing the size of the types of electrode described above would

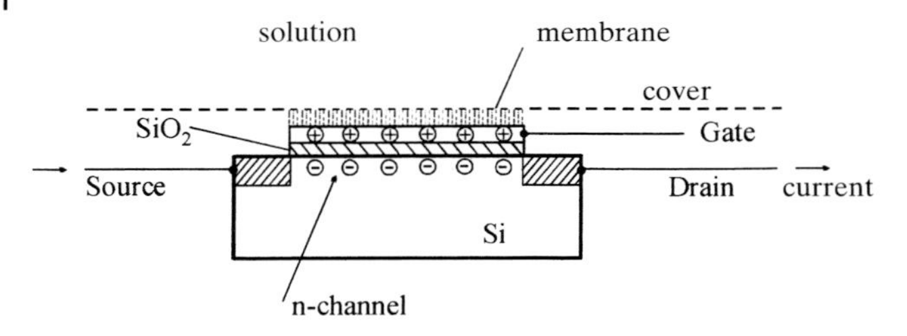

Fig. 10.14a (Solid lines): Field-effect transistor (highly schematic); in the example the gate is shown as positively charged. (Dotted lines): Extension of the principle to form an ion-sensitive FET.

substantially worsen the signal/noise ratio. This is because no amplification of the signal takes place is such sensors, but sensitivity to pick-up increases. To circumvent this problem, devices have been developed that permit amplification to take place at the membrane itself. Such devices are based on the operational principles of the *Field Effect Transistor* (FET), and one such device is shown in schematic form in Fig. 10.14a. The basic construction consists of a channel of *n*-type Si separated from a metallic electrode (the gate) by a thin insulating layer of SiO_2. Current flows from *source* to *drain* in the device, but the flow of this current can be altered, as in any transistor, by altering the potential on the gate. The main advantage of the FET type of transistor is that the control of the gate is purely through its potential; the resistance between the gate and the source/drain channel is extremely high. This allows us to replace the electronic gate in a conventional FET with an ion-selective membrane. The potential across this membrane is sensitive to ion concentration in solution; changes in the latter then change the current-voltage characteristics of the source-drain channel which can be detected as, for example, changes in current at constant drain-source voltage. It is evident that such *Ion-Selective* FETs (ISFETs) can be used with more selective membranes, such as those with immobilised enzymes, allowing a wide variety of applications.

10.3.4
Sensors for the Analysis of Gases

Electrochemical techniques for the determination of gas concentrations in solution have a multiplicity of applications, including CO_2 in blood, oxygen in streams or even the oxygen content of molten steel. Partial pressures of gases in a mixture can also be determined by dissolution, taking advantage of Henry's law, which relates gas solubility to its partial pressure. Gas sensors based on conductivity, and on potentiometric and amperometric principles have all been developed and are described in outline below: all these types of sensor, however, have been found to require careful calibration, to show drift and interference effects, and to need temperature compensation built in.

10.3.4.1 Conductimetric Sensors

The most straightforward type of conductimetric sensor is a solution conductivity cell into which the gas to be analysed is admitted. Provided an appropriate reactant is present already in the solution before the gas is admitted, and this can react with the gas either to form ions or alter their concentration, the liquid will show a conductivity difference that can be used analytically. A simple example is the determination of traces of SO_2 in air: SO_2 reacts with an aqueous solution of H_2O_2 to form sulphuric acid, the latter giving rise to a marked change in ionic conductivity in the cell. The sensitivity of this type of cell lies as low as 0.1 ppm, with an upper limit of ca. 15% by volume in the gas stream. Similar detectors have been constructed for HCl, NH_3, PH_3, AsH_3 and CS_2, as well as for traces of organic sulphur-containing compounds.

Gas sensors based on the conductivity of semiconducting materials such as transition-metal oxides have also been developed. These sensors do not require electrolyte: adsorption of the gas on the semiconductor surface alters the conductivity of the semiconductor in ways not radically different from the operation of the ISFETs treated above. Their mode of operation is, therefore, not primarily electrochemical, but their simplicity of fabrication and operation make them strong competitors.

10.3.4.2 Potentiometric Gas Sensors

In the simplest form of this type of sensor, the gases are dissolved in solution and the equilibrium potential established at an appropriate electrode is then proportional, from the Nernst equation, to the logarithm of the pressure. This type of approach can be used directly for H_2, and indirectly for such gases as SO_2. In the latter case, dissolution in bromine water gives rise to the reaction:

$$SO_2 + 2H_2O + Br_2 \rightarrow 4H^+ + SO_4^{2-} + 2Br^- \qquad (10.20)$$

and the reaction can be monitored through the Br_2/Br^- couple or the pH.

This type of process underlies the CO_2 sensor shown in Fig. 10.15, in which CO_2 diffuses through a porous membrane as shown; dissolution in water gives rise to HCO_3^- and H^+ ion, the latter being detected by a classical glass electrode as shown. This principle has also been used to detect traces of NH_3, SO_2, NO_x, H_2S and HCN, and miniaturisation through the use of ISFETs is possible.

Oxygen can also be detected in this way, though the poor electrochemical kinetics of O_2 at room temperature mean that much higher temperatures are needed. The measurement cell is therefore fabricated using an oxide-ion conductor with porous platinum electrodes on the two sides. By maintaining a standard O_2 pressure, p_S, on one side, the oxygen partial pressure on the other side can be determined from the Nernst equation:

$$E = \frac{RT}{4F} \ln\left\{\frac{p_{O_2}}{p_S}\right\} \qquad (10.21)$$

The cell operates at temperatures above 800°C, and an example of its use is shown in Fig. 10.16, where the oxygen content of molten steel is being measured. Here p_S is the

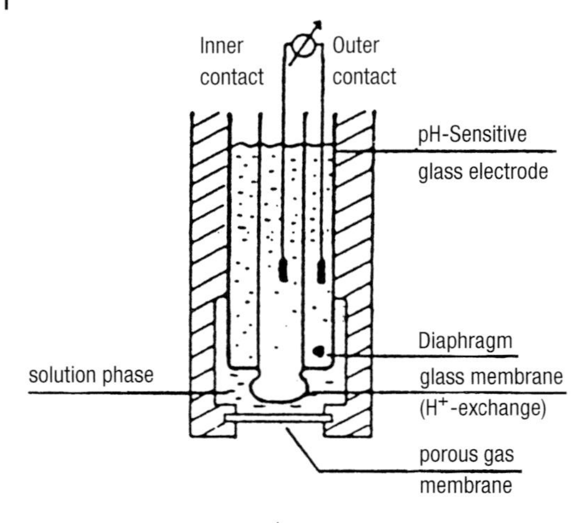

Fig. 10.15 Construction of a CO_2 sensor (see text).

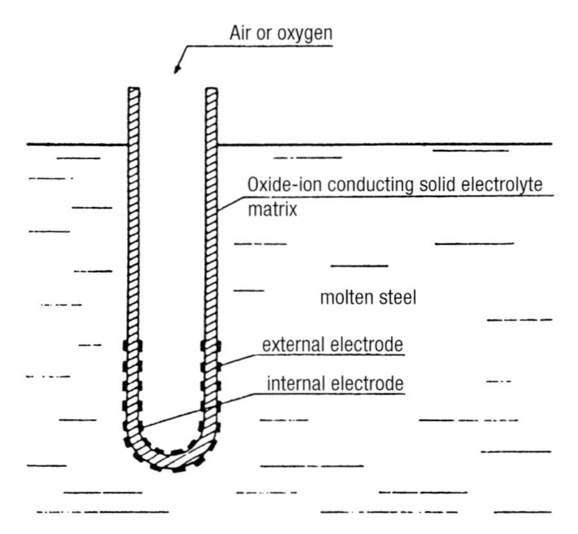

Fig. 10.16 Determination of oxygen content in steel melts using an oxide ion-conducting membrane; not shown are the electrode connections.

partial pressure of O_2 in air. A similar principle is used to determine the oxygen content of the hot exhaust stream from internal combustion engines: the λ-Sonde principle. The value of λ is given by:

$$\lambda = \frac{\textit{The amount of air admitted to the motor}}{\textit{Stoichiometric amount of air needed completely to burn fuel}} \tag{10.22}$$

If $\lambda > 1$, the partial oxygen pressure in the exhaust gas from the internal combustion engine rises to $10^{-4} - 10^{-2}$ bar, whereas for $\lambda < 1$, the pressure falls to below 10^{-10} bar. Clearly, λ should be close to unity for ideal functioning of the engine, and feedback from the sensor is used to control the fuel/air mixture. Amperometric oxygen detectors have also been developed, as described below.

10.3.4.3 Amperometric Gas Sensors

Amperometric sensors for the determination of traces of gases depend on the oxidation or reduction of electroactive species which are generated by dissolution of the gas, and both direct and indirect processes can be used. The sensitivity of this type of sensor can be very high, in the ppb range, and they frequently exhibit good selectivity as well. Such sensors usually rely on a thin electrolyte layer to ensure rapid transport of dissolved gas to the electrode surface; this thin layer may be incorporated directly into the cell design or may be part of a three-phase boundary, the latter being particularly suited to direct measurement in the gas phase.

The magnitude of the gas reduction or oxidation current is determined by the partial pressure of the gas in the gas stream and the effective thickness of the Nernst diffusion layer for the 3-phase gas/electrolyte film/catalyst boundary. We have, for the limiting current:

$$j_{\text{lim}} = nFDc_0/\delta_{\text{N}} \tag{10.23}$$

For a sensor directly measuring partial pressures in the gas phase, we have, from Henry's law,

$$j_{\text{lim}} = nFDKp_{\text{gas}}/\delta_{\text{N}} \tag{10.24}$$

where c_0 is the equilibrium concentration of gas in the liquid electrolyte and p_{gas} the corresponding gas partial pressure in the gas phase, with K the value of Henry's constant.

A simple example of a gas sensor is the Clark cell, which is used for the detection of dissolved oxygen. This cell consists of a thin electrolyte layer in contact with the silver electrode and a gas-permeable membrane, as shown in Fig. 10.17a. Dissolved oxygen diffuses through the membrane and the thin electrolyte layer, and is reduced at the silver electrode.

Direct amperometric sensors are frequently based on fuel-cell technology and incorporate familiar gas-diffusion electrodes. A schematic of the sensor version of a H_2/O_2 fuel cell is shown in Fig. 10.17b: both the H_2-working electrode and the O_2-counter electrode are gas-diffusion electrodes as described above (Section 8.6), and the electrolyte is sulphuric acid in the form of a gel. Given that the working and counter-electrodes are connected, the potential of the hydrogen electrode is maintained in the limiting current region, and the hydrogen concentration at the surface is, therefore, effectively zero. Typical current response is a few nanoamps per ppm hydrogen, and a current follower/amplifier is used as shown in Fig. 10.17b. The desired selectivity in an amperometric sensor can be introduced by suitable choice of catalyst, modification of the electrolyte to discriminate between different gases

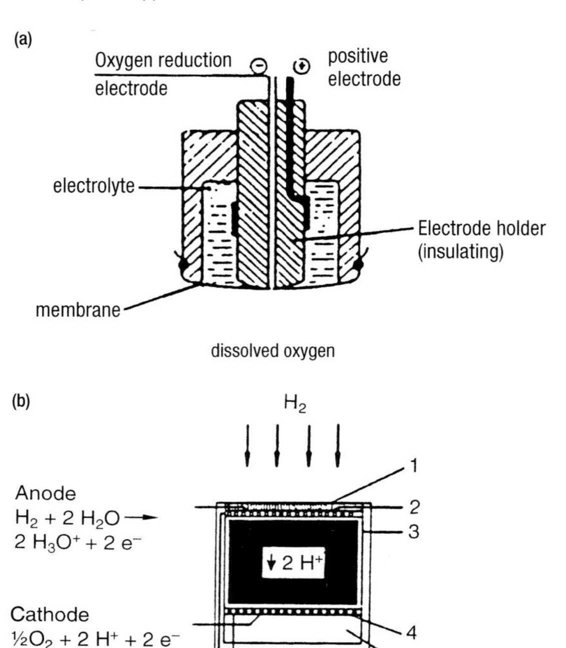

Fig. 10.17 (a) Clark electrode for detection of dissolved oxygen (b) Schematic representation of an amperometric 2-electrode sensor (after D. Kitzelmann and C. Gottschalk: tm-Technisches Messen **62** (1995) 152). (1): PTFE membrane as diffusion barrier; (2): working electrode (PTFE-bonded platinised carbon layer); (3): H_2SO_4 as electrolyte; (4): O_2-diffusion counter electrode; (5) Current follower/amplifier; (6): Air supply for counter electrode; *i* is the output signal.

or the use of a potential range in which only the species to be determined is electroactive.

Indirect amperometric sensors exploit an auxiliary chemical step which is coupled to the charge-transfer reaction at the electrode. A typical example is the amperometric sensor for H_2S shown in Fig. 10.18. A finely porous silver layer serves as the working electrode, and an organic electrolyte is used to form a three-phase boundary. Ag^+ ions dissolved in the electrolyte react very selectively in the three-phase region with H_2S to form $Ag_2S + 2H^+$. If the silver electrode is held at a constant potential with respect to a suitable reference electrode, the silver ions will be formed at the electrode surface, and the resultant current can be measured. As the counter electrode, an oxygen-diffusion electrode can be used, and Fig. 10.19 shows the time response to a pulse of H_2S at the 20 ppm level.

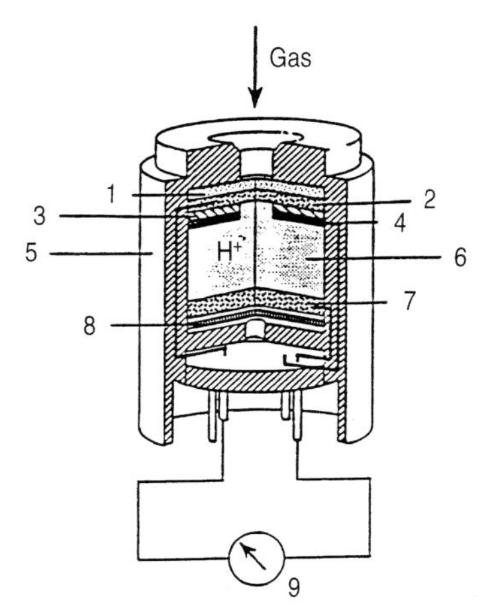

Fig. 10.18 Indirect amperometric 3-electrode sensor for H_2S (after *loc. cit.*, Fig. 10.17); (1): membrane as diffusion barrier; (2): working electrode (porous silver); (3): separator; (4): Reference electrode; (5): outer housing; (6): Organic electrolyte; (7): O_2-diffusion electrode; (8): membrane for pressure equalisation; (9): nano-amperometer.

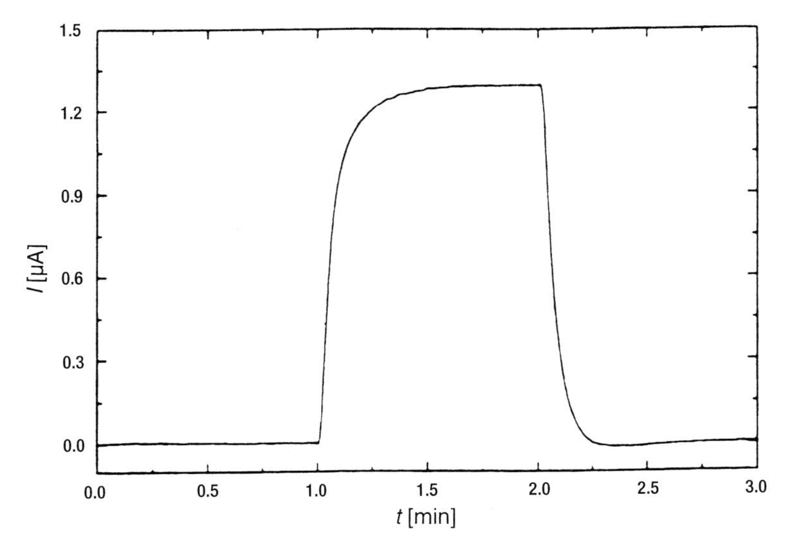

Fig. 10.19 Signal-time transient of the H_2S sensor of fig. Fig. 10.18 following introduction of 20 ppm H_2S.

Other examples of selective sensors with preceding reactions are given in Table 10.1; in addition to the equations for the equilibria used for coupled chemical reactions, some direct amperometric sensor reactions are also listed, and the sensitivities of the different reactions given. It should be noticed that interferences, particularly the reduction of adventitious oxygen, can be troublesome.

Table 10.1 Detection limits and electrochemical reactants for amperometric sensors.

Gas	Chemical Formula	Reaction limit [ppm]
Ammonia	NH_3	3
Arsine	AsH_3	0.01
Bromine	Br_2	0.01
Bromide	HBr	0.5
Carbonyl Chloride (Phosgene)	$COCl_2$	0.005
Chlorine	Cl_2	0.01
Chlorine Dioxide	ClO_2	0.01
Hydrogen Chloride	HCl	0.5
Diborane	B_2H_6	0.02
Ethanol	CH_3CH_2OH	0.5
Ethylene Oxide	C_2H_4O	0.5
Fluorine	F_2	0.01
Hydrogen Fluoride	HF	0.3
Formaldehyde	HCHO	0.5
Germane	GeH_4	0.01
Hydrazine	N_2H_4	0.02
Carbon Dioxide	CO_2	range 0–5 %
Caron Monoxide	CO	3
Ozone	O_3	0.05
Phosphine	PH_3	0.01
Nitric Acid	HNO_3	1
Sulphur Dioxide	SO_2	0.5
Hydrogen Sulphide	H_2S	0.1–1
Nitrogen Dioxide	NO_2	0.5
Nitrogen Monoxide	NO	3
Silane	SiH_4	0.5

Gas sensors can also be used to monitor biological systems. One example is the use of Clark electrodes to study toxicity. One such electrode is placed in contact with a bacterial sample in an oxygen-containing culture medium and a second electrode is in contact with a second bacterial sample in an identical medium to which has been added the toxic substance to be tested. The more toxic this substance, the less will be the growth of the bacteria and the more slowly will oxygen be consumed, leading to an increasing difference in the response of the two electrodes. This process is far faster than conventional tests. A second, similar, setup can be

Reactions with Direct electron Transfer

Gas	Electrochemical Equation	Reaction Type
H_2	$H_2 \rightarrow 2H_{ad} \rightarrow 2H^+ + 2e^-$	Oxidation
SO_2	$SO_2 + 2H_2O \rightarrow H_2SO_4 + 2H^+ + 2e^-$	Oxidation
O_2	$O_2 + 2H_2O + 4e^- \rightarrow 4OH^-$	Reduction
N_2H_4	$N_2N_4 \rightarrow N_2 + 4H^+ + 4e^-$	Oxidation
NO	$NO + H_2O \rightarrow NO_2 + 2H^+ + 2e^-$	Oxidation
NO_2	$NO_2 + 2H^+ + 2e^- \rightarrow NO + H_2O$	Reduction
NH_3	$2NH_3 \rightarrow N_2 + 6H^+ + 6e^-$	Oxidation
H_2S	$H_2S + 4H_2O \rightarrow H_2SO_4 + 8H^+ + 8e^-$	Oxidation
O_3	$O_3 + 2H^+ + 2e^- \rightarrow O_2 + H_2O$	Reduction
HCN	$2HCN + 2H_2O \rightarrow 2HCHO + N_2 + 2H^+ + 2e^-$	Oxidation
SiH_4	$SiH_4 + 3H_2O \rightarrow H_2SiO_3 + 8H^+ + 8e^-$	Oxidation

Reactions with coupled chemical processes

Gas	Electrochemical and Chemical Reactions	Reaction Type
CO	$Pt - O + CO \rightarrow Pt + CO_2$	Oxidation
	$Pt + H_2O \rightarrow Pt - O + 2H^+ + 2e^-$	
AsH_3, PH_3	$AsH_3 + 8Ag^+ + 4H_2O \rightarrow H_3AsO_4 + 8H^+ + 8Ag^0$	Oxidation
	$8Ag^0 \rightarrow 8Ag^+ + 8e^-$	
Cl_2	$Cl_2 + 2Br^- \rightarrow Br_2 + 2Cl^-$	Reduction
	$Br_2 + 2e^- \rightarrow 2Br^-$	
O_3	$O_3 + 2Br^- + 2H^+ \rightarrow O_2 + Br_2 + H_2O$	Reduction
	$Br_2 + 2e^- \rightarrow 2Br^-$	
HCl	$4HCl + Au^{3+} \rightarrow H[AuCl_4] + 3H^+$	Complexation
	$Au \rightarrow Au^{3+} + 3e^-$	
H_2S	$H_2S + 2Ag^+ \rightarrow Ag_2S + 2H^+$	Precipitation
	$2Ag^0 \rightarrow 2Ag^+ + 2e^-$	
HCN	$HCN + Ag^+ \rightarrow AgCN + H^+$	Precipitation
	$Ag^0 \rightarrow Ag^+ + e^-$	

used to monitor the rate of back-mutation in genetically engineered Salmonella strains designed to be unable to grow in histidine-free environments. If mutation in such a medium takes place back to the wild-type, then oxygen consumption will show a marked increase, and this experiment adapted as an electrochemical version of the well-known Ames-Test for mutagenicitiy and carcinogenicitiy.

References to Chapter 10

General References to Electroanalytical Techniques

"Comprehensive Treatise of electrochemistry" eds. J.O'M. Bockris, B.E. Conway, E. Yeager and R.E. White, Plenum Press, New York, vol. **8**, 1984.

J.J. Lingane: *"Electroanalytical Chemistry"*, Wiley-Interscience, New York, 1958.

E.A.M.F. Dahmen: *"Electroanalysis: Theory and Applications in Aqueous and Non-aqueous Media and in Automated Chemical Control"*, Elsevier, Amsterdam, 1986.

More specific texts

J.T. Stock: *"Amperometric Titrations"*, Wiley, New York, 1965; reprinted by R.E. Krieger Publishing Co., Huntington, New York, 1975.

R.A. Robinson and R.H. Stokes: "Electrolyte Solutions", Butterworth, London, 1959.

L. Meites: *"Polarographic Techniques"*, 2nd. Ed., Interscience Publishers, New York, 1965.

J. Heyrovsky and J. Kuta: *"Principles of Polarography"*, Academic Press, New York, 1966.

P. Zuman and I. Kolthoff (eds.): *"Progress in Polarography"*, 3 volumes, Interscience Publishers, New York, 1962 – 1972.

P.J. Elving: *"Voltammetry in Organic Analysis"*, in H.W. Nürnberg (ed.), *"Electroanalytical Chemistry"* p. 197, Wiley, New York, 1974.

A.A. Vlcek: *"Polarographic Behaviour of Coordination Compounds"* in F.A. Cotton (ed.): *"Progress in Inorganic Chemistry"*, vol. 5, p.211, Interscience Publishers, New York, 1963.

H. Hoffman and J. Volke: *"Polarographic Analysis in Pharmacy"*, loc. cit. p. 287.

P. Zuman (ed.): *"Topics in Organic Polarography"*, Plenum Press, London, 1970.

B. Breyer and H. Brauer: *"Alternating Current Polarography and Tensammetry"*, Interscience Publishers, New York, 1963.

D.E. Smith: *"AC Polarography and Related Techniques: Theory and Practice"* in A.J. Bard (ed.): *Electroanalytical Chemistry* vol. 1, p.1, Marcel Dekker, New York, 1966.

D.E. Smith: *"Recent Developments in Alternating Current Polarography"* in: Crit. Rev. Anal. Chem. **2** (1971) 247.

A.M. Bond: *"Modern Polarographic Methods in Analytical Chemistry"*, M. Dekker, New York, 1980.

D.G. Davis: *"Applications of Chronopotentiometry to Problems in Analytical Chemistry"*, in *"Electroanalytical Chemistry"*, ed. A.J. Bard, vol. 1, p. 157 Marcel Dekker, New York (1966).

J.J. Lingane: *"Analytical Aspects of Chronopotentiometry"*, The Analyst **91** (1966) 1.

D. Jagner: *"Potentiometric Stripping Analysis"*, The Analyst **107** (1982) 593.

V. Vydra, K. Stulik and E. Julakova: *"Electrochemical Stripping Analysis"*, Ellis Horwood, Chichester, 1976.

J.J. Wang: *"Stripping Analysis; Principles, Instrumentation and Applications"*, VCH Publishers, Florida, 1985.

P.K. Bailey: *"Analysis with Ion-selective Electrodes"*, Heyden, London, 1976.

J. Koryta and K. Stulik: *"Ion-selective Electrodes"*, Cambridge University Press, 1983.

H. Freiser: *"Ion-selective electrodes in Analytical Chemistry"*, Plenum Press, New York, 1981, 2 vols.

J. Janata: *"Principles of Chemical Sensors"*, Plenum Press, New York, 1989.

W. Göpel and K.D. Schierbaum: "*Chemical Sensors Based on Catalytic Reactions*" in G. Ertl, H. Knözinger abd J. Weitkamp (eds.), "*Handbook of Heterogeneous Catalysis*", Wiley-VCH, Weinheim, 1997, p. 1283.

A.M. Azad: "*Solid-state Gas Sensors: A Review*" in J. Electrochem. Soc. **139** (1992) 3690.

W.E. Morf, "*The Principles of Ion-Selective Electrodes and of Membrane Transport*", Elsevier, Amsterdam, 1981.

P. Fabry and E. Siebert: "*Electrochemical Sensors*" in "*The CRC Handbook of Solid-State Electrochemistry*", eds. P.J. Grellings and H.J.M. Bouwmeaster, CRC Press, Boca Raton, Florida, 1997.

Subject Index

Electrochemistry. Carl H. Hamann, Andrew Hamnett, Wolf Vielstich

ISBN: 978-3-527-31069-2

C

F

G

H

I

K

L

P

Q

R

Further Titles of Interest

Holze, R.

Experimental Electrochemistry

A Student's Lab Course

2007
Softcover
ISBN 3-527-31098-3

Bard, A. J., Faulkner, L. R.

Electrochemical Methods

Fundamentals and Applications

2001
Hardcover
ISBN 0-471-04372-9